FUNDAMENTAL CONSTANTS

Symbol	Name	Value
c	Velocity of light	2.9979×10^{10} cm sec^{-1}
e	Electronic charge	1.6021×10^{-19} coulomb
N_0	Avagadro's number	6.0225×10^{23}
h	Planck's constant	6.6256×10^{-27} erg sec
F	Faraday constant	96,487 coulomb eq^{-1}
R	Gas constant	82.056 cm^3 atm mole^{-1} deg^{-1}
		1.9872 cal mole^{-1} deg^{-1}
		8.3143 joule mole^{-1} deg^{-1}
k	Boltzmann constant	1.3805×10^{-16} erg deg^{-1}

GREEK ALPHABET

Greek letter	Greek name	Greek letter	Greek name
A α	Alpha	N ν	Nu
B β	Beta	Ξ ξ	Xi
Γ γ	Gamma	O o	Omicron
Δ δ	Delta	Π π	Pi
E ε	Epsilon	P ρ	Rho
Z ζ	Zeta	Σ σ	Sigma
H η	Eta	T τ	Tau
Θ θ	Theta	Υ υ	Upsilon
I ι	Iota	Φ ϕ	
K κ	Kappa	X	
Λ λ	Lambda		
M μ	Mu		Omega

Modern Methods of Chemical Analysis

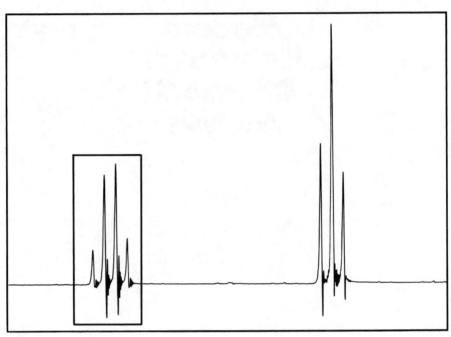

The n.m.r. spectrum of diethylether, reproduced from the film "Nuclear Magnetic Resonance," *produced by the Chemistry Department, UCLA, and distributed by John Wiley and Sons.*

Modern Methods of Chemical Analysis

SECOND EDITION

ROBERT L. PECSOK
University of Hawaii

L. DONALD SHIELDS
California State University, Fullerton

THOMAS CAIRNS
U.S. Food & Drug Administration, Los Angeles

IAN G. McWILLIAM
Swinburne College of Technology, Melbourne

JOHN WILEY & SONS

New York • Chichester • Brisbane • Toronto • Singapore

This book was written by Thomas Cairns in his private
capacity. No official support or endorsement by the Food
and Drug Administration is intended or should be inferred.

Library of Congress Cataloging in Publication Data
Main entry under title:

Modern methods of chemical analysis.

    Edition for 1968 by R. L. Pecsok and
L. D. Shields.
    Includes bibliographical references and index.
    1. Chemistry, Analytic. I. Pecsok, Robert L.
Modern methods of chemical analysis. II. Title.
QD75.2.M65 1976   543   76-13894
ISBN 0-471-67662-4

Printed in the United States of America

10 9 8

PREFACE

The first edition of this book represented a significant departure from the usual textbook on quantitative analysis. Because of the well-established trend of including gravimetric and volumetric analysis in the general chemistry course, we replaced much of the standard elementary discussions of stoichiometry and equilibria with more comprehensive descriptions of chromatography, spectrophotometry, nuclear magnetic resonance and mass spectrometry, emphasizing applications in the organic and biochemical areas.

As we noted in our first preface, "perhaps the only constant feature of science is that of change." Over the last 10 years an increasing majority of students enrolled in the analytical course are nonchemists who are preparing for careers that demand a working knowledge of *modern* analytical methods. For most of these students, this is their last formal exposure to the wealth of techniques currently used by all chemists. Some of these techniques are used primarily to determine the composition of samples. Others are used to separate complex mixtures or to purify components. Still others are used to obtain detailed characteristic information leading to identification and structure of compounds. Our list also includes methods for handling vast quantities of data in a systematic fashion, and for following and controlling reactions in the laboratory or in the industrial chemical plant.

In this second edition we have reorganized several of the major areas to improve clarity and to save space for new topics. For example, the principles of separation are treated more generally and completely in the introductory chapter on chromatography. The interpretation of various types of absorption spectra is expanded with additional chapters on spark source mass spectrometry, flame emission, atomic absorption, and X-ray methods. The discussion of acid-base equilibria has been thoroughly reorganized and compacted to make room for the related topic of complex ion equilibria. Two new sections have been added—one on methods of data handling, and another on methods used for automatic and process analyzers, including auto-analyzers, process monitoring and process control.

The organization of the book into independent parts makes it possible to select topics to fit a variety of course lengths and purposes. Our objective is to reach the middle level of compromise between an elementary survey and a sophisticated treatment in depth. It will be most useful to students who have had the background in chemical principles covered in most general chemistry courses and, at least, an elementary exposure to organic chemistry. Sufficient material is given for a two-semester integrated course covering most of the topics of the more traditional courses in quantitative analysis, qualitative organic analysis and instrumental methods of analysis.

Our primary objective has been to write a book for students to use and enjoy; it is not a monograph or a compendium of procedures. We hope our readers will find our explanations of theory plausible and understandable—we have sometimes sacrificed formality and rigor where it didn't seem necessary for reasonable comprehension and practical use. In deference to an almost unanimous request, we have provided answers for numerical problems in the expanded sets following most chapters.

Students and instructors, too numerous to mention individually, have volunteered suggestions and thoughtful comments that have been most helpful to us. In particular, we wish to acknowledge the contributions of Dr. Norman Juster, Dr. Quintus Fernando, Dr. Christopher Foote, Dr. David Lightner, Dr. Royce Murray, Dr. Ian Bowater and Dr. John Rowe. We are most grateful to Dr. Thomas Isenhour for his detailed criticism of the preliminary draft of this edition, and to Marcella Hughes for her consistent and excellent work at the typewriter.

Robert L. Pecsok
L. Donald Shields
Thomas Cairns
Ian G. McWilliam

CONTENTS

List of Symbols   xv

1 Introduction   1

PHASE CHANGES AND SEPARATIONS

2 Analytical Uses of Phase Changes   5

    2–1   Introduction   5
    2–2   Phase Changes for Pure Compounds   5
    2–3   Phase Changes for Mixtures   13
    2–4   Fractional Distillation   19
    2–5   Nonideal Solutions   24

3 Extraction   28

    3–1   Introduction   28
    3–2   Distribution Law   29
    3–3   Successive Extractions   30
    3–4   Separation of Mixtures by Extraction   33
    3–5   The Craig Method of Multiple Extraction   34
    3–6   Continuous Countercurrent Extraction   38

4 Chromatography   41

    4–1   Introduction   41
    4–2   Types of Chromatography   42
    4–3   Chromatographic Theory   44

5 Liquid Column Chromatography   54

    5–1   Introduction   54
    5–2   Liquid–Solid Adsorption Chromatography   54

5–3   Liquid–Liquid Partition Chromatography   59
5–4   High Performance Liquid Chromatography   60
5–5   Ion Exchange Methods   62
5–6   Gel Chromatography   70

6 Plane Chromatography   76

6–1   Introduction   76
6–2   Paper Chromatography   77
6–3   Thin-Layer Chromatography   81
6–4   Zone Electrophoresis   82

7 Gas Chromatography   85

7–1   Introduction   85
7–2   Basic GLC Apparatus   85
7–3   Carrier Gas   86
7–4   Sample Introduction   86
7–5   Columns   88
7–6   Column Performance   88
7–7   Solid Supports   89
7–8   Liquid Phases   90
7–9   Detectors   92
7–10 Qualitative Analysis: Retention Parameters   97
7–11 Quantitative Analysis   102
7–12 Temperature Effects   105
7–13 Gas–Solid Chromatography   106
7–14 Applications   108

ELECTROMAGNETIC RADIATION

8 Electromagnetic Radiation and its Interaction with Matter   115

8–1   Introduction   115
8–2   Nature of Electromagnetic Radiation   115
8–3   Absorption and Emission of Radiation   117
8–4   Luminescence Phenomena   120
8–5   Refraction and Refractive Index   122
8–6   Interference and Diffraction   125
8–7   Light Scattering   127
8–8   Rotation of Polarized Light   128

9 Quantitative Analysis by Absorption of Electromagnetic Radiation   133

9–1   Introduction   133
9–2   Quantitative Laws   133
9–3   Deviations from Beer's Law   138
9–4   Photometric Error   142
9–5   Intensity of Luminescence   143

10 Instrumentation for Spectrometry   147

    10–1 Introduction   147
    10–2 Sources of Radiant Energy   147
    10–3 Monochromators   148
    10–4 Sample Handling   153
    10–5 Detection Devices   154
    10–6 Amplification and Readout of Detector Signals   158
    10–7 Single-Beam vs. Double-Beam Operation   160
    10–8 Typical Instruments   161

11 Infrared Spectroscopy   165

    11–1 Introduction   165
    11–2 Infrared Theory   166
    11–3 Sample Preparation   170
    11–4 Infrared Absorption Spectra   171
    11–5 Infrared Correlation Charts   173
    11–6 Interpretation of Infrared Absorption Spectra   173
    11–7 Representative Spectra   177
    11–8 Elucidation of Structure   184
    11–9 Tables of Absorption Characteristics   201

12 Ultraviolet Spectroscopy   226

    12–1 Introduction   226
    12–2 Theory   226
    12–3 The Chromophores   228
    12–4 Solvation and Substitution   235
    12–5 Interpretation and Use of Ultraviolet Spectra   237

13 Flame Emission and Atomic Absorption Spectroscopy   243

    13–1 Introduction   243
    13–2 Emission and Absorption in Flames   243
    13–3 Atomization and Ionization   245
    13–4 Flames   246
    13–5 Burners and Nebulizers   248
    13–6 Nonflame Atomization   251
    13–7 Radiation Sources and Optical Systems   253
    13–8 Flame Spectra and Interferences   256
    13–9 Quantitative Analysis   259
    13–10 Typical Applications   262

14 X-ray Methods   267

    14–1 Introduction   267
    14–2 Emission, Dispersion and Detection of X rays   267
    14–3 Absorption of X rays   273
    14–4 Diffraction of X rays   275
    14–5 Electron Microprobe Analysis   278

15 Nuclear Magnetic Resonance Spectroscopy   279

15–1   Introduction   279
15–2   Magnetic Properties of the Nucleus   279
15–3   Nuclear Resonance   281
15–4   Nuclear Relaxation   282
15–5   N.M.R. Instrumentation and Technique   283
15–6   The Chemical Shift   285
15–7   Spin–Spin Splitting   292
15–8   Interpretation of N.M.R. Spectra   299
15–9   Additional Topics   305

MASS SPECTROMETRY

16 Mass Spectrometry of Organic Compounds   316

16–1   Introduction   316
16–2   Methods of Ion Production   316
16–3   Ion Separators   321
16–4   Ion Collection and Recording   325
16–5   Sample Handling Techniques   325
16–6   Resolution   328
16–7   The Mass Spectrum   328
16–8   Interpretation of Mass Spectra   331
16–9   Behavior of Classes of Compounds   334
16–10 Identification of Unknowns   343
16–11 Analysis of Mixtures   346
16–12 Some Additional Applications   347

17 Spark Source Mass Spectrometry   353

17–1   Introduction   353
17–2   Instrumentation   353
17–3   Sample Handling   354
17–4   Interpretation of Spectra   356
17–5   Qualitative Identification of Elements   356
17–6   Quantitative Analysis   358

ELECTROANALYTICAL CHEMISTRY

18 Electrochemical Cells and Potentiometry   361

18–1   Introduction   361
18–2   Electrochemical Cells   362
18–3   Electrode Potentials   365
18–4   Types of Electrodes   377
18–5   Cell Voltage Measurements   383

18–6   Definition of pH and the pH Meter   384
18–7   Equilibrium Constants   386
18–8   Potentiometric Titrations   389
18–9   Coupled Reaction Systems   394

19   **Some Other Electrochemical Techniques**   400

19–1   Electrodeposition   400
19–2   Controlled Current Electrolysis   401
19–3   Controlled Cathode (or Anode) Potential Electrolysis   403
19–4   Secondary Coulometric Titrations   403
19–5   Polarography   406

ACIDS, BASES AND THEIR SALTS AND COMPLEXES

20   **Monoprotic Systems**   413

20–1   Introduction   413
20–2   Acid–Base Concepts   413
20–3   Role of the Solvent   414
20–4   Relative Strengths of Acids and Bases   416
20–5   Equilibria in Monoprotic Systems   419
20–6   Acid–Base Titrations in Aqueous Solution   427
20–7   Acid–Base Titrations in Nonaqueous Solvents   431

21   **Polyprotic Systems**   438

21–1   Introduction   438
21–2   Phosphoric Acid System   438
21–3   Distribution of Species Diagrams   439
21–4   Calculations for Polyprotic Systems   442
21–5   Titrations with Polyprotic Systems   445
21–6   Polyamine and Amino Acid Systems   446
21–7   Amino Acid Titrations   451

22   **Metal Ion Complexes**   454

22–1   Introduction   454
22–2   Monodentate Ligands   454
22–3   Metal Chelates   456
22–4   Conditional Formation Constants   458
22–5   Titration Equilibria   460
22–6   Compleximetric Indicators   462
22–7   Scope of EDTA Applications   463

RADIOCHEMISTRY

23 Radiochemical Methods   466

    23–1   Introduction   466
    23–2   Radioactive Disintegration   466
    23–3   Interaction of Radiation with Matter   469
    23–4   Detection and Measurement of Radioactivity   471
    23–5   Counting Statistics, Errors and Corrections   475
    23–6   Some Applications of Radioisotopes   480
    23–7   Radiation Safety   482

EVALUATION AND PROCESSING OF ANALYTICAL DATA

24 Statistical Treatment of Data   485

    24–1   Introduction   485
    24–2   Significant Figures   485
    24–3   Types of Errors   486
    24–4   Normal Distribution of Random Errors   487
    24–5   Statistical Treatment of Small Data Sets   490
    24–6   Control Charts   492

25 Data Processing   494

    25–1   Introduction   494
    25–2   Indicating and Recording Devices   494
    25–3   Integrators   497
    25–4   Digital Computers   501
    25–5   Computerized Gas Chromatography   503
    25–6   Computerized Mass Spectrometry   505
    25–7   Computerized N.M.R. Spectroscopy   509
    25–8   Other Analytical Applications of Digital Computers   512
    25–9   Operational Amplifiers   513
    25–10 Analog and Hybrid Computers   515

AUTOMATIC AND PROCESS ANALYZERS

26 Automatic Analyzers   519

    26–1   Introduction   519
    26–2   Basic Automatic Analyzer   520
    26–3   Beckman "DSA 560"   521
    26–4   Ljundberg "Autolab"   522
    26–5   DuPont "ACA"   522
    26–6   Technicon "AutoAnalyzer"   524
    26–7   Electro-Nucleonics "GeMSAEC System"   528

27 Process Analyzers   531

   27–1 Introduction   531
   27–2 Specific and Nonspecific Analyzers   531
   27–3 Typical Process Analyzer   532
   27–4 Sample Handling System   533
   27–5 Analyzer Calibration   534
   27–6 Process Gas Chromatograph   535
   27–7 Oxygen Analyzers   538
   27–8 Infrared Analyzers   540
   27–9 Moisture Analyzers   542
   27–10 Colorimetric Analyzers   544

28 Process Control   547

   28–1 Introduction   547
   28–2 Feedback Control   547
   28–3 Dynamic Response of the System   549
   28–4 Cascade and Feedforward Control   551
   28–5 Controller Characteristics   553
   28–6 Applications of Control Systems   558

Index   564

LIST OF SYMBOLS
AND PAGE WHERE FIRST USED

A	absorbance, 135
A	activity of a radioisotope, 468, 480
A	amplifier gain, 498
A	Van Deemter coefficient, 49
A_S	area of stationary phase, 77
a	absorptivity, 135
a_i	activity of ith species, 64, 371
B	Van Deemter coefficient, 49
b	cell path length, 134
C	capacitance of a capacitor, 498
C	Van Deemter coefficient, 49
C_i	concentration of ith component, 28
c	velocity of light in vacuum, 115
D	distribution ratio, 29
D_M	diffusivity coefficient in mobile phase, 88
D_S	diffusivity coefficient in stationary phase, 88
$D(t)$	output peak shape, 512
DAT	derivative action time, 555
$\bar{d}$	average deviation, 488
d	line spacing in a grating, 126
d_f	thickness of liquid layer, 88
d_p	diameter of a packing particle, 88
dme	dropping mercury electrode, 407
E	energy of a photon, 117
E	electrode potential, 365
E_{cell}	voltage of a cell, 365
E_{ep}	electrode potential at equivalence point, 391
E_j	junction potential, 374
E_{ov}	overvoltage, 401
$E_{1/2}$	half-wave potential, 407
$E°$	standard electrode potential, 367
E'	formal potential, 375
E'_7	biochemical formal potential, 376
$E*$	cell constant of a pH meter, 384
$E(t)$	true peak shape, 512
e	charge on an electron, 268
e	voltage at a point in electric circuit, 498
F	Faraday constant, 367
F	flow rate of mobile phase, 98
F	magnetic force, 321
F	weight factor of a detector, 103
f	fraction titrated, 393

f_i	activity coefficient of ith species,	64, 371
G	gain of controller,	549
G_u	ultimate gain,	556
g_i, g_0	statistical weights,	243
H	height equivalent of a theoretical plate,	23
H_0	intensity of applied magnetic field,	281
ΔH_f	heat of fusion,	6
ΔH_s	heat of sublimation,	11
ΔH_v	heat of vaporization,	8
h	Planck's constant,	116
I	intensity of radiation,	119
I_F	intensity of luminescence,	143
I_0	intensity of incident radiation,	133
I_t	intensity of transmitted radiation,	134
I	ion current in mass spectrometry,	346
I	retention index,	101
I	spin number of a nucleus,	279
IAT	integral action time,	555
i	electric current,	401
i_d	diffusion current in polarography,	407
J	coupling constant,	292
j	pressure correction factor,	99
K	equilibrium constant,	28
K_a	acid dissociation constant,	29, 416
K_b	base dissociation constant,	416
K_D	distribution coefficient,	28, 44
K_d	dimerization constant,	29
K_f	molal freezing point depression constant,	14
K_f	formation constant,	455
K'_f	conditional formation constant,	458
K_s	autoprotolysis constant,	413
K_w	ion product of water,	413
k	Boltzmann constant,	243
k	force constant, Hooke's law,	166
k	Henry's law constant,	18
k'	capacity factor,	45
L	dead time,	556
L	length of column,	45
l	path length of a polarimeter,	131
M	molecular rotation,	131
m	mass of a particle,	167
m	mass flow rate of mercury in dme,	408
m	molality,	14
N	number of counts,	478
N_0	number of atoms in ground state,	243
N_j	number of atoms in excited state,	243
n	index of refraction,	115
n	number of electrons in half-reaction,	367
n	number of extractions,	33
n	number of scans	
n	number of theoretical plates,	21, 47

n	number of transfers, 35	
P	radiant power, 119	
P	pressure of a gas (vapor), 8	
P_i	inlet pressure, 98	
P_o	outlet pressure, 98	
$\bar{P}$	average pressure, 99	
$P°$	vapor pressure of a pure liquid, 14	
P_u	ultimate period, 556	
p	fraction of solute in stationary (lower) phase, 36	
p_o	initial controller output pressure	
Q	range quotient, 491	
Q_L	limit of linear response of a detector, 92	
Q_0	limit of detection, 92	
q	charge on a capacitor, 498	
q	fraction of solute in mobile (upper) phase, 36	
R	counting rate, 476	
R	electrical resistance, 401	
R	gas constant, 9	
R	number of unsaturated sites, 333	
R	reflux ratio, 22	
R	resolution of peaks, 51	
R_f	retardation factor, 76	
R_M	molar response of a detector, 103	
R_N	noise level of a detector, 92	
r	radius of ion path, 321	
s	scissoring vibration, 169	
s	standard deviation, 490	
T	transmittance, 135	
T	temperature, 7	
T_b	boiling point, 10	
T_m	melting point, 5	
t	Student's t value, 490	
t_a	elution time for an air peak, 46	
t_d	drop time of dme, 408	
t_M	retention time of a nonsorbed substance, 46	
t_R	retention time, 45	
$t_{1/2}$	half-life of a radioisotope, 467	
u	linear velocity of mobile phase, 45	
V	voltage at a point in an electric field, 267	
V	volume of titrant added, 390	
V	volume of a gas, 9	
V_e	elution volume, 64	
V_g	specific retention volume, 99	
V_i	volume of liquid in gel column inside beads, 71	
V_o	volume of liquid in gel column outside beads, 71	
V_M	volume of mobile phase in a column, 46	
V_N	net retention volume, 99	
V_R	retention volume, 46, 98	
$V_{R°}$	corrected retention volume, 99	
V'_R	adjusted retention volume, 99	
V_S	volume of stationary phase in column, 46	

V_{ep} volume of titrant added at equivalence point, 390
v velocity of a particle, 318
W_i weight of ith component, 31
W_i width of ith peak, 48
W_L weight of liquid phase in column, 99
X_i mole fraction of ith component, 14
$\bar{x}$ arithmetic mean of a set of values, 488
Y_i mole fraction of ith component of a vapor, 18
Z atomic number, 270
z number of carbon atoms in a n-paraffin, 101
α relative volatility, 18
α separation factor, 37
α relative retention, 46, 100
α fraction of total concentration, 439
α specific rotation, 130
γ bending vibration (out of plane), 169
γ magnetogyric ratio, 281
δ bending vibration (in the plane), 169
δ chemical shift, 286
ϵ molar absorptivity, 135
ζ damping factor, 550
λ decay constant of radioisotope, 467
λ wavelength of radiation, 115
μ ionic strength of a solution, 372
μ linear absorption coefficient for X rays, 273
μ nuclear magnetic moment, 281
μ reduced mass, 167
μ true value, 486
μ_M mass absorption coefficient for X rays, 273
ν frequency of radiation, 115
ν stretching vibration, 169
$\bar{\nu}$ wavenumber (reciprocal centimeters), 116
ρ_L density of liquid phase, 100
ρ rocking vibration, 169
σ shielding constant, 285
σ standard deviation, 477, 488
τ chemical shift, 287
τ dead time of counter, 476
τ time constant, 512
τ twisting vibration, 169
ω wagging vibration, 169

1

INTRODUCTION

If chemistry is aptly described as "what chemists do," it must be that "analytical chemistry is what analytical chemists do." However, it is no easier to pinpoint the subject matter of analytical chemistry than it is to describe chemistry as a whole. Biologists, geophysicists, engineers—all practice chemistry to some extent, and nearly all chemists practice analytical chemistry; for, among other things, chemistry is the study of the composition and behavior of the natural world. Anyone wishing to know more about the composition of substances must employ analytical methods to determine the kinds and amounts of compounds, elements, atoms, and sub-atomic particles present in a given sample, as well as to examine the detailed arrangements of the various species. And to study the behavior of materials, analytical methods must be used before, during, and after certain reactions or "changes."

The recent rapid advances in our knowledge of the physical world parallel a similar rapid advance in the science of analytical methods. However, neither the balance nor buret, which provided most of our analytical measurements prior to 1940, has been relegated to the museum—both are still indispensable in any chemical laboratory, and often are the ultimate weapons in calibrating a fancier instrumental method. They were good enough to allow Nobel Laureate T. W. Richards to prove the existence of isotopes by very precise determinations of the atomic weight of lead from various sources. But now mass spectrometry permits atomic weights to be determined more precisely and with a fraction of the effort. Similarly, the American Petroleum Institute invested many thousands of man-hours in attempting to isolate the hundreds of components in petroleum and gasoline by tedious fractional distillations. Today the same results can be obtained by one man in a few hours with a gas chromatograph. Most of our present knowledge about the detailed structure of compounds and atoms has been gained by studying the interaction of electromagnetic radiation with matter—through various forms of spectroscopy.

Even the beginning student soon becomes aware of the importance of modern analytical methods in all branches of science. This is not to say that breakthroughs in chemistry must wait for the analytical chemist to invent new methods, though most new techniques and instruments have been invented by scientists with analytical problems to solve. Chromatography was first introduced by botanists. Nuclear magnetic resonance spectroscopy and mass spectroscopy were first studied by physicists. Yet, for the most part, scientists who are not analytical chemists tend to consider analytical methods as tools of the trade. They are primarily interested in the results they can provide.

The analytical chemist, on the other hand, is concerned with methods *per se*. He must become familiar with the theoretical principles and instrumentation of all methods. In so doing, he often becomes involved in bio-, organic, inorganic, and physical chemistry. He seeks to improve methods, searches for new areas of application, and points out the limitations so that the technique will not be misused. Accuracy, precision, reproducibility, and reliability are everyday words in his vocabulary. In a real sense, the analytical chemist is the keeper of chemistry's tools and techniques. This is no small assignment, and in this book we attempt merely to introduce some of the methods which all chemistry students will use in the more specialized branches of chemistry.

The determination of a melting point is hardly a modern analytical technique, but it illustrates how an extremely simple experiment can identify a substance or measure its purity. The melting process involves a two-phase system and is best understood from a study of a "phase diagram." The same principles apply to distillation, a somewhat more sophisticated separation technique where the concept of a column process is introduced. In distillation, there is a distribution of material between a gas phase and a liquid phase, whereas in extraction there is a distribution of material between two immiscible liquid phases. Otherwise the two processes have much in common. In extraction we develop the concept of a distribution coefficient which describes how a component is distributed between two phases. The concepts of distribution (from extraction) and flow through columns (from distillation) form the basic framework for chromatographic methods in which a complex mixture is separated into its components by selective retardation as the sample is washed through a packed column or over a specially prepared paper surface. Chromatographic methods include some of the most powerful and useful separating techniques yet devised.

As a second major topic, the study of the interaction of electromagnetic radiation with matter serves as a background for the great variety of spectroscopic methods. At one end of the spectrum γ rays and X rays tell us much about the details of nuclear, atomic, and molecular structure. At the other end of the spectrum there is relatively low-energy radiation in the radio frequency range. Normally these low-level frequencies are of concern to communication engineers, but when combined in the proper way with an intense magnetic field, this kind of energy can interact with certain nuclei, giving us a new tool called "nuclear magnetic resonance."

At intermediate frequencies are ultraviolet and infrared radiation—both capable of yielding much information about the nature of atoms, ions, and molecules. The most familiar part of the spectrum, visible radiation, lies between the ultraviolet and infrared. While the inorganic chemist may use visible radiation to study the structure and stability of transition metal complexes, the layman may use it for a simple "eyeball" determination of the strength of his cup of tea.

In most spectroscopic methods, the sample is exposed to some kind of radiation. In mass spectrometry, the sample is bombarded with a beam of electrons which ionizes and "cracks" the molecules into many pieces. The spectrometer then sorts the electrically charged fragments in order of their mass-to-charge ratios.

The third major topic is the more traditional study of chemical equilibria. Stoichiometry and equilibria are the foundations of classical gravimetric, titrimetric, and electrometric analytical methods. The treatment here will extend the elementary concepts to more complex systems including the study of electrode reactions under carefully controlled conditions and the use of solvents other than water.

The next topic concerns some radiochemical techniques which are becoming more important with the search for additional energy sources. For many analytical problems, the use of radioactive isotopes provides the ultimate in trace analysis.

The final section introduces some of the concepts used in the operation of a large-scale commercial or government analytical laboratory. To a great degree, automation has replaced the drudgery of routine analyses. At the push of a button, ingenious devices can weigh a sample, dissolve it, add reagents, mix and filter, heat or cool it, titrate it or measure its absorbance, and print the result on a card. Process analyzers will give an immediate and continuous monitoring of a chemical process. Such information entered into a feedback loop can control the operation of an entire plant.

The methods we have selected by no means exhaust the list available to the modern chemist. No matter what list is chosen, it is soon outdated. Rather than completeness, our objective is to present a variety of principles and concepts which underlie most of the instruments and techniques in current use.

We should point out that most of the instruments the chemist now uses routinely were not available 25 years ago or were available only as laboratory curiosities in a few research laboratories. There is no doubt that in the next 25 years we will see an even more spectacular change in the types of analytical methods used routinely and the kind of information they are able to unfold.

Given a sound background, the student will be better prepared and more capable of remaining current. The first of the general references listed below is one of the best sources of information on new developments as they occur.

GENERAL REFERENCES

Annual Reviews—a special issue of *Analytical Chemistry*. In even years the literature concerning fundamental techniques is reviewed in all areas of analytical chemistry. In odd years, the literature in selected applied areas is reviewed. These reviews are usually the quickest and easiest way to begin a literature search.

I. M. Kolthoff and P. J. Elving, *Treatise on Analytical Chemistry*, Wiley-Interscience, New York, 1959 ff. (a continuing collection). An encyclopedic work in many volumes in which experts discuss all fields of analytical chemistry by methods and by applications. This work is the best single source for analytical methods.

C. N. Reilley and R. W. Murray, eds., *Advances in Analytical Chemistry and Instrumentation*, Wiley-Interscience, New York, 1960 ff. A continuing series of reviews of specific areas.

A. Weissberger, *Physical Methods of Organic Chemistry*, Wiley-Interscience, New York, 1959 ff. Another multivolume work with discussions in depth. Parts of it have been revised several times.

F. L. J. Sixma and H. Wynberg, *A Manual of Physical Methods in Organic Chemistry*, Wiley, New York, 1964. A condensed version of the theoretical principles of most of the methods covered in this text, as well as experimental procedures.

H. H. Willard, L. L. Merritt, Jr., and J. A. Dean, *Instrumental Methods of Analysis*, 5th ed., Van Nostrand Reinhold, New York, 1974. Theory, instrumentation, and experimental procedures.

H. A. Laitinen and W. E. Harris, *Chemical Analysis*, 2nd ed., McGraw-Hill, New York, 1975. A definitive text at the advanced level on noninstrumental methods.

PHASE CHANGES AND SEPARATIONS

2

ANALYTICAL USES OF PHASE CHANGES

2–1 INTRODUCTION

Our study of chemical analysis begins with one of the earliest observations—substances melt and boil at particular temperatures, and in so doing change in form from a solid to a liquid, or from a liquid to a gas, or the reverse. The transitions between these states of matter, or *phases*, provide information which helps to identify the substance and determine its purity. Furthermore, since the mechanical separation of two phases is relatively straightforward, a study of the distribution of a component between two phases provides the basis for all separation techniques.

The phase changes represented below may occur with either a single pure compound or element, or a mixture of two or more substances.

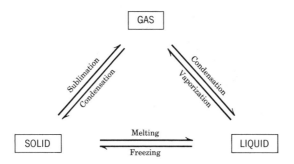

2–2 PHASE CHANGES FOR PURE COMPOUNDS

Solid–Liquid Equilibrium. When a solid sample is heated its temperature rises as shown in Figure 2–1. At a characteristic temperature, the solid begins to melt and there is a discontinuity in the heating curve. If the heating is continued, the solid continues to melt at a constant temperature, known as the *melting point, T_m*.

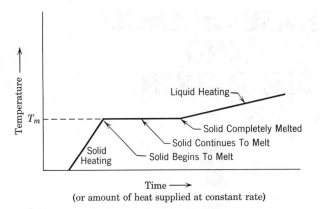

Figure 2–1 Heating curve for a typical pure solid substance.

The additional heat is absorbed as *latent heat of fusion*, ΔH_f, which is the energy required to disrupt the crystal lattice. When the solid is completely melted, the temperature of the liquid begins to rise again, but usually not at the same rate as for the solid because the heat capacity of the liquid differs from that of the solid. We have assumed that during the phase transition the system is well-mixed, so that an equilibrium is maintained.

$$\boxed{\text{SOLID}} \quad \overset{+\Delta H_f}{\underset{-\Delta H_f}{\rightleftarrows}} \quad \boxed{\text{LIQUID}}$$

If we reverse the process just described and cool a pure liquid, we obtain a cooling curve that is not quite a mirror image of the heating curve, Figure 2–2. The temperature may fall below the equilibrium freezing (melting) point before the first crystals appear, a phenomencn known as *supercooling*. The extent of supercooling varies with the nature of the substance, the rate of cooling,

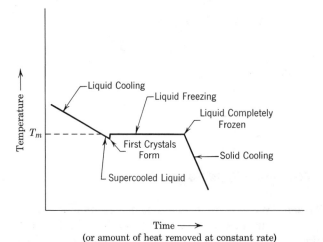

Figure 2–2 Cooling curve for a typical pure liquid substance.

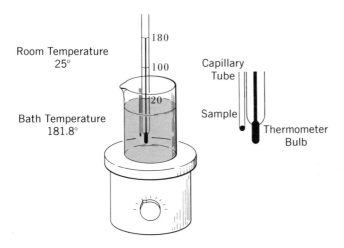

Figure 2–3 Simple apparatus for determining the melting point. Close-up of position of sample.

mechanical actions such as stirring and vibration, and with the presence of impurities which can act as nuclei in the formation of the crystals.

The melting point of a *pure* solid is always sharp (except for liquid crystals) and serves to identify it. If the substance is not pure, it will melt over a range of temperature, as we will discuss in a later section. The change in volume on melting or freezing is relatively small; therefore the effect of pressure is almost always negligible.

Melting points are often determined in the very simple apparatus shown in Figure 2–3. The sample, contained in a thin-walled capillary, and the thermometer should be in close proximity and heated at a rate not exceeding 1° per minute as the melting point is approached (final 10°). More sophisticated apparatus is available for increased accuracy, speed, and convenience.

There are several sources of error in this determination, some of which are easily corrected. Adequate heat transfer must be provided by using a well-stirred bath, a thin-walled capillary, and a very slow heating rate in the region of the melting point. The thermometer should not be trusted unless it has been properly calibrated. Calibration in an ice slush bath and in boiling water at a known pressure is usually adequate. If the bath temperature is well above room temperature, a thermometer correction may be necessary to allow for the variation in the density of mercury along the exposed column. For mercury thermometers, the correction, ΔT, is

$$\Delta T = 0.000154 l (t - n) \tag{2–1}$$

where 0.000154 is the coefficient of linear expansion of mercury in a glass tube, l the length of the exposed column in degrees, t the observed temperature, and n the average temperature of the exposed stem.

Example/Problem 2–1. The thermometer in Figure 2–3 reads 180° and is immersed to the 20° mark. The correction is:

$$\Delta T = 0.000154 \times (180 - 20) \times \left(180 - \frac{180 + 25}{2} \right)$$

$$= 1.9°$$

The corrected reading is therefore 181.9°C.

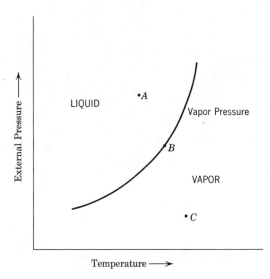

Figure 2–4   Vapor pressure curve for a typical pure liquid substance.

Liquid–Gas Equilibrium.   All molecules in a liquid are in constant motion, thus generating an internal pressure and a tendency to escape from the liquid. The *vapor pressure* of a liquid is a measure of this escaping tendency and is the pressure that would be observed in the vapor phase if the pure liquid were placed in an evacuated chamber and allowed to come to equilibrium. If the chamber also contains another gas, the total pressure includes the vapor pressure of the liquid, and we speak of the *partial pressure* of each component. The vapor pressure of the liquid is a property of the liquid and does not depend on the composition of the vapor phase in contact with it. However, it does depend on the composition of the liquid phase (see below). Because vapor pressure is the result of molecular motions, we would expect it to increase with an increase in temperature. Figure 2–4 shows how the vapor pressure of a typical liquid changes with temperature. If the vapor pressure is equal to or greater than the external pressure, the liquid will boil as fast as the *latent heat of vaporization,* ΔH_v, can be supplied until the liquid is completely evaporated, or until an equilibrium is established between the liquid and the vapor.

$$\boxed{\text{LIQUID}} \xrightleftharpoons[-\Delta H_v]{+\Delta H_v} \boxed{\text{GAS}}$$

The temperature at which the vapor pressure is one atmosphere (760 torr) is the *normal boiling point*; however, the boiling point at any pressure is given by the curve in Figure 2–4. This curve also separates two regions: any combination of external pressure and temperature to the left of the line (e.g., point A) is in a region in which the liquid phase is the stable form. Point C is in a region where the gas phase is the stable form. Point B and other points on the line give the only conditions for which the two phases can co-exist in equilibrium.

For pure liquids, the vapor pressure–temperature curve is expressed by the *Clapeyron equation*, developed by Emile Clapeyron in 1834.

$$\frac{dP}{dT} = \frac{\Delta H_v}{T \Delta V} \tag{2–2}$$

where ΔV is the difference between the volume occupied by a mole of substance in the gas phase and in the liquid phase. The liquid volume is very small compared to the gas volume, therefore $\Delta V \approx V$, the gas volume. If we also assume that the vapor is an ideal gas, then

$$V = RT/P \ (R = 1.987 \text{ cal mole}^{-1} \text{ deg}^{-1})$$

Thus, we can substitute RT/P for ΔV in Equation 2–2

$$\frac{dP}{dT} = \frac{\Delta H_v}{RT^2} \times P$$

or

$$\frac{dP}{P} = \frac{\Delta H_v}{RT^2} dT$$

Integration of this equation yields

$$\log P = \frac{-\Delta H_v}{2.3RT} + \text{const} \tag{2-3}$$

or

$$\log \frac{P_2}{P_1} = -\frac{\Delta H_v}{2.3R} \left(\frac{1}{T_2} - \frac{1}{T_1} \right) \tag{2-4}$$

Over a considerable range of temperature, ΔH_v is constant and can be determined from the slope of a plot of the data (e.g., Table 2–1) according to Equation 2–3, as shown in Figure 2–5. Once we have a value for ΔH_v, Equation 2–4 is useful for computing the vapor pressure, P_2, at temperature T_2 from the known value, P_1, at another temperature T_1.

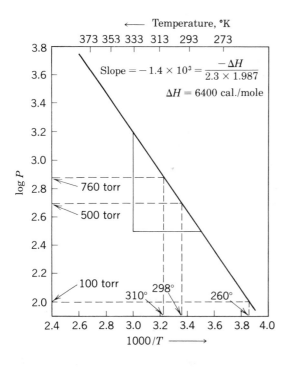

Figure 2–5 Vapor pressure–temperature plot for *n*-pentane according to Equation 2–3.

Example/Problem 2–2. What is the heat of vaporization and the vapor pressure of *n*-pentane at 25°C? The data from Table 2–1 are plotted in Figure 2–5, from which the slope is -1.4×10^3 deg, giving a value for ΔH_v of 6400 cal/mole. From the graph, the vapor pressure at 25°C ($1/T = 0.00336$) is 500 torr, or, using Equation 2–4 and values for 20°C, we have

$$\log \frac{P_2}{420} = \frac{-6400}{2.303 \times 1.987} \left(\frac{1}{298} - \frac{1}{293} \right)$$

$$\log P_2 - \log 420 = 0.0801$$

$$P_2 = 505 \text{ torr}$$

Table 2–1 Vapor Pressures of Paraffins in Torr[a]

Temperature	n-Pentane	n-Hexane	n-Heptane	n-Octane
0°C	183	40	11	3
10	282	70	21	6
20	420	120	36	10
30	611	170	58	18
40	873	280	92	31
50	1193	410	141	49
60	1605	565	209	78
70	2119	770	302	118
80	2735	1050	427	175
90	3498	1390	589	253
100	4410	1840	795	354
110	--	—	1047	482
120	—	—	1367	646
130	—	—	—	859

[a] 1 torr = 1 mm Hg.

Example/Problem 2–3. What is the boiling point of *n*-pentane at 760 torr? At 100 torr? From Equation 2–4 and values at 40°C, we have

$$\log \frac{760}{873} = \frac{-6400}{2.3 \times 1.987} \left(\frac{1}{T_b} - \frac{1}{313} \right)$$

$$T_b = 309°\text{K} \ (36°\text{C}) \text{ at } 760 \text{ torr}$$

and for the reduced pressure (100 torr), using values at 0°C, we have

$$\log \frac{100}{183} = \frac{-6400}{2.3 \times 1.987} \left(\frac{1}{T_b} - \frac{1}{273} \right)$$

$$T_b = 260°\text{K} \ (-13°\text{C}) \text{ at } 100 \text{ torr}$$

From Figure 2–5 the corresponding boiling points are 310° and 260°.

At lower pressure the boiling point is not only at a lower temperature, but also changes faster with a change in pressure. Compounds which are sensitive to heat are often distilled at temperatures well below their normal boiling points by reducing the pressure inside the apparatus as shown in Figure 2–6. A mixture of compounds A and B, whose boiling points are also given in Figure 2–6, would be more easily separated by distillation at low pressure because the difference in

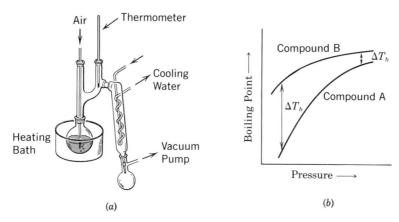

Figure 2–6 (a) Apparatus for distilling under reduced pressure. (b) The effect of pressure on the boiling points of two compounds to be separated.

their boiling points is larger. There is a practical limit to conducting distillations at reduced pressures, namely, the time required because of the reduced rate of evaporation. Typical times required to distil 1 g of substance at various pressures are given in Table 2–2.

Table 2–2 Time Required to Distil 1 g of Material

Pressure, Torr	Time
10	0.5 min
1	1 hr
10^{-1}	1 wk
10^{-2}	2 yr
10^{-3}	40 yr

Solid–Gas Equilibrium. The escaping tendency of a molecule in a solid is usually very low compared to a liquid. Nevertheless, a vapor pressure does exist (approx 10^{-26} torr for lead metal at room temperature) and it depends on temperature in the same manner as the Clapeyron equation.

$$\log P = \frac{-\Delta H_s}{2.3RT} + \text{constant} \qquad (2\text{–}5)$$

where ΔH_s is the *latent heat of sublimation.* The vapor pressure of a typical solid as a function of temperature is shown in Figure 2–7. In general, the vapor pressure of a solid is less than that of its liquid phase, and the pressure–temperature curve is steeper.

Phase Diagram. It is convenient to summarize the three types of phase transitions in a single graph, called a *phase diagram.* As a familiar example, the phase diagram for water is given in Figure 2–8. Curve *A*, separating the regions in which the solid and vapor are stable, gives the vapor pressure of the solid as a function of temperature. If the pressure above the solid is less than this, the solid will sublime. If the pressure is greater, the vapor will condense. Similarly curve *B* between the liquid and vapor regions, depicts the vapor pressure of the liquid as a

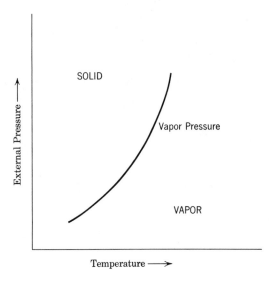

Figure 2–7 Vapor pressure curve for a typical pure solid substance.

function of temperature. The dashed portion of this line (curve *D*) gives the vapor pressure of the supercooled liquid, if it exists. Curve *C* shows the melting point as a function of pressure. It is very nearly a vertical line with a negative slope because water expands slightly on freezing.

For a given temperature and pressure, the phase diagram indicates which phase(s) will exist at equilibrium. If a particular combination of pressure and temperature lies on a curve, two phases will co-exist in equilibrium. There is one unique combination where the curves intersect, the *triple point*, at which all three phases can co-exist.

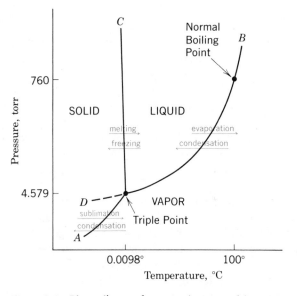

Figure 2–8 Phase diagram for water (not to scale).

2-3   PHASE CHANGES FOR MIXTURES

Solid–Liquid Equilibrium.   The melting point of a pure substance is described as "sharp," that is, the entire transition from solid to liquid occurs at a single temperature. (In practice, a melting temperature range of 0.5 to 1.0° is considered to be "sharp.") If the substance is not pure (a mixture), the melting occurs over a range of temperatures as shown in Figure 2–9. Compare this figure with Figure 2–1. In general, the melting point of an impure substance is lower than that of the pure substance and the melting range is greater.

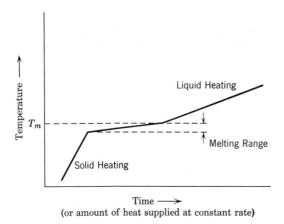

Time ⟶
(or amount of heat supplied at constant rate)

Figure 2–9   Heating curve for a solid mixture.

Freezing Point of Mixtures.   Odd as it may seem, we will approach freezing points through vapor pressure curves. Let us consider an isolated system consisting of solid and liquid camphor at its melting point, 179°C. The vapor pressure curves for this system of *pure* camphor are the solid lines A and B in Figure 2–10. A small amount of naphthalene added to the system will dissolve in

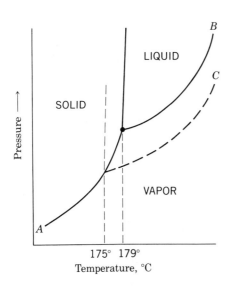

Figure 2–10   Phase diagram for pure camphor and for camphor with a small amount of naphthalene added (curve C).

the liquid camphor. The concentration of camphor molecules at the liquid surface is now less than before adding the naphthalene. As a consequence, the rate of vaporization of liquid camphor is decreased proportionately and its vapor pressure is decreased. The dashed line, C, in Figure 2–10 gives the vapor pressure of liquid camphor containing a small amount of naphthalene. The actual displacement of curve C from curve B was first investigated by F. M. Raoult in 1887, and is given by *Raoult's law* which states that the vapor pressure, P_c, of a solvent (in this case, camphor) in a solution is equal to the vapor pressure of the pure solvent P_c° multiplied by the mole fraction, X_c, of the solvent in the solution:

$$P_c = X_c P_c^\circ \tag{2–6}$$

Mole fraction is a measure of concentration and is equal to the number of moles of the given substance divided by the total number of moles of all substances in the same phase. The addition of a small amount of naphthalene does not alter the vapor pressure of solid camphor because the naphthalene dissolves entirely in the liquid phase, leaving the solid pure.

As in Figure 2–8, the intersection of the two vapor pressure curves, A and B, or A and C, in Figure 2–10, must define the freezing point. Thus the addition of an impurity lowers the freezing point of the liquid. If more impurity is added, there will be a still greater change in the freezing point. For *dilute* solutions, all changes will be small and we can represent the curves of Figure 2–10 as straight lines as shown in Figure 2–11. The obvious conclusion is that the change in freezing point, ΔT_f, is directly proportional to the amount of naphthalene added. If we measure concentration as molality (moles of solute per 1000 g of solvent), the proportionality constant is called the *molal freezing point depression constant, K_f*. For dilute solutions a simple equation describes this behavior:

$$\Delta T_f = K_f m \tag{2–7}$$

where m is the molality of the solute. The value of K_f depends only on the nature of solvent and not of the solute. Values of K_f for a few solvents are given in Table 2–3. If a known weight of a solute is dissolved in a known amount of solvent, a determination of the freezing point depression will yield the molality of the

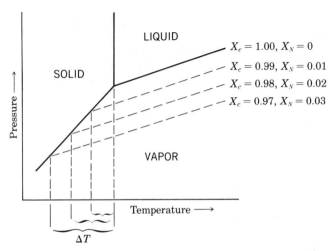

Figure 2–11 Expanded portion of phase diagram for camphor with several amounts of naphthalene added.

Table 2–3 Molal Freezing Point Constants

Solvent	Freezing Point, °C	K_f, deg/mole
Acetic acid	17	−3.9
Benzene	5.4	−5.12
Camphor	179	−38.0
Ethylene dibromide	10.1	−11.8
Naphthalene	80	−6.8
Water	0	−1.86

solution, from which the molecular weight of the solute can be computed. Although simple in principle, exact determination requires great care.

Example/Problem 2–4.   Determine the molecular weight of an unknown from the following data: 4.35 g of the unknown was dissolved in 200 g of liquid camphor, giving a solution with a freezing point of 173.5°C.

$$\Delta T_f = 179.0° - 173.5° = 5.5°$$

$$\Delta T_f = K_f m$$

$$5.5 = 38.0 \times m$$

$$m = 5.5/38.0 = 0.145 \text{ mole}/1000 \text{ g solvent}$$

$$0.145/5 = 0.029 \text{ mole}/200 \text{ g solvent}$$

$$\text{Molecular wt.} = \frac{4.35 \text{ g}}{0.029 \text{ mole}} = 150$$

In treating freezing points of mixtures, we generally assume that the liquid phase is homogeneous; that is, the two liquids are miscible. But the two solid substances may or may not form a homogeneous phase. A solid phase whose composition is the same throughout is, for the present discussion, called a *solid solution*. Some mixtures of solids are able to form homogeneous solid solutions while others remain as mixtures of two pure solid phases.

Mixtures with No Solid Solutions.   If we start with pure naphthalene, m.p. 80°C, and add camphor, the argument is the same as for adding naphthalene to pure camphor. The two effects are combined in Figure 2–12, in which the melting point is plotted as a function of the mole % composition of the liquid phase. How can we use this graph?

1. The melting points of the pure compounds are given on the respective axes at 100% camphor and 100% naphthalene, points A and B.
2. Curve AE gives the initial freezing points of camphor containing various amounts of naphthalene, for example, a liquid mixture containing 20% naphthalene begins to freeze at 117°C, point C.
3. Curve BE gives the initial freezing points of naphthalene containing various amounts of camphor.
4. Curves AE and BE intersect at point E which gives the composition of the mixture having the lowest possible freezing point, called the *eutectic point* (from the Greek words meaning "easily melted").
5. If a liquid mixture containing 20% naphthalene (point D) is cooled (curve DC), crystals of pure camphor will appear when the temperature reaches 117°C. Further cooling will cause additional pure camphor to solidify. The

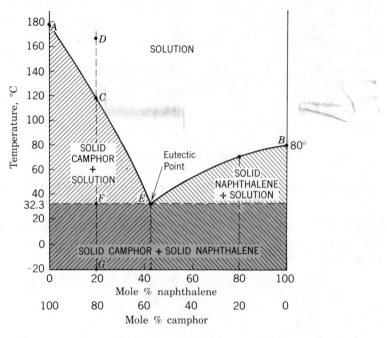

Figure 2–12 Melting point–composition diagram for the camphor–naphthalene system.

liquid phase thus becomes enriched in naphthalene and its freezing point decreases along curve *CE*. When point *E* is reached, the remaining liquid (now at the eutectic composition) freezes. When all the liquid has frozen, the composition of the solid must, of course, be the same as the composition of the original liquid mixture, 20% naphthalene. Further cooling follows line *FG*.

6. Heating a mixture of solids containing 20% naphthalene would follow the reverse course from (5) above. At a temperature of 32.3°C, eutectic proportions (58% camphor–42% naphthalene) of the solid will melt until no solid naphthalene remains. The temperature remains fixed at 32.3°C during the melting of the eutectic. Then the remaining camphor melts, gradually decreasing the concentration of naphthalene in the liquid melt and thus raising the melting temperature along *EC*. The heating curve for this process is given in Figure 2–13.

7. A eutectic mixture has a sharp melting point and a sharp freezing point. Thus a sharp melting point doesn't necessarily prove a pure compound. However, the addition of a small amount of either pure compound to a eutectic mixture will raise its melting point, whereas the addition of a small amount of either pure compound to the other pure compound will lower its melting point.

Identification by Melting Point. The sharp melting point of a pure compound serves as a means of identification. However the usual "pure" organic compound seldom has a melting range of less than 0.3 to 0.5°. Thus, several compounds may have essentially the same melting points and could be confused. Pure samples of the suspected compounds permit a more positive identification. Portions of the unknown are mixed with each of the known compounds and the melting point of each mixture is taken. The "mixed" melting point of all such mixtures will be lowered except when the two compounds are identical.

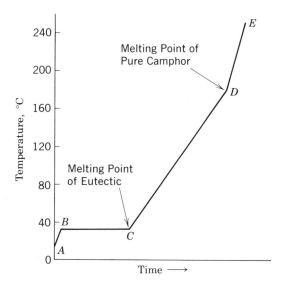

Figure 2–13 Heating curve for a solid sample with an average composition of 20 mole % naphthalene–80 mole % camphor.

Liquid–Vapor Equilibrium for Binary Mixtures. Let us consider a mixture of n-hexane (A) and n-heptane (B). The molecules of A will exert a certain vapor pressure, P_A, which will be less than if the liquid were pure A, having a vapor pressure of P_A°. Likewise, molecules of B exert a vapor pressure of P_B which is less than P_B°. As given by Raoult's law (Equation 2–6):

$$P_A = X_A P_A^\circ \tag{2–8}$$

$$P_B = X_B P_B^\circ \tag{2–9}$$

$$P = P_A + P_B = X_A P_A^\circ + X_B P_B^\circ \tag{2–10}$$

where X_A and X_B are the mole fractions of A and B in the liquid solution and P is the total pressure in the vapor phase above the liquid. This law is valid only for an ideal solution, or to put it the other way around, an *ideal solution* is one that obeys Raoult's law. Ideal behavior is most likely to be observed if the two components are chemically similar. For example, our mixture of two paraffins, n-hexane and n-heptane, has ideal behavior as shown in Figure 2–14.

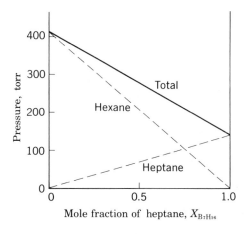

Figure 2–14 Vapor pressure–composition curves for the hexane–heptane system at 50°C.

More often the linear relations expressed in Equations 2–8 and 2–9 are valid only for highly dilute solutions, and even then only for the major component—the solvent. In this instance, the vapor pressure of the solute, P_A, often obeys Henry's law formulated by William Henry in 1804.

$$P_A = kX_A \qquad\qquad (2\text{–}11)$$

where k is an empirical constant determined in the region where X_A approaches zero.

The data for Figure 2–14 pertain to a constant temperature. In Figure 2–15 the total vapor pressure for a mixture of n-hexane and n-heptane is plotted against the composition of the mixture. The family of curves shows the effect of temperature. The intersection of each curve with the horizontal dashed line at 760 torr gives the boiling temperature of the corresponding mixture; that is, a solution which is 2.5 mole % heptane will boil at 70°C, 46 mole % heptane will boil at 80°C, and 78 mole % heptane at 90°C. If we had more lines on the graph, we could find the boiling point of any mixture of n-hexane and n-heptane.

It is more informative to plot the boiling point itself as a function of composition at a constant pressure of 760 torr. Also, it is equally important to consider the composition of the vapor which is escaping from the solution. Let X_A and X_B represent the mole fractions of hexane and heptane in the liquid phase, respectively, and Y_A and Y_B the mole fractions of hexane and heptane in the vapor phase, where there is also a simple relationship between partial pressure and composition:

$$\frac{P_A}{P_T} = Y_A \qquad \text{and} \qquad \frac{P_B}{P_T} = Y_B \qquad\qquad (2\text{–}12)$$

Now the ratio of P_A/P_B is readily obtained from Equations 2–8, 2–9 and 2–12:

$$\frac{P_A}{P_B} = \frac{Y_A}{Y_B} = \frac{X_A P_A^\circ}{X_B P_B^\circ} = \alpha\,\frac{X_A}{X_B} \qquad\qquad (2\text{–}13)$$

where α is called the *relative volatility* of A with respect to B and is equal to the

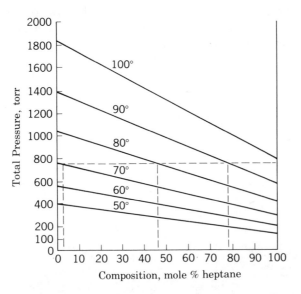

Figure 2–15 Effect of temperature on the total vapor pressure for the hexane–heptane system.

ratio of their saturation vapor pressures, P_A^o/P_B^o. If component A is the more volatile component, then $P_A^o > P_B^o$; therefore, Y_A/Y_B must be greater than X_A/X_B. This is *the fundamental principle of fractional distillation*—namely, that in the process of boiling a solution, the vapor becomes enriched in the more volatile component.

Example/Problem 2–5. What is the composition of the vapor which is in equilibrium with a liquid containing 46 mole % heptane–54 mole % hexane at its boiling point?

From Figure 2–15, the mixture boils at 80°C.

The vapor pressure of hexane is $X_A P_A^o = 0.54 \times 1050 = 567$ torr.

The vapor pressure of heptane is $X_B P_B^o = 0.46 \times 427 = 193$ torr.

The vapor composition is $\dfrac{567}{760} \times 100 = 74.5$ mole % hexane

$$\text{and} \quad \frac{193}{760} \times 100 = 25.5 \text{ mole \% heptane}$$

Thus, an enrichment from 54 to 74.5 mole % hexane.

Suppose, for example, that we condense the first vapors from the distillation given in Example/Problem 2–5. The resulting liquid phase (condensate) now contains 74.5 mole % hexane. If this condensate is distilled, a calculation similar to that in the example shows that the vapor will be further enriched to 87.5 mole % hexane. Many more similar calculations would give the two lines on Figure 2–16 in which pairs of points at the ends of horizontal tie-lines give the composition of the vapor and liquid which are in equilibrium at the boiling point of the given mixture.

The enrichment of the vapor so obtained is a function of the relative vapor pressures of the two components. A single distillation of a solution cannot yield pure products. However, an understanding of the enrichment process is basic to the study of fractional distillation.

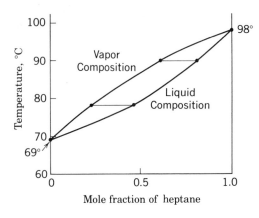

Figure 2–16 Boiling point–composition diagram for the hexane–heptane system. Upper curve is the vapor composition and lower curve is the liquid composition of the boiling mixtures. Horizontal tielines connect vapor and liquid composition in equilibrium at various temperatures. Lower curve is also the boiling point as a function of composition.

2–4 FRACTIONAL DISTILLATION

We have shown that, except for azeotropes which will be discussed later, simple distillation results in a partial separation of two components—the vapor phase is enriched in the more volatile component. Neither phase is pure, however. In the technique of fractional distillation, we repeat this partial separation process many times, each time obtaining a further separation.

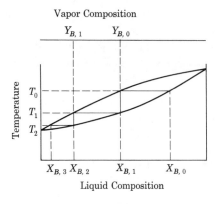

Vapor Composition

Figure 2–17 The enrichment process during the fractional distillation of a mixture of hexane and heptane.

We next consider the following very effective, but very impractical, experiment for which the data are given in Figure 2–17. We begin with a large amount of solution of composition $X_{B,0}$—so large an amount that subsequent removal of vapor will not change its composition materially. When it is heated to T_0 it will begin to boil, producing a vapor of composition $Y_{B,0}$. This vapor is condensed in a separate part of the apparatus producing a liquid of the same composition, $X_{B,1}(= Y_{B,0})$, with a boiling point T_1. This condensate is now brought to temperature T_1 and a small amount of its vapor is collected. The second condensate has a composition $X_{B,2}$ and a boiling point of T_2. This stepwise process can be repeated until the last vapor is essentially pure hexane. But we can remove only a small amount of vapor each time. Unless we have an infinite amount of starting material, the yield is infinitesimally small.

Bubble-Cap Column. The experiment just described is better carried out in a bubble-cap column, as illustrated in Figure 2–18. The original mixture is heated in the still pot 0 to its boiling point. The vapor passes through plate 1 and is deflected and condensed by the bubble cap. Plate 1 is maintained at the boiling point of the mixture contained therein, which is, of course, somewhat lower than that of the original mixture in the still pot. The vapors formed in plate 1 are condensed at the

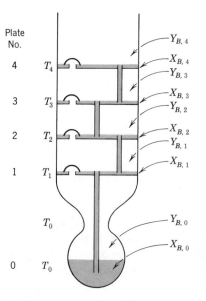

Figure 2–18 Operation of a bubble-cap distillation column.

bubble cap in plate 2, and so forth up to the top of the column. The excess liquid in each plate flows through the overflow tube down to the next plate. A condenser is placed at the top of the column. When the system has reached equilibrium, the composition of the vapor and liquid in each plate corresponds to the steps drawn in Figure 2–17. Accordingly, the vapors in each plate are progressively enriched in the more volatile component.

By adding a sufficient number of plates, it is possible to separate two components to any required degree of purity. However, the addition of plates requires that more of the mixture will be "held up" within the column. Furthermore, as soon as any product is removed from the top of the column, the equilibrium existing in the top plate is disturbed and this effect is shortly reflected down the column. Any departure from equilibrium results in a less effective separation. Therefore, it is necessary to make a compromise between desirable yield per unit time, or *throughput*, and purity. The *effective number of plates* is equal to the number of theoretical enrichment steps, and this is always less than the actual number of plates in the column.

The number of theoretical plates, n, required to enrich a binary mixture of A and B originally containing concentrations $X_{A,0}$ and $X_{B,0}$ to a given degree of purity, $X_{A,f}$, can be computed from the relative volatility, α. The vapor from the first vaporization in the still pot gives

$$\frac{Y_{A,0}}{Y_{B,0}} = \alpha \frac{X_{A,0}}{X_{B,0}} \tag{2-14}$$

or

$$\frac{Y_{A,0}}{1 - Y_{A,0}} = \alpha \frac{X_{A,0}}{1 - X_{A,0}} \tag{2-15}$$

When the vapor from the still pot condenses in the first plate, its concentration does not change; that is, $X_{A,1} = Y_{A,0}$, and when it evaporates from the first plate

$$\frac{Y_{A,1}}{1 - Y_{A,1}} = \alpha \frac{X_{A,1}}{1 - X_{A,1}} = \alpha \frac{Y_{A,0}}{1 - Y_{A,0}} = \alpha^2 \frac{X_{A,0}}{1 - X_{A,0}}$$

After n such plates, giving $n + 1$ repetitive evaporation-condensations

$$\frac{Y_{A,n}}{1 - Y_{A,n}} = \alpha^{n+1} \frac{X_{A,0}}{1 - X_{A,0}} \tag{2-16}$$

or

$$\log \frac{Y_{A,n}}{1 - Y_{A,n}} = (n + 1) \log \alpha + \log \frac{X_{A,0}}{1 - X_{A,0}} \tag{2-17}$$

When the vapor in the nth plate condenses, the composition of the final distillate will be $Y_{A,n} = X_{A,f}$. Therefore

$$\log \frac{Y_{A,n}}{1 - Y_{A,n}} = \log \frac{X_{A,f}}{1 - X_{A,f}} \tag{2-18}$$

and

$$n + 1 = \frac{\log \dfrac{X_{A,f}(1 - X_{A,0})}{X_{A,0}(1 - X_{A,f})}}{\log \alpha} \tag{2-19}$$

Equation 2–19 is called the *Fenske equation*.

Example/Problem 2–6. How many theoretical plates are required to enrich an equi-molar mixture of benzene and toluene ($\alpha = 2.47$) so that the final distillate contains $X_{A,f}$ (benzene) $= 0.995$?

$$n = \left[\left(\log \frac{0.995 \times 0.500}{0.500 \times 0.005}\right)\Big/ \log 2.47\right] - 1$$

= 4.9 plates, not including the still pot and final condenser.

In practice, at least six plates would be required for the separation just described. As a rough approximation, let us assume (1) that a "good" separation should yield a product at the top of the column with a composition of at least 95% of the more volatile component while leaving a still-pot composition of 95% of the less volatile component, (2) that a typical mixture has an average boiling point of 150°C, and (3) that the column is operated with total reflux. Then for a binary mixture whose pure components have boiling points differing by ΔT_b, the minimum number of plates required for a "good" separation is given in Table 2–4. Simple unpacked columns rarely have more than one theoretical plate, although the still pot itself also furnishes one plate.

Table 2–4 Number of Theoretical Plates Required to Make a Good Separation

ΔT_b, °C	No. of Plates
108	1
72	2
36	5
20	10
10	20
2	100

[After K. B. Wiberg, *Laboratory Technique in Organic Chemistry*, McGraw-Hill, New York, 1960, p. 44.]

Reflux Ratio. Condensation of the vapors takes place at the top of any column, and the condensate is either withdrawn as product or returned to the column. The ratio of the amount returned to that withdrawn is called the *reflux ratio*, R, which can vary from zero to infinity. For large-scale industrial purposes, a low value of R (often less than unity) is desirable in order to increase the yield of distillate. For analytical uses, large values of R are needed (10 to 50 are common) in order to maintain conditions close to equilibrium and thus to obtain a better separation. The reflux ratio may be fixed by the geometry of the column head or may be varied in some columns by adjusting the position of the take-off, or by altering the position of a stopcock, possibly with an automatic timing device. As a rule of thumb, the reflux ratio should be approximately equal to the number of plates in the column.

Fractionating Columns. In practice, bubble-cap columns are not convenient for laboratory work. The yield is too small relative to the large amount of material contained in the column; in other words a bubble-cap column has a small *throughput* with a large *holdup*. An approximation of the phenomenon occurring in the bubble-cap column can be achieved by inserting baffles, projections, or various kinds of loose or porous packing materials in an open column. Vapors are thus condensed and partially revaporized many times as they pass up the column, and there is a continuous flow of condensate back down the column. If the column is properly insulated, its temperature will gradually decrease toward the top of the column. The effectiveness of such a column depends on many factors, such as the design of the packing, temperature control, the length of the column and the rate at which the product is removed. In order to measure the column's efficiency under operating conditions, we compare the result (degree of separation) obtained from the given column with that to be expected from an ideal bubble-cap column giving the same degree of separation. The number of actual plates in the corresponding bubble-cap column defines the *number of "theoretical plates"* in the given column. An equivalent description of the number of theoretical plates is the number of steps required in the boiling point-composition diagram to reach the same degree of separation; one equilibrium separation is achieved in each theoretical plate. The fundamental unit of efficiency is the *height equivalent to a theoretical plate*, HETP or *H*, which is equal to the length of the column divided by the number of theoretical plates. It should be noted that the number of theoretical plates, and *H*, depends on the nature of the mixture to be separated. We will return to the theoretical plate concept many times in the discussions of extraction and chromatography.

Types of Fractionating Columns. A few types of columns in common use will be described briefly (Figure 2-19).

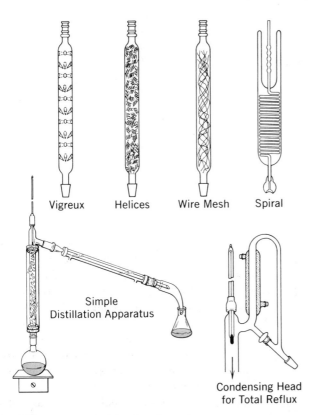

Vigreux   Helices   Wire Mesh   Spiral

Simple
Distillation Apparatus

Condensing Head
for Total Reflux

Figure 2-19 Typical fractionation columns.

A Vigreux column consists of a glass tube which has been indented in a regular fashion with projections extending inward and slightly downward. It may be insulated with asbestos tape or sealed in a vacuum jacket. It is inexpensive and permits a relatively high throughput with a low holdup. A Vigreux column is much better than a plain empty tube, but at best it is still rather inefficient.

A glass tube packed with irregularly shaped pieces of material is one of the most common forms of fractionating columns. Glass or metal helices offer a large surface for good equilibration between the vapor and liquid. Metal helices give more efficient columns than glass, but cannot be used with corrosive mixtures. A column packed with a copper or stainless steel sponge is a surprisingly good and very inexpensive variation. It is adequate for many simple laboratory distillations.

A concentric-tube column consists of a straight, uniform diameter inner tube exactly centered within a precision-bore outer tube. The vapors pass through the annular space (~ 0.75 mm) between the tubes while the descending liquid flows down the walls. The high area-to-volume ratio gives extremely good efficiency with a very low holdup.

The spinning-band column is designed for optimum performance. A twisted strip of wire gauze is inserted into a tube and spun at 2000 to 3500 rpm. A very low holdup is achieved because excess liquid is thrown against the wall of the tube, allowing it to drain freely.

Typical performance data are given in Table 2–5. The limits given are approximate for ordinary operation.

Table 2–5 Performance Data of Fractionating Columns

Type	Throughput, ml/min	Holdup, ml/plate	HETP, cm
Vigreux	5–10	0.5–2	7–12
Glass helices	2–7	0.7–1	3–5
Metal helices	1–5	0.2–0.5	1.0–1.5
Concentric-tube	0.5–2	0.02–0.03	0.5–1.0
Spinning-band	3–5	0.01–0.03	1–3

2–5 NONIDEAL SOLUTIONS

When a solution displays a total vapor pressure greater than that predicted by Raoult's law, it is said to show a *positive deviation*. The molecules of each substance seem to prefer their own kind and have an abnormally large tendency to escape from a mixture. The solution thus has an abnormally low boiling point. Perhaps the best known example is ethanol–water; its boiling point–composition diagram is shown in Figure 2–20. A minimum boiling point is observed at a composition of 95.6% (w/w) ethanol. To the left of this minimum, water is the more volatile component, and to the right, ethanol is the more volatile. A solution containing 95.6% ethanol has the lowest boiling point and is known as an *azeotropic mixture* because it cannot be separated by distillation alone.

If the vapor pressure of a mixture is lower than that predicted by Raoult's law, the solution has a *negative deviation*. These mixtures exhibit an abnormally high

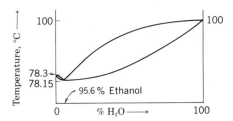

Figure 2–20 Boiling point–composition diagram for the ethanol–water system.

boiling point, as exemplified by the acetone–chloroform system shown in Figure 2–21. This behavior is to be expected whenever the two components react in some way with each other (formation of loose molecular complexes). The composition of the solution remaining in the still pot always approaches that corresponding to the maximum boiling point and then boils with a constant composition. This, too, is an azeotrope; a better known example of this kind is constant-boiling hydrochloric acid which boils at 108.6°C at 1 atmosphere with a composition of 20.2% ($6N$) HCl.

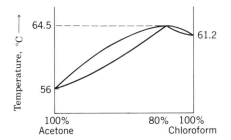

Figure 2–21 Boiling point–composition diagram for the acetone–chloroform system.

There are no doubt many more azeotropes than are normally recognized because they frequently occur with a composition of nearly 100% of one of the components. There is no way to predict their occurrence *a priori*. Some typical cases are given in Table 2–6.

Table 2–6 Typical Azeotropic Mixtures

	A. With Minimum Boiling Point	
Component, %	*Boiling Point of Pure Component,* °C	*Minimum Boiling Point,* °C
66.7 Benzene	80.2 ⎱	71.9
33.3 Isopropanol	82.5 ⎰	
11.7 *tert*-Butanol	82.8 ⎱	79.9
88.3 Water	100.0 ⎰	
44 Methanol	64.7 ⎱	62.3
56 Ethyl acetate	77.1 ⎰	

Table 2–6 (*continued*)

<table>
<tr><td colspan="3" align="center">B. With Maximum Boiling Point</td></tr>
<tr><td>Component, %</td><td>Boiling Point
of Pure Component, °C</td><td>Maximum
Boiling Point, °C</td></tr>
<tr><td>31.3 Acetic acid
68.7 Triethylamine</td><td>118.1⎫
89.4⎭</td><td>162</td></tr>
<tr><td>77 Formic acid
23 Water</td><td>101⎫
100⎭</td><td>107.1</td></tr>
<tr><td>42 Phenol
58 Aniline</td><td>181.5⎫
184.4⎭</td><td>186.2</td></tr>
</table>

QUESTIONS AND PROBLEMS

2–1. The "observed" melting point of some impure camphor is 170.0°C. The thermometer is immersed in the melting point bath (See Figure 2–3) to the 120° mark. A second thermometer shows that the average temperature of the exposed stem is 120°C. What is the corrected melting point of the impure camphor? The melting point of pure camphor is 178.9°C. What is the molality of the impurity? The impurity is known to be naphthalene. Calculate the percentage by weight of naphthalene in the impure camphor. *Ans.* 170.4°C, 0.224 m, 2.78% w/w

2–2. Plot the stem correction as a function of observed reading for a thermometer which is always immersed to the $-10°$ mark, assuming that the average temperature of the exposed stem is 30°C.

2–3. Calculate the expected freezing point of an aqueous solution containing 10 g of glycerol in 50 g of water. *Ans.* $-4.0°C$

2–4. A 7.50-g sample of a paraffin hydrocarbon was dissolved in 50.0 g of benzene. The freezing point of the solution was 0.0°C. Identify the hydrocarbon. *Ans.* $C_{10}H_{22}$

2–5. If you were to choose a solvent for the determination of molecular weight by the freezing point depression, which solvent (from those in Table 2–3) would you select for maximum accuracy? Why?

2–6. Describe the composition of the system containing an intimate mixture of 2 moles of naphthalene and 2 moles of camphor as it is heated from 0° to 200°C.

2–7. A 1.00-g sample of a polymer with the general formula $(CH_2)_n$ dissolved in 7.5 g of benzene depressed the freezing point of benzene by 0.50°. What is the value of n? *Ans.* 98

2–8. Draw the temperature–composition curve for mixtures of cinnamic acid (m.p. 136.8°C) and benzoic acid (m.p. 121.5°C). The melting point of the eutectic mixture containing 57 mole % benzoic acid is 82°C. There are no solid solutions. Predict the shape of the cooling curve (temperature vs. time) for:
(a) Pure benzoic acid.
(b) A mixture containing 20% cinnamic acid.
(c) A mixture containing 20% benzoic acid.
(d) The eutectic mixture.

2–9. The vapor pressure of ether is 442 torr at 20°C, 647 torr at 30°C, 760 torr at 34.6°C. What is its heat of vaporization? If the pressures are expressed in atmospheres, the value of R is 1.987 cal mole^{-1} deg^{-1}. *Ans.* 7000 cal/mole

2–10. The heat of vaporization of water at its normal boiling point is 9718 cal mole^{-1}. What is its vapor pressure at 75°C? *Ans.* 299 torr

2–11. What is the vapor pressure of a benzene-toluene solution at 30°C? State the answer in the form of an equation involving mole fraction of benzene, and the two vapor pressures of the pure compounds—118 torr for benzene and 36 torr for toluene, both at 30°C. Assume Raoult's law is valid for this system.

2–12. From the data in Problem 2–11, calculate the mole fraction of benzene in the vapor which is in equilibrium with an equi-molar solution of benzene and toluene at 30°C.

Ans. 0.765

2–13. The vapor pressure of *n*-octane at 100°C is 351 torr, while that of *iso*-octane is 777 torr. If a fuel contains 90 mole % *iso*-octane and 10 mole % *n*-octane what is its vapor pressure at 100°C? What is the composition of the vapor in equilibrium with the liquid at 100°C? What additional information is needed in order to compute the boiling point of this fuel?   *Ans.* 734 torr, $Y_{iso} = 0.952$

2–14. The vapor pressures of some paraffins are given in Table 2–1.

(a) What is the vapor pressure of a 1:1:1 (by moles) mixture of *n*-pentane, *n*-heptane, and *n*-octane at 30°C?   *Ans.* 229 torr

(b) What is the vapor pressure of a 1:1:1 (by weight) mixture of *n*-pentane, *n*-heptane, and *n*-octane at 30°C?   *Ans.* 282 torr

(c) What mixture of *n*-pentane and *n*-heptane boils at 90°C under atmospheric pressure?   *Ans.* 5.9 mole % pentane

(d) *n*-Heptane boils at 98°C. What would be the composition of a mixture of *n*-pentane and *n*-octane which boils at the same temperature? How could you distinguish between these two liquids (heptane and the pentane–octane mixture) which have the same boiling point by distillation alone?

(e) What is the composition of a mixture of *n*-hexane and *n*-heptane which boils at 80°C?   *Ans.* $X_{hex} = 0.535$

(f) What is the composition of the vapor over a mixture of *n*-hexane and *n*-heptane (mole fraction = 0.5) at 100°?   *Ans.* $Y_{hex} = 0.70$

(g) Does the composition of the vapor calculated in (f) depend on temperature?

2–15. A liquid mixture of 3 moles of *A* and 2 moles of *B* boils freely at 100°C when atmospheric pressure is 760 torr. Calculate the vapor pressure of pure *A* if the vapor pressure of pure *B* is 400 torr (all at 100°C).   *Ans.* 1000 torr

2–16. Calculate the total pressure of the system and the mole fraction of *A* in the vapor of a mixture of *A* and *B* ($X_A = 0.3$) at a temperature where the vapor pressure of pure *A* is 700 torr and that of pure *B* is 300 torr.   *Ans.* 420 torr, 0.5

2–17. Why should the reflux ratio be increased when the number of plates in the column is increased?

2–18. Why is the effective number of plates in a bubble-cap column less than the actual number?

2–19. How would you expect *H* to vary with the reflux ratio?

REFERENCES

G. R. Robertson, W. E. Truce, and T. L. Jacobs, *Laboratory Practice in Organic Chemistry*, 5th ed., Macmillan, New York, 1974; Chapters 5–8.

J. Cason and H. Rapoport, *Laboratory Text in Organic Chemistry*, 2nd ed., Prentice-Hall, Englewood Cliffs, N.J., 1962; pp. 9–17, 30–43, 270–311.

K. B. Wiberg, *Laboratory Technique in Organic Chemistry*, McGraw-Hill, New York, 1960; Chapters 1 and 2.

3

EXTRACTION

3–1   INTRODUCTION

Extraction methods have much in common with distillation methods. In fractional distillation, a separation of components is possible because of the difference in vapor pressure or volatility of the components. At a given temperature and pressure, the equilibrium concentrations of a component in the liquid phase, C_L, and vapor phase, C_G, are expressed by the equation:

$$K = \frac{C_L}{C_G} \qquad (3-1)$$

where K is an equilibrium constant. For a two-component system, K is greater for the less volatile component than for the more volatile. Therefore, in the process of vaporization we have achieved a partial separation of the original mixture.

Extraction is an analogous separation process in which a solute is distributed between two immiscible solvents. A similar law defines the ratio of the concentrations of the solute in the two solvents, 1 and 2:

$$K_D = \frac{C_1}{C_2} \qquad (3-2)$$

where K_D is the *distribution coefficient* or *partition coefficient*—a special type of equilibrium constant which is related to the relative solubilities of the solute in the two solvents. Often one solvent is water and the other is an organic solvent, so that inorganic ionic species as well as polar organic compounds are found largely in the aqueous phase while nonpolar organic compounds are largely in the organic phase. This is another way of saying "like dissolves like." In dilute solution, to a first approximation, the distribution coefficient is independent of concentration. More precisely, activities should be used in Equation 3–2.

Example.   Suppose that we wish to separate the excess fatty acids from a sample of toilet soap. A pair of solvents such as ether and water would be very effective because fatty acids are far more soluble in ether than in water, while the opposite is true for soap. If ether is arbitrarily defined as "solvent 1" in Equation 3–2, then K_D is very large for fatty acids and very small for soap.

In Equation 3–2, phase 1 appears in the numerator and phase 2 in the denominator. Often the organic phase is placed in the numerator, but just as often, the lighter phase (which may or may not be organic) is placed in the numerator. Thus, the assignment of phase numbers is arbitrary and must be explicitly stated to avoid ambiguity in quoting values of the distribution coefficients.

3–2 DISTRIBUTION LAW

The distribution coefficient pertains only to a single species and does not include possible products of side reactions. For example, let us consider the extraction of benzoic acid, HB, from water (acidified with HCl to suppress dissociation of the benzoic acid) into an organic solvent such as ether. Figure 3–1a illustrates the equilibrium situation, for which

$$K_D = \frac{[HB]_{et}}{[HB]_{aq}} \tag{3-3}$$

Complications arise, however, if the aqueous layer is not acidified and the benzoic acid dissociates

$$HB \rightleftharpoons H^+ + B^-$$

$$K_a = \frac{[H^+][B^-]}{[HB]} \tag{3-4}$$

Now there are two independent equilibria as illustrated in Figure 3–1b. Note that the partition equilibrium pertains only to the undissociated benzoic acid molecules in the two phases, and that the dissociation equilibrium pertains only to the species in the aqueous phase (benzoic acid does not dissociate in an ether solution). A second complication arises if benzene is used as the organic solvent in place of ether, as shown in Figure 3–1c. Benzoic acid is partially dimerized in benzene

$$2HB \rightleftharpoons HB \cdot HB$$

$$K_d = \frac{[HB \cdot HB]}{[HB]^2} \tag{3-5}$$

In extracting benzoic acid, we may want to know how much benzoic acid, regardless of its form, is in each phase. The expression which takes all forms into account is called the *distribution ratio, D*

$$D = \frac{\text{total concentration of benzoic acid in organic phase}}{\text{total concentration of benzoic acid in aqueous phase}} \tag{3-6}$$

or

$$D = \frac{[HB]_{org} + 2[HB \cdot HB]_{org}}{[HB]_{aq} + [B^-]_{aq}} \tag{3-7}$$

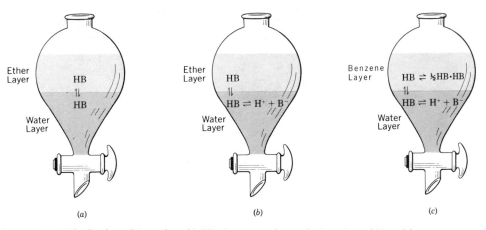

Figure 3–1 Distribution of benzoic acid, HB, between ether and water (a and b) and between benzene and water (c) in an extraction.

In this particular case, we can make substitutions as follows.
From Equation 3–4:

$$[B^-] = K_a \frac{[HB]}{[H^+]} \tag{3-8}$$

From Equation 3–5:

$$[HB \cdot HB] = K_d [HB]^2 \tag{3-9}$$

Substituting from Equations 3–8 and 3–9 into Equation 3–7, we obtain

$$D = \frac{[HB]_{org} + 2K_d[HB]^2_{org}}{[HB]_{aq} + K_a \dfrac{[HB]_{aq}}{[H^+]_{aq}}} = \frac{[HB]_{org}(1 + 2K_d[HB])}{[HB]_{aq}(1 + K_a/[H^+])} = \frac{K_D(1 + 2K_d[HB])}{1 + K_a/[H^+]} \tag{3-10}$$

Equation 3–10 shows that the distribution ratio can be changed by the simple expedient of changing the pH of the aqueous solution. In acid solution (high $[H^+]$, low pH), D will be large and benzoic acid will be found largely in the organic layer. In alkaline solution (low $[H^+]$, high pH), D will be small and benzoic acid will be found in the aqueous layer (almost entirely as benzoate ions).

Example/Problem 3–1. One gram of benzoic acid originally dissolved in 100 ml of water is to be equilibrated with 100 ml of ether. The distribution coefficient, K_D is 100 and the dissociation constant, K_a, is 6.5×10^{-5}. Calculate the distribution ratio, D, if the aqueous layer is at pH 3, pH 5 and pH 7. Calculate D at several other pH values and plot D versus pH for this system.

$$D = \frac{K_D}{1 + K_a/[H^+]}$$

At pH 3:
$$D = \frac{100}{1 + 6.5 \times 10^{-5}/10^{-3}} = \frac{100}{1.065} = 93.9$$

At pH 5:
$$D = \frac{100}{1 + 6.5 \times 10^{-5}/10^{-5}} = \frac{100}{7.5} = 13.3$$

At pH 7:
$$D = \frac{100}{1 + 6.5 \times 10^{-5}/10^{-7}} = \frac{100}{651} = 0.15$$

Expressions similar to Equation 3–10 can be derived for other types of side reactions, for example, formation of ion pairs or metal complexes in either or both phases. If the side reactions involve ionic (charged) species, one can assume that ions exist only in the aqueous layer, or in a very polar organic solvent (high dielectric constant). The characteristics of some useful solvents are listed in Table 3–1.

3–3 SUCCESSIVE EXTRACTIONS

If the distribution coefficient is very large (> 1000), a single extraction in a simple separatory funnel will probably remove essentially all of a solute from one phase to another. However, it can be shown that for a given amount of extracting solvent, it is more effective to divide it into several small portions and use each portion successively rather than to make a single extraction with all of the solvent at one time.

For example, assume that 4 g of butyric acid is to be extracted from 500 ml of water with 500 ml of ether. The distribution coefficient for this system is 3.0 at

Table 3–1   Some Useful Solvents

Solvent	Boiling Point, °C	Freezing Point, °C	Dielectric Constant, Debye units
Diethyl ether	35	− 116	4.3
Carbon disulfide	46	− 111	2.6
Acetone	56	− 95	20.7
Chloroform	61	− 64	4.8
Methanol	65	− 98	32.6
Tetrahydrofuran	66	− 65	7.6
Di-isopropyl ether	68	− 60	3.9
Carbon tetrachloride	76	− 23	2.2
Ethyl acetate	77	− 84	6.0
Ethanol	78	− 117	24.3
Benzene	80	5.5	2.3
Cyclohexane	81	6.5	2.0
Isopropanol	82	− 89	18.3
Water	100	0	78.5
Dioxane	102	12	2.2
Toluene	111	− 95	2.4
Acetic acid (glacial)	118	17	6.2
N,N-Dimethylformamide	154	− 61	34.8
Diethylene glycol	245	− 10	37.7

25°C. If the ether is used in a single batch:

$$K_D = C_{et}/C_{aq} = 3.0 = \frac{(4-x)/0.5}{x/0.5}$$

where x is the weight of butyric acid remaining in the water layer. Thus $x = 1$, and 3 g is extracted into the ether layer, as shown in Figure 3–2. However, if the ether is used in two successive 250-ml portions, for the first extraction:

$$K_D = 3.0 = \frac{(4-x_1)/0.25}{x_1/0.50}$$

In this case, $x_1 = 1.60$ g remains in the water layer, and 2.40 g is found in the ether layer, which is then removed. In the second extraction with the remaining 250-ml portion of ether:

$$K_D = 3 = \frac{(1.6-x_2)/0.25}{x_2/0.50}$$

After the second extraction, $x_2 = 0.64$ g remains in the water layer, and an additional 0.96 g is extracted by the ether. Thus a total of $2.40 + 0.96 = 3.36$ g has been extracted, a significant improvement.

A similar calculation shows that had the ether been divided into five 100-ml portions, only 0.23 g of the original 4 g of butyric acid would remain in the aqueous phase after the fifth extraction, and the combined ether extracts would contain 3.77 g.

For the general case, assume that W_0 g of a solute originally present in V_A ml of solvent A is to be extracted with successive portions of V_B ml of solvent B, as

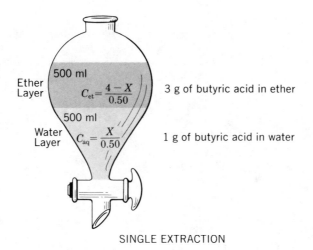

SINGLE EXTRACTION

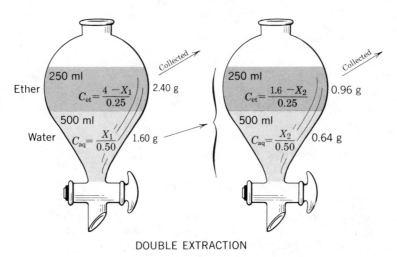

DOUBLE EXTRACTION

Figure 3–2 Comparison of extraction procedures for butyric acid: 4 g of acid originally in 500 ml of water extracted with one 500-ml portion of ether (top), or two 250-ml portions of ether (bottom).

shown in Figure 3–3.

$$K_D = C_B/C_A; \qquad W_0 = W_A + W_B$$
$$W_A = C_A V_A; \qquad W_B = C_B V_B$$

After the first equilibration:

$$\text{fraction in } A = \frac{W_{A,1}}{W_0} = \frac{C_{A,1} V_A}{C_{A,1} V_A + C_{B,1} V_B} = \frac{V_A}{V_A + \dfrac{C_{B,1}}{C_{A,1}} V_B} = \frac{V_A}{V_A + K_D V_B}$$

or

$$W_{A,1} = W_0 \left(\frac{V_A}{V_A + K_D V_B} \right)$$

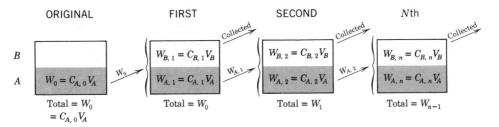

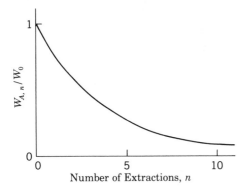

Figure 3–3 Successive extractions of sample of W_0 g originally in V_A-ml of water with V_B-ml portions of ether.

The second extraction is the same as the first, except that $W_{A,1}$ g of solute is present instead of W_0 g. Thus, by analogy (repeating the steps above):

$$W_{A,2} = W_{A,1}\left(\frac{V_A}{V_A + K_D V_B}\right) = W_0\left(\frac{V_A}{V_A + K_D V_B}\right)^2$$

Repeating the procedure through n equilibrations (collecting and combining the B layers) gives:

$$W_{A,n} = W_0 \left(\frac{V_A}{V_A + K_D V_B}\right)^n \tag{3–11}$$

where $W_{A,n}/W_0$ is the fraction of the solute remaining in the A layer after n extractions with n equal portions of solvent B. As n increases, $W_{A,n}$ decreases, but the relationship is exponential and diminishing returns soon set in, as shown in Figure 3–4.

More than five successive extractions are seldom worthwhile. If this number has not resulted in an adequate extraction, it is more practical to look for a better extractant (more favorable K_D). It should be noted that in deriving Equation 3–11, it was tacitly assumed that the two phases are completely separated after each equilibration. In practice, this is not always easy to do.

Figure 3–4 Fraction of solute remaining in water layer after a number of successive extractions with fresh portions of ether.

3–4   SEPARATION OF MIXTURES BY EXTRACTION

The extraction of a single substance from one solvent to another is of little interest. The great value of extraction is the possibility of separating two or more substances based upon a difference in their distribution coefficients. If one solute has a K much greater than 1, and the other much less than 1, a single extraction will cause nearly complete separation. This fortunate circumstance will arise only if two solutes are very different chemically, in which case the pair could no doubt be separated easily by some other method. If the two solutes have similar, but not identical, distribution coefficients, a single extraction will cause only a partial separation with an enrichment of one solute in one solvent and an enrichment of the other solute in the other solvent. If we are to make an adequate separation, we must repeat the process many times. After the first extraction, each phase may be further equilibrated with a fresh portion of the opposite solvent. By a systematic recycling of the various intermediate fractions, a satisfactory separation can eventually be achieved at a cost of considerable amounts of solvent and numerous manipulations. Lyman Craig has developed a machine to perform these operations semiautomatically. The mathematical relations describing the Craig process are helpful in understanding many column operations because the distribution profile of a substance as it passes through this apparatus approximates that obtained in a chromatographic separation to be discussed in the following chapters.

3–5   THE CRAIG METHOD OF MULTIPLE EXTRACTION

Craig Apparatus.   The apparatus consists of a series of separatory vessels connected so that the outlet of one vessel flows into the inlet of the next. Each vessel consists of two chambers connected to each other as shown in Figure 3–5. The operation is begun by introducing through inlet A an amount of the heavier solvent which will fill chamber B somewhat less than half full. Each of the vessels in the train is filled in a like manner. The sample to be separated is introduced as a solution in the lighter solvent into chamber B of the first vessel. The assembly is rocked back and forth through an angle of about 35° around pivot P. After equilibration has been achieved and the solvents separated into two layers, the assembly is rotated 90° clockwise. The lighter solvent flows through connecting tube C into chamber D while the heavier solvent is trapped in the lower part of chamber B. When the assembly is rotated back to its original position, the lighter solvent now in D flows through outlet E into chamber B of the *next* stage. Hundreds of these assemblies can be mounted side by side or in banks, and all of them rocked and rotated simultaneously by a motor timed to operate as indicated.

The Craig Process.   The Craig machine has been a very powerful and practical tool in biochemistry for extremely difficult separations of substances that are chemically very similar. In order to describe the operation mathematically, we will

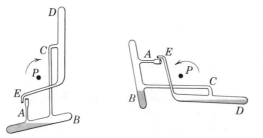

Figure 3–5 Extraction vessel of the Craig apparatus.

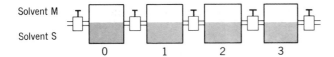

Solvent M

Solvent S

       0   1   2   3

Figure 3–6 Apparatus for the ideal Craig process.

use a schematic representation of the machine shown in Figure 3–6. Consider a series of vessels numbered consecutively from 0. Each vessel is of the same size and is initially half filled with the heavier solvent to be used (solvent S). There is also a series of connecting tubes and valves so that the solvent to be contained in the upper half of the vessels can be transferred from one vessel to the next when desired. No mixing is allowed during the transfer. We will follow the course of a single solute, although whether or not it is present in a mixture is immaterial since each solute should behave independently of all others. The arithmetic will be greatly simplified if we assume that each phase occupies one-half the volume of the vessel and that the K_D of the solute is 1.

To start the operation, we introduce the sample dissolved in the first portion of the lighter solvent, M, into vessel 0. After equilibration (shaking and settling), one-half of the solute is in the upper phase and one-half in the lower phase, S. The upper layer, solvent M, is then transferred to vessel 1 and a fresh portion of solvent M is added to vessel 0. After equilibration, one-quarter of the solute is now found in each phase of each vessel, 0 and 1. Next solvent M in vessels 0 and 1 is transferred to vessels 1 and 2, respectively, along with a fresh batch of solvent M to vessel 0. The pattern of the operation has now been established and the distribution of the solute develops as in Figure 3–7 in which the vessels are labeled across the top and the number of transfers labeled down the side. To continue the process, it is easier to tabulate the fraction of the solute found in each vessel (including both layers) after n transfers and equilibrations as shown in Table 3–2.

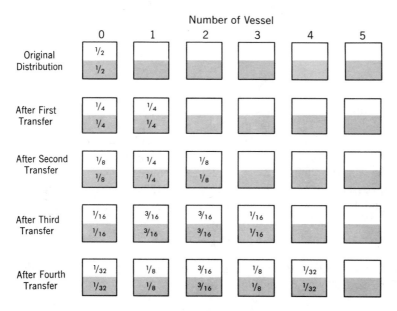

Figure 3–7 Successive distribution of solute in the Craig process. Fraction of total solute in each layer of each vessel for $K_D = 1$ and $V_M = V_S$.

Table 3–2   Distribution of Solute in the Craig Process: Fraction in Each Vessel

No. of Transfers	Vessel Number								
	0	1	2	3	4	5	6	7	
0	1								$\times 2^0$
1	1	1							$\times 2^{-1}$
2	1	2	1						$\times 2^{-2}$
3	1	3	3	1					$\times 2^{-3}$
4	1	4	6	4	1				$\times 2^{-4}$
5	1	5	10	10	5	1			$\times 2^{-5}$
6	1	6	15	20	15	6	1		$\times 2^{-6}$
7	1	7	21	35	35	21	7	1	$\times 2^{-7}$

The numbers on each line of the table are the successive terms in the expansion of the binomial $(p + q)^n$, that is,

$$(p + q)^n = p^n + np^{n-1}q + \frac{n(n-1)}{2!} p^{n-2}q^2 + \frac{n(n-1)(n-2)}{3!} p^{n-3}q^3 + \cdots + q^n$$

$$(3\text{–}12)$$

where p is the fraction of total solute which is in phase S of any vessel, q is the fraction of total solute in phase M of the same vessel, and n is the number of transfers. We define the distribution coefficient* as

$$K_D = \frac{C_S \text{ (the lower, or stationary phase)}}{C_M \text{ (the upper, or moving phase)}}$$

$$(3\text{–}13)$$

Then

$$p = \frac{C_S V_S}{C_S V_S + C_M V_M} = \frac{K_D V_S}{K_D V_S + V_M}$$

$$(3\text{–}14)$$

and

$$q = \frac{C_M V_M}{C_S V_S + C_M V_M} = \frac{V_M}{K_D V_S + V_M}$$

$$(3\text{–}15)$$

In the example we have been discussing, $K_D = 1$ and $V_S = V_M$; therefore, from Equations 3–14 and 3–15, both p and q are 1/2. We can now generate the numbers in Table 3–2 by substituting $p = 1/2$ and $q = 1/2$ into Equation 3–12, carrying out the binomial expansion for any given number of transfers. For example, after 7 transfers, vessel 0 contains $(1/2)^7$ or 1/128 of the solute, vessel 1 contains $7(1/2)^7$ or 7/128 of it, vessel 2 contains $((7 \times 6)/2)(1/2)^7$ or 21/128 of it, etc. In the general case, $V_S \neq V_M$ and $K_D \neq 1$, and a combination of Equations 3–12 to 3–15 yields

$$(p + q)^n = \left(\frac{K_D V_S}{K_D V_S + V_M} + \frac{V_M}{K_D V_S + V_M} \right)^n$$

$$(3\text{–}16)$$

The calculations become tedious for many transfers, but the general approach is straightforward following Equation 3–16.

* Note that K_D defined here is the inverse of K_D defined for Equation 3–11.

Example/Problem 3-2. Calculate the distribution of a substance after three transfers in a Craig apparatus for which $V_S = 2$ ml and $V_M = 4$ ml. The distribution coefficient as defined by Equation 3-13 is 3.0.

$$p = \frac{K_D V_S}{K_D V_S + V_M} = \frac{3 \times 2}{3 \times 2 + 4} = \frac{6}{10}$$

$$q = \frac{V_M}{K_D V_S + V_M} = \frac{4}{3 \times 2 + 4} = \frac{4}{10}$$

$$(p + q)^n = \left(\frac{6}{10}\right)^3 + 3\left(\frac{6}{10}\right)^2\left(\frac{4}{10}\right) + 3\left(\frac{6}{10}\right)\left(\frac{4}{10}\right)^2 + \left(\frac{4}{10}\right)^3$$

$$= \frac{216}{1000} + \frac{432}{1000} + \frac{288}{1000} + \frac{64}{1000}$$

Thus, the first four vessels contain 21.6%, 43.2%, 28.8%, and 6.4%, respectively.

So far we have considered only a single solute which is transferred a number of times, and we have shown that the distribution among the vessels is clearly a function of several variables: K_D, V_S/V_M and n. Let us now consider a sample which contains two components (solutes), each with a different distribution coefficient; for example, $K_A = 1/2$ and $K_B = 1$. If our sample is small so that none of the solutions become saturated, each of the solutes will behave independently. That is, we can calculate the distribution of each solute, as if it were the only one present. Curve A in Figure 3-8 shows the fraction of total solute A ($K_A = 0.5$) in each vessel (0 to 8) after eight transfers ($n = 8$). Curve B shows the distribution for solute B ($K_B = 1.0$). If the experiment is carried through 200 transfers, the distributions are as shown in Figure 3-8, curves A' and B'. Several important observations are worth noting:

1. With a separation factor of 2 ($\alpha = K_B/K_A$), 8 transfers give only a very partial separation of the solutes. All vessels, (except the first) contain significant amounts of both A and B.

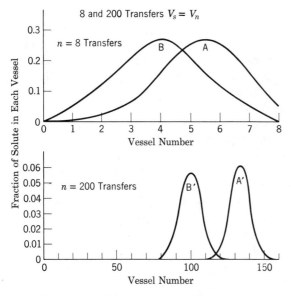

Figure 3-8 Distribution of solutes A ($K_A = 0.5$) and B ($K_B = 1.0$) in Craig separation after 8 and 200 transfers. $V_M = V_S$.

2. After 200 transfers, the separation is still not complete but probably adequate for most purposes. Vessels 80 to 117 contain nearly all of solute B and vessels 118 to 155 contain nearly all of solute A. Vessels 112 to 123 could be discarded to obtain a higher purity but with a lower yield.

3. Another 100 transfers would improve the separation but would increase the time and the amount of solvents required.

4. The concentration of the solutes decreases with an increase in number of transfers. With 8 transfers, the vessel at the center of the peak (maximum concentration) contains about 25% of the solute. With 200 transfers, the maximum amount of solute in one vessel is only about 6% of the total.

5. After 8 transfers, the solutes are spread over 8 vessels. After 200 transfers, each solute is spread over approximately 50 vessels. The "peaks" become broader with an increase in transfers; however, the solute occupies a smaller fraction of the vessels used.

6. The solute is never completely removed from even the first vessel, although after the first few transfers the amount remaining in the first vessel may be safely neglected. For example, if $K_D = 1$ and $V_S = V_M$, then $p = 1/2$ and the amount remaining in the first vessel is p^n, or 3% ($n = 5$), 0.1% ($n = 10$), 10^{-4}% ($n = 20$).

7. If the separation factor had been 1.1 instead of 2, 10,000 transfers would be required to achieve the same degree of separation—not very practical even with an automated Craig machine. Separations of this type are readily accomplished by chromatography, to be discussed in the next few chapters.

3–6 CONTINUOUS COUNTERCURRENT EXTRACTION

In the Craig apparatus, one solvent moves with respect to the other in a discontinuous fashion. Imagine now that the size of each equilibration vessel is reduced until the entire system becomes a single column, and that the extracting solvent is passed through the system continuously. If both solvents are restricted to thin layers, it is possible to approach an equilibrium state at all points in the apparatus even though it is not exactly attained anywhere. Instead of a series of discrete vessels, it is convenient to construct a column packed with some porous material which will hold one of the solvents stationary on its surface, while allowing the other solvent to percolate through it. The "stationary" solvent need not be fixed in position; it is only necessary that the two solvents "pass through each other," exposing a large interface. Thus the term "countercurrent" is appropriate. In this way an extremely large number of equilibrations can be achieved without expanding the apparatus unduly.

The relationship of continuous countercurrent extraction to the discontinuous Craig process is much the same as the relationship of continuous fractional distillation with a packed column to the ideal bubble-cap plate distillation. Hence, it is common to refer to "theoretical plates" in a continuous extraction column. Here this expression refers to the number of separate equilibrations and transfers which would have to be done in order to achieve the same degree of separation. Actually, countercurrent extraction is identical to one form of chromatography.

QUESTIONS AND PROBLEMS

3–1. What is the minimum value of K_D which would allow the extraction of 99.9% of a solute from 50 ml of water with five successive 50-ml portions of ether?   *Ans. K* = 3

3–2. If five extractions with 100-ml portions of ether extract 90% of a solute from an aqueous solution, what fraction of the solute will ten similar extractions remove?
Ans. 99%

3–3. Derive an expression similar to Equation 3–10 for the distribution of pyridine between water and benzene:

$$C_5H_5N + H^+ = C_5H_5NH^+ \text{ (in water)}$$

3–4. In the Craig process, some of the solute remains in the first vessel no matter how many transfers are made. Assuming that $V_S = V_M$, and that $K_D = 1$, what fraction of the solute will remain in the first two vessels after ten transfers?
Ans. $1/1024 + 1/102 \approx 1\%$

3–5. A sample containing approximately 1 g of an organic substance A is dissolved in 100 ml of water. The aqueous solution is then shaken with 100 ml of ether. Analysis of the ether layer shows that 0.7 g of A is extracted.
(a) What is the value of $K_D = C_{et}/C_{aq}$ for this system?   *Ans.* 2.33
(b) How many additional extractions with 100-ml portions of ether would be required to extract 95% of the substance?   *Ans.* 2 (additional)
(c) What fraction of A would have been extracted by four 25-ml portions of ether?
Ans. 84%

3–6. The distribution coefficient of an organic compound is $K_D = C_{aq}/C_{pent} = 5.0$. A 0.55-g sample is placed in vessel 0 of a Craig apparatus. Each vessel holds 4 ml of water and 4 ml of pentane (the lighter solvent which is transferred). Calculate the fraction of the compound which is in each vessel after four transfers.
Ans. 48.2%, 38.6%, 11.6%, 1.5%, 0.1%

3–7. Three extractions with 50-ml portions of chloroform removed 97% of a solute from 200 ml of an aqueous solution. Calculate the distribution coefficient of the solute: $K_D = C_{chlor}/C_{aq}$.   *Ans.* 8.9

3–8. Two weak bases (organic amines) have basic dissociation constants of $K_{bA} = 1 \times 10^{-4}$ and $K_{bB} = 1 \times 10^{-8}$. Both bases have distribution coefficients of approximately 10 between chloroform and water. Describe a procedure, making use of the difference in K_b, for separating the two by extraction.

3–9. Two weak acids, *HX* and *HY*, have distribution coefficients, $K_D = C_{et}/C_{aq}$, of 5 and 50, respectively, and dissociation constants of 1×10^{-4} and 1×10^{-8}, respectively. Calculate the distribution ratios of the two acids at integral values of pH of the aqueous solution from pH 4 to pH 11, assuming equal volumes of the two solvents. What is the minimum pH for which the ratio of *D* values is at least 10^5?
Ans. At pH 4 $D_{HX} = 2.5, D_{HY} = 50$
At pH 7 $D_{HX} = 0.005, D_{HY} = 45$
Minimum pH = 10

3–10. Formaldehyde has a distribution coefficient, $C_{et}/C_{aq} = 0.111$ at 25°. How many liters of water will be required to remove in one extraction 95% of the formaldehyde from 1 liter of ether containing 0.5 mole of formaldehyde?   *Ans.* 2.1 liters

3–11. How much formaldehyde would remain in 50 ml of ether initially containing 5 g of formaldehyde after five successive extractions with 25-ml portions of water?
Ans. 1 mg

3–12. Suppose you were given 100 ml of an aqueous solution containing 1 mg of LSD. You wish to extract the LSD into ether. The distribution coefficient is given by

$$K_D = \frac{C_{et}}{C_{aq}} = 10$$

(a) How much LSD would be extracted with one 150-ml portion of ether?

(b) How much LSD would be extracted with three successive 50-ml portions of
   ether?   *Ans.* (a) 0.935 mg (b) 0.995 mg

3–13. The dissociation constant for propionic acid, HOPr, in water is $K_a = 1.00 \times 10^{-5}$, and
its distribution coefficient between ether and water is $K_D = 2.85$. You are given a
solution of 8.00 g of HOPr in 500 ml of water which has been adjusted to pH 4.00 by
the addition of base. The HOPr is extracted from the aqueous solution with two
250-ml portions of ether. What is the pH of the water layer after the second
extraction?   *Ans.* 4.72

REFERENCES

L. C. Craig and D. Craig, in *Technique of Organic Chemistry*, 2nd ed., A. Weissberger, Editor, Vol. III,
   Part I, Wiley-Interscience, New York, 1956; pp. 150–392. A comprehensive and definitive
   discussion of extraction and distribution.

E. W. Berg, *Physical and Chemical Methods of Separation*, McGraw-Hill, New York, 1963; Chapter 3.
   Intermediate level discussion of extraction, probably more useful to the student than Weissberger.

H. Irving and R. J. P. Williams in *Treatise on Analytical Chemistry*, Part I, Volume 3, edited by I. M.
   Kolthoff and P. J. Elving, Wiley-Interscience, New York, 1959; pp. 1309–65.

G. H. Morrison and H. Freiser, *Solvent Extraction in Analytical Chemistry*, Wiley, New York, 1957.

B. L. Karger, L. R. Snyder and C. Horvath, *An Introduction to Separation Science*, Wiley-
   Interscience, New York, 1973.

4

CHROMATOGRAPHY

4-1   INTRODUCTION

A chemical analysis is of doubtful value if the property measured cannot be specifically related to the substance in question. In most cases involving real samples, interfering constituents must be removed and/or the desired constituent must be isolated or concentrated before it can be identified and measured. Numerous separation techniques are available, but of all of them, chromatography is the most widely and frequently used. Most chromatographic separations of mixtures are routinely accomplished in a few minutes with relatively simple and inexpensive equipment.

Although similar separations, which could now be called chromatography, were observed in ancient times and were used by nineteenth century chemists, the first detailed description of chromatography is generally credited to Michael Tswett, a Russian botanist, working in Warsaw. In 1906 he published a description of the separation of chlorophylls and other pigments in a plant extract using an apparatus similar to that shown in Figure 4-1. He placed a petroleum ether solution containing the sample at the top of a narrow glass tube which was packed firmly with powdered calcium carbonate. When he washed the sample through the column with additional petroleum ether, he observed that the pigments were resolved (separated) into various colored zones. After carefully removing the packing, he was able to extract the individual constituents from the zones and identify them. Tswett explained that the more strongly adsorbed pigments "displaced" the more weakly adsorbed ones. The appearance of the colored bands (zones) suggested the name "chromatography" from the Greek words for "color" and "to write," although the colors are incidental and have no bearing on the principles of the method.

Chromatography now embraces a variety of processes which are based on differential distributions of the sample components between two phases. One phase remains fixed in the system and is called the *stationary phase*. The other phase, called the *mobile phase*, percolates through the interstices, or over the surface, of the fixed phase. The movement of the mobile phase causes a differential migration of the sample components. The diverse mechanisms responsible for the differential migration are the subjects of the next few chapters.

In many respects, chromatography can be likened to the Craig multiple extraction process. Both utilize two phases, one of which moves with respect to the other. In chromatography, however, the mobile phase moves continuously. At each point along the column, equilibration between the phases is very fast, although never completely achieved. The Craig process is limited to the use of two immiscible liquid solvents, whereas in chromatography we will also make use of other types of phases. Compared to the elaborate apparatus required for the

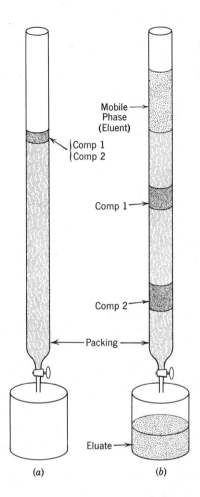

Figure 4-1 Separation of a two-component mixture in a simple chromatographic column: (a) after addition of sample; (b) after partial elution.

Craig process, a chromatographic separation can be achieved with a piece of paper and a beaker, or a buret partially filled with the stationary phase.

4–2   TYPES OF CHROMATOGRAPHY

There are several ways to classify the collection of chromatographic techniques; the most fundamental is based on naming the types of phases used (mobile–stationary), as for example: gas–liquid (GLC), gas–solid (GSC), liquid–liquid (LLC) and liquid–solid (LSC). In another classification scheme, the operational technique is stressed; for example, elution chromatography, in which a small, discrete sample is introduced and then washed (carried or eluted) through the column by the mobile phase, as described by Tswett. All forms of chromatography that we will discuss here are variations of elution chromatography, although other operational techniques are known.

In practice, chromatographic methods are grouped partly on the basis of the types of phases and partly by the mechanism of establishing the phase distribution. Several types are of current importance and will be discussed in detail in the next few chapters.

Liquid–solid or adsorption chromatography, discovered by Tswett and reintroduced by Kuhn and Lederer in 1931, has been used extensively for organic and biochemical analysis. Typically, silica gel or alumina, which have a very large ratio of surface area to volume as well as chemically active surfaces, are used as column-packing materials. Unfortunately there are only a few suitable adsorbents, so the choice is limited. An even more serious limitation is the fact that the distribution coefficient for adsorption often depends on the total concentration. It will be shown that this behavior often results in incomplete separations.

Liquid–liquid or partition chromatography, introduced by Martin and Synge in 1941, was a major advance; in fact, they were later awarded the Nobel Prize for this work. The stationary phase consists of a thin layer of liquid held in place on the surface of a porous inert solid. A wide variety of liquid combinations are suitable so that this method is very versatile. Furthermore, the distribution coefficient for these systems is more likely to be independent of concentration, giving sharper separations.

Gas–solid chromatography was used before 1800 to purify gases, but did not become popular until much later. Hesse, Claesson, and Phillips made major contributions in the 1940s. In the past it has suffered from the same defects as liquid adsorption chromatography, but current research with new types of solid phases is extending the application of this technique.

Gas–liquid chromatography (sometimes called vapor-phase chromatography or VPC by organic chemists) is perhaps the most powerful and most versatile of all methods of separation. It has caused a revolution in organic chemistry since it was first introduced by James and Martin in 1952. At the moment, the most serious remaining obstacle seems to be that the sample components (or their derivatives) must have a vapor pressure of at least a few torr at the temperature of the column. Sample sizes from less than a microgram to more than 100 g can be handled, and traces of the order of 10^{-15} g have been detected. There are problems still to be solved, but Nobel Laureate A. J. P. Martin once predicted that the day would come when the analytical laboratory will consist of a master gas chromatograph which can separate any sample automatically and send the various components to the appropriate slave device for identification and measurement.

Ion exchange chromatography is a special field of liquid–solid chromatography. As the name implies, it is specifically applicable to ionic species—not the case with most other forms of chromatography. The introduction of synthetic resins with ion exchange properties just before World War II has revolutionized the difficult separations of rare earth metals and amino acids. At one time, the entire world's supply of mendelevium (about 17 atoms) was concentrated in a single bead of an ion exchange resin.

Paper chromatography is a special field of liquid–liquid chromatography in which the stationary liquid is a film of water adsorbed on a paper mat. Other stationary liquids can be used as well. This technique is the ultimate in simplicity. All one needs to do is to spot the sample near the edge of a piece of paper, and dip the edge of the paper into the eluting solvent. With sensitive developing reagents, it is particularly suited to separating and identifying traces of components.

Thin-layer chromatography is similar to paper chromatography except that the paper is replaced by a glass or plastic plate coated with a thin layer of alumina, silica gel, or other powdered material. The properties of this "thin layer" are far more reproducible than for paper, therefore TLC has just about replaced paper chromatography in the laboratory.

Gel filtration is a separation process performed with a gel consisting of modified dextran—a three-dimensional network of linear polysaccharide molecules which have been cross-linked. The material swells in water to form a sieve-like structure which can sort out molecules by their size. In addition to geometrical effects, some surface effects are involved. Molecules in the molecular weight range from 100 to several million can be concentrated and separated. Gel permeation chromatography is a similar technique utilizing a polystyrene-type gel suitable for separating polymers.

Continuous-zone electrophoresis consists of a double-barreled approach. During the course of an elution by paper chromatography, an electrical field is applied across the paper perpendicular to the flow of solvent. Thus ionic species are deflected at an angle from the mainstream, depending on their charge and mobility. This procedure, using crossed gradients, makes continuous chromatography possible. Otherwise, elution chromatography is essentially a batch process.

4–3   CHROMATOGRAPHIC THEORY

The rapid advances and widespread use of chromatography are largely due to the theory first presented by A. J. P. Martin and R. L. M. Synge for liquid partition chromatography. Their work provided the basis for a more refined and detailed treatment of gas–liquid chromatography. However, the theoretical principles presented here are applicable to all types of chromatography.

Let us consider first a small section of a column to be used for liquid–liquid partition chromatography, as in Figure 4–2. A given sample molecule finds itself either in the stationary phase, in which case it is making no forward progress; or in the mobile phase, in which case it is being carried along the column at the same velocity as the mobile phase. The mobile phase acts as a carrier—it provides the transportation. The stationary phase acts as a retarder.

Distribution Equilibria.   The distribution of sample molecules between the two phases is governed by an equilibrium constant, known as the *distribution coefficient, K* (sometimes called the *partition coefficient*). As always, equilibration is a dynamic process, and a given molecule passes rapidly back and forth between the two phases, so that on the average the concentrations obey the distribution law:

$$K = \frac{C_S}{C_M} \tag{4–1}$$

where the subscripts S and M pertain to the stationary and mobile phases, respectively.

The distribution coefficient, K, defined in Equation 4–1, applies to many kinds of mechanisms depending on the nature of the phases and the types of interaction of the sample components with each phase. Nevertheless, whether the process is ion

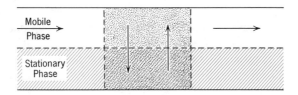

Figure 4–2   Section of a chromatographic column. The stationary phase is arbitrarily located in the lower half of the column for clarity.

exchange, solution, adsorption, or vaporization from solution, the value of K determines the relative populations in the two phases. If the value of K is large, the population in the stationary phase is larger than that in the mobile phase and the molecules of that component will spend more time, on the average, in the stationary phase. For a truly dynamic equilibrium, the fraction of the total time that an average molecule spends in the mobile phase is directly related to the fraction of the total population which is found in the mobile phase.

$$\begin{aligned}\frac{\text{Fraction of time spent}}{\text{in mobile phase}} &= \frac{\text{No. of molecules in mobile phase}}{\text{Total no. of molecules}} \\[2mm] &= \frac{C_M V_M}{C_M V_M + C_S V_S} \\[2mm] &= \frac{1}{1 + K V_S / V_M} \\[2mm] &= \frac{1}{1 + k'}\end{aligned} \qquad (4\text{--}2)$$

where $k' = K V_S / V_M$, a new variable called the *capacity factor*. k' is also equal to $C_S V_S / C_M V_M$, or the ratio of the moles of sample component in the stationary phase divided by the moles in the mobile phase.

Rate of Travel.   Equation 4–2 gives the fraction of time spent in the mobile phase, and if we multiply this by the linear velocity, u, of the carrier (mobile phase), we have the rate of travel of an average sample molecule:

$$\text{Rate} = u \frac{1}{1 + k'} \qquad (4\text{--}3)$$

Thus, the rate of travel of an average molecule is determined by:

1. Carrier velocity: same for all components.
2. Ratio of volume of stationary phase to volume of mobile phase: same for all components.
3. Distribution coefficient: unique for each component.

In a chromatographic experiment, a sample containing several constituents is introduced at the inlet of the column. All sample molecules start from the same place at the same time, but each component moves at a different rate. Two experimental techniques are possible:

1. After a fixed time, each component has traveled a different distance and the zones are separated in space, as in Tswett's experiment or in paper chromatography.
2. At a fixed distance (the outlet of the column), each component appears after a different time interval and is "sensed" by a detector. A plot of detector response vs. time is called a *chromatogram*. Such a plot resembles a distribution diagram within the column (concentration vs. distance at a fixed time) but the two types of plots should not be confused.

Retention Time.   The time it takes a component to travel the length of the column, L, is known as the *retention time*, t_R. From this definition and Equation 4–3, we obtain

$$t_R = \frac{\text{length}}{\text{rate}} = \frac{L}{u}(1 + k') = t_M (1 + k') \qquad (4\text{--}4)$$

where t_M is the time required for a carrier molecule to traverse the column.

Equation 4–4 is one of the basic retention equations for all forms of chromatography. It is most often applied to gas chromatography, in which case a small amount of air may be introduced along with the sample. The air normally passes through the column at the same rate as the carrier gas because oxygen and nitrogen are not sorbed (retarded) by most stationary phases. Hence, t_M is the time required for the *air peak*, and is sometimes given the symbol t_a.

Retention Volume. It is often more convenient to measure the volume of mobile phase required to *elute* the sample component (i.e., move it completely through the column). If the volume flow rate of mobile phase, F, is constant, then the *retention volume*, V_R, is obtained.

$$\text{volume} = \text{time} \times \text{flow rate}$$

$$V_R = t_R F \tag{4–5}$$

Substituting Equation 4–5 into Equation 4–4, we obtain

$$V_R = V_M(1 + k') = V_M + K V_S \tag{4–6}$$

where V_M is the volume of mobile phase contained within the column at a given instant. Alternatively, V_M is the void or empty space in the column not occupied by the stationary phase. Equation 4–6 is another form of the basic retention equation and is applicable to all types of chromatography. If the stationary phase is a solid, V_S may be converted to its area (adsorption) or its ion exchange capacity, or any other appropriate unit.

Relative Retention. The measured retention time or volume is of little value in identifying a component, because t_R is determined in part by how the column is constructed and operated; and, therefore, it is not characteristic of the component. We can eliminate the effects of the operational variables if we compare the retention time of a component to that of a known compound run on the same column under the same conditions. (The standard may be added to the sample, if convenient.) Before taking the ratio of the two retention times, it is important to subtract the time required to elute a nonretarded component. This step is essential because all components must travel the length of the column and therefore spend the same amount of time in the mobile phase. We should compare only the times they spend in (or on) the stationary phase. The same argument applies to *relative retention volume*, etc. The *relative retention* is often given the symbol "α".

$$\alpha = \frac{t_R - t_M}{t_R^* - t_M} = \frac{V_R - V_M}{V_R^* - V_M} = \frac{K}{K^*} \tag{4–7}$$

where the asterisks refer to the standard substance.

Ideal Chromatography. If all molecules behaved identically to the average molecule which we have been discussing, we would have an "ideal" situation. In *ideal chromatography*, the component zones (also called bands or peaks) would have no tendency to spread and would elute from the outlet of the column in the same volume or time interval as they were introduced at the inlet. Since the zone of introduction can be very sharp, an ideal chromatogram would appear as a series of very sharp, completely separated peaks, even for a sample containing many components as shown in Figure 4–3a.

Nonideal Chromatography. The real world is, of course, never quite ideal, and one of the aspects of nonideality is that in a collection of molecules there are random variations in behavior. The net result of nonideal behavior is that zones which are compact and sharp at the inlet become diffuse and broad as they pass through the column, as shown in Figure 4–3b.

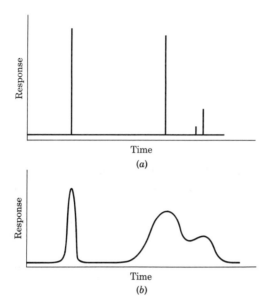

Figure 4–3 Band broadening: (a) ideal behavior; (b) nonideal behavior.

The phenomenon of band spreading has important consequences for chromatography. Closely spaced zones (K values close together) which could be separated ideally, do not separate completely because the width of the zones exceeds the space between them. A rigorous treatment of band spreading mechanisms is a complex mathematical exercise which we will not treat in detail here. The first step is to design a physical model, and, as usual, this ends in a compromise between a simple, easily handled model and one that attempts to take all factors into account.

Plate Theory.   Martin and Synge recognized the similarity between the processes occurring in a chromatographic column and a fractional distillation column and adapted the "theoretical plate concept" which had been used so successfully in treating separations by distillation. However, the analogy to a Craig machine (discussed in Chapter 3) is even more direct. We can imagine that a chromatographic column consists of a large number of identical segments, or *theoretical plates*, in each of which an equilibration is achieved. Although this model is not precise, the deviations are not important with a very large number of plates. In the Craig process, we can predict the distribution of a component after a given number of transfers (equilibrations). Conversely, in chromatography, we observe the distribution of a component along the column, or, alternatively, measure its concentration as a function of time as it reaches the column outlet and passes through a detector. The mathematical relationship between the distribution along the column and number of transfers (or theoretical plates) is analogous for the two processes. To sum it up, a chromatographic column contains the same number of theoretical plates as the number of vessels in a Craig machine giving the same separation behavior (providing a corresponding Craig machine could be constructed). In chromatographic terms, the *number of theoretical plates, n,* in a column is determined from the observed chromatogram, as shown in Figure 4–4.

$$n = \left(\frac{4t_R}{W}\right)^2 \qquad (4\text{–}8)$$

Equation 4–8 requires that the peak width be measured at the intersections of the

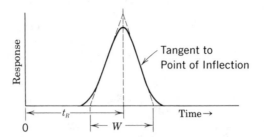

Figure 4–4 Schematic chromatogram.

tangents with the base line. Since n is dimensionless, both t_R and W must be measured in the same units; for example, cm on the chart itself, although time or volume units are equally satisfactory as long as the two units are consistent.

The number of theoretical plates in a column, computed by Equation 4–8 is not solely a function of how the column is constructed, but also of the nature of the solute, flow rate, temperature, method of sample introduction, and other variables. Therefore n is only an approximate number useful for comparative purposes as a description of *column efficiency.* (A similar quantity is the miles per gallon given by an automobile, which depends on both the details of construction as well as on how it is operated.) For a given retention time, the larger the n value, the narrower the peaks, and the closer the approach to ideal behavior.

There are two ways of increasing the number of theoretical plates in a column. First, since each theoretical plate occupies the same space in the column, the number of plates is directly proportional to the column length, L. Therefore, if we double L, we double n. This also must double t_R, according to Equation 4–4. Thus, all retention times (and the separations between peaks) will be doubled. However, the peaks will also become broader, in this case by a factor of $\sqrt{2}$ as determined from Equation 4–8. In other words, the actual gain in separation is roughly a square root function of the column length.

A better approach to ideal behavior is to increase the number of equilibrations within the same time span. In other words, we should increase the number of theoretical plates per unit column length. We have seen that the plate theory tells us the total number of plates in the column under given operating conditions, but says nothing about how to improve this number. In working on this problem, we find that a more useful measure of column performance is the space occupied by each theoretical plate, that is, the column length divided by the total number of plates it contains:

$$H = \frac{L}{n} = \frac{L}{16}\left(\frac{W}{t_R}\right)^2 \tag{4–9}$$

H, sometimes called HETP, is the *height equivalent to a theoretical plate.* Small values of H are highly desirable and we next turn our attention to those factors under the operator's control which influence the value of H.

Rate Theory.   The behavior of a collection of molecules in a chromatographic column is indeed chaotic; however, we recognize three important deviations from ideality which cause H to be greater than its ideal value of zero.

1. As the molecules move from inlet to outlet of the column, they take random paths of different lengths. This is true whether the column is packed or empty. If the column is packed with irregularly shaped particles, there may be gross variations in the streamlines. The differences in path lengths mean that molecules which started together at the inlet will arrive at the outlet over a spread of time.

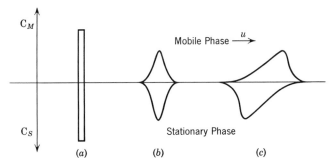

Figure 4–5 Concentration profiles through a zone in a chromatographic column: (a) ideal chromatography; (b) band spreading caused by diffusion from the zone; (c) band spreading caused by slow equilibration between the phases.

2. Molecules tend to diffuse from regions of high concentration to low. As the zone passes through the column, it is inevitable that this diffusion will cause it to spread both forwards and backwards as indicated in Figure 4–5b.
3. A dynamic equilibrium may be established rapidly, but never quite instantaneously. The continuous flow of the mobile phase causes a departure from equilibrium: the ratio C_S/C_M will always be smaller than K at the leading edge of a zone and larger than K at the trailing edge, as shown in Figure 4–5c. This time lag also causes the zone to spread in both directions. Each of these three factors operates more or less independently, and their total effect on H (which is a direct measure of band width according to Equation 4–9) is a summation of three terms:

$$H = \begin{matrix} \text{Contribution from} \\ \text{non-equal paths} \end{matrix} + \begin{matrix} \text{Contribution from} \\ \text{diffusion along} \\ \text{column} \end{matrix} + \begin{matrix} \text{Contribution from} \\ \text{nonequilibrium} \end{matrix}$$

(Eddy diffusion) (Longitudinal diffusion) (Mass transfer)

A group of Dutch petroleum chemists first derived a mathematical expression for the above statement, known as the *Van Deemter equation*:

$$H = A + B/u + Cu \tag{4–10}$$

Each of the three terms on the right side of Equation 4–10 is more complex than shown here. A, B and C can be regarded as constants for a given column, but each constant includes several experimental parameters specific to each type of chromatography. An expanded version of the Van Deemter equation will be given in Chapter 7. Here we will discuss only a few general observations.

1. The A term is a function of the size and uniformity of the particles of packing in the column. Small particles, tightly packed, give small A terms; however, in most well-packed columns, the value of the A term is usually close to zero and may be neglected.
2. The B term is related to diffusion along the length of the column. Diffusion in liquids is approximately 10^5 slower than in gases, so the B term is far less important with a liquid mobile phase. Note that the B term is inversely proportional to carrier velocity. At low velocities, elution times are longer and there is more time for diffusion to spread the band.
3. The C term represents the broadening due to nonequilibrium. It is a complex

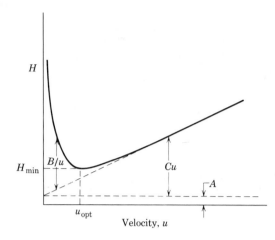

Figure 4–6 Van Deemter plot showing the relative contributions of each term as a function of mobile phase velocity.

function of the geometry of the stationary phase, the distribution coefficient, and the rates of diffusion in both phases. Note that the C term is directly proportional to carrier velocity. At high velocities, there is less time available for equilibration.

4. For the best separation, H should be minimized which will maximize n and give the narrowest peaks.

5. To reduce H most effectively, one should determine which of the three terms is the largest and attempt to reduce its value first. Values of A, B, and C are determined as follows:

   (a) Run a chromatogram for a particular peak at three different flow rates, reasonably far apart.

   (b) Determine H for each flow rate from the chromatogram and Equation 4–9.

   (c) Set up three simultaneous equations similar to Equation 4–10, substituting flow rate for velocity if more convenient.

   (d) Solve the equations for A, B, and C.

   The relative contributions of the three terms for a typical gas–liquid chromatogram are plotted as a function of velocity in Figure 4–6.

6. The one variable affecting H which is most easily controlled by the operator is the carrier velocity. At low velocities the B term is very large and controls the value of H. At high velocities, the C term predominates. Both extremes should be avoided. There is an intermediate, *optimum velocity* which can be obtained by differentiating Equation 4–10 if values for A, B, and C are known, or graphically as indicated in Figure 4–6.

7. The optimum velocity as determined above may be very slow, giving inconveniently long retention times. Note that the Van Deemter plot rises fairly slowly to the right of the minimum so that velocities up to twice the optimum still give reasonable separations in shorter times.

8. The Van Deemter equation, as given in Equation 4–10 is based on many approximations. Numerous workers have modified and expanded it to arrive at a closer fit to experimental data. Nevertheless, even the simplified version is extremely useful in improving column performance.

Resolution. In practice, the most important objective is to be able to separate the sample components—not necessarily to build the world's most efficient column. We are often concerned with the degree of separation, or *resolution*, of

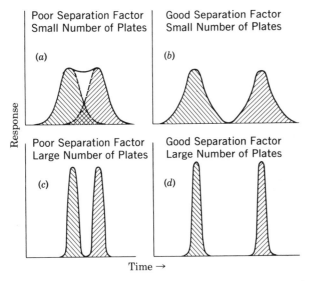

Figure 4–7 Effect of separation factor and number of theoretical plates in the column on the resolution of two peaks.

compounds of similar character. The two components in Figure 4–7a are poorly resolved and overlap to an extent that makes it impossible to identify or determine the amount of either. To improve the resolution, we can either (a) change the temperature or the nature of the phases to give a greater separation between the peaks as in Figure 4–7b, or (b) reduce the width of the peaks by improving the column efficiency (decrease H, increase n) as in Figure 4–7c, or (c) a combination of both (a) and (b) as in Figure 4–7d, where the resolution is now more than satisfactory. Quantitatively, the resolution, R, of two peaks having retention time t_1 and t_2 and peak widths W_1 and W_2 is:

$$R = \frac{2(t_2 - t_1)}{W_1 + W_2} \tag{4–11}$$

The resolution of the two peaks shown in Figure 4–8 is about 1.05. A resolution of $R > 1.5$ gives so-called *baseline resolution* or essentially complete separation of two peaks of equal size. A resolution of $R = 1.0$, giving a peak overlap of approximately 2%, is usually considered to be adequate for analytical purposes.

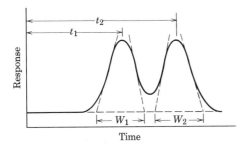

Figure 4–8 Measurements used in calculating the resolution of two peaks.

Adequate resolution may be difficult to obtain for complex samples containing many components of similar properties or for trace components. The best procedures for improving resolution vary with the type of chromatography; however, some useful generalities are in order.

To improve resolution:

1. Increase $\Delta t_R = t_2 - t_1$
   (a) Increase column length, L
   (b) Increase the amount of stationary phase, V_S or V_L
   (c) Employ a better separation factor, $\alpha = t_2'/t_1'$
       i. Decrease temperature, T
       ii. Select a different stationary phase
       iii. Select a different mobile phase (if liquid).
2. Decrease the band width, W
   (a) Employ more uniform packing
       i. Pack more carefully
       ii. Use smaller particles
   (b) Increase interfacial area between phases
   (c) Optimize flow rate
   (d) Reduce sample size
   (e) Reduce dead space in system
   (f) Reduce time constant of detector
   (g) Decrease diameter of column

In developing or improving a chromatographic procedure, one or more of the following objectives may be important:

1. High resolution: for complex mixtures
2. Fast analysis time: for kinetic studies or process control
3. Large throughput: for preparatory work or concentrating trace components.

All three of these objectives cannot be achieved simultaneously; for example, large samples and/or high flow rates give poor resolution. Inevitably, the operator must decide on the best compromise to meet the requirements. In making the best choices, chromatographic theory can be a great help.

QUESTIONS AND PROBLEMS

4–1. Substances P and Q have distribution coefficients of 490 and 460, respectively, on a particular column. Which compound will be eluted first in a chromatographic separation?

4–2. Component Y is eluted in 15.0 min. Component Z requires 25.0 min and a non-retarded substance X requires 2.0 min.
  (a) What is the relative retention time of Z with respect to Y?   *Ans.* 1.77
  (b) What is the relative retention time of Y with respect to Z?   *Ans.* 0.57
  (c) What is the capacity factor for Y in this column?   *Ans.* 6.5
  (d) What fraction of the total time does Y spend in the mobile phase?   *Ans.* 0.133
  (e) Of the 25 minutes required to elute Z, how much time does the average Z molecule spend in the stationary phase?   *Ans.* 23.0 min

4–3. The width of a certain peak (measured in time units) is 50 sec and its retention time is 50 min. How many theoretical plates does the column contain under these conditions?   *Ans.* 57,600

4–4. Iso-octane and n-octane give retention times of 800 and 815 seconds on a column known to have 8100 theoretical plates.
  (a) What resolution will be obtained if a sample containing both compounds is run on this column?   *Ans.* 0.42
  (b) Assuming the retention times to be unchanged, how many plates would be required to achieve a resolution of 1.00?   *Ans.* 45,500

(c) Assuming the retention times to be unchanged, how many plates would be required to achieve baseline resolution ($R = 1.5$)? *Ans.* 102,400

4-5. The retention times of several compounds, measured from sample injection, are: air, 45 sec; propane, 1.5 min; pentane, 2.35 min; acetone, 2.45 min; butyraldehyde, 3.95 min; xylene, 15.0 min. What are the relative retention times of the organic compounds using pentane as the standard? *Ans. α for acetone = 1.06*

4-6. The constants in the Van Deemter equation (4-10) for a particular column at 150°C were found to be: $A = 0.08$ cm, $B = 0.15$ cm^2/sec and $C = 0.03$ sec. What is the optimum gas (mobile phase) velocity for this column? What is the corresponding minimum H? *Ans. 2.24 cm/sec, 2.14 mm*

4-7. Obtain an expression for H_{min} and u_{opt} in terms of A, B, and C by differentiating Equation 4-10 and setting $dH/du = 0$. *Ans. $H_{min} = A + 2\sqrt{BC}$; $u_{opt} = \sqrt{B/C}$*

4-8. State the requirements for ideal chromatography and show why it is impossible to achieve.

4-9. Predict the effect on resolution of adding the sample over a long period of time.

4-10. Starting from Equations 4-7, 4-8, and 4-11 and the approximations that $W_1 = W_2$ and that $t_a \ll t_R$, derive an expression for R in terms of n and α.

$$Ans.\ R = (\sqrt{n}/4)((\alpha - 1)/\alpha)$$

4-11. Using the answer given for Problem 4-10, predict the resolution which will be obtained for a two-component mixture in a 50-m column, if a 20-m column (otherwise identical) gives a resolution of 0.80 for the same mixture. *Ans. 1.27*

4-12. What length of column is required to give a resolution of 1.50 for the mixture in Problem 4-11? *Ans. 70.3 m*

REFERENCES

H. Purnell, *Gas Chromatography*, Wiley, New York, 1962.

E. Heftmann, *Chromatography*, 2nd ed., Van Nostrand, Reinhold, New York, 1967.

B. L. Karger, L. R. Snyder, and C. Horvath, *An Introduction to Separation Science*, Wiley-Interscience, New York, 1973.

J. M. Miller, *Separation Methods in Chemical Analysis*, Wiley, New York, 1975.

5

LIQUID COLUMN CHROMATOGRAPHY

5–1   INTRODUCTION

The common features of the kinds of chromatography to be discussed in this chapter are: (a) a column packed with a porous stationary phase, and (b) a liquid mobile phase to elute the sample components through the column. Four kinds of chromatography are studied: (1) *adsorption*, in which the components are selectively adsorbed on the surface of the packing; (2) *partition*, in which the components are selectively partitioned between the eluent and a thin liquid film held stationary on an inert solid support; (3) *ion exchange*, in which ionic constituents of the sample are selectively retarded by exchange with replaceable ions of the packing; and (4) *gel*, in which the column is packed with a permeable gel, causing separations by a sieving action based on molecular size.

Liquid chromatography is particularly useful for samples containing large molecules or ionic substances with low vapor pressures, and for thermally unstable substances which cannot be vaporized without decomposing. Liquid eluents are seldom inert; therefore, the distribution coefficients depend on the chemical nature of both the stationary and the mobile phase. Liquids have much higher viscosities and greater resistance to flow than do gases. In addition rates of diffusion of solutes are five orders of magnitude slower in liquids than in gases; thus, the C term in the Van Deemter equation (4–10) is much larger than the A and B terms at normal flow rates. Van Deemter plots for gas and liquid chromatography are compared in Figure 5–1. As conventionally performed, liquid chromatography is a time consuming procedure, because low flow rates are indicated for best efficiency. There are some new developments discussed in Section 5–4 that speed up the process to a point where it is now competitive with gas chromatography.

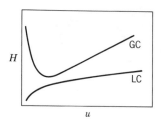

Figure 5–1   Typical Van Deemter plots for gas and liquid chromatography.

5–2   LIQUID–SOLID ADSORPTION CHROMATOGRAPHY

Adsorption Processes.   The properties of atoms, ions, or molecules located at the surface of a solid particle are somewhat different from the properties of the same species in the interior. The bonds within the surface layer are perturbed by the lack of an overlaying structure. Hence the surface layer is at a higher energy level and is described as having *surface activity*. If the solid particle is immersed in a fluid (gas or liquid), the active surface attracts and tends to *adsorb species* from the fluid. The attractive forces may be ionic (electrostatic), dipole–dipole, dipole–induced dipole, London forces, or a combination of these. If the fluid is a solution, any or all of the solutes as well as the solvent may be adsorbed. A good adsorbent should possess a large surface area containing many chemically active sites. In general, there is a competition by the various species in the fluid phase for the active sites on the surface. Usually, the surface loses its activity when it is covered by a monolayer of adsorbed species.

Adsorption Equilibrium. When a solution flows over the active surface of a solid, an equilibrium is established for the adsorption or desorption of the species present. The relationship between the concentration of a given species in solution and the amount which is adsorbed can be expressed as an equation or as a plot of C_S (adsorbed) versus C_M (mobile phase), as in Figure 5–2. The line so obtained is called the *adsorption isotherm* (constant temperature).

Three types of isotherms are illustrated in Figure 5–2.

1. *Linear.* This is the desired situation; it indicates that the surface does not become "saturated" with the species in question. The slope of a linear isotherm gives the distribution coefficient, and, in this case, it is independent of total concentration:

$$K = C_S/C_M \quad \text{or} \quad C_S = KC_M \quad (5\text{–}1)$$

2. *Convex.* Because of the variation in the activity of the available adsorption sites, most systems depart from linearity. Many liquid–solid systems follow a relationship known as the *Freundlich isotherm*:

$$K = C_S/C_M^{1/n} \quad \text{or} \quad C_S = KC_M^{1/n} \quad (5\text{–}2)$$

where the value of n varies with the nature of the system and the temperature. In general, $n > 1$ which leads to a convex shape for the

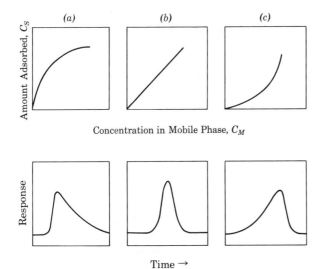

Figure 5–2 Typical isotherms for adsorption and their effect on peak shape.

isotherms as shown in Figure 5–2a. In many gas–solid systems, the rate of adsorption is directly proportional to the number of vacant sites until the surface is completely covered with a monolayer of the adsorbed species. Thus, the rate (and the distribution coefficient) must decrease as the concentration increases. Let us represent adsorption as a chemical reaction:

$$A + S \rightleftharpoons AS$$

where A is a molecule to be adsorbed from the gas phase, S is a vacant adsorption site and AS is an adsorbed molecule of A. The equilibrium expression for this reaction is:

$$K = \frac{X_{AS}}{X_A X_S} = \frac{X_{AS}}{P_A X_S}$$

where X is a mole fraction and P_A is the pressure. To account for all sites, it must be true that

$$X_S + X_{AS} = 1$$

Therefore:

$$K = \frac{X_{AS}}{P_A(1 - X_{AS})} \tag{5–3}$$

Converting mole fractions to concentrations, we obtain

$$C_S = \frac{k_1 C_M}{1 + k_2 C_M} \tag{5–4}$$

where the small k's represent constants for each system. Equation 5–3 (or 5–4) is called the *Langmuir isotherm* and it also leads to a convex isotherm reaching a limiting value at a monolayer.

3. *Concave.* Additional reactions that take place on adsorption occasionally enhance the overall adsorption process. Such cases are uncommon, but not unknown. The adsorption isotherm curves upward as in Figure 5–2c.

Peak Distortion. Recall that the rate of travel of a component through a chromatographic system is a function of the fraction of the component found in the mobile phase (Equation 4–2). As the band spreads, there is inevitably a region of high concentration in the center of the band relative to the edges. If the isotherm is convex (Figure 5–2a) the center of the band must travel at a faster rate than both the leading and trailing edges. In the extreme case the center will nearly overtake the leading edge giving a very sharp front and long trailing edge. This behavior is called *band tailing,* and is obviously undesirable. The opposite case (Figure 5–2c) gives a long leading edge and a sharp back to the peak.

Peak tailing is more pronounced with active adsorbents. One way to reduce its deleterious effects is to partially deactivate the solid by covering the most active sites with another substance or by raising the temperature. Another approach is to reduce the size of the sample so that the system remains within the nearly linear portion of the isotherm found at very low concentrations.

Adsorbents. Relatively few materials are used as chromatographic adsorbents. Silica gel and alumina are by far the most popular. The adsorbing power of the material depends on the chemical nature of the surface, the area available, and its pretreatment. These properties are easily modified and difficult to control; therefore, the order given in Table 5–1 is not always observed.

The reproducibility of surfaces is an ever-present problem in adsorption

Table 5–1 Some Common Adsorbents in Order of Decreasing Adsorptive Power

Alumina	Calcium carbonate
Charcoal	Sucrose
Silica gel	Starch
Magnesia	Powdered cellulose

chromatography. The surface activity of most common adsorbents varies from site to site on the same particle, and from batch to batch. Pretreatment including exactly prescribed washing and drying is always advisable.

Alumina. The adsorptive activity of alumina can be controlled by varying the amount of water it contains. Thus we can prepare a tailored adsorbent by dehydrating at 360°C for 5 hours and then allowing the dehydrated material to adsorb a suitable amount of water. The *Brockmann scale of activity* is based on the amount of water that the alumina contains: $I = 0\% \ H_2O$, $II = 3\%$, $III = 6\%$, $IV = 10\%$, $V = 15\%$. The scale values can be obtained empirically by observing the chromatographic behavior of certain azo dye mixtures under carefully controlled conditions.

Example. A column, 1.5 cm diameter × 10 cm long, is filled to a height of 5 cm with alumina. A 10-ml aliquot of a mixture containing 20 mg azobenzene (A) and 20 mg p-methoxyazobenzene (B) in 50 ml benzene-petroleum ether (1:4) is added to the top of the column. The solvent is allowed to flow out at a rate of 20 to 30 drops per minute. Just as the top of the dye solution reaches the top of the alumina, another 20 ml of the solvent is added to develop the chromatogram. If, at the end of this development, dye A has reached the bottom of the column while dye B remains at the top, the alumina used has activity I. If the dye A passes completely through the column while dye B is found at the bottom, then the alumina has activity II. Lower activities (higher Brockmann numbers) are determined with other dyes in the series which are more firmly adsorbed.

The surface area of alumina is approximately 150 m^2/g. About 5% by weight of water is sufficient to coat the alumina with a monolayer. For most purposes, alumina treated with about 3% water provides the most useful surface.

Silica Gel. Although silica gel has a higher surface area (ca. 500 m^2/g), it is less chemically active than alumina and is preferred for separating organic compounds that are sensitive to rearrangement on active catalytic surfaces.

Solvents. The choice of an eluting solvent is just as important as the choice of adsorbent. A liquid mobile phase not only provides transportation as the carrier, but also influences the distribution coefficients through its solvent power. In addition to the relative solubilities of the solutes in the eluting solvent, it is necessary to consider the competition between the solutes and solvent for adsorption sites on the surface of the stationary phase. A solvent which elutes the solutes too fast will not separate them, whereas a solvent which elutes the components too slowly will give inconveniently long retention times. Long retention times also cause excessive band broadening and unnecessary dilution of the sample. It is fortunate that a long list of solvents is available. If necessary, mixtures of solvents are used, or even a series of mixtures (*gradient elution*) with an increasing fraction of the more polar eluent. It is advantageous to introduce the

Table 5–2 Some Common Solvents in
Order of Increasing Eluting Power from
Alumina

Fluoroalkanes	Acetone
Petroleum ether	Ethyl acetate
Carbon tetrachloride	Dioxane
Trichloroethylene	Acetonitrile
Isopropyl ether	Ethanol
Toluene	Methanol
Benzene	Water
Chloroform	
Ether	
Methyl ethyl ketone	

sample in a solvent in which it is highly soluble in order to keep the sample volume
at a minimum. However, this is not necessarily the solvent which will give the
optimum resolution in a reasonable time. A short list of solvents is given in Table
5–2; the order given is known as the *eluotropic series.*

The eluting power of a solvent depends on the solid phase and also on the
nature of the solutes. An examination of Tables 5–1 and 5–2 indicates that the
substances are listed rather closely in order of polarity as measured roughly by the
dielectric constant (see Table 3–1). Essentially nonpolar hydrocarbon solutes will
be held more tightly on nonpolar column materials than polar solutes will be. On
the other hand, the surface of alumina is very polar, and the order of retention will
be quite the opposite. The effect of substituent groups on adsorption is very
important and can be predicted qualitatively from the list in Table 5–3.

Table 5–3 Some Groups of Solutes in Order of
Increasing Adsorbability on Alumina Columns

Perfluoro carbons	Aldehydes and ketones
Saturated hydrocarbons	Alcohols and thiols
Unsaturated hydrocarbons	Acids and bases
Halides and esters	

Packing and Operating the Column.   Irregularly shaped zones, which are one of
the most annoying features of chromatography, are caused by nonuniform
packing. Columns to be used for liquid chromatography are usually of large
enough diameter so that they can be mechanically vibrated, or tamped with a long
plunger during packing. Alternatively, the packing can be poured into the column
as a slurry, and the particles allowed to settle before adding more slurry. If
columns develop channels in use, they can be rejuvenated by back-flushing; that
is, a reversal of the liquid flow to stir up the particles after which they are allowed
to settle more uniformly.

A sintered glass disk or a small plug of glass wool is used at the bottom of the
column to support the packing. Similarly, the top of the packing can be protected
from disturbances when pouring in sample or eluent by covering it with a piece of

filter paper or other innocuous material. Once the column is packed properly, the level of the liquid must never be allowed to go below the top of the packing. If this happens, small air bubbles will be entrapped and cause channeling in further use.

The flow rate of the eluent must be kept constant so that meaningful data can be easily recorded. The flow rate depends on the particle size of the packing, the dimensions of the column, the viscosity of the liquid, and the pressure applied to force the liquid through (or the position of a stopcock at the outlet of the column). Satisfactory results are obtained if the average linear velocity of the eluent is of the order of 1 cm/min.

Detection Methods. Most common stationary phases are white or nearly colorless, making it possible to observe the components if they are colored. A number of organic compounds fluoresce in ultraviolet light. In a special technique, called development chromatography, the elution is stopped when the first band appears at the end. Then the column packing is carefully extruded and streaked with color-developing reagents to indicate the position of the bands. The separated components can be extracted from the packing material if desired.

Compared with the simple, inexpensive, nearly universal detectors available for gas chromatography, those available for liquid chromatography are often highly selective, cumbersome, expensive, and useful over only very narrow ranges of concentration. Flow-through micro-cells are available for many instruments such as refractometers, ultraviolet and infrared spectrophotometers, fluorimeters, liquid scintillation counters (for components containing radio-isotopes), and various electrochemical devices. Most of these detectors will be described in later chapters. In any event, fractions of the effluent can be collected automatically and examined further by any method, or the components can be collected after evaporation of the solvent.

5–3 LIQUID–LIQUID PARTITION CHROMATOGRAPHY

Partition chromatography offers great advantages over adsorption chromatography. It is far more reproducible and predictable from solubility data. Even more important, the distribution coefficient is constant over a much greater range of concentration, yielding sharper, more symmetrically shaped bands. It is the method of choice whenever a suitable solvent pair can be found.

Some care must be given to the choice of solvent pairs. They must, of course, be immiscible. The more polar of the two is normally coated on a solid support. Optimally, the fraction of a solute which is dissolved in the mobile phase should be in the range from 0.05 to 0.5; otherwise the retention time will be too long or too short. Most often, a thin film of water supported on silica gel is used as the stationary phase. Up to 50% water can be adsorbed by silica gel before it becomes too moist to handle easily. For some purposes, it may be desirable to include a buffer system in the aqueous phase to change and/or control the solubility of the solutes. The list given in Table 5–2 will be helpful in choosing the best solvent.

Reverse-Phase Chromatography. There are some situations in partition chromatography in which it is advantageous to use the nonpolar organic solvent as the stationary phase in order to get a more favorable value for the partition coefficient. Obviously, a nonpolar solid support is required. Powdered rubber coated with benzene is a very satisfactory stationary phase with an aqueous mobile phase.

5–4   HIGH PERFORMANCE LIQUID CHROMATOGRAPHY

In the introduction to this chapter, we noted that in liquid chromatography, maximum efficiency is achieved at low flow rates. This unhappy state of affairs results from the low rates of diffusion which are characteristic of liquid phases. The only practical way to increase the *rate* of diffusion is to raise the temperature, but there is little to be gained in the temperature range available. The alternative is to reduce the *distance* through which the molecules must diffuse. The answer, then, lies in finding ways to make and use smaller particles for column packings. This apparently simple expedient, combined with a number of other improvements in apparatus and technique, has evolved into a new practice of liquid chromatography which is competitive with gas chromatography in speed and resolution of complex mixtures.

Packing Materials.   Closely sized particles as small as 10 μm are now becoming available. Particles smaller than 10 μm are extremely difficult to handle and give an almost impermeable column. To alleviate this problem, a solid glass bead of 30 to 50 μm in diameter can be coated with a layer (thin skin of 1 to 2 μm thickness) of porous material. These coated beads are called *pellicular* beads. The porous layer may serve as a solid stationary phase or be coated with a very thin layer of liquid stationary phase with an extremely large surface area.

The thin film of stationary phase may be rinsed away by mobile phase under the high pressures used (see below). One method of overcoming this problem is to chemically bond the liquid phase to the solid support. Porous silica beads are esterified with various alcohols forming the corresponding silicate esters. The organic parts stand on end, projecting from the bead surface, and resemble a *brush*—a name used to describe this type of packing. These esterified, siliceous packings are not thermally stable and are subject to hydrolysis and exchange with lower alcohols. Another type of chemically bonded packing utilizes silicone polymers which are more stable because of their three-dimensional cross-linked structure.

High Pressures.   Columns packed with small particles require high inlet pressures in order to give a reasonable flow rate. Pressures up to 10,000 psi are not difficult to handle in the small columns used (2 to 3 mm diam.). Improved, pulse-free pumping systems are incorporated in modern liquid chromatographs.

Detectors.   The smaller columns and faster flow rates place rigid requirements on the detection system. Flow-through detectors with low dead volumes and high sensitivity are a necessity. For example, a component at the 0.01% level in a 10 μg sample requires a system that can detect 1 ng of the component. Furthermore, the detector should not introduce extracolumn band broadening because of its size. The dead volume within the detector should be no more than one-tenth of the peak volume. Since highly efficient *LC* columns can give peak volumes of the order of 50 μl, a detector volume of $<5 \mu$l is desirable.

Two types of detectors are currently popular. One is based on a flow-through micro-cell placed in an ultraviolet spectrophotometer (see Chapter 10). A low pressure mercury lamp is commonly used as a source and the absorption measured with the 254-nm line where most organic compounds having double bonds or aromatic groups cause at least some absorption. Measurements are not restricted to 254 nm if one employs a standard spectrophotometer. Since most common solvents do not absorb radiation in the ultraviolet and visible region, these detectors are extremely sensitive and free from interferences.

A second popular detector, the differential refractometer (see Section 8–5), continuously monitors the difference in refractive index between the pure mobile phase (reference stream) and the column effluent. These devices respond to essentially all sample components, but the refractive index differences are minute and fluctuate with small changes in temperature or composition of the mobile phase (gradient elution is difficult).

Another detector utilizes the flame ionization detector developed for gas chromatography (see Section 7–8). A continuous wire or fine chain moves through the stream emerging from the column outlet where it becomes coated with a film of the effluent. The wire then passes through an oven to evaporate the mobile phase, and finally through an oxidation oven in which the components are burned in a stream of oxygen at about 850°C. The gaseous products are passed over a nickel catalyst bed at 350°C which converts the carbon dioxide to methane, a substance that is readily measured by the flame ionization detector. Many variations have been tried in an effort to overcome the difficulties of the transport system; however, these devices offer the unique advantage of responding directly to the number of carbon atoms in the component molecule.

A judicious combination of these improvements has led to separations of closely related compounds in very short times, such as the impressive separation of five hydroxylated aromatics in less than 60 sec described in Figure 5–3.

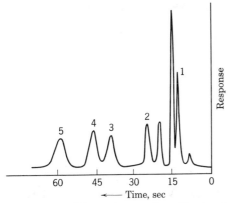

Figure 5–3 High speed separation of hydroxylated aromatics by liquid–solid chromatography. Column: 250 mm × 3.2 mm; packing: 8–9 μ porous silica microspheres, ca. 75 Å; mobile phase: dichloromethane, 50% water-saturated; pressure: 2000 psi; flow rate: 10.5 ml/min; temperature: 27°C; sample: 4 μl of mixture in hexane; identification of peaks:

(Other peaks are impurities in standards.)

[Reprinted by permission of editor, *J. Chromatog. Sci.*, **10**, 593 (1972).]

5–5   ION EXCHANGE METHODS

Many natural clays are able to function as ion exchangers because they are insoluble polymeric materials with a loose structure and many replaceable metal ions. A continual exchange of ions takes place between these clays and the waters passing through them. The detailed process is modified and complicated by the pH and carbonate content of the water. Clays have varied and unpredictable properties, and it was only with the introduction of synthetic ion exchangers that a reproducible, scientific technique became possible.

Modern ion exchange resins were first used in 1935 by Adams and Holmes. These resins consist of three-dimensional networks of polymeric chains cross-linked with short chains containing ionizable functional groups. Thus, there is an insoluble phase with fixed ionic sites of one charge, while the oppositely charged species are free to move about in the solvent and be replaced by other ions of like charge, provided that electroneutrality is maintained. These resins can be used either in a batch process for a wholesale exchange of ions with those in solution, or packed into a column and used in a chromatographic fashion. They have proven invaluable in the separation of ionic substances with closely similar chemical behavior.

Types of Resins.   A typical resin is prepared by a polymerization of styrene and divinylbenzene:

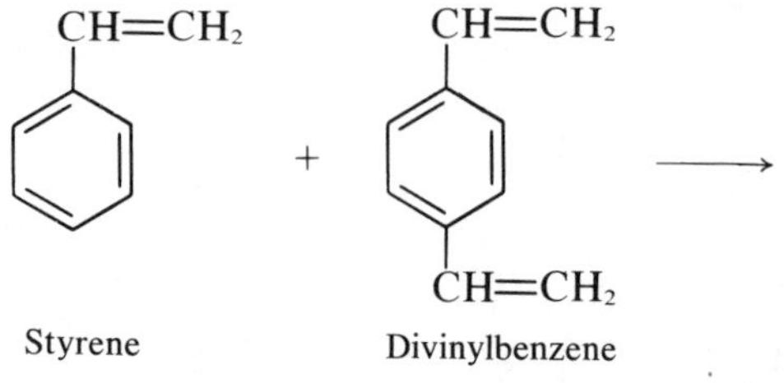

Styrene   Divinylbenzene

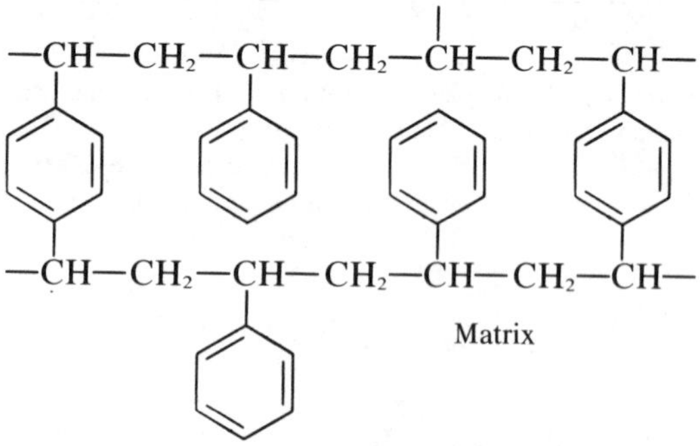

Matrix

The above structure is repeated in three dimensions, with the number of cross-linkages determined by the ratio of divinylbenzene to styrene. Increasing the cross-linking increases the rigidity, reduces swelling (see below), reduces porosity, and reduces the solubility of the polymeric structure. Ordinarily about 10% divinylbenzene is used, but both higher and lower percentages are available in commercial resins. The particle size of the product is determined largely by the extent of mechanical agitation during the reaction, so that resin beads of numerous graded sizes are available.

Cationic Exchangers.   Acidic functional groups are easily introduced, for example, by sulfonation in which a sulfonic acid group is attached to nearly every aromatic nucleus. Sulfonic acids are strong acids with essentially completely dissociated protons, although these protons are not free to leave the resin unless replaced by other positive ions. Weakly acidic cation exchangers are prepared by introducing carboxylic acid functional groups on a polyacrylic acid matrix. The total number of equivalents of replaceable protons per unit volume of resin determines the *exchange capacity* of the resin. For weak ion exchange resins, the capacity is a function of pH.

Anionic Exchangers. If basic functional groups are introduced, the resin can exchange anions rather than cations. Strong anion exchangers are prepared with a tertiary amine, yielding a strongly basic quaternary ammonium group. Weaker anionic exchangers can be prepared with secondary amines, yielding a weakly basic tertiary amine.

A few common resins are listed in Table 5–4.

Table 5–4 Selected Ion Exchange Resins

Type	Nature	Capacity meq./ml	Trade Name
Strong cation	Sulfonated polystyrene	1.9	Dowex 50 Amberlite IR 120
Weak cation	Condensed acrylic acid	4.2	Amberlite IRC 50
Strong anion	Polystyrene with —CH$_2$NMe$_3$Cl	1.2	Dowex 1 Amberlite IRA 400
Weak anion	Polystyrene with *sec*-amine	2.0	Dowex 3 Amberlite IR 45

Swelling. The high proportion of polar groups within the resin gives it a hygroscopic character. In a sulfonated resin, one can consider that the —SO$_3^-$ and H$^+$ ions dissolve in the adsorbed water yielding a solution of high concentration. Since the ions cannot diffuse out, there is a tendency for water to diffuse in to equalize the concentration, but the amount of water that can diffuse in is restricted by the space available in the interstices of the resin structure. Even so, large amounts of water, up to a gram per gram of dry resin, are able to penetrate the interior causing distortion of the structure and swelling of the resin bead. The osmotic forces which drive the water from the exterior to the interior of the resin beads cause an enormous internal pressure. Dry beads must never be allowed to swell while tightly packed in a column, and even a change in the nature of the solvent may result in an "explosion" of a tightly packed glass column. Before the sample is introduced, the column is usually converted entirely to the acid form, HR, by washing with a solution of hydrochloric acid.

The important properties which determine the behavior of a resin can be summarized:

1. Size of particles—rate of exchange and permeability of the packed column.
2. Degree of cross-linking—rigidity, porosity, swelling.
3. Nature of functional group—kind of ion exchanged.
4. Strength of functional group—distribution coefficient.
5. Number of functional groups—capacity of resin.

Theoretical Principles

Ion Exchange Equilibria. The net result of an ion exchange reaction can be expressed as a replacement of equivalent quantities of like-charged ions:

$$HR + Na^+ \rightleftharpoons NaR + H^+$$
$$2HR + Ca^{+2} \rightleftharpoons CaR_2 + 2H^+$$
$$RCl + OH^- \rightleftharpoons ROH + Cl^-$$

where R^+ or R^- represents the resin matrix. The law of mass action applies, and therefore the equilibrium constant (the *selectivity coefficient*) takes the usual form:

$$K = \frac{a_{NaR} a_{H^+}}{a_{HR} a_{Na^+}} = \frac{[NaR][H^+]}{[HR][Na^+]} \times \frac{f_{NaR} f_{H^+}}{f_{HR} f_{Na^+}}$$

K is a constant only if the *activities* of the various species are used, otherwise it will vary with relative and total concentrations due to changes in activity coefficients. There is no adequate theory for treating activity coefficients in concentrated solutions of strong electrolytes such as exist within the resin, where concentrations of the order of 2 to 8 M are encountered. The activity coefficient of the resin phase is even more difficult to treat. Consequently, there is little choice but to use a pseudo-equilibrium constant, or concentration ratio. Selectivity coefficients, defined by neglecting activity coefficients, are determined empirically and are reasonably constant for given conditions. Experimental observations have been formulated into a number of useful rules:

1. Selectivity coefficients approach unity as the cross-linking is decreased.
2. The exchange of ions that causes expansion of the resin is less favored than those exchanges that do not; or, the smaller the (hydrated) ion, the greater the affinity for the resin.
3. The greater the charge on the ion, the greater the affinity for the resin.
4. The affinity of high molecular weight organic ions and some anionic complexes of metal ions are unusually high, probably because the electrostatic forces are augmented by short range adsorption (Van der Waal's) forces.

To a first approximation, these rules predict the observed order of affinity for groups of ions:

$$Li^+ < H^+ < Na^+ < NH_4^+ < K^+ < Rb^+ < Cs^+ < Ag^+ < Tl^+$$
$$Be^{+2} < Mn^{+2} < Mg^{+2} < Zn^{+2} < Co^{+2} < Cu^{+2} < Cd^{+2} < Ni^{+2} < Ca^{+2} < Sr^{+2}$$
$$< Pb^{+2} < Ba^{+2}$$
$$Na^+ < Ca^{+2} < La^{+3} < Th^{+4}$$
$$OH^-, F^- < CH_3CO_2^- < HCO_2^- < H_2PO_4^- < HCO_3^- < Cl^- < NO_2^- < HSO_3^-$$
$$< CN^- < Br^- < NO_3^- < HSO_4^- < I^-$$

The above orders often show inversions due to changes in pH, relative concentrations, nature of resin, complex formation, ionic strength, etc.

The *plate theory*, developed by Martin and Synge for partition chromatography, can be applied directly to an ion exchange column with only a change of terminology. We will define a distribution ratio, D, as:

$$D = \frac{\text{quantity of sample in resin of a given plate}}{\text{quantity of sample in interstitial solution of same plate}} \quad (5-5)$$

Then the volume of eluent required to move the sample through the column, V_e, measured to the peak maximum is

$$V_e = V_M + DV_M \qquad (5-6)$$

where V_M is the interstitial volume of the column. Let us assume the sample contains cations A^+ and B^+ which are to be separated from each other. When the sample is introduced, it is first retained at the top of the column by exchange of cations:

$$A^+ + HR \rightleftharpoons AR + H^+$$
$$B^+ + HR \rightleftharpoons BR + H^+$$

The corresponding selectivity coefficients are:

$$K_A = \frac{[AR][H^+]}{[HR][A^+]} \qquad (5-7)$$

and

$$K_B = \frac{[BR][H^+]}{[HR][B^+]} \qquad (5-8)$$

A separation factor, $\alpha_{A/B}$, can be derived from Equations 5–7 and 5–8

$$\frac{K_A}{K_B} = \alpha_{A/B} = \frac{[AR][B^+]}{[A^+][BR]} \qquad (5-9)$$

In Equations 5–7 to 5–9 we have neglected the activity coefficients, but, to a first approximation, the last equation tells us that the separation factor is independent of the concentration of *other* ions (unless they react with A^+ or B^+). Now the cations A^+ and B^+ can be eluted from the column only if they are replaced by another cation contained in the eluent. In this case, let us assume that the eluent contains a high concentration of H^+. Equations 5–7 and 5–8 show that increasing $[H^+]$ must also increase $[A^+]$ and $[B^+]$. In other words, we can change the distribution ratios for A^+ and B^+ (Equation 5–5) and thus the elution volume (Equation 5–6) manyfold by altering the concentration of eluting ion. A high concentration of the latter will lead to faster elution with sharper but less well resolved peaks as shown in Figure 5–4.

Effect of pH of Eluent. The extent of dissociation of weak acids and bases, and the hydrolysis of salts and metal ions is controlled by the pH of the medium. Thus the electrical charge on a species may be increased, decreased, or even reversed by a change in pH. In this manner, we have a delicate but powerful means of influencing the distribution, or of preventing exchange altogether. This behavior is especially important in the separation of amino acids which can carry a positive, negative, or no net charge depending on the pH of the eluent. Buffered eluents are obviously indicated for separations of this kind, but one must not forget that the ionic constituents of the buffers are also subject to exchange with the resin so that the pH within the column may bear no relation to that which was prepared. The effect of pH on the elution of a typical weak acid is shown in Figure 5–5.

Effect of Complexing Agents. Ligands (defined in Section 22–1) which are neutral molecules have no effect on the charge of an ion, but do alter the exchange equilibrium constant. Many metal ions are complexed by anions yielding negatively charged complex ions. Thus, the rare earth metal cations, which are poorly separated by cationic exchangers, can be complexed and separated quite well by anionic exchangers.

Most of the useful complexing agents are themselves weak acids, weak bases, or anions or cations thereof. Thus the pH and complexing effects are often

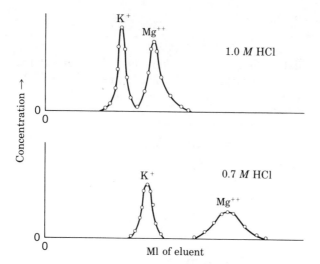

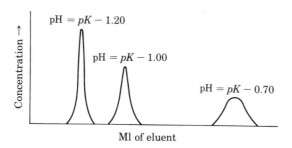

Figure 5–4 Elution of potassium and magnesium ions with two different concentrations of hydrochloric acid. [From Kolthoff and Elving, *Treatise on Analytical Chemistry*, Wiley-Interscience, 1961, Part I, Vol. 3, p. 1544.]

Figure 5–5 Effect of pH on the elution of a weak acid anion on an anion exchanger. [From Kolthoff and Elving, *Treatise on Analytical Chemistry*, Wiley-Interscience, 1961, Part I. Vol. 3, p. 1545.]

interdependent. Although there is a sensitive control of the separation, the resulting multiple equilibria are so complicated that theoretical predictions are of little value because of the necessary approximations.

Technique of Ion Exchange Chromatography

Selection and Preparation of the Resin. A wide variety of resins are available commercially from which one must select an appropriate mesh size, cross-linking, and quality. Analytical-grade (AG) resins are preferred because they have been more carefully sized and washed to remove foreign organic and inorganic materials. It may be necessary to convert the resin from one form to another; for example, the hydrogen form can be converted to the sodium form by extensive washing in the column with a strong sodium chloride solution until the effluent is neutral. Used resins are "regenerated" in this way.

If it is important to know the weight of resin used, it must be dried or brought to a known moisture content in a *hygrostat* (a vessel with controlled, constant humidity). In any event, before packing it must be equilibrated with water by prolonged soaking. After the resin has settled, the "fines" which float in the water are poured off.

Packing the Column. Simple columns are constructed from glass tubing with a reservoir at the top for the eluent and a fritted disk or glass wool plug at the bottom to support the resin bed. Usually the bottom of the column is drawn to a narrower diameter and bent in a double-U shape so that the outlet is higher than the top of the resin bed. This prevents air bubbles from leaking into the column. The resin is packed as an aqueous slurry and allowed to settle with occasional tapping. Once packed, the level of liquid should never be allowed to drain lower than the top of the resin bed or air bubbles will be entrapped. The best way to remove air bubbles or channeling is to back-flush the column with an upflow of water. A paper or glass fiber disk placed on top of the bed will minimize disturbance of the resin when adding the sample.

Total Capacity of the Column. The total exchange capacity influences the maximum sample size and is used to check the long term stability of the resin. The capacity of a resin in milli-equivalents per gram of dry resin is normally marked on the bottle by the manufacturer. Experimentally, it is most readily determined by converting the resin entirely to the hydrogen form (if it is cationic), and then eluting with a sodium chloride solution until it is completely converted to the sodium form. The effluent then must contain hydrochloric acid in an amount equivalent to the capacity of the column—easily determined by titration with sodium hydroxide. Common resins have a capacity of 1 to 5 meq/ml or roughly 1 to 5 N in acid or base.

Detection Methods. The difficulty of detecting small amounts of sample components in the presence of a large concentration of eluting ion is one of the major disadvantages of ion exchange methods. Continuous recording is not common, although in specific applications, light absorption, refractive index, pH, radioactivity, or polarographic measurements have been utilized. The most common practice is to collect numerous small, equal volume fractions and analyze each fraction for the species sought.

Some Applications

Removal of Ions. The household water "softener" is perhaps the most common example of ion exchange. Calcium, magnesium, iron, and all other multiply charged cations are replaced with sodium ion. The softened water then contains sodium salts which are innocuous in plumbing systems and for most home uses. Sodium is chosen because it is harmless in the water and because the resin can be readily regenerated with a strong solution of common salt.

Completely de-ionized water is prepared by passing the raw water through a cation exchanger which replaces all cations with hydrogen ion, and then through an anion exchanger which replaces all anions with hydroxide ion. In effect, the salts are replaced with the ions of water. The two resins can be combined in a single mixed bed so that the water never becomes too acidic or basic as it might if passed through the two resins separately. De-ionized water having a conductivity of less than 10^{-6} mho/cm is prepared more conveniently by ion exchange than by distillation. However, the de-ionization process does not remove nonelectrolytes, and thus the water may still be quite impure. Ion exchange offers a convenient and effective method for de-salting solutions of organic or biochemical mixtures.

The removal of one or more interfering ions by replacement with an innocuous ion for a given process or procedure is an obvious application. The determination of total salt content of a solution is simplified by conversion of the cations to hydrogen ion, or the anions to hydroxide ion, followed by a simple acid–base titration.

Concentration of a Trace Constituent.   Whenever a trace amount of an ion must be isolated or concentrated from a large volume of aqueous solution, one of the methods of choice is to remove it with an ion exchanger followed by elution into a small volume of eluent. This is a common step in the determination of trace metals in water, copper in milk, or the recovery of precious metals. Perhaps the most spectacular example occurred in the isolation and identification of the first sample of mendelevium. Ten thousand atoms of einsteinium on a gold foil were bombarded with high energy alpha particles. The target was quickly dissolved with aqua regia and the gold extracted with ethyl acetate. The aqueous phase, which contained the einsteinium and any mendelevium produced, was separated with miniature ion exchange columns. At one point the entire world's supply (about 17 atoms) of this newly discovered element was contained in a single resin bead.

Preparation of Reagents.   Determinate solutions of strong acids and bases are not easily prepared because of the lack of primary standard reagents. On the other hand, primary standard sodium or potassium chloride is readily available and their solutions are stable indefinitely. Aliquots of these solutions, when passed through a resin in the hydrogen or hydroxide form, will produce equivalent amounts of acid or base. Many other solutions which are difficult to prepare or standardize can be made in a similar fashion.

Separation of Metals.   Ion exchange is especially advantageous for the separation of metal ions with very similar properties for which specific methods are not available. For example, the alkali and alkaline earth metals are always difficult to determine in mixtures, but can be readily separated in an ion exchange column. The separation of rare earths is a classic problem formerly accomplished only by numerous and tedious fractional crystallizations. The selectivity coefficients for the rare earth metal ions are nearly identical to each other, but with the addition of chelating agents such as EDTA the selectivity coefficients are altered to varying degrees. Thus the separation becomes much simpler. Ion exchange columns now provide pure rare earth compounds on a commercial scale.

Separation of Amino Acids.   Perhaps the most impressive example of the versatility and potency of ion exchange methods is the separation of the complex mixtures of amino acids encountered in biochemistry. The amphoteric nature of this group of acids makes it possible to change the sign of the charge or to remove the net charge so that a given acid is amenable to exchange on a cationic or anionic resin, or neither, by controlling the pH of the solution. Thus, at a given pH, a mixture of amino acids can be separated into three groups according to their isoelectric points by passing it through the two types of resins successively. After changing the pH, the groups can be further subdivided as many times as desired.

Alternatively, the original mixture can be resolved in a single pass through an appropriate resin by a graded elution technique in which the pH of the eluent is gradually increased in a stepwise fashion. Moore and Stein were able to separate 50 amino acids by this technique. The conditions for this remarkable separation are given below the diagram in Figure 5–6. Both the pH and temperature were increased during the elution.

More recently amino acids have been separated with resins containing immobilized metal ions (e.g., Cu^{+2}, Cd^{+2}). The metal ions then act as exchange sites for

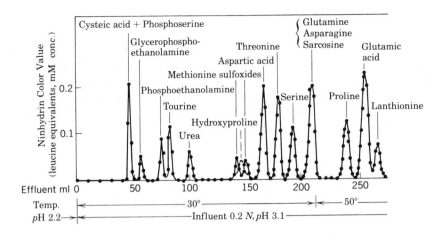

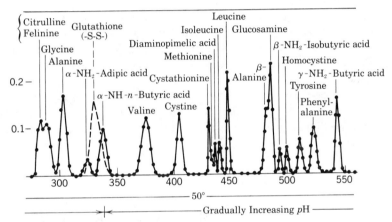

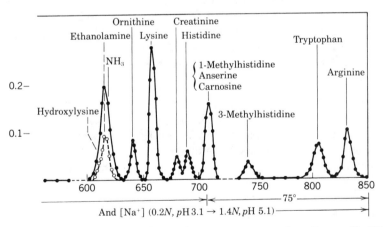

Figure 5–6 Separation of amino acids and related compounds on Dowex 50 − X4 column 0.9 × 150 cm. [After Moore and Stein, *J. Biol. Chem.*, **211**, 893 (1954).]

ligands, in particular for amino acids. If a mixture of amino acids is passed through such a resin, the various amine groups compete in the formation of complexes with the metal ions and emerge in order of increasing stability of the complexes. Although an ion exchange column is used, the sample components are not necessarily ionic and the process is called *ligand exchange.*

5–6   GEL CHROMATOGRAPHY

The separation of very high molecular weight substances is most readily accomplished by the use of columns packed with gels. Several varieties of gels separate molecules primarily on the basis of their sizes by a "sieving" or "filtering" process. Hence, the names "gel filtration," used by biochemists, and "gel permeation chromatography," used by polymer chemists, describe the same general technique. Adsorption and electrical charge effects may also play a role in these separations.

The gels have very open, three-dimensional networks formed by cross-linking long polymeric chains. Instead of ion exchange sites, most of these gels have polar groups capable of adsorbing water or other polar solvents. A few are able to adsorb nonpolar solvents. In either case, the adsorption causes an opening of the structure, or "swelling," leaving interstices within the gel. Depending on the extent of cross-linking, there will be a critical size (*exclusion limit*) of a molecule that can just penetrate the interior. Larger molecules will pass through the column with no retardation because they cannot enter the gel. Smaller molecules will penetrate the interior to a degree determined by their size. Exclusion limits from about 1000 to several million (molecular weight) are available by selecting the appropriate gel.

Gel chromatography is another variation of the general chromatographic method. Again, the same principles are applicable, although the mechanism of the separation, the handling of the sample and the detection of the components will differ in detail.

Types of Gels

Sephadex. For proteins and most of the large molecules of biochemical interest, Sephadex is the most popular of the gel materials. Sephadex is prepared from polysaccharide dextrans which have been synthesized by the action of a bacterium on sucrose. Each glucose residue contains three hydroxyl groups giving the dextran a polar character. The cross-linking reaction is accomplished with epichlorhydrin, CH_2—$CHCH_2Cl$. By adjusting the conditions, the amount of

cross-linking, and thus the size of the pores, can be carefully controlled. Sephadex gels are insoluble in water, and are stable in bases, weak acids, and mild reducing and oxidizing agents. Prolonged exposure to 0.1 N hydrochloric acid or strong oxidizing agents will cause breakdown of the gel granules. Temperatures above 120°C should also be avoided. In addition to water in which they are commonly used, Sephadex gels swell in glycol, dimethylsulfoxide, and formamide, but not in glacial acetic acid, methanol, or ethanol. These gels are classified by the amount of "water regain," which is a function of the looseness of the structure or the extent of cross-linking. Several types and their properties are listed in Table 5–5.

Bio-Gel. A more chemically inert series of gels, called Bio-Gel P, is prepared by copolymerization of acrylamide and N,N'-methylene-bis-acrylamide. The polymer is insoluble in water and common organic solvents and is useful in the pH range 2 to 11. Strong bases may hydrolyze the amide groups. The inert polyacrylamide matrix minimizes the possibility of adsorption of polar materials. Ten types of Bio-Gel P are commercially available with exclusion limits varying from molecular weight 1800 to 400,000.

Table 5–5 Types of Sephadex Gels

Type	Water Regain, g/g Dry Sephadex	Exclusion Limit, Mol. Wt.	Fractionation Range, Mol. Wt. Limits
G–10	1.0	700	up to 700
G–25	2.5	5,000	100–5,000
G–50	5.0	10,000	500–10,000
G–75	7.5	50,000	1,000–50,000
G–100	10.0	100,000	5,000–100,000
G–200	20.0	200,000	5,000–200,000

[Courtesy of Pharmacia, Inc.]

Agarose. For molecular weights over 500,000, agarose gels (Sepharose and Bio-Gel A) are effective. These gels are prepared from a polygalacto-pyranose in the form of rather soft beads that cannot withstand high pressures.

Porous Glass. For high pressures, the rigidity of a porous glass bead is advantageous. Porous glass beads of accurately controlled pore sizes are available commercially; for example, Porasil and Bio-Glas.

Styragel. For completely nonaqueous separations, a gel that will swell in an organic solvent is required. Styragel is a rigid cross-linked polystyrene gel which can be prepared in a range of porosities similar to the other gels. It is useful at temperatures up to 150°C with the following solvents: tetrahydrofuran, benzene, trichlorobenzene, perchloroethylene, cresol, dimethylsulfoxide, chloroform, carbon tetrachloride, aromatics, and others, but not with water, acetone, or alcohols. Exclusion limits vary from 1600 to 40,000,000. The rigidity of Styragels permits their use under high pressure in high performance columns.

Theoretical Principles

The Ideal Process. In its simplest form, we shall imagine that the interior of the gel bead consists of conically shaped holes all having identical shapes and sizes. The large molecules to be separated will either be too large to enter the hole at all, or be able to penetrate the interior of the cones to a depth determined by the size of the molecule. If they are reasonably small molecules or simple ions, they will be able to move freely into all parts of the interior of the gel. With this model it is easy to see that the fraction of the interior volume available to any given molecule varies from 0 to 1 and is a function of the diameter of the molecule relative to the diameter of the pores.

There are two kinds of solvent in a gel column: that within the gel having a volume V_i, and that outside the beads having a volume V_o. Let us assume that the only differentiating feature of the column results from the variation in the fraction of the volume V_i available to different sized molecules. A molecule too large to enter the holes will be swept through the outside volume without ever entering the gel. In this case:

$$V_e = V_o$$

where V_e is the elution volume. Smaller molecules must be swept through V_o plus some additional volume which is a fraction of V_i, or

$$V_e = V_o + K V_i \qquad (5\text{–}10)$$

where K is a "distribution" coefficient which can be defined in several ways. It is the ratio between the "average" concentration of the solute in the interior and the concentration outside the gel. One must remember that in that part of the interior volume available to the solute, its concentration is the same as that outside, but in the remaining interior volume the concentration is zero. K can also be defined as the volume fraction of the interior solvent available to the solute, or in terms of experimentally measurable quantities:

$$K = (V_e - V_o)/V_i \qquad (5\text{--}11)$$

V_i can be obtained from the weight of the dry gel, a, and the water regain value, W_R (Table 5–5):

$$V_i = aW_R \qquad (5\text{--}12)$$

If the sieving action just described were the only mechanism, K values should lie between 0 and 1. An interesting consequence of this mechanism is that in gel chromatography the elution volume is usually less than the void volume within the column, whereas in other forms of chromatography the elution volume is always larger than the void volume because of the retention mechanism. In gel chromatography, retention is replaced by partial exclusion. A simplified illustration of what happens in the column is given in Figure 5–7.

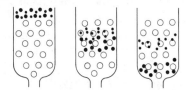

Figure 5–7 Separation by exclusion in a gel chromatography column.

Two solutes having K values equal to K_A and K_B will have elution volumes of $V_o + K_A V_i$ and $V_o + K_B V_i$ and will be separated by the difference $(K_B - K_A)V_i$. If the eluted bands are no broader than the sample volume, V_s, which was used, then the maximum sample volume for complete separation is:

$$V_s < (K_B - K_A)V_i \qquad (5\text{--}13)$$

The same band broadening processes operate in gel chromatography as in all other forms. For optimum efficiency (minimum H) one should use small beads uniformly packed, and slow flow rates as shown in Table 5–6.

Table 5–6 Efficiency of a Sephadex G–200 Column $(4 \times 42$ cm$)$ with Serum Albumin

Particle Size, Mesh	Flow Rate, ml/hr	H, cm
100–200	75	0.62
100–200	50	0.48
100–200	25	0.23
200–270	25	0.08

[Courtesy of Pharmacia, Inc.]

Nonideal Conditions. So far, we have treated gel separations from a geometric approach; however, the simple theory based on partial exclusion is only an approximation of the true nature of the process. There are, of course, often other mechanisms operating which cause a departure from the simple behavior attributed to sieving alone. The solvent in the primary adsorption layer on the interior surface appears not to be available to the solutes no matter what their size. This presumably explains why many small species have K values of about 0.9 rather than the expected 1.0. Other species have K values exceeding unity which must result from adsorptive forces retarding the molecules on the interior surfaces. Adsorption is effective with solutes which can react or form complexes with the hydroxyl groups of Sephadex, for example, sodium hydroxide, sodium tetraborate, and aromatic and heterocyclic compounds. K values for strongly adsorbed substances can be as high as 20. Still another mechanism appears to be a restricted diffusion within the pores. No doubt the predominant mechanism varies from case to case.

Technique

Column Preparation. Gel columns are prepared in the same way as ion exchange columns. The gel beads must first be swelled by equilibrium with the solvent—a few hours for dense, highly cross-linked gels, and a day or more for the loosely cross-linked varieties which take up many times as much solvent. Fines can be removed by decantation.

The gel bed is supported in the column on a glass wool plug or nylon net and the previously swollen gel is added in the form of a slurry and allowed to settle. No air bubbles must be entrained within the bed and the level of liquid must never be allowed to go lower than the top of the bed.

Operation. Samples are introduced as carefully as possible to avoid disturbing the bed itself. Since larger samples are used than in most forms of chromatography, it may be difficult to begin the elution without mixing the sample and the eluent. Sometimes a nondisturbing substance such as sodium chloride is added to the sample to increase its density. Elution is usually carried out under a constant hydrostatic pressure head to achieve a constant flow rate.

Detection. The common detection methods include: collecting and analyzing fractions and continuous methods with flow-through cells in which ultraviolet absorption, refractive index, or radioactivity is measured.

Applications

Desalting. One of the common separation problems in biochemistry is the removal of salts and small molecules from macromolecules. The large difference in distribution coefficients for this separation makes it possible to use simple columns with high flow rates. For example, with Sephadex G–25, solute molecules having molecular weights over 5000 are eluted in a volume V_o. Smaller molecules with molecular weights below 1000 will be eluted following the macromolecules inversely in order of their size. One of the advantages of this method of desalting is that the macromolecules are eluted with essentially no dilution.

Concentrating. Dilute solutions of macromolecules with molecular weights higher than the exclusion limit may be readily concentrated by utilizing the hygroscopic nature of the dry gel. Sephadex G–200 absorbs 20 times its weight of water, although G–25 is preferred because of its more rapid action. Since salts and small molecules are imbibed to the same extent as the water, their concentrations in the supernatant solvent remain essentially unchanged, and the ionic strength and pH of the macromolecular solution remain essentially unaltered. This is an important advantage in the separation of proteins which are easily denatured by changes in solution composition and temperature.

Fractionation. The complete separation of mixtures of closely related molecules having small differences in K values will require long columns, slow flow rates, and long times. However, the distribution of molecular weights in a complex mixture is often sufficient information to characterize the sample or to give a rough idea of its composition. It has been observed, and can be predicted from theory, that the elution volume is related to molecular weight in a simple fashion:

$$V_e = A - B \log \text{Mol. Wt.} \qquad (5\text{–}14)$$

where A and B are constants whose values are determined from a plot of V_e versus $\log \text{Mol. Wt.}$ for several known compounds. The equation is valid over a considerable range of molecular weights, provided all compounds are closely related. Values for A and B must be redetermined for each column, each set of operating conditions, and for different classes of compounds. Some representative data are given in Figure 5–8.

Gel chromatography has been used primarily by biochemists to separate proteins, peptides, nucleic acids, polysaccharides, enzymes, hormones, and the like, and by polymer chemists to characterize molecular weight distributions in polymeric mixtures.

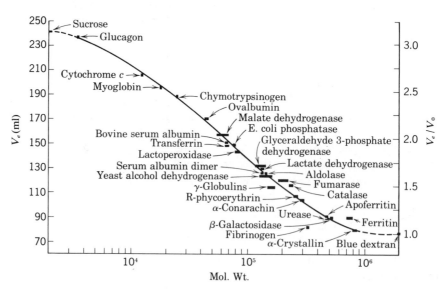

Figure 5–8 Relation between elution volume and molecular weight for proteins on a Sephadex G-200 column, 2.5×50 cm, at pH 7.5. [After Andrews, *Biochem. J.*, **96**, 595 (1965).]

QUESTIONS AND PROBLEMS

5–1. If a monolayer of water occupies 2.86×10^3 m^2/ml, estimate the number of monolayers of water corresponding to each step on the Brockmann scale for a sample of alumina with a surface area of 175 m^2/g.   *Ans.* for II, 0.49 monolayer

5–2. A 3.00-g sample of sugar and potassium nitrate is dissolved in exactly 100 ml of water and passed through a cation exchange column in the H-form. The effluent requires 5.30 ml of 0.0100 N sodium hydroxide for titration. Calculate the percentage of KNO_3 in the sample. *Ans.* 0.179%

5–3. Describe a procedure for analyzing a solution of sodium chloride in hydrochloric acid using ion exchange resins and acid–base titrations. Indicate the method of calculating the concentration of each of the principal ions present.

5–4. One gram of dry cation exchange resin has a capacity of 5.0 meq. After swelling and packing in a column, the same amount of resin occupies 7.5 ml. If the total column contains 25 ml, how many mg of calcium ion could it exchange? *Ans.* 334 mg

5–5. Commercial ion-exchange water softeners are commonly regenerated with a concentrated solution of rock salt. Why is it not feasible to regenerate mixed bed water deionizers?

5–6. A common expression for the distribution coefficient in ion exchange studies is

$$K = \frac{\text{amount of ion/g of dry resin}}{\text{amount of ion/ml of solution}}$$

Calculate K for Mg^{+2} ion for the following case. A 1.000-g portion of dry resin is shaken with 30.0 ml of 0.1000 M $Mg(NO_3)_2$. After equilibration, a 10-ml aliquot of the supernatant solution required 2.00 ml of 0.0111 F EDTA solution for titration of the magnesium:

$$Mg^{+2} + EDTA \rightleftharpoons Mg\cdot EDTA \qquad \qquad Ans.\ 1320$$

5–7. Show that Equation 5–6 is equivalent to

$$V_e = V_M + KV_S$$

where $K = C_S/C_M$ and V_S is the volume of resin.

5–8. The elution volume of Zn^{+2} on a certain ion exchange column is 400 ml for which it is known that the interstitial volume is 7.0 ml. What is the value of D (as defined in Equation 5–5) for Zn^{+2} under these conditions? *Ans.* 56.1

5–9. In the separation of amino acids shown in Figure 5–6, the pH was held constant at 3.1 for the first 340 ml of eluent. Would you expect acids with the largest dissociation constants to elute first or last during this interval? What other factors influence the order of elution of these acids? The last portion of eluent (700 to 850 ml) is at pH 5.1. Do you expect that the acids eluted in this interval are stronger or weaker than those eluted first? Explain your answers.

5–10. Sucrose and Blue dextran are often used to measure V_o and V_i for gel columns (see Figure 5–8). On a certain gel column the elution volumes of these two substances are 55.5 and 9.0 ml, respectively.
(a) Predict the elution volume of a substance with a K value of 0.40. *Ans.* 27.6 ml
(b) What is the K value for a substance for which the elution volume is 25.0 ml? *Ans.* 0.344

REFERENCES

A. J. P. Martin and R. L. M. Synge, *Biochem. J.*, **35**, 91, 1358 (1941). The classic papers on partition chromatography.

S. Moore and W. H. Stein, *J. Biol. Chem.* **192**, 663 (1951); **211**, 893 (1954). Classic papers on the separation of amino acids.

L. R. Snyder, *Principles of Adsorption Chromatography*, Dekker, New York, 1968.

F. Helfferich, *Ion Exchange*, McGraw-Hill, New York, 1962.

O. Samuelson, *Ion Exchange Separations in Analytical Chemistry*, 2nd ed., Wiley, New York, 1963.

H. Determann, *Gel Chromatography*, Springer-Verlag, New York, 1968.

J. J. Kirkland, *Modern Practice of Liquid Chromatography*, Wiley-Interscience, New York, 1971.

6

PLANE CHROMATOGRAPHY

6–1   INTRODUCTION

A drop of liquid spotted on a piece of paper or cloth spreads in a circular pattern, and if the liquid contains colored substances, concentric rings are easily observed. This simple analytical technique was used by the ancient Romans to test dyes and pigments. About 100 years ago, the German chemists Runge, Schoenbein, and Goppelsroeder introduced improvements to make the technique more nearly reproducible and quantitative. It is only within the last 30 years, however, that paper chromatography has been accepted as a reliable method. A. J. P. Martin, who also was the pioneer in liquid partition and gas chromatography, and his co-workers, Consden and Gordon, published a classical description of paper partition chromatography (PC) in which the moisture in the paper served as a stationary phase. Within a few years, this utterly simple scheme of analysis helped to revolutionize research in biochemistry.

More recently, and primarily due to the work of Stahl, thin-layer chromatography (TLC) has found widespread use. The "thin layer" of adsorbent is spread on a glass or plastic plate and is used much the same as a piece of paper. For convenience, the operational details of plane chromatography are discussed under the heading of PC, but the discussion applies to TLC as well. Because it is more versatile and reproducible, TLC has largely replaced PC in the laboratory.

In addition to their obvious simplicity, PC and TLC require only minute amounts of samples. Chromatography on a plane surface, rather than with a column, offers the unique advantage of two-dimensional operation. Thus the selective properties of two different solvents can be utilized in developing a single chromatogram.

Theoretical Principles

Most of the theoretical principles described in column chromatography apply to plane chromatography as well. The concept of a "theoretical plate" may be more difficult to visualize in a piece of paper, but it is clear that the separation is accomplished by successive equilibrations of the sample components between two phases, one of which moves over the other. Likewise the same kinds of nonideal processes must cause zone spreading on a plane surface as in a column.

The degree of retention in plane chromatography is customarily expressed as the *retardation factor*, R_f:

$$R_f = \frac{\text{distance solute moved}}{\text{distance solvent moved}} \qquad (6\text{–}1)$$

The "front" of the solvent is the only point at which the distance it moved can be measured. However, the distance the sample moved is measured to the center of the spot, or to its point of maximum density. We may also define a distribution coefficient, K, in terms of the concentration of solute in the moving phase, C_M and in the fixed phase, C_S:

$$K = C_S/C_M \qquad (6\text{–}2)$$

There is a simple relationship between K and R_f value. The distance an average solute molecule moves is directly proportional to its velocity, which in turn is equal to the velocity of the solvent times the fraction of time the solute spends in the moving phase. The latter can be expressed in terms of the number of molecules in each phase, or as the equilibrium distribution of solute between the two phases:

$$R_f = \frac{\text{number of moles of solute in moving phase}}{\text{total moles of solute in both phases}} = \frac{C_M A_M}{C_M A_M + C_S A_S} \qquad (6\text{–}3)$$

where A_M and A_S represent the cross-sectional areas of the two phases (perpendicular to the plane of the paper). Dividing numerator and denominator by C_M and substituting from Equation 6–2, we obtain:

$$R_f = \frac{A_M}{A_M + A_S C_S/C_M} = \frac{A_M}{A_M + KA_S} \qquad (6\text{–}4)$$

The cross-sectional areas are difficult to determine, so that Equation 6–4 is of little practical use, but it does show that R_f is a modified form of an equilibrium constant, and therefore R_f values can be expected to depend on the same kinds of parameters as other measures of retention more commonly used in column chromatographic methods. R_f values are also subject to many minor influences such as: variations in the paper, method, and direction of development, size and concentration of the sample, and even the distance traveled by the spot. For these reasons, it is more convenient and accurate to use a relative R_f, or R_{std}, value, for which a standard compound is added to the sample (or measured separately under identical conditions). The R_{std} value is simply the ratio of the distances traveled by the two spots during the same time of development.

6–2 PAPER CHROMATOGRAPHY

Nature of the Paper. The paper commonly used consists of highly purified cellulose. The polymeric cellulose structure contains several thousand anhydroglucose units linked through oxygen atoms. Theoretically, there are three hydroxyl groups on each glucose unit, but many of these have been partially oxidized during manufacture to aldehyde, ketone, or carboxyl functional groups. In addition to these variations, the paper will contain traces of many impurities, including inorganic substances held as exchangeable cations on the hydroxyl groups, adsorbed salts, or mineral matter left on the paper during processing. Cellulose has a great affinity for water and other polar solvents and holds them tenaciously through the formation of hydrogen bonds. These solvents penetrate the fibers and cause the paper to swell, changing its dimensions. In water, the cellulose is highly polar and becomes electronegative. This effect is decreased in solvents of lower dielectric constants or with increasing salt concentrations. The paper exhibits weak ion exchange properties as well as adsorptive properties. Furthermore, it is a mild reducing agent, and on prolonged contact it will react with many oxidizing agents.

Most of the properties just described vary from paper to paper even within the

same lot. Acid-washing, rinsing, and drying aid in obtaining reproducible behavior, but, at best, variations in R_f values can be expected. Modified forms of
paper are now available in which the paper has been impregnated with alumina,
silica gel, ion exchange resins, etc. While these modifications lead to different
mechanisms for the separation, the technique is the same.

Apparatus. The apparatus required for paper chromatography consists of a
support for the paper, a solvent trough, and an air-tight chamber in which to
develop the chromatogram. The closed chamber is necessary to prevent the
evaporation of the volatile solvents from the large exposed area of the paper. The
size of the chamber may vary from an ordinary test-tube to a large aquarium
depending on the size of paper. If a great deal of paper chromatography is to be
done, it may be simpler to control the temperature of the entire room, say to 18°C
or so. The basic components illustrated in Figure 6–1 are sufficient for simple
paper chromatography, but a number of modifications are available for convenience or for special methods of development to be described below.

The sample is applied prior to dipping into the eluting solvent as a small spot
with any device that will transfer a small volume; e.g., a toothpick, loop of
platinum wire, capillary tube, micro-pipet, etc. For some methods, the sample may
be applied in a narrow line perpendicular to the flow of solvent. If the sample is
applied as a solution, its solvent should be evaporated before proceeding.

The methods of detection commonly used are: (1) inherent visible colors of the
components (whenever possible), (2) reactions with color-producing reagents, (3)
ultraviolet absorbance, (4) infrared absorbance, (5) fluorescence, (6) radioactivity,
(7) bioautography, or (8) extraction and further chemical or physical tests. Test
reagents are easily applied by spraying, immersing, or by exposing to vapors. For
example, a ninhydrin reagent sprayed on the paper reacts with amines and amino
acids to form a blue or purple color.

Bioautography involves placing the paper in contact with a culture medium for
a period of time followed by examination of the relative growth rates of bacteria
along the paper strip. Sample compounds may have either a positive or negative
effect on the growth rate of the bacteria in the culture medium.

Development. For all types of development, the paper should be equilibrated
with solvent vapors in the chamber before development is begun.

Ascending Development. This is the simplest and most popular type. The paper
is suspended vertically with about an inch immersed in the solvent and the sample
spot initially located about an inch above the surface of the solvent to prevent
diffusion of the sample downwards into the solvent reservoir. The solvent ascends
through the paper by capillary action. The rate of ascent is slow and decreases
with time because of gravity. However, the slow rate enhances the possibility of

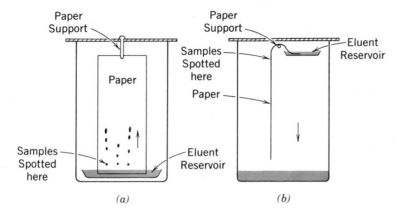

Figure 6–1 Basic apparatus for paper chromatography: (a) ascending solvent; (b) descending solvent.

achieving partition equilibrium and often results in compact, sharply defined spots.

Descending Development. With a solvent trough at the top of the chamber, the direction of flow is downwards. To prevent rapid siphoning, the paper is folded into a U-shape with a short initial rise from the solvent tank. Although the apparatus is somewhat more elaborate, the method is much faster than the previous one, longer pieces of paper may be used, and large amounts of solvent can be used if necessary for slow-moving spots.

Two-Dimensional Development. In all previously described methods of chromatography, the eluting solvent (or carrier gas) could flow in one direction only—through the column or across the paper. On a plane surface, such as a piece of paper, it is possible to carry out a sequential development in two directions. The sample is applied as a spot close to a corner. This is developed in the normal fashion by ascending or descending procedure until the fastest moving spot approaches the end of the paper. Then the paper is removed, and after evaporating the solvent, it is turned 90° and developed a second time with another solvent having different eluting properties. In this manner, samples which could be only partially separated with either solvent alone, may be completely separated by the combination of solvents as shown in Figure 6–2.

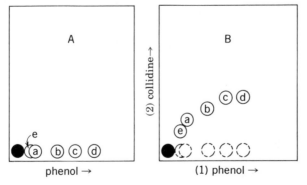

Figure 6–2 Two-dimensional paper chromatography. Eluents: A, phenol only; B, phenol followed by collidine. Solutes: a, glutamic acid; b, serine; c, threonine; d, alanine; e, aspartic acid.

Radial Development. In this method the sample is spotted at the center of a horizontally placed disk of paper with a wick to supply solvent to the center from a supply trough below. The components move outward along radial paths forming circles of increasing diameters. The bands are more or less self sharpening because the solvent moves the trailing edge faster than the leading edge. A covered Petri dish is an adequate container, or the paper can be sandwiched between two glass plates, with a hole in the center of the upper one for the introduction of solvent.

For fast development, one can rotate the disk at high speed so that the solvent flows by centrifugal force as well as by capillary action.

Choice of Solvent Systems. The more polar phase, usually water, is preferentially adsorbed by the paper and is held stationary while the less polar phase flows over it. In some cases it may be desirable to remove the water from the paper in a preliminary drying operation and use an alcohol, glycol, formamide, or other polar solvent as a stationary phase. If water is used as a stationary phase, it is not necessary that the mobile solvent be immiscible because the water phase is so tenaciously held by the paper. The aqueous phase may, of course, be buffered if desired. The mobile phase may be a mixture of solvents, for example, alcohols, acids, ketones, esters, amines, phenols, hydrocarbons, etc., selected so as to achieve maximum separation of the sample components.

In addition to optimum separation, there are other factors to be considered in choosing a solvent system. The more components a solvent system contains, the more difficult it will be to maintain a saturated atmosphere in the chamber. Partial evaporation which changes the composition of the solvent will affect the R_f values. Likewise, temperature change will be more critical in a mixed system. The solvent system must not interfere with the detection procedure (or must be readily and completely vaporizable so that it can be eliminated before detection).

Another very important consideration which is easily overlooked is that minute, nonvolatile impurities in the developing solvent can be easily confused with traces in the sample, or may interfere in the detection method.

Ideally, one should choose solvent systems such that the two phases are immiscible, and in which the sample components have a high, but different, solubility in the two phases. This situation will lead to maximum separation with a minimum of spreading in the shortest time.

Reversed-Phase Chromatography. If the paper is coated with a hydrophobic substance, such as rubber latex, mineral oil, or silicone oil, and a polar phase used to elute, a reversal of the normal distributions is obtained. Such systems may be advantageous for the separation of fatty acids and nonpolar compounds which may move too fast because of low solubility with a polar stationary phase.

Advantages and Limitations. Paper chromatography is usually used for analyses that would be difficult, time-consuming, or expensive by more conventional methods. Though some of the procedures may require several hours, the operator time is indeed very small. It is particularly suitable for samples in the microgram region, and for mixtures of substances that are closely related chemically.

The small sample limitation may well be a disadvantage if detection requires an auxiliary technique. The variability of the paper is a serious problem although this situation is improving. Most of the procedures have been arrived at in an empirical fashion, and could hardly be called quantitative.

Applications. Paper chromatography has been successfully applied to problems in inorganic and organic chemistry as well as in biochemistry. Prospectors have used it to determine metals in ore samples. Isomeric metal complexes can be resolved. Metals with similar chemical properties are easily resolved. Nearly any mixture of organic chemicals can be separated and organic chemists often use it as a quick check for purity or as a pilot method to determine the optimum conditions for a larger scale column chromatographic separation.

The method has made its greatest impact in biochemistry where difficult analytical problems with vanishingly small samples are legion. The literature expands continually: the control of purity of pharmaceuticals and food products, the study of ripening and fermentation, the detection of drugs and dopes in animals and humans, the analysis of cosmetics, the detection of adulterants and contaminants in food and drink, and, of course, the analysis of most of the reaction mixtures studied by biochemists.

6–3 THIN-LAYER CHROMATOGRAPHY

The operations performed in thin-layer chromatography are essentially the same as in paper chromatography, and most of the previous discussion applies to this remarkably simple form of chromatography as well. Instead of a piece of paper, a thin layer of finely divided adsorbent supported on a glass or plastic plate is used.

Nature of the Thin Layer. The most commonly used adsorbents are silica gel, alumina, diatomaceous earth, and powdered cellulose, but other materials such as Sephadex, ion exchange resins, or inorganics have been used for special purposes. Silica gel is acidic and has a high capacity. It is useful for both adsorption and partition chromatography. Alumina is basic and used primarily for adsorption. Diatomaceous earth is nearly neutral and is used as a support for partition phenomena. Any of these materials can be used in pure form, but it is more convenient to combine them with a binder (adhesive) such as plaster of Paris to make a more cohesive layer.

Preparation of the Layer. The glass plate is usually flat and smooth, or occasionally lightly ground or ridged. The size and shape may be dictated by the apparatus to be used—microscope slides are inexpensive, readily available, and adequate for small scale work. In any event, the glass surface must be scrupulously cleaned with a detergent and/or an organic solvent to remove any grease.

The thickness of the layer will determine the capacity of the system. Layers of 0.15 to 2.0 mm thickness are satisfactory. The primary concern is to obtain a layer of uniform thickness. Most thin layers are produced by spreading a film of an aqueous slurry of the adsorbent over the entire surface. The slurry must be neither too thick (viscous) nor too thin, or it will not spread properly. Silica Gel G, which contains 10 to 15% calcined calcium sulfate, mixed with about twice its weight of water, makes a very satisfactory slurry.

In commercial spreading machines a slotted trough travels over the glass and deposits a uniform layer. Alternatively, we can apply tape along opposite edges of the glass and use a glass rod for a trowel, producing a layer the thickness of the tape. For occasional use, a pair of microscope or lantern slides can be dipped back-to-back into a non-aqueous slurry containing 35 g Silica Gel G in 100 ml of chloroform, raised slowly, and allowed to drain and dry.

After spreading, the binder requires about 30 minutes to "set." For adsorption chromatography, the layer is activated by heating at 110°C for several hours. For partition chromatography, no drying is required and the residual water acts as the stationary phase. The edges should be cleaned and the plates stored in a desiccator or an appropriate cabinet. Remember that an activated plate will adsorb water vapor and other vapors from the air, making drastic changes in its chromatographic behavior.

Precoated TLC plates (glass or plastic) are available from many laboratory supply firms. For occasional use, they are much more convenient and probably more reproducible than the homemade variety.

Methods of Development and Detection. The choice of solvent depends on the same factors as already discussed in other forms of liquid chromatography, and the methods of elution are much the same as for paper chromatography. The procedure must, of course, be conducted in a closed chamber to prevent evaporation of the solvent. In order to detect the spots, iodine vapor is used extensively as a "universal" reagent for organic compounds. The iodine spot disappears rapidly but can be made more permanent by spraying with a 0.5% benzidine solution in absolute ethanol. Another common detecting reagent is a spray of sulfuric acid which upon warming chars the sample components, leaving black spots. There are a host of more specific reagents available, and, of course, the spot can be scraped off, eluted, and investigated by any available method.

Advantages and Applications. Compared to paper chromatography, thin-layer is more versatile (wider range of stationary phases and solvents), faster, and more reproducible. It is often used as a pilot technique for a quick look at the complexity of a mixture or as an aid in establishing the best conditions for large scale column chromatography. Because of its speed and simplicity, it is often used to follow the course of reactions or to monitor more elaborate and complex separation techniques. Thin-layer chromatograms often serve to identify drugs, plant extracts, and biochemical preparations, or to detect contaminants or adulterants.

6–4 ZONE ELECTROPHORESIS

Electrophoresis pertains to the transport of electrically charged particles—ions, colloids, macromolecular ions, or particulate matter—in an electrical field. It is not a form of chromatography; however, differences in the rates of migration provide a powerful means of separating biocolloids such as proteins, polysaccharides, and nucleic acids as well as the characterization of their components. Although electrophoresis can take place in gaseous media (electrostatic smoke precipitators), we are concerned here only with differential migration in a liquid phase. Electrode reactions must occur in order to sustain the current, but the reactions per se are not important to the separation and need not be considered.

Moving Boundary Electrophoresis. The liquid phase may be a bulk solution contained in a U-tube, or a thin film on paper or porous support. The former technique was introduced by Arne Tiselius in 1937 for the analysis of protein mixtures. He placed the sample in the center part of a square-shaped U-tube provided with electrodes at the ends (Figure 6–3). On passage of a current the charged species began to migrate toward the appropriate electrode. Complete separations do not occur, but if mechanical disturbances are carefully avoided (no mixing), *moving boundaries* can be observed as discontinuities in concentration (changes in color or refractive index). This technique requires considerable experimental skill and is slow and not very sensitive; however, it was exceedingly important in the early development of protein chemistry.

Zone Electrophoresis. In this form of electrophoresis, the migration takes place

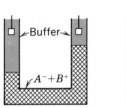
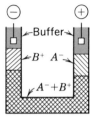

Figure 6–3 Basic apparatus for electrophoresis. Resolution of two components under ideal conditions.

in a thin film of liquid coated on paper or porous material. The apparatus may consist of a strip of filter paper supported on a horizontal glass plate with the ends of the paper dipping into reservoirs of buffer solution. The electrodes are pressed against the paper near the ends and the sample spotted near the middle, as shown in Figure 6–4a. Upon passage of current, sample components travel in one direction or the other (Figure 6–4b), forming spots or zones. Cellulose acetate membranes have largely replaced filter paper as a support medium. Various gels, resins, and powdered materials are also useful in special cases. *Immunoelec- trophoresis* is carried out with a layer of agar gel 1 to 2 mm in thickness, formed on a microscope slide. After electrophoretic separation of a mixture of proteins, a trough is cut parallel to the edge of the plate (Figure 6–4c) and filled with a suitable antiserum. Incubation for 24 to 48 hours in a humid atmosphere allows the antiserum to diffuse through the agar layer. When the antiserum reaches the protein spots, visible "arcs" are formed corresponding to pairs of antigen- antibody proteins (Figure 6–4d).

    Continuous-Flow Electrophoresis.   Electrophoresis on paper (or thin layer) makes possible one of the few forms of truly continuous elution chromatography. A diagram of the apparatus is shown in Figure 6–5. The sample is applied

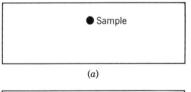

(a)

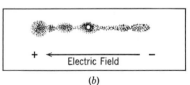

(b)

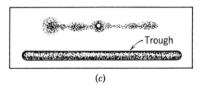

(c)

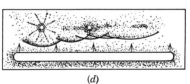

(d)

Figure 6–4 Separation by electrophoresis: (a) sample spotted on a layer of agar; (b) components are separated by electric current; (c) immuno- electrophoresis with antiserum introduced into longitudinal trough alongside electrophoretic migration path; (d) arcs formed where antigens and antibodies meet by diffusion.

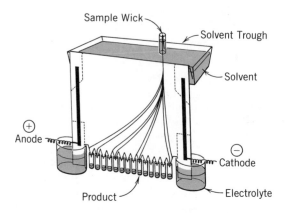

Figure 6–5 Continuous electrophoresis on paper. [From Durham, *J. Amer. Chem. Soc.*, **73**, 4875 (1951).]

continuously from a wick at the top of the paper and is eluted in a downward direction. With no electrical current flowing, the sample would of course travel in a straight vertical line and emerge unseparated, although each component would be traveling at a different velocity depending on its distribution coefficient. With an electrical current flowing, however, charged particles will travel to the left or to the right depending on their charge with a horizontal velocity depending on their mobility. The resultant direction of travel is the vector sum of the two perpendicular components. Thus each charged component should emerge at a different place at the bottom of the paper while uncharged components will emerge directly below the point of application. Continuous electrophoresis makes possible the preparation of relatively large amounts of purified substances in a short time.

REFERENCES

J. Sherma and G. Zweig, *Paper Chromatography*, Academic Press, New York, 1971.

E. Stahl, *Thin-Layer Chromatography*, Springer-Verlag, New York, 1965.

J. G. Kirchner, *Thin-Layer Chromatography* (Technique of Organic Chemistry, Vol. 12), Wiley-Interscience, New York, 1967.

M. Bier, ed., *Electrophoresis*, Academic Press, New York, Vol. I, 1959; Vol. II, 1967.

B. L. Karger, L. R. Snyder, and C. Horvath, *An Introduction to Separation Science*, Wiley-Interscience, New York, 1973.

7

GAS CHROMATOGRAPHY

7-1 INTRODUCTION

The separation of benzene (b.p. 80.1°C) from cyclohexane (b.p. 80.8°C) by fractional distillation is virtually impossible, yet it can be accomplished with the simplest of GLC equipment by an unskilled technician within a few minutes. The more difficult separations—for example, the detection of a hundred or more hydrocarbons in gasoline, the analysis of a fruit peel for traces of pesticide residues, or the separation of steroids at 100° below their boiling points—push the GLC method to its limits, and require intelligent application of theoretical principles. There is always a delicate compromise among speed, completeness of separation, and the size of sample that can be handled, not to mention cost and complexity of the apparatus required. As with most instruments, the time spent examining "what's inside the black box" will be well repaid.

In this chapter we will utilize the general principles of chromatography developed in Chapter 4, plus a few others that are relevant to the gaseous mobile phase. In liquid chromatography we are restricted to temperatures mostly in the range from 0°C to 150°, in fact most LC is performed at or near room temperature, whereas in GC the range can be extended from about $-196°$ to $+500°$C or even higher. It is not surprising, then, that GC equipment appears in a variety of types, including a model that has been operated successfully by remote control on the surface of the moon.

7-2 BASIC GLC APPARATUS

Unlike a liquid chromatographic system, a gas chromatograph requires a completely closed system (except for the gas outlet at the end). The essential components are shown in Figure 7-1. The carrier gas, supplied from a pressurized tank, passes through one or more pressure regulators which control the flow rate through the apparatus. The sample is introduced into a heated chamber either through a silicone rubber septum with hypodermic syringe if it is a liquid, or by means of a special sampling valve if it is a gas. From here the carrier gas carries the sample components through the column where they are separated and one after the other pass through a detector which sends a signal to a recorder. A thermostatted oven is provided for the column, injector, and detector, although the last two may be heated separately.

There are innumerable variations of each of these basic components of a gas chromatograph, and scores of commercial instruments are available in a wide price range for various specific applications. Process gas chromatographs are discussed in Chapters 27 and 28. Whereas one pH meter is much like another, gas chromatographs should be selected on the basis of their applicability to the

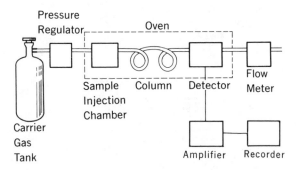

Figure 7-1 Essential features of a gas chromatograph.

problem at hand, their versatility, accuracy, ease of service, and price. Often, a $500 instrument used intelligently can do a better job than a $5000 instrument improperly used. We will examine some of the important considerations in setting up a gas chromatographic system.

7-3   CARRIER GAS

At ordinary pressures and temperatures, the common carrier gases are regarded as chemically inert: for example, helium, nitrogen, hydrogen, and argon. These gases are relatively inexpensive, readily available, and are not hazardous to handle other than the flammability of hydrogen which does require adequate safety precautions. Also, because they are inert, interactions between sample molecules and carrier molecules can be neglected (except at high pressures) and the distribution coefficient of a component is determined entirely by its volatility from the stationary phase. The choice of carrier gas is usually based on availability in a high purity grade, or on the requirements of the detector which must sense the component in an extremely high dilution of carrier gas. For example, thermal conductivity detectors work best with hydrogen or helium. For most purposes, helium is the popular choice. Since even trace impurities in the carrier gas can cause noise in the detector signal, the purity is critical. Passage through a cold molecular sieve trap is very effective in removing the last traces of water vapor frequently found in commercial gases.

Flow through the column is caused by the difference in pressure between inlet and outlet. Pressure regulating valves (See Section 28–6) maintain an inlet pressure, P_i, of 10 to 50 psig (pounds per square inch, gauge-above atmospheric) or occasionally higher. The outlet pressure, P_o, is normally atmospheric pressure although it can be increased by restricting the outlet orifice. On some instruments, flow controllers automatically vary P_i to maintain a constant flow rate, but on most instruments it is assumed that a constant P_i is adequate. In contrast to liquids, gases are compressible and this leads to complications in determining the flow rate and volume of gas passing through the column (See Section 7–10).

7-4   SAMPLE INTRODUCTION

As in other types of chromatography it is important to introduce the sample in the shortest time and in the smallest volume possible. The sample chamber may be heated for rapid vaporization of liquid samples and should be designed so that the carrier gas sweeps the sample directly into the column. If these requirements are not met, the sample will be unnecessarily spread out before the separation process has begun. In some instruments, the sample is injected directly into the column at the inlet (*on-column injection*). This is preferable for most samples unless they have a very high boiling point.

The size of the sample is dictated by several factors: the amount available, the capacity of the column, and the sensitivity of the detector. The ordinary laboratory chromatograph can handle liquid samples in the range of 0.1 to 10 μl and gaseous samples in the range from 1 to 10 ml. A capillary column (see below) requires much smaller samples, of the order of 10^{-3} to 10^{-2} μl. Samples of this size must be introduced by a splitting technique, as indicated in Figure 7–2. In this way only a small fraction of the injected sample is used, the remainder is vented to the atmosphere. Accurate measurement of samples of this size is questionable and may account for much of the error in quantitative analysis. For reliable thermodynamic data, retention values should be obtained with several sample sizes and extrapolated to zero sample (infinite dilution).

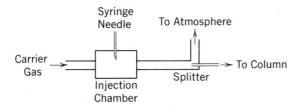

Figure 7–2 Splitter for obtaining very small samples.

Liquid samples are injected from a microsyringe of appropriate capacity through a rubber septum. A swift, neat motion of the plunger is necessary. Gas samples are easier to measure accurately because of their larger volume. For routine gas analysis, a two-position, hexport valve, shown in Figure 7–3, is convenient and accurate. Solid samples can be dissolved in a solvent, or sealed in a thin-walled glass vial which is then inserted into the sample chamber and finally crushed from the outside.

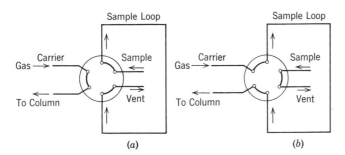

Figure 7–3 Schematic of typical hexport sampling valve: (a) carrier gas is bypassed directly to column, sample gas feeds through sample loop; (b) carrier gas is diverted through sample loop, sample gas is bypassed to vent.

7–5   COLUMNS

There are two distinct types of columns in common use, packed and open tubular (capillary). The packed columns are easier to fabricate, less expensive, last longer, have a higher capacity, and suffice for all but the most difficult separations. The open tubular columns have less pressure drop and therefore can be made much longer (more plates).

Packed columns are usually 1 to 20 m long and 3 to 10 mm (1/8 in. to 3/8 in.) in diameter, although columns up to 10 cm or more in diameter are used for preparative scale work. Open tubular columns are usually 10 to 50 m long and 0.2 to 1.2 mm in diameter. The geometrical configuration of the column is of little concern as long as the bends are not sharp. Columns are bent in a U- or W-shape or coiled to fit the oven. Short columns are often made of glass, but longer columns are made of copper, aluminum, or stainless steel tubing to facilitate bending which may be done after the column is filled. It is possible to draw glass capillary tubing in lengths of several hundred meters—capillary columns up to a mile in length have been fabricated.

7–6   COLUMN PERFORMANCE

Successful resolution of complex mixtures often requires careful attention to column efficiency which is indicated by n, the number of theoretical plates in the column, or by H, the height equivalent of a theoretical plate. In Section 4–3 we introduced the Van Deemter equation, originally developed for GLC to account for the experimental factors which influence H. In a more expanded version, the Van Deemter equation for a packed column is given as

$$H = 2\lambda d_p + \frac{2\gamma D_M}{u} + \frac{8k' d_f^2 u}{\pi^2 (1+k')^2 D_S} + \frac{\omega d_p^2 u}{D_M} \qquad (7–1)$$

$$A \text{ term} \qquad B \text{ term} \qquad C_l \text{ term} \qquad C_g \text{ term}$$

where λ, γ and ω are functions of the packing structure. The A and B terms are as given previously but the C term has been divided into two parts: C_l (liquid phase) and C_g (gas phase)—both of which contribute to the mass transfer effect. In the C_l term, the thickness of the liquid film appears as d_f^2. Note that H will be different for each solute, because C_l depends on k' (and therefore on K). Remember that k' is determined by the choice of liquid phase, the temperature, and the ratio of the volumes of the two phases, $k' = KV_L/V_G$. D_S is the diffusion coefficient of the solute in the stationary (liquid) phase. The C_g term is often ignored with packed columns because of the short distances that solute molecules have to travel in the gas phase in order to reach the stationary phase and the much larger value of the diffusion coefficient in the mobile phase, D_M, compared to D_S. However, its effect becomes relatively more important with very thin liquid films. The particle diameter, d_p, appears in both the A term and the C_g term; in fact, Giddings has proposed that these two effects (nonuniform paths and cross-column transport in the mobile phase) are interactive, or *coupled*, adding another term to Equation 7–1. The effect of the coupling is that a large A term tends to decrease the effect of the C_g term.

7–7   SOLID SUPPORT

The ideal solid support is not yet available. It should have a high specific surface ($1 \text{ m}^2/\text{g}$), and the surface must be chemically inert although wettable by the liquid phase so that the latter will spread in a thin layer of uniform thickness. In addition, the solid support must have thermal stability, mechanical strength, and be available in uniformly sized, near-spherically shaped particles.

The most commonly used supports are derived from diatomaceous earth, a spongy siliceous material consisting of the skeletal residues of diatoms, a form of microscopic algae. (Diatomaceous earth is also used as a filter aid, as the main constituent of firebrick, and as a mild abrasive.) The raw material has a surface area of about $20 \text{ m}^2/\text{g}$, but it is very fragile and, of course, loaded with impurities. In one process the diatomaceous earth is mixed with clay and baked at 900°C. It is then crushed and graded according to size. The final material has a surface area of about $4 \text{ m}^2/\text{g}$. The surface is fairly active toward polar compounds and is pink; hence, designations like Chromosorb-P. In a second process, the raw material is mixed with a sodium carbonate flux before baking (calcining). The product is more fragile, less chemically active, has a smaller surface area (0.5 to $1 \text{ m}^2/\text{g}$), and is white; hence designations like Chromosorb-W.

These support materials are interlaced with a network of fine pores, requiring about 0.5% (by weight) of a liquid to cover the entire surface with a monolayer. As a heavier coating is applied, the finer pores fill up first, but even with a 20% coating many free passageways still exist, and the average thickness of the layer is still only a few hundred angstroms.

The surface of commercial firebrick (Chromosorb-P) has the general structure:

$$\begin{array}{cc} \text{OH} & \text{OH} \\ | & | \\ -\text{Si}-\text{O}-\text{Si}- \\ | & | \end{array}$$

The —OH groups are acidic and somewhat polar, and tend to react with polar solutes, especially those with basic functional groups. The surface activity can be partially reduced and some impurities removed by washing in acids or bases. For the separation of amines, it is desirable to leave a thin coating of sodium hydroxide on the surface. A very effective treatment consists of silanizing the surface with hexamethyldisilazane (HMDS):

$$\begin{array}{cc} \text{OH} & \text{OH} \\ | & | \\ -\text{Si}-\text{O}-\text{Si}- \\ | & | \end{array} \quad + \quad (\text{Me}_3\text{Si})_2\text{NH} \longrightarrow$$

$$\begin{array}{cc} \text{Me} & \text{Me} \\ | & | \\ \text{Me}-\text{Si}-\text{Me} & \text{Me}-\text{Si}-\text{Me} \\ | & | \\ \text{O} & \text{O} \\ | & | \\ -\text{Si} - \text{O} - \text{Si}- \\ | & | \end{array}$$

In this way the polar —O—H is replaced by a relatively inert trimethyl silyl group.

Powdered Teflon is useful as a support for very polar solutes, but it has a low surface area. It is so inert that it is difficult to coat evenly. It is more easily handled if first cooled in a refrigerator. Some other supports which have been suggested are micro glass beads, graphitized carbon, and carborundum.

Coating the Support. The column packing is prepared by mixing the solid with the correct amount of liquid phase dissolved in a suitable low-boiling solvent such as pentane, dichloromethane, or acetone. The solvent is then evaporated with judicious heating; stirring as necessary to obtain a uniform coating. The last traces of solvent may be removed under vacuum. Columns are usually filled by pouring the packing into the straight tube with gentle shaking or tapping. Both ends are plugged with glass wool and the column bent to the appropriate shape to fit the oven. Alternatively, bent columns can be packed by loading the packing material in a high velocity gas stream. Great care is necessary in packing large columns to avoid channeling and segregation of the particles according to size. A newly packed column must be conditioned by passing carrier gas through it at elevated temperature for several hours.

Open tubular columns, of course, contain no packing. The thin coat of liquid phase is applied by forcing a dilute solution through the column at a slow rate. The solution remaining on the wall is evaporated by passing through carrier gas, leaving a layer of liquid phase. Experience helps in getting a satisfactory coating.

7–8 LIQUID PHASE

The versatility of GLC is in large part due to the wide variety of liquid phases available. Hundreds of liquids have been tried, but perhaps a dozen or so will suffice for most purposes.

The requirements for a good liquid phase are: (1) it should be essentially nonvolatile (vapor pressure <0.1 torr) at the temperature it is to be used; (2) it must be thermally stable; (3) it should yield appropriate K values for the components to be studied—neither too small nor too large; (4) it should be readily available in a reproducible form, preferably as a single pure compound of known molecular weight; and (5) it should be inert toward the solutes, or if it reacts, it should do so fast and reversibly.

No single liquid meets all requirements for all possible solutes. Some are needed for low temperatures, others for high temperatures. For some studies a nonselective liquid is desirable, for others a highly selective phase is needed. A few liquids commonly used are listed in Table 7–1. The choice is often based on availability and/or habit, but the difference between a successful and a poor separation may well lie in the choice of the best liquid phase.

The factors which determine K, and thus indirectly influence the retention time, are not completely understood. The kinds of solutions used in gas chromatography often involve solutes and solvents of widely different chemical properties, and always of different molecular dimensions. Such solutions rarely obey Raoult's law, but often obey Henry's law provided they are highly dilute. The distribution coefficient is directly related to Henry's law constant, or the volatility of the solute. The volatility, in turn, is a function of the cohesive forces acting between the molecules of the solute and solvent. In this regard, some qualitative concepts will help us to understand the factors which determine retention at the molecular level.

1. *Dipole–Dipole Interactions.* Molecules containing electronegative or electropositive atoms possess a permanent electrical dipole, and will interact

Table 7–1 Some Common Liquid Phases

Liquid Phase	Typical Samples	Polarity[a]	Max. Temp., °C
Squalane	Hydrocarbons	N	125
Apiezon L	High boiling hydrocarbons, esters, ethers	N	300
Methyl silicone	Steroids, pesticides, alkaloids, esters	N	300
Dinonyl phthalate	All types	I	175
Silicone oil	All types	I	275
Diethyleneglycol succinate	Esters	P	200
Carbowax 20M	Alcohols, aromatics amines, ketones	P	250
Polyamid Resin	Amino compounds	P	300
β,β-Oxydipropio-nitrile	Olefins, alcohols, aldehydes	P	100
AgNO₃ in propylene glycol	Olefins, cyclic hydrocarbons	P	50
Inorganic eutectics	Volatile inorganics	P	

[a] N, nonpolar; I, intermediate polarity; P, polar.

strongly with other molecules possessing a dipole. Thus a polar solute will have an abnormally low volatility (high solubility) in a polar solvent. On the other hand, in a nonpolar solvent, the dipole–dipole interactions between the polar solute molecules themselves are decreased by dilution, resulting in an abnormally high volatility (low solubility). Dipole–dipole interactions are decreased as the temperature is increased.

2. *Induction Forces.* If either the solvent or solute contains a permanent dipole, it can induce a temporary dipole in the other. The magnitude of the force depends on the polarizability of the other molecule and is generally rather small.

3. *Dispersion Forces.* The vibrations of nonpolar molecules often produce temporary dipoles by slight separation of the electrical charges within the molecules. The oscillating dipoles can induce similar temporary dipoles in neighboring polarizable molecules. A small force of attraction is thus generated. These forces are present in all solutions.

4. *Hydrogen Bonds.* An especially strong dipole–dipole interaction is possible when one molecule contains a polarized hydrogen atom and the other a strong electronegative atom, such as a fluorine or oxygen atom. The extra strength results from the closeness of approach afforded by the small size of the proton.

5. *Formation of Metal Complexes.* Solutions of silver nitrate in glycols or benzyl cyanide selectively absorb olefins because of the weak organo-metallic complexes formed. Thus the olefins are retained far longer than corresponding paraffins in these columns. Similarly, heavy metal salts of fatty acids retard amines because of complex formation.

In general, nonpolar liquid phases are nonselective; that is, in the absence of special forces between solute and solvent, the volatility of the solute is determined primarily by its vapor pressure. Thus separations will be in the order of increasing boiling points of the solutes. On the other hand, with polar liquid phases, the volatility is not so simply determined because of the complicating factors just mentioned.

Obviously, the polarity of the liquid phase is of primary concern, however, an acceptable definition of "polarity" has not yet been developed. In a general sense, polarity is measured by the dielectric constant, but, in the chromatographic sense, "polarity" is a function of any type of interaction between solute and solvent. Because these forces vary with the type of solute, it is not possible to designate the "polarity" of a solvent with a single number. Several schemes have been proposed for characterizing the polarity of a solvent with a series of numbers relating to its tendency to donate electrons, donate protons, participate in dipole–dipole interactions, form covalent or hydrogen bonds, etc. We can expect that such a systematic designation will make it easier to choose the best liquid phase for a given problem.

7–9   DETECTORS

The remarkable separations performed in the column must somehow be sensed and recorded. All of the components are highly diluted in the carrier gas with concentrations of 1 part per thousand at best, ranging down to zero. Furthermore, sharp peaks may pass through the detector in less than a second, while the last peaks may not emerge for hours and be barely discernible above the base line. Somehow the detector must ignore the large amount of carrier gas and find the trace amounts of sample components contained therein.

The universal detector has not yet been invented, for it must meet all of the following requirements: low limit of detection, linear response over an extreme range of concentrations, uniform response to all possible substances, simple calibration, short response time, small internal volume, low noise, long term stability; and it must be simple, inexpensive, robust, and safe to operate. There may be some occasions when selective response to a few components is desirable as an aid to identification, or as a means of finding a trace component which is incompletely separated from a major component.

Detector Evaluation.   Comparison of detectors is not always meaningful because their performance may depend on how it is measured. In general terms, it is useful to plot the response or signal, R, versus the quantity measure, Q, as in Figure 7–4. The limit of detection, Q_0, is the quantity of substance that gives a readable signal. It is determined by the noise level, R_N, of the detector, and it is assumed that a signal equal to $2R_N$ can be distinguished from the background

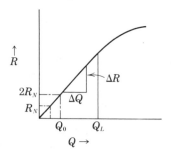

Figure 7–4   Response curve of a detector.

noise. Q_0 corresponds to $2R_N$. The slope of the curve, $\Delta R/\Delta Q$, defines the *sensitivity*. The region where the curve begins to deviate from a straight line defines the *limit of linear response*, with Q_L being the upper limit. The measurements are a function of the construction of the detector, but also of the way it is operated.

A "sensitive" detector may not have the lowest "limit of detection" because it may also have a high noise level. These terms must be carefully defined and understood, as they apply to our recent concern for the impact of trace contaminants in the environment. Legislation requiring "zero" tolerance, say of a particular carcinogen, is meaningless and impossible to enforce unless the method for detection is also stated. To the analyst, "zero" means less than a detectable amount by the best technique available. Obviously, the zero level changes as improvements in technology become available.

Types of Detectors. Scores of detectors have been proposed, and we could classify them ad infinitum. *Differential detectors* measure instantaneous concentration or instantaneous rate of emergence of the component. *Integral detectors* accumulate the instantaneous signal and give the total amount which has been measured up to a given instant. Signals from differential detectors are usually integrated for quantitative analysis, and signals from integral detectors are often differentiated to make them easier to interpret for qualitative analysis. Either signal in Figure 7–5 can be derived from the other.

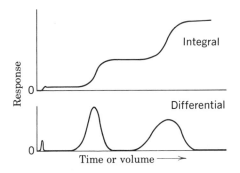

Figure 7–5 Differential and integral response for the same chromatogram.

Detectors can also be classified as *destructive* or *nondestructive*, depending on whether or not the sample components can be collected unchanged for further study.

Hydrogen Flame Detector. One of the simplest detectors is the hydrogen flame detector, shown in Figure 7–6 with an exquisitely simple chromatograph. Hydrogen must be used as a carrier gas. It is burned as it emerges from the column through a hollow needle, yielding a nearly colorless flame. When an organic component emerges, the flame becomes yellow. The retention time can be measured with a stopwatch. The amount of the component is roughly proportional to the height and/or luminosity of the flame. This may sound very crude, yet it is not difficult to install a thermocouple to measure the flame temperature or a photocell to measure the luminosity and increase the accuracy manyfold. Better still, since most organic compounds are ionized in the flame, an ion current can be collected between two oppositely charged electrodes, as in Figure 7–7. This is the principle of the *flame ionization detector*, one of the most sensitive and popular

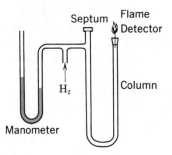

Figure 7–6 Simple chromatograph with hydrogen flame detector. [After Cowan and Sugihara, *J. Chem. Educ.*, **36**, 246 (1959).]

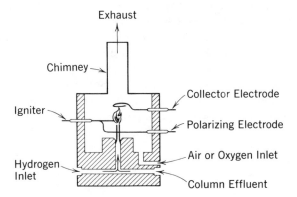

Figure 7–7 Hydrogen flame ionization detector.

detectors in current use. The ions produced are collected between two electrodes, one of which may be the jet itself. Because the electrical resistance of the flame is very high (about 10^{12} ohms) and the current is extremely small (about 10^{-10} amp), the associated electronics are complicated and moderately expensive.

The ionization processes occurring in the flame are not completely understood. The ion current, however, is approximately proportional to the number of carbon atoms entering the flame. More precisely, the response depends also on the state of oxidation of the carbon atoms—those which are already completely oxidized do not lead to ion formation. The relative response per carbon atom, or "effective carbon number," of several organic functional groups is listed in Table 7–2. The

Table 7–2 Contributions to Effective Carbon Numbers

C-aliphatic, aromatic, olefinic	1.0
C-acetylenic	1.3
C-carbonyl	0
C-nitrate	0.3
O-ether	−1.0
O-primary alcohol	−0.6
O-secondary alcohol	−0.75
O-tertiary alcohol, esters	−0.25
N-amines	(same as O in alcohols)

[After J. C. Sternberg, W. S. Gallaway, and D. T. L. Jones, *Gas Chromatography* (Edited by Brenner, Callen, and Weiss), Academic Press, New York, 1962; p. 265.]

detector is insensitive to most inorganic compounds. It is especially worth noting that it does not "see" water vapor or air. The flame ionization detector is relatively simple, extremely sensitive, and has a wide range of linear response.

Thermal Conductivity Detector. The measurement of the thermal conductivity of a gas is based on the transfer of heat from a hot filament to a cooler surface. Thus the gas conducts heat from the filament to the wall. If a constant amount of electrical energy is supplied to the filament, its temperature will be a function of the thermal conductivity of the gas. Rather than determining the temperature of the filament, it is easier to determine its electrical resistance which increases with temperature. As applied to gas chromatography, a dual detector is used to minimize the effect of the thermal conductivity of the carrier gas and to minimize fluctuations in the temperature, pressure, and power supply. A schematic diagram of the detector is shown in Figure 7–8, and the associated electrical circuit in Figure 7–9. In some detectors, thermistors are used in place of filaments. Thermistors are somewhat more sensitive that filaments below about 100°C, but the reverse is true above about 150°C.

The temperature of the filament (and thus the signal) is a function of the bridge current, the geometry of the cell, and the thermal conductivity and flow rate of the gas. Increasing the bridge current will increase the running temperature of the

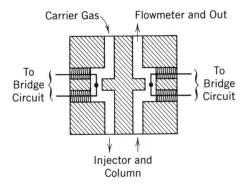

Figure 7–8 Thermal conductivity detector.

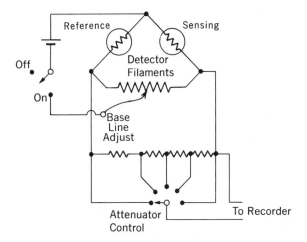

Figure 7–9 Circuitry for thermal conductivity detector.

Table 7–3 Thermal Conductivities of Substances at 100°C

Substance	$k \times 10^5$, cal/°C/mole	RMR[a]	RWR[b]
Hydrogen, H_2 (Mol. Wt. = 2)	53.4	—	—
Helium, He (At. Wt. = 4)	41.6	—	—
Methane, CH_4 (Mol. Wt. = 16)	10.9	36	2.25
Nitrogen, N_2 (Mol. Wt. = 28)	7.5	42	1.50
Ethane, C_2H_6 (Mol. Wt. = 30)	7.3	51	1.70
Propane, C_3H_8 (Mol. Wt. = 44)	6.3	65	1.47
Ethyl propyl ether, $C_2H_5OC_3H_7$ (Mol. Wt. = 88)	5.4	121	1.37
Carbon dioxide, CO_2 (Mol. Wt. = 44)	5.3	48	1.09
Ethanol, C_2H_5OH (Mol. Wt. = 46)	5.3	72	1.56
Argon, Ar (At. Wt. = 40)	5.2	42	1.05
n-Hexane, C_6H_{14} (Mol. Wt. = 86)	5.0	123	1.43
Benzene, C_6H_6 (Mol. Wt. = 78)	4.4	100	1.28
Ethyl acetate, $CH_3COOC_2H_5$ (Mol. Wt. = 88)	4.1	111	1.27
Carbon tetrachloride, CCl_4 (Mol. Wt. = 154)	2.2	108	0.70

[a] RMR—Relative molar response with helium as carrier gas, Benzene = 100.
[b] RWR—Relative weight response = RMR/Mol. Wt.
[From W. A. Dietz, *J. Gas Chromatog.*, **5**, 68 (1967).]

filaments and increase the sensitivity, but the filaments will burn out sooner. Since the detector is measuring thermal conductivity, it is very important to keep the temperature of the detector walls constant; often this is the limiting factor in quantitative analysis.

The thermal conductivities of a number of vapors are listed in Table 7–3, from which it is seen that both hydrogen and helium have very high thermal conductivities relative to most organic compounds. If either of these gases is used as a carrier gas, there will be a decrease in the thermal conductivity of the gas mixture whenever an organic compound is eluted. The temperature of the filament and its electrical resistance increase. The bridge circuit becomes unbalanced and a signal is sent to the recorder. With nitrogen as a carrier, the sensitivity is reduced and the signal may be negative (inverse peak) because its thermal conductivity is closer to that of organic compounds.

The thermal conductivity of a mixture of gases is not easy to calculate, and this detector must be calibrated for each compound for highest accuracy. Some generalizations have been observed; for example, the *relative* response (signal per mole of compound/signal per mole of standard, in this case, benzene) for many compounds follows the relation

$$\text{Relative Response} = A + BM \qquad (\text{benzene} = 100) \qquad (7\text{–}2)$$

where M is the molecular weight, and A and B are constants for a given homologous series of compounds, a series in which a member is derived from the previous member by the insertion of a CH_2 group. Some values of A and B are listed in Table 7–4.

Many forms of geometry of the cell cavities have been proposed to reduce the effects of flow fluctuations without decreasing the sensitivity or increasing the time constant unduly. The internal volume of the detector, and especially the dead space, should be at a minimum in order not to remix the components. The thermal conductivity detector is simple, rugged, inexpensive, moderately sensitive, non-

Table 7–4 Values of A and B in Equation 7–2, with Helium Carrier Gas in the Temperature Range 30° to 160°C

Type	Carbon No.	A	B
n-Alkanes	1–3	20.6	1.04
n-Alkanes	3–10	6.7	1.35
Methyl alkanes	4–7	10.8	1.25
Olefins	2–4	13.0	1.20
Methyl benzenes	7–9	9.7	1.16
n-Ketones	3–8	35.9	0.86
Primary alcohols	2–7	34.9	0.81
n-Ethers	4–10	43.3	0.89

[After A. E. Messner, D. M. Rosie and P. A. Argabright, *Anal. Chem.*, **31**, 230 (1959).]

selective, essentially nondestructive, very accurate if properly calibrated, and more widely used than any other.

Electron Capture Detector. In this detector, the effluent gas is ionized by a stream of particles emanating from a radioactive source, typically ^{3}H or ^{63}Ni. Thus, the carrier gas produces a steady supply of positive ions and free electrons which can be measured as a standing current between two charged electrodes, much like the hydrogen flame detector. When sample components pass through the detector, the standing current may be perturbed. Compounds containing highly electronegative atoms capture free electrons very efficiently, and they are detected by the *decrease* in the standing current. This occurs because of the increased rate of recombination of positive and negative ions, compared with positive ions and electrons. Other species have little effect on the ion current, thus the electron capture detector is highly selective and very sensitive for compounds containing halogens, phosphorus, lead, nitro groups, and polynuclear aromatic ring systems. It is ideally suited to the detection of minute traces of many pesticides and defoliants.

Other Detectors. Of the many other detectors proposed, some of the more important are: (a) the *gas density balance* for which the response is a precise function of molecular weight; (b) *coulometric titrator* in which the column effluent is burned to give HCl, H_2S, etc., and then passed through a solution to be titrated with electrolytically generated silver ion (specific for halogens and sulfur); (c) *thermionic detector* (sensitized flame ionization detector) of which there are several versions, each incorporating a FID provided with a screen or porous block coated with an alkali halide at or just above the flame. (Depending on which salt is used, these detectors are highly selective and several orders of magnitude more sensitive to phosphorus-, sulfur-, or halogen-containing compounds); and (d) *flame photometer detector* in which the flame of an FID serves as a radiation source of a flame photometer. (It is also selective and sensitive to phosphorous-, sulfur-, and chlorine-compounds.)

7–10 QUALITATIVE ANALYSIS: RETENTION PARAMETERS

The gas chromatograph, as a separator, has no equal, but as an analytical tool it leaves much to be desired. It is easy to find impurities in a supposedly chemically pure reagent, or to detect 200 constituents in petroleum, but the identification is left to the ingenuity of the analyst.

Comparison of retention behavior with a known sample using several columns at more than one temperature is almost certain proof of identity. But this method would, in general, require infinite patience if not an infinite supply of pure chemicals. In a negative sense, the absence of extraneous peaks is now considered to be one of the best indications of purity of a reagent.

Ideally, it should be possible to identify a compound from the retention time which is read directly on the recorder chart. However, the retention time alone is of little value because it depends on so many other factors in addition to the nature of the compound. The objective is to find the most convenient parameter which will identify a compound. We will consider several of these parameters.

The *distribution coefficient*, K, is an equilibrium constant which is characteristic of the component and the liquid phase at a given temperature; therefore, it could serve as an identifying parameter. However, K values are seldom determined in practice (except for theoretical studies) because there are easier ways to identify components.

The *retention volume*, V_R, is independent of flow rate, F (whereas t_R is inversely proportional to F). Flow rates are most easily measured with a soap bubble flow meter at the end of the apparatus. Thus the carrier gas at that point is at room temperature and pressure, and presumably saturated with water vapor from the flow meter. Therefore, the true flow rate at column temperature, T_c, is

$$F_c = F_{\text{meas}} \times \frac{T_c}{T_{\text{room}}} \times \frac{P_o - P_{\text{H}_2\text{O}}}{P_o} \tag{7-3}$$

where $P_{\text{H}_2\text{O}}$ is the vapor pressure of water at the temperature of the flow meter. It follows that:

$$V_R = t_R \times F_c \tag{7-4}$$

But this is the retention volume measured at the pressure of the column outlet (normally atmospheric) whereas the carrier gas in the column is under a higher pressure. As the gas passes through the column, the pressure decreases continuously (but not linearly) from the inlet pressure, P_i to the outlet pressure, P_o. Therefore its volume (and flow rate in ml/min) must increase in inverse fashion to the pressure. If the column is operated with a large pressure drop and if the flow rate is measured at the column outlet, the average flow rate is much lower than the measured rate. As a consequence only a part of the column can be operating at maximum efficiency, and that part may be only the last portion before the outlet, as shown in Figure 7-10. In order to determine meaningful values of retention

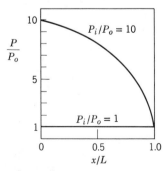
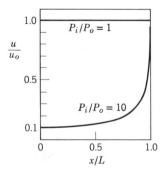

Figure 7-10 Pressure and gas velocity profiles through a gas chromatographic column where $P_i/P_o = 1$ and 10.

volumes, we must determine the average pressure, $\bar{P}$, in the column. By a suitable combination of the gas law, the relation between velocity and pressure gradient, and the requirement for a constant throughput, it can be shown that:

$$\bar{P} = P_o \times \frac{2[(P_i/P_o)^3 - 1]}{3[(P_i/P_o)^2 - 1]} = P_o/j \qquad (7-5)$$

where j is the *pressure drop correction factor*. Selected values of j are given in Table 7–5.

Table 7–5 Values of the Pressure Drop Correction Factor

P_i/P_o	j	P_i/P_o	j
1.00	1.000	1.80	0.695
1.10	0.952	1.90	0.668
1.20	0.907	2.00	0.643
1.30	0.865	2.10	0.619
1.40	0.826	2.20	0.597
1.50	0.790	2.30	0.576
1.60	0.756	2.40	0.557
1.70	0.725	2.50	0.539

The "*corrected*" *retention volume*, V_R^o, is thus:

$$V_R^o = t_R F_c j \qquad (7-6)$$

The next operation is called an "adjustment." Recall from Equation 4–6, that the measured retention volume includes the volume of mobile phase contained within the column (dead space), V_M. This volume is the same for all components and should be subtracted from the measured retention volume. The "*adjusted*" *retention volume*, V_R', is thus:

$$V_R' = (t_R - t_M)F_c \qquad (7-7)$$

The two corrections just given are combined in defining the *net retention volume*, V_N:

$$V_N = (t_R - t_M)F_c j \qquad (7-8)$$

Finally, since the retention volume also depends on the volume of stationary phase in the column (Equation 4–6), the net retention volume per gram of liquid phase, called the *specific retention volume*, V_g, is given as

$$V_g = \frac{V_N}{W_L} = \frac{d_R - d_a}{CS} \times F_c \times j \times \frac{273}{T_c} \times \frac{1}{W_L} \qquad (7-9)$$

where d_R and d_a are distances (cm) measured on the chart from injection to the component peak maximum and the air peak, respectively. CS is the chart speed (cm/min) and W_L is the weight (g) of liquid phase in the column.

V_g, like K, is a true equilibrium constant, dependent only on the nature of the solute and solvent, and the temperature of the column. The two constants differ only because of the different units used.

$$V_g = \frac{\text{adjusted and corrected retention volume}}{\text{weight of liquid in column}} \times \frac{273}{T_c} \qquad (7-10)$$

$$K = \frac{\text{conc of solute in liquid phase}}{\text{conc of solute in gas phase}} \qquad (7\text{--}11)$$

Example/Problem 7–1. Retention data for pentane on a typical column are given below. Calculate the specific retention volume and distribution coefficient.

Retention time, t_R	4.50 min
Air peak time, t_a	30 sec
Flow rate at 100°C, F_c	75 ml/min
Column temperature, T_c	100°C
Weight of liquid phase, W_L	4.2 g
Density of liquid phase at 100°C, ρ_L	0.95 g/ml
Inlet pressure, P_i	1185 torr
Outlet pressure, P_o	740 torr

From Table 7–6 and $P_i/P_o = 1.60$, $j = 0.756$
From Equations 7–8 and 7–9:

$$V_g = (4.50\text{--}0.50) \times 75 \times 0.756 \times \frac{273}{373} \times \frac{1}{4.2} = 39.5 \text{ ml/g}$$

From Equation 7–6:

$$V_R^\circ = 4.50 \times 75 \times 0.756 = 255.2 \text{ ml}$$

$$V_M = 0.50 \times 75 \times 0.756 = 28.4 \text{ ml}$$

From Equation 4–6 ($V_R = V_M + KV_S$):

For GLC: $\qquad\qquad V_R = V_R^\circ \quad$ and $\quad V_S = V_L = W_L/\rho_L$

$$255.2 = 28.4 + K \times \frac{4.2}{0.95}$$

$$K = 51.3$$

The relationship between V_g and K is easily derived by combining Equations 4–6 and 7–10:

$$V_R^\circ - V_M^\circ = V_N = KV_L$$

$$V_g = \frac{V_N}{W_L} \times \frac{273}{T_c} = \frac{K}{\rho_L} \times \frac{273}{T_c} \qquad (7\text{--}12)$$

The *specific retention volume*, or the partition coefficient, is a reliable means of identification, but vast tables of data are required for the innumerable combinations of compounds, liquid phases, and temperatures. Many of these data are available, but are well scattered in the literature. A large fraction of the data is unreliable, or important details are missing. There are, however, some useful theoretical considerations and empirical observations which remove some of the witchcraft from the problem of identification.

Relative Retention. A re-examination of Equation 7–9 will show that in the course of running a chromatogram with several components, if all experimental conditions remain the same, then:

$$\frac{d_1 - d_a}{d_2 - d_a} = \frac{t_1 - t_a}{t_2 - t_a} = \frac{V_{R_1} - V_a}{V_{R_2} - V_a} = \frac{V_{g_1}}{V_{g_2}} = \frac{K_1}{K_2} = \alpha_{1,2} \qquad (7\text{--}13)$$

where the subscripts refer to the two components and air, and $\alpha_{1,2}$ is the *relative retention* of component 1 with respect to component 2. The latter may be a standard compound to which all others are referred. Like K, α is a form of equilibrium constant, but is readily determined from two measurements of distance on the chart. No other measurements need be made—they all cancel out. This highly delightful situation is marred by the fact that we must choose and agree upon a standard substance. The standard should be readily available, should have a retention time close to (but easily resolved from) the sample components, and be chemically compatible with the sample. Obviously, no single substance will meet all requirements.

Retention Index.   In this system, introduced by Kovats, the relative retention is referred to the series of normal paraffin hydrocarbons as standard substances. The Retention Index, I, is defined as:

$$I = 100 \frac{\log \alpha_{i,p_z}}{\log \alpha_{p_{z+1},p_z}} + 100z \qquad (7\text{--}14)$$

where α_{i,p_z} is the relative retention of compound i compared to the n-paraffin with z carbon atoms, and α_{p_{z+1},p_z} is the relative retention of the two n-paraffins with $z+1$ and z carbon atoms. By definition, I for all n-paraffins is $100z$ for all liquid phases at all temperatures. The I value for all other compounds depends on the nature of the liquid phase as well as the nature of the compound. $I/100$ is the (fractional) number of carbon atoms in a hypothetical n-paraffin having the same retention time as the compound in question.

It has been observed (supported by sound thermodynamic arguments) that for the homologous series of n-paraffins a plot of $\log (t_R - t_a)$ versus I is a straight line, unique for a given set of operating conditions for a particular column. The data for just two n-paraffins are sufficient to establish this straight line; therefore I values for other compounds can be read from such a graph as shown in the following example.

Example/Problem 7–2.   Retention times for several substances in a particular column are:

Substance	t_R, sec	$t_R - t_a$, sec
Air	1.2	—
Propane	32.8	31.6
Cpd X	51.2	50.0
n-Pentane	80.7	79.5

(a) What is the retention index, I, of n-butane on this column? The I value of n-butane is 400 on all columns, by definition.

(b) What is the retention index of Cpd X on this column?
   i. Solution using Equation 7–14.

$$I = 100 \frac{2 \log (50/31.6)}{\log (79.5/31.6)} + 100 \times 3$$
$$= 99 + 300 = 399$$

   (Note that the two paraffins given are C-3 and C-5; therefore $\log \alpha_{p_{z+2},p_z}$ is divided by 2.)
   ii. Graphical solution. Construct a table and the corresponding plot, Figure 7–11.

Substance	$t_R - t_a$	$\log(t_R - t_a)$	I
Propane	31.6	1.50	300 (By definition)
Cpd X	50.0	1.70	400 (From plot)
Pentane	79.5	1.90	500 (By definition)

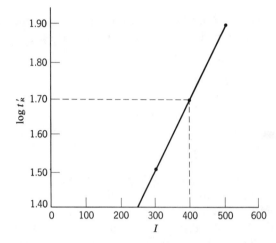

Figure 7–11 Retention Index plot in Example/Problem 7–2.

(c) Is Compound X butane? It is eluted from this column as if it were butane. This behavior is consistent with butane, but not proof. The same compounds should be chromatographed on several other columns of varying polarity. Consistent I values of 400 would constitute reasonable proof.

This system has a number of advantages: (1) a series of well-defined, readily available standards having a wide boiling range is used; (2) the I values are relatively insensitive to temperature; and (3) additional structural information can be obtained by noting the *change* in retention index, ΔI, when a compound is measured on a very polar and a nonpolar phase. ΔI values are related to structure (nature of functional groups and their positions). Kovats and others have prepared tables giving partial ΔI values for common functional groups and positional features. (See, for example, the Schomberg reference at the end of this chapter.) The partial ΔI values appear to be additive for the molecule as a whole, but there are many exceptions.

At best, chromatographic data leave an element of uncertainty in the identification of sample components unless the sample is already characterized by type. In totally unknown cases, or when more certain identification is required, the safest procedure is to analyze the separated components by another technique, such as mass spectrometry or infrared spectrophotometry.

7–11   QUANTITATIVE ANALYSIS

Integral detectors give a signal which is directly proportional to some bulk property of the amount of a component which has been eluted, for example, a gas volume or titrant volume. Calibration presents no problem for this type of detector. With differential detectors, the interpretation is not so simple.

For those detectors which give a signal proportional to the instantaneous concentration, we must sum the amount found in each volume increment, or, in other words, we must find the area under the peak—more precisely the area between the curve and the hypothetical baseline.

$$\text{Amount} = \int C\, dV = \frac{\text{Area}}{R_M} = \int \frac{R\, dt}{R_M} \qquad (7\text{-}15)$$

where R is the actual instantaneous response (signal) of the detector and R_M is the molar response (area/mole). If the amount is to be expressed in grams rather than moles, it may be more convenient to use a weight factor, F, which is usually defined in inverse fashion to the response

$$F = \frac{\text{Mol. Wt.}}{R_M} \text{ (gram/unit area)} \qquad (7\text{-}16)$$

Note that the area can be determined in any units (squares of any dimension on the chart paper, weight of the chart paper, millivolts × time, millivolts × volume, etc.), provided that the detector response is determined in the same units.

Measurement of Area. In practice, the area is measured by any of several techniques:

1. Cut out the peak with a scissors and weigh the paper. Determine the weight of paper per unit area in a separate experiment.
2. Use a planimeter, a mechanical device incorporating interconnecting levers which accumulate the area on a dial as the boundary is traversed by an attached marker.
3. Estimate the area by triangulation. Draw tangents to the points of inflection on the peak sides, and compute the area of the triangle formed with the base line, as in Figure 7–12.

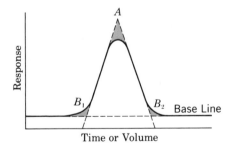

Figure 7–12 Measurement of peak area by triangulation. $A \approx B_1 + B_2$.

4. Measure the peak height and peak width at half-height, then multiply these two values.
5. Use a mechanical or electronic accessory to the recorder which will automatically integrate the area. (Some devices operate directly from the detector—see Chapter 25.)

Of the several comparisons of these techniques that have been made, a study made by Varian Aerograph is most useful. An eight-component mixture was run ten times and the areas integrated by several methods. The results, given as precision (relative standard deviation), and times required are shown in Table 7–6.

Table 7–6   Comparison of Integration Methods

Method	Time/Trace, min	Precision Rel. Std. Dev.
Triangulation	45–60	4.06%
Planimeter	45–60	4.06
Height × width at half-height	50–60	2.58
Cut and weigh	100–200	1.74
Disc integrator	15–30	1.29
Digital integrator	5–10	0.44

[Adapted from H. M. McNair *Basic Gas Chromatography*, Varian Aerograph, Walnut Creek, Calif., 1971.]

Any of these methods may be satisfactory, provided the calibration factors (R_M or F) are determined in the same way. The choice is determined by the convenience desired and the price we are willing to pay. Incompletely resolved (overlapping) peaks are usually treated by dropping a perpendicular from the minimum signal to the base line and dividing the area accordingly.

It is probably more difficult to determine the proper calibration factor than the peak area. From the discussion of detectors, it is evident that a different factor is required for each component and that the conditions for operating the detector must remain constant. Absolute calibration of a detector is indeed difficult and requires carefully prepared samples of known composition or precisely measured amounts of pure substances.

Normalization.   When the chromatogram represents the entire sample, a normalization method is satisfactory. To a first approximation, the fraction of the total area under each peak is the same as the fractional composition of the sample.

$$\% \text{ of Component } B = \frac{A_B}{\Sigma A_i} \times 100 \tag{7–17}$$

where A_i is the area of the ith component. However, if the molar responses or weight factors are known, these can be taken into account for better accuracy.

$$\text{Mole } \% \text{ of } B = \frac{A_B/R_{M,B}}{\Sigma(A_i/R_{M,i})} \times 100 = \frac{A_B/RMR_B}{\Sigma(A_i/RMR_i)} \times 100 \tag{7–18}$$

$$\text{Wt.}\% \text{ of } B = \frac{A_B F_B}{\Sigma(A_i F_i)} \times 100 = \frac{A_B/RWR_B}{\Sigma(A_i/RWR_i)} \times 100 \tag{7–19}$$

Note that in the normalization method (Equation 7–18 or 7–19) the use of relative values for molar response (RMR) or weight factors (RWR) is just as good as the absolute values. The relative values are much easier to determine (Equation 7–2 and Tables 7–3 and 7–4).

Example/Problem 7–3.   A sample containing benzene, heptane, and 2-methylhexane gives a three-peak chromatogram with areas of 35, 58, and 13 "squares" for the respective peaks. Calculate the percentage composition. (See data table at the top of page 105.)

If the peak is so sharp (narrow) that the area cannot be measured accurately, it is permissible to use the peak height as a measure of the amount, provided a proper calibration factor is determined. This technique is especially applicable to repetitive analysis such as those required for process control.

Compound	Area	Relative Molar Response, Area/Mole, (Table 7–3)	Mole/Area, Relative	Weighted Area, A/RMR	Mole %
Benzene	35	100	1.00	35.0	41.1
Heptane	58	143	0.70	40.6	47.8
2-Methylhexane	13	136	0.735	9.5	11.1

7–12 TEMPERATURE EFFECTS

There are three important parts of the chromatograph in which the temperature must be controlled. First, the injector temperature determines the rate at which the sample is vaporized. Since the fastest rate possible is desired in order to get the sample into the column in a very small volume, the injector is kept at a relatively high temperature consistent with the thermal stability of the sample, usually about 50° above the component boiling points. Second, the detector must be hot enough so that the constituents do not condense in it. On the other hand, the sensitivity of a thermal conductivity detector decreases with temperature, so that some optimum temperature is selected just above the column temperature. The thermal stability of the electrical insulation must also be considered. Temperature control is discussed in Section 28–6.

Finally, the temperature of the column itself is an important factor in determining retention and resolution. At high temperatures, components will tend to spend most of the time in the gas phase because of the decrease in solubility with increasing temperature. Thus they will elute quickly and close together—resolution is poor. At low temperatures, components will spend most of the time in the liquid phase, thus will elute slowly and usually farther apart—resolution is improved at the expense of time.

Programmed Temperature. The effects just discussed are shown dramatically in the chromatograms of Figure 7–13a and b showing the same mixture chromatographed at 45° and 120°C. The lower temperature yields good results only for the first four peaks, while the higher temperature is adequate for those beyond the first four. It is possible to get the best of both worlds by *programming* the temperature during the run. In Figure 7–13c the temperature was continuously increased from the initial 30° at a rate of 5° per min. The early peaks, 1 through 4, remain nicely resolved at low temperature, while the later peaks are sharpened and speeded up so that even peak 9 is eluted in just over 30 min. Temperature programming automatically provides a convenient column temperature for each component in a wide boiling mixture. Note also that in isothermal operation, the peaks for a homologous series are spread unevenly (log t_R is proportional to carbon number), whereas with temperature programming they are evenly spaced (t_R is proportional to carbon number).

Programmed temperature operation is advantageous whenever the boiling point range of sample components is greater than 100°. It requires more elaborate heating devices, a positive control of the flow rate, and a liquid phase useful over the entire temperature range.

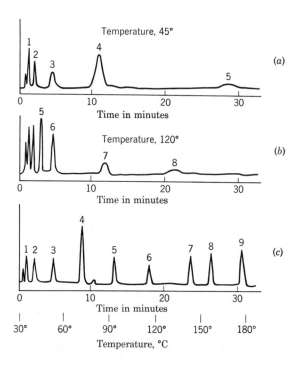

Figure 7–13   Isothermal (a and b) and programmed temperature (c) chromatograms of a mixture containing:

Peak No.	Compound	B.P., °C
1	n-propane	−42
2	n-butane	−0.5
3	n-pentane	36
4	n-hexane	69
5	n-heptane	98
6	n-octane	126
7	bromoform	150
8	m-chlorotoluene	162
9	m-bromotoluene	184

[From Habgood and Harris, *Anal. Chem.*, **32**, 450 (1960).]

7–13   GAS–SOLID CHROMATOGRAPHY

Historically, gas–solid chromatography (GSC) preceded gas–liquid chromatography (GLC) by many years. Charcoal filters to purify air have long been in common use. This is a form of GSC, but the development of the technique as a scientific method of separation was plagued by a number of difficulties.

Separation by GSC is very similar to that by GLC: the primary difference is that the partition within the column is caused by partial and selective adsorption on a solid surface rather than solubility in a liquid phase. The apparatus required is identical to that for GLC except for the column packing which is a finely divided, porous solid with a high specific surface.

Comparison of GSC and GLC.   Although there are many similarities between the two techniques, GSC has some unique advantages as well as some problems not associated with GLC.

1. GSC distribution coefficients, K, relating amount of component adsorbed per unit area of solid surface or per gram of packing to the concentration or partial pressure in the gas phase tend to be much larger than the partition coefficients for GLC. Thus retention times will be inconveniently long, except for the "fixed" gases like hydrogen, nitrogen, argon, oxygen, etc., which give retention times too short to be resolved by GLC.
2. Since there is no liquid phase to be retained in the column, the upper temperature limit of operation can be higher in GSC than in GLC. At higher temperatures, K values are lower and retention times shorter, but the temperature limit may be determined by thermal instability of the sample or the materials of construction of the apparatus itself.
3. Adsorption isotherms (a plot of amount adsorbed vs. partial pressure at constant temperature) are generally curved. In other words, the distribution coefficient is not independent of concentration. Curved isotherms result in skewed peaks. Most isotherms, however, tend toward linearity at very low concentrations, indicating that extremely small samples are to be preferred.
4. Useful solid adsorbents have areas of about $100 \text{ m}^2/\text{g}$, nearly 100 times that of solid supports most often used for GLC. Solids with such large surfaces are generally very effective catalysts whether or not they are chemically reactive in bulk. Thus samples may be readily pyrolyzed or otherwise converted into other substances, or even irreversibly adsorbed never to be seen again.
5. Solid surfaces are very difficult to reproduce—the size and shape of the area are certain to vary as well as the exact and detailed composition of the surface layer. Thus retention behavior is difficult to reproduce and a theoretical interpretation can be only approximate.

Some Common Adsorbents. Compared to the vast number of liquid phases used in GLC, relatively few adsorbents are used for GSC.

Carbon. There are many varieties available, differing in surface area, porosity, and surface activity. The surface can be treated either chemically or physically to make it more or less reactive. Carbon columns have been used in permanent gas analysis and for separations of the low boiling hydrocarbons.

Molecular Sieves. Some compounds, such as the calcium zeolites, have a well-defined, open structure with holes of molecular dimensions. Molecules having dimensions less than a critical size, depending on the size of the holes, are able to penetrate the interior of the porous structure and may be adsorbed on the interior surfaces or effectively "trapped." Larger molecules which do not penetrate the porous structures may pass through the column without significant retention. The size of the holes is carefully controlled in commercial preparations such as Linde sieves 4A, 5A, and 13X. Again, the surface activity is a function of adsorbed water and, in fact, molecular sieves are very effective drying agents for gas streams. In addition to their advantageous use for the analysis of the permanent and other inorganic gases, molecular sieves exhibit unusual behavior toward hydrocarbons. Linde 5A sieve retains *normal*-paraffins which are able to penetrate the holes, but not branched or cyclic hydrocarbons.

Porous Polymers. Spherical beads of uniform structure and pore size are prepared from the polymerization of styrene cross-linked with divinylbenzene, much like ion exchange resins. The composition can be modified by introducing various polar monomers to make beads of varying polarity. A particular series, known as Porapak, is widely used in gas chromatography. Solute molecules partition directly from the gas phase into the amorphous polymeric material; thus, the beads serve both as a solid support and a liquid phase. Porapak Q is especially useful for the separation of aqueous solutions of polar compounds.

Inorganic Salts. Some complex inorganic salts, for example, hydrates or ammoniates, can be decomposed thermally to yield open structures with very polar and often very selective adsorbing surfaces. Other salts can be deposited by evaporation from solution onto the surface of Chromosorb, silica gel, or alumina giving the characteristic adsorbing properties of the salt used. These techniques are relatively new, but would seem to add greatly to the versatility of GSC.

Up to the present, GSC has been used mostly for the separation of permanent gases and light hydrocarbons. However, with the development of improved adsorbents and more sensitive detectors, the versatility of the method is enormously extended.

7–14   APPLICATIONS

During the last 25 years, the literature of gas chromatography has jumped from one paper in 1952 to nearly 2500 per year. It would be hopeless to attempt to list all the areas of applications which by now have reached into all branches of chemistry. Gas chromatography cannot, by itself, solve all problems, but a few examples will indicate its scope.

Fuel Gases. The energy crisis has stimulated efforts to manufacture fuel gas from sewage and other wastes. The analysis of the products is best handled by a combination of two columns containing Porapak Q and Linde Molecular Sieve 5A. Figure 7–14 shows a complete analysis of a mixture containing H_2, CO_2, C_2H_4, C_2H_6, C_2H_2, O_2, CO, and CH_4 in about 18 min.

Auto Exhaust Gas. Air pollution caused in large part by the automobile engine is a continuing major concern. The analysis of the exhaust gases by gas

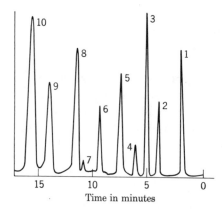

Figure 7–14   Separation of light gas mixture containing:

1 Hydrogen	6 (Disregard)
2 Carbon dioxide	7 Oxygen
3 Ethylene	8 Nitrogen
4 Ethane	9 Carbon monoxide
5 Acetylene	10 Methane

[From Marchio, *J. Chromatog. Sci.*, **9**, 432 (1971).]

chromatography has helped us to understand the combustion processes and to improve all parts of the fuel system.

High Purity Metals. Most metals and their compounds are not sufficiently volatile to be determined directly by gas chromatography. However, volatile complexes can be prepared for essentially all metals. The most useful complexes for this purpose are those formed by trifluoro- and hexafluoroacetylacetone:

$$CF_3C—CH_2—C—CF_3$$
$$\|\qquad\quad\|$$
$$O\qquad\quad O$$

The high sensitivity of the electron-capture detector for fluoro compounds makes this technique ideal for trace impurities in metals. For example, trace amounts of 0.1 ppm ($\sim 10^{-5}\ \mu g$) of Cr and Al are easily determined in uranium metal after extraction with trifluoroacetylacetone, as shown in Figure 7–15.

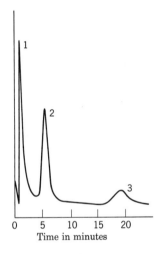

0   5   10   15   20
Time in minutes

Figure 7–15 Chromatogram of a solution of trifluoroacetyl-acetone complexes of (1) Be, (2) Al, and (3) Cr present at the 0.1 ppm level in uranium metal. Column: silanized glass beads coated with 0.2% silicone oil; Temperature: 115°C. Detector: electron-capture. [From Gentz, Hovin, Malherbe, and Schott, *Anal. Chem.*, **43**, 235 (1971).]

Identification of Natural Products. The flavors and aromas of natural products (foods, flowers, perfumes, etc.) are the result of a unique combination of trace quantities of hundreds of organic compounds. Strawberry flavor was one of the first to be investigated by gas chromatography. One of the objectives is to be able to duplicate rare or expensive flavors and aromas with synthetic chemicals. Another objective is to make a positive identification of a particular variety. In a recent study, the oil obtained from a number of cheeses by centrifugation was chromatographed. Distinctive patterns, as in Figure 7–16, are obtained for each variety of cheese. Some of the peaks change with aging or adulteration.

Preparative Scale. If it were easy to scale up the chromatograph, the remarkable separative powers of gas chromatography would be a bonanza for manufacturers of high purity chemicals. Unfortunately, resolution is drastically reduced as the sample size and column diameter are increased, because it is difficult to maintain a uniform flow profile across the column (i.e., the A term in the Van Deemter equation becomes very large). The high cost of designing and constructing large scale apparatus must be compared with the cost of repetitive small scale operation in conventional apparatus. To date, a number of columns one foot or more in diameter have been constructed for special applications. Figure 7–17 is the chromatogram of the separation of a 1.5-liter sample in a one-foot column.

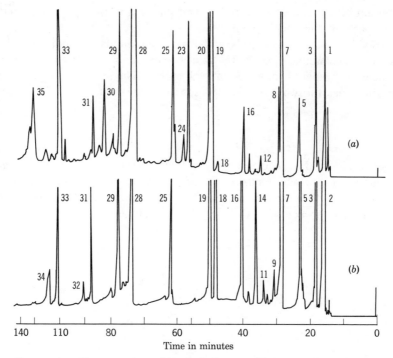

Figure 7-16 Chromatograms of 10 μl of oil expressed from (a) Roquefort cheese; (b) blue cheese. Column: 500 ft, 0.02 in. diam., nonyl phenol polyethylene glycol ether (Dowfax 9N15); temperature: programmed from 25 to 125°C at 2°/min; detector: flame ionization. Peaks not identified, but numbers represent the same compound on both graphs. [From Liebich, Douglas, Boyer, and Zlatkis, *J. Chromatog. Sci.*, **8**, 351 (1970).]

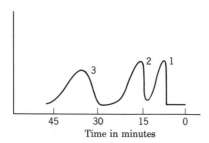

Figure 7-17 Preparative trace of a 1475 ml sample containing equal volumes of (1) *n*-hexane, (2) *n*-heptane and (3) *n*-octane. Column: 8 ft, 12 in. diameter, celite with 25% asphalt; temperature: 150°C. [From Carel, Clement, and Perkins, Jr., *J. Chromatog. Sci.*, **7**, 219 (1969).]

Process Control.   Chromatography is not an ideal technique for controlling a process in which fast action is desirable. It does not give an immediate and continuous signal, as for example in spectrophotometry or potentiometry. However, in situations involving complex mixtures that would require separation of interferences before other techniques are applicable, gas chromatography has much to offer. For this application, speed is the primary consideration—in favorable cases the data can be acquired in a few seconds. The detector signal can be sent directly to a computer system which may be responding to many chromatographs and/or other instruments and controlling addition of reagents, temperature, pressure, and so on, to maximize the production. (See Section 28–5.)

Biomedical Applications.   Perhaps the chemical system most difficult of all to monitor and control is the human body. As our knowledge of medicine and our

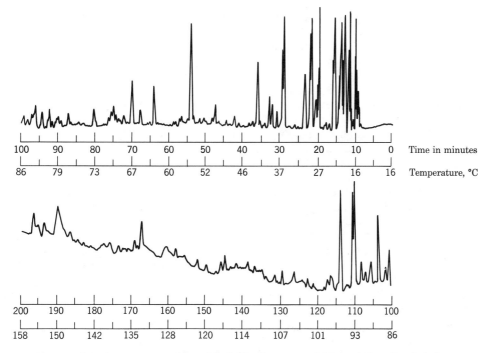

Figure 7–18 Gas chromatogram of breath volatiles from normal human being, 10 exhalations. Column: 1000-ft, 0.03-in. i.d., open tubular column coated with methylsilicone oil; temperature: programmed from 16°C at 0.05°/min; detector: Flame ionization. [From Teranishi, Mon, Robinson, Cory, and Pauling, *Anal. Chem.*, **44**, 18 (1972).]

analytical technology have advanced, our doctors have come to rely on an ever increasing battery of chemical analyses of body fluids. At some future time you may walk into a medical laboratory, leave a sample of breath, blood, and urine, and within a few minutes obtain a complete diagnosis of the state of your health along with prescriptions to cure any problems. Linus Pauling and his co-workers have concluded that the gas chromatograph is the best instrument for this purpose. Using dual 1000-ft capillary columns coated with methyl silicone in a programmed temperature gas chromatograph, they have obtained chromatograms showing 280 substances in human urine and 250 substances in human breath, Figure 7–18. Quantitation is achieved by a computer program; for persons on a standard diet, the day-to-day deviations are less than 10%. The ready availability of such detailed information should lead to fantastic breakthroughs in health care.

QUESTIONS AND PROBLEMS

7–1. At the beginning of this chapter is a statement that the separation of benzene and cyclohexane, which have almost identical boiling points, is easy. How would you do it?

7–2. A two-component mixture is separated on a certain column with a resolution of 1.25 when the inlet pressure is 1.25 atm. How will a change in inlet pressure to 2.50 atm affect the retention times, H, and the resolution?

7–3. Predict the effect on peak shape of:
   (a) Introducing the sample over a long period of time (say, 10 sec) rather than as a sharp plug.

  (b) Using too low a temperature at the injector so that the sample does not vaporize quickly.

  (c) Increasing the temperature of the column.

  (d) Increasing the flow rate.

  (e) Doubling the length of the column.

  (f) Doubling the recorder chart speed.

7–4. For a particular experiment the column inlet pressure is held at 5 atm with the outlet at 1 atm. What is the average pressure in the column?   *Ans.* 3.44 atm

7–5. Define the following terms used in chromatography by giving an equation, labeled diagram, or description:

  (a) Resolution

  (b) Retention volume

  (c) Specific retention volume

  (d) Liquid phase

  (e) Theoretical plate

  (f) Van Deemter plot

  (g) Optimum flow rate

7–6. The following data are obtained in a chromatographic analysis: $t_R = 5.0$ min, $t_a = 1.0$ min, $V_L = 2.0$ ml, $F_c = 50$ ml/min. Calculate:

  (a) k' *Ans.* 4.0

  (b) V_M *Ans.* 50 ml

  (c) K *Ans.* 100

  (d) V_R *Ans.* 250 ml

7–7. What is the value of H for a 2.00-m column if a peak with a retention time of 10.0 min has a width of 30 sec?   *Ans.* 0.031 cm

7–8. Van Deemter constants are given for two columns of equal length:

	A	B	C
Col 1	0.18 cm	0.40 cm^2/sec	0.24 sec
Col 2	0.05	0.50	0.10

  (a) Which of these two columns gives the larger number of theoretical plates if the carrier gas velocity is 0.50 cm/sec?

  (b) What is the optimum velocity for Column 1?   *Ans.* 1.3 cm/sec

7–9. Retention times (corrected for the air peak) are given for the following compounds on a particular column. What is the retention index of each of these compounds on this column?

Ethane	0.25 min	2-Methylbutane	1.20 min
Propane	0.45	Butene-1	0.80
n-Butane	0.95	Hexene-1	2.95
n-Pentane	1.80	Ethylene	0.15
n-Hexane	3.50	Benzene	3.75
n-Heptane	6.95	n-Butanol	8.4
n-Octane	13.7	Water	3.50

Ans. Ethane, 200; Hexene-1, 574

7–10. Retention times in minutes (corrected for air peak) for several straight-chain paraffin hydrocarbons on a silicone oil column at 90°C are given below. The flow rate was 85 ml/min.

C_1	C_2	C_3	C_4	C_5	C_6	C_7	C_8
0.00	0.12	0.60	1.44	2.80	5.90	11.6	22.8

  (a) Calculate the retention volume for each paraffin.

  (b) Calculate the relative retention volume for each paraffin (pentane = 100).

  (c) Plot log relative retention time vs. carbon number.

7–11. Describe how you could use a table of V_g values to identify the unknown compounds in a sample.

7-12. Glass micro-beads pack very uniformly in a column, but they have a small surface area relative to Chromosorb. Considering these two properties of glass beads, compare the efficacy of glass beads versus Chromosorb as a solid support.

7-13. For propanol on a silicone oil column, the specific retention volume, V_g, varies with temperature as follows:

T, °C	40	60	72	84	97	111
V_g, ml/g	169	85.0	64.6	44.5	31.6	22.0

(a) Plot both V_g and log V_g vs. both T and $1/T$ (in absolute degrees).

(b) What conclusion can you draw from these curves?

7-14. A certain detector gives a signal which depends on the molecular weight in a linear fashion:

$$\text{Rel. sens (area/unit wt)} = (1.52 + 1.10 \times \text{mol wt}) \text{ cm}^2/\mu g$$

A two-component mixture containing acetone and decane is analyzed using this detector. The peak areas are 2.50 and 7.50 cm², respectively.

(a) What is the weight % composition of the sample?   *Ans.* acetone = 44.6%

(b) What is the mole % composition of the sample?   *Ans.* acetone = 66.4%

7-15. A 2.00-mg sample gave the following peaks with areas given in parentheses (arbitrary units): hexane (35), 1-heptanol (20), methyl butyl ketone (15), toluene (80).

(a) Calculate the mole % composition of the sample assuming the molar response of the detector (area/mole) is the same for each component.

Ans. mole % hexane = 23.3%

(b) Calculate the weight % composition of the sample assuming the molar response of the detector (area/mole) is the same for each component.

Ans. weight % hexane = 21.2%

(c) Calculate the weight % composition given that the weight factors (weight/unit area) of the detector for each of these compounds is: 0.97, 1.25, 1.41 and 0.93, respectively.   *Ans.* weight % hexane = 21.9%

7-16. Two columns, one a packed column and the other a capillary, have the following specifications:

	Col 1	Col 2
Dead volume, t_a, ml	15.0	3.0
Adjusted retention time, $t_R - t_a$, ml		
Component P	160	1.1
Component Q	170	1.2
Number of theoretical plates	6,400	25,600

Which of these columns gives the better resolution (assuming the peaks are triangular in shape). Explain why.

7-17. In one form of detection system, the effluent is passed through a reactor which oxidizes organic compounds to carbon dioxide and water. The water vapor is removed and the CO_2 concentration is measured with a thermal conductivity detector. Typical data are given below:

Compound	Area	Compound	Area
n-Pentane	25.0	n-Heptane	49.0
iso-Pentanol	35.0	Toluene	70.0

(a) Calculate the mole % composition of the sample.

(b) Calculate the weight % composition of the sample.

*Ans. n-*pentane, 17.2 mole %, 13.8 weight %

REFERENCES

H. Purnell, *Gas Chromatography*, Wiley, New York, 1962.

A. B. Littlewood, *Gas Chromatography*, 2nd ed., Academic Press, New York, 1970.

L. S. Ettre and A. Zlatkis, *The Practice of Gas Chromatography*, Wiley-Interscience, New York, 1967.

W. E. Harris and H. W. Habgood, *Programmed Temperature Gas Chromatography*, Wiley, New York, 1966.

Journal of Chromatographic Science, May 1973, contains several review articles on detectors.

G. Schomburg and G. Dielmann, "Identification by Means of Retention Parameters," *J. Chromatographic Science*, **11**, 151 (1973).

H. M. McNair and E. J. Bonelli, *Basic Gas Chromatography*, Varian Aerograph, Walnut Creek, Calif., 1965.

D. A. Leathard and B. L. Shurlock, *Identification Techniques in Gas Chromatography*, Wiley, New York, 1970.

ELECTROMAGNETIC RADIATION

8

ELECTROMAGNETIC RADIATION AND ITS INTERACTION WITH MATTER

8–1   INTRODUCTION

Electromagnetic radiation is a form of radiant energy which exhibits both wave and particle properties. The phenomena of refraction, reflection, constructive and destructive interference are examples of wave properties. At the other extreme, Einstein's explanation of the photoelectric effect suggests that electromagnetic radiation consists of discrete particles called *photons* which have definite energies and travel through space with the velocity of light. Although waves and particles seem to be incompatible, we must invoke the "particle-wave" duality to explain both electron behavior and the nature of electromagnetic radiation.

8–2   NATURE OF ELECTROMAGNETIC RADIATION

Wave Properties.   As indicated in Figure 8–1, and as the name implies, an electromagnetic wave has an electric component and a magnetic component. The two components oscillate in planes perpendicular to each other and perpendicular to the direction of propagation of the radiation. Only the electric component is active in ordinary energy transfer interaction with matter. Henceforth, in our discussion of wave behavior we will consider only the electric component. In Figure 8–1, wavelength, λ, is the distance between two corresponding points on the wave. The square of the amplitude, A, of the wave is a measure of the intensity of the wave.

Another important property of an electromagnetic wave is its frequency, ν, or the number of complete wavelength units which pass a fixed point per unit of time. The units of frequency are cycles per second or hertz. The wavelength and frequency are related to the velocity of light by the expression:

$$\lambda\nu = c/n \qquad (8\text{–}1)$$

where c = the velocity of light in a vacuum (2.9976×10^{10} cm/sec) and n is the refractive index (the ratio of the velocity of light in a vacuum to its velocity in the

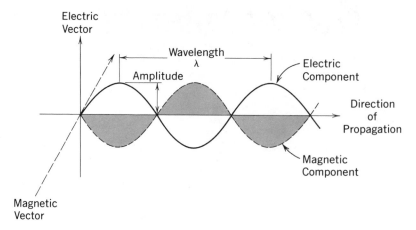

Figure 8–1   An electromagnetic wave.

medium in question). The frequency of a given radiation is the same in every medium. Only the velocity and wavelength of the radiation change from medium to medium. The frequency may also be defined as the number of wavelength units in 2.9976×10^{10} cm, a very large number. For many purposes it is more convenient to use the number of wavelength units in one centimeter. This number is called the wavenumber, $\bar{\nu}$:

$$\bar{\nu} = 1/\lambda = \nu n/c \text{ (units, cm}^{-1})$$

Example/Problem 8–1. Green light has a wavelength of approximately 530 nanometers (nm) in a vacuum. Calculate the wavelength, λ, and the wavenumber, $\bar{\nu}$, for green light in water (1 nm = 10^{-7} cm).

$$n_{\text{vacuum}} = 1.0000; n_{\text{water}} = 1.332$$

Solution.   For green light in a vacuum,

$$\nu = \frac{c}{n\lambda} = \frac{3.00 \times 10^{10} \text{ cm/sec}}{(1.000)(5.30 \times 10^{-5} \text{ cm})} = 5.66 \times 10^{14} \text{ Hz}$$

Consequently, in water, $\nu = 5.66 \times 10^{14}$ Hz. However, $n = 1.332$. Therefore for green light in water,

$$\lambda_{\text{water}} = \frac{c}{n\nu} = \frac{3.00 \times 10^{10} \text{ cm/sec}}{(1.332)(5.66 \times 10^{14} \text{ sec}^{-1})} = 3.98 \times 10^{-5} \text{ cm}$$

Since $\bar{\nu}_{\text{water}} = \dfrac{1}{\lambda_{\text{water}}}$, $\bar{\nu}_{\text{water}} = 2.51 \times 10^{4} \text{ cm}^{-1}$.

Particle Properties.   To describe how electromagnetic radiation interacts with matter, it is useful to think of a light beam as a train of photons. The energy of *each* photon is proportional to the frequency of the radiation and is given by the relationship:

$$E = h\nu = hc/n\lambda \tag{8–2}$$

where E = the energy of the photon in ergs, ν = the frequency of the electromagnetic radiation in hertz, and h = Planck's constant, 6.624×10^{-27} erg-sec. A photon of high frequency (short wavelength) has a higher energy content than one of lower frequency (longer wavelength). The intensity of a light beam is proportional to the number of photons and is independent of the energy of each photon.

The Spectrum. Electromagnetic radiation which is of interest in chemistry varies from the highly energetic gamma rays to the very low energy radiowaves. The entire range of radiation is commonly referred to as the electromagnetic spectrum; the portion of chemical interest is presented in Table 8–1. The various regions of the spectrum are defined by the type of apparatus used to generate or detect the radiation, for example, the human eye, infrared spectrophotometer, X-ray machine, etc. The limits indicated in Table 8–1 are arbitrary and diffuse— the regions overlap. Table 8–2 summarizes some of the units commonly used to describe the properties of electromagnetic radiation.

8–3 ABSORPTION AND EMISSION OF RADIATION

Electromagnetic radiation can interact with matter in a number of ways. If the interaction results in the transfer of energy from a beam of radiant energy to the matter, it is called *absorption*. The reverse process in which a portion of the internal energy of matter is converted into radiant energy is called *emission*. Part of the radiation which passes into matter, instead of being absorbed, may be scattered or reflected, or may be re-emitted at the same wavelength or a different wavelength upon emerging from the sample. Radiation which is neither absorbed nor scattered may undergo changes in orientation or polarization as it passes through the sample. We will return to a discussion of these "other" interactions in later sections of this chapter.

Absorption of Radiation. To explain the absorption of radiation, it is convenient to invoke the particle nature of light, which is to say that the energy of a photon is "quantized" and is given by Planck's equation

$$E = h\nu \tag{8–2}$$

Quantum theory proposes that if there is a "collision" between a photon and a receptor (atom, ion, or molecule), there is a finite probability that this energy may be transferred to the receptor in a discontinuous process. In simple terms, the receptor either absorbs the complete quantum of energy, or it does not.

$$M + h\nu \rightarrow M^*$$

where M^* is an "excited" receptor.

The energy levels of an atom or molecule are also quantized. Consider the energy levels for a typical atom shown in Figure 8–2. Monatomic substances normally exist in the gaseous state and their atoms are largely in the lowest energy level (ground state). They are able to absorb radiant energy only through an increase in their electronic energy, that is, by promotion of an electron to a higher energy orbital. The vertical lines labeled ΔE_1, ΔE_2, and so on, indicate "allowed" transitions between energy levels. Such transitions can occur only if photons of exactly the same energy are available. Otherwise, no absorption can occur.

Energy levels of a typical molecule are shown in Figure 8–3. The total energy state of a molecule includes electronic, vibrational, and rotational components, all of which are quantized. For each electronic level, there will be several sub-levels corresponding to vibrational states, and the latter are further subdivided into rotational levels. Transitions between these levels will be discussed in more detail in Sections 11–2 and 12–2. Apparently the absorption or radiation by a molecule is

Table 8–1 The Electromagnetic Spectrum

	0.01 Å	0.1 Å	1 Å	10 Å	100 Å / 10 nm	1,000 Å / 100 nm	1,000 nm / 1 μm	10 μm	100 μm / 0.01 cm	1,000 μm / 0.1 cm	1 cm	10 cm
E (electron volts)				1240	124	12.4	1.24	0.124	0.012	0.001		
E (kcal/mole)					2850	285	28.5	2.85				
E (cal/mole)								2850	285	28.5	2.85	
$\bar{\nu}$ (cm^{-1})							10,000	1,000	100	10	1	0.1
ν (Hz)	3×10^{20}		3×10^{18}		3×10^{16}		3×10^{14}		3×10^{12}		3×10^{10}	
λ	0.01 Å	0.1 Å	1 Å	10 Å	100 Å / 10 nm	1,000 Å / 100 nm	1,000 nm / 1 μm	10 μm	100 μm / 0.01 cm	1,000 μm / 0.1 cm	1 cm	10 cm

Spectral regions:

- Gamma rays
- X-rays — Inner Shell, Middle Shell
- Ultraviolet — Vacuum UV, Near UV — Valence Shells
- Visible
- Infrared — Near IR, Fund. IR, Far IR
- Microwave
- Radio

Transitions:

- Nuclear Transitions
- Electronic Transitions
- Molecular Transitions — Vibrations, Rotations
- Spin Orientations

Table 8–2   Units and Symbols Used To Describe Electromagnetic Radiation

Quantity	Unit	Symbol	Conversions
Wavelength	Micron (micrometer)	μm	$1\ \mu m = 10^{-4}\ cm$
λ	$\lceil$Nanometer	nm	$1\ nm = 10^{-3}\ \mu m = 10^{-7}\ cm$
	$\lfloor$Millimicron[a]	mμ	$1\ m\mu = 1\ nm$
	Angstrom	Å	$1\ Å = 10^{-8}\ cm$
Frequency	Cycles per sec	cps	
ν	Hertz	Hz	$1\ Hz = 1\ cps$
	Megahertz	MHz	$1\ MHz = 10^{6}\ cps$
Wavenumber	Reciprocal centimeter	cm^{-1} or $\bar{\nu}$	$cm^{-1} = 1/\lambda$
Intensity[b]	Energy per sec per unit solid angle	I	
Radiant power[b]	Energy per sec	P	

[a] Note the difference between mμ (millimicron = 10^{-9} m) and μm (micrometer or micron = 10^{-6} m). Both symbols are in common use, along with μ for micron.

[b] Radiant power and intensity are often used interchangeably in spectrophotometry. In this text we will retain the more common usage of intensity, I.

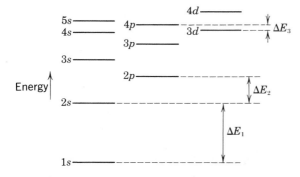

Figure 8–2   Energy level diagram for subshells in polyelectron atoms.

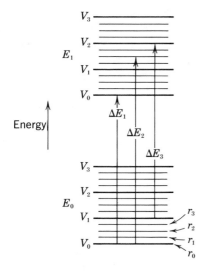

Figure 8–3   Schematic representation of molecular electronic, vibrational, and rotational energy levels.

far more complex than absorption by individual atoms. A given electronic transition may also involve a vibrational change as well as a rotational change resulting in several photon energies that can be absorbed. For example, in Figure 8–3 ΔE_1, ΔE_2 and ΔE_3 all represent electronic transitions involving the same two electronic levels, but different vibrational and rotational levels. Quantum mechanical selection rules indicate that some of these possibilities are not allowed and that some are more probable than others. Each absorption thus corresponds to energy transfer from photons of a given energy (radiation of a given frequency or wavelength).

In summary, an atom or a molecule cannot accept energy indiscriminately, only in quanta of the appropriate value to cause an excitation from one energy level to another. In absorption spectrophotometry, we observe which energies the sample can absorb by varying the wavelength of the incident radiation and measuring the decrease in intensity of the transmitted radiation (energy absorbed). A plot of the energy absorbed (or of the intensity of the transmitted beam) as a function of wavelength or frequency is called an *absorption spectrum*. Figure 8–4 shows an absorption spectrum in which many closely spaced transitions overlap giving a broad absorption band typical of many complex molecules and ions.

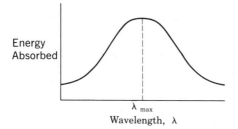

Wavelength, λ **Figure 8–4**   Typical broad absorption band.

Emission of Radiation.   The interaction of electromagnetic radiation with matter is a reversible phenomenon. Species which happen to be in an excited state can emit photons of characteristic energies by returning to lower energy states or to the ground state. The initial excitation can be produced from a variety of energy sources, including electric arcs or sparks, flames, electron bombardment, X-ray beams, and so on. Most of these sources are sufficiently energetic to break chemical bonds so that atomic emission spectra are observed. Such spectra consist of relatively few sharp lines at frequencies characteristic of the element (*line spectra*). On the other hand, the emission spectra of solids and liquids appear to be continuous because the individual wavelengths are so numerous and closely spaced that they overlap each other and cannot be resolved with ordinary instruments. Heated solids produce continuous spectra because of the unquantized changes in the kinetic energies of the individual particles. Continuous spectra are of little use in analytical spectroscopy except, of course, as sources for absorption studies.

The production and interpretation of emission spectra will be discussed in greater detail for flame photometry (Section 13–2) and X-ray spectrometry (Section 14–2).

8–4   LUMINESCENCE PHENOMENA

Energy level diagrams for real molecules are generally more complicated than the schematic diagram given in Figure 8–3. A molecule in a *singlet state* has all its

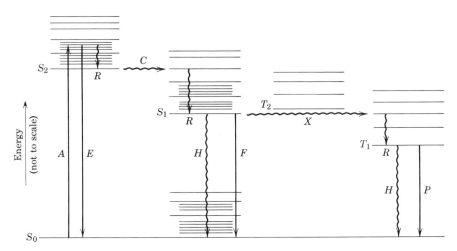

Figure 8–5 Schematic energy level diagram of singlet and triplet states of a molecule showing luminescence phenomena.

electrons spin-paired—the ground state of most molecules is a singlet state. A molecule in a *triplet state* has two electrons for which the spins are unpaired. The triplet state is slightly lower in energy than the corresponding singlet state because two paired electrons have a higher energy than two unpaired ones. An energy diagram showing these various states is given in Figure 8–5. Let us examine the events which might follow the absorption of a photon, raising the energy of the molecule from its ground state, S_0, to an excited vibrational level of an excited singlet state, S_2 (Arrow A).

The molecule may lose this acquired energy through one of several alternate pathways. The process might be immediately reversed, and a photon emitted of the same energy as that absorbed (Arrow E). This process is called *resonance fluorescence* and is the basis for a technique known as *atomic fluorescence*. If the excited molecule is in solution with many close neighbors, it is far more likely that it will lose its vibrational excitation energy through collisions and fall to the lowest vibrational level of the S_2 state, a process called *vibrational relaxation* (Arrow R). In general, there will be excited levels of the next lower singlet state (S_1 in this case) that have energies comparable to lowest S_2 state. A transition from S_2 to S_1 is thus highly favored and is called an *internal conversion* (Arrow C). The molecule then rapidly loses energy through additional collisions until it reaches the lowest level of the lowest excited singlet state, S_1.

At this point, one of several things may happen. It may simply return to the ground state by further collisions, dissipating the energy as nonradiated heat (Arrow H). It may return directly from the S_1 level to the ground state by emitting a photon, termed *normal fluorescence* (Arrow F). The frequency of this fluorescence will be lower than the resonance fluorescence. The duration of fluorescence is equal to the lifetime of the excited singlet state (10^{-9} to 10^{-7} sec). Many organic and some inorganic compounds fluoresce in the visible region when they are irradiated with ultraviolet light. Fluorescence is also important in the production of low energy, longer wavelength X rays by irradiation of a sample with shorter wavelength, higher energy X rays.

A third possibility is that the molecule can shift from the singlet state to the corresponding triplet state ($S_1 \rightarrow T_1$), a phenomenon called *intersystem crossing* (Arrow X). The crossing involves unpairing of two electrons and leaves the molecule in an excited vibrational level. Vibrational relaxation will quickly bring it to the lowest T_1 level. Very few molecules exhibit intersystem crossing, but, for those that do, the lifetime of the T_1 state is relatively long (10^{-6} to 10 sec). The energy of the T_1 state is lower than that of the S_1 state, therefore a triplet molecule is more likely to lose energy through collisions. However, some substances do return from the triplet state to the ground state via photon emission (Arrow P), called *phosphorescence*. The duration of phosphorescence depends on the lifetime of the T_1 state and may last as long as 10 sec. Only a few types of molecules exhibit phosphorescence and its measurement is not yet a widely used analytical technique.

8–5   REFRACTION AND REFRACTIVE INDEX

In addition to absorption and emission there are other interaction phenomena which are useful for both identification and structural studies. Refraction is one of these important nonabsorptive processes. When radiation passes from one medium to another, it is partially reflected and partially transmitted. The transmitted radiation retains its characteristic frequencies in the new medium; however, both the velocity and direction of propagation may change. Interfacial characteristics for radiation passing from air into glass are illustrated in Figure 8–6. In order to explain this behavior, it is helpful to consider a "wave front" perpendicular to the direction of travel. If the ray strikes the surface of the glass at an angle (ϕ_1 is the angle of incidence), then one side of the wave front reaches the interface and enters the glass before the other. Therefore, while one side of the wave front is traveling in glass, the other side is still traveling in air. We have already defined the ratio of the velocity of light in vacuum to that in another medium as the index of refraction, n, that is,

$$n = v_{\text{vac}}/v_{\text{medium}} \qquad (8\text{–}3)$$

The index of refraction of air is so close to unity ($n_{\text{air}} = 1.00027$), that for ordinary purposes

$$n = v_{\text{air}}/v_{\text{medium}} \qquad (8\text{–}4)$$

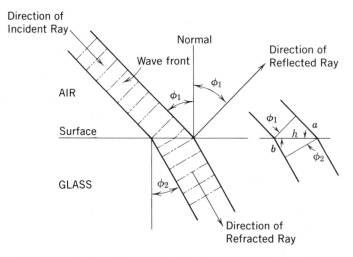

Figure 8–6   Explanation of diffraction of a light ray passing from air into glass.

In Figure 8–6 there are two right triangles with a common hypotenuse, h. Side a is the distance traveled by one side of the wave front in air while the other side of the wave front travels a distance b in glass. Both of these distances must be traveled in the same time in order for the oscillations to remain in phase across the wave front. Therefore,

$$\text{time} = \frac{\text{distance}}{\text{velocity}} = \underbrace{\frac{a}{c}}_{\text{air}} = \underbrace{\frac{b}{c/n_{\text{glass}}}}_{\text{glass}} \qquad (8\text{--}5)$$

or

$$\frac{h\,\sin\phi_1}{c} = \frac{h\,\sin\phi_2}{c/n_{\text{glass}}} \qquad (8\text{--}6)$$

or

$$n_{\text{glass}} = \frac{\sin\phi_1}{\sin\phi_2} \qquad (8\text{--}7)$$

The above treatment holds as well for the transmission of radiation across the interface between any two media, although two values of n will appear. (We have taken $n_{\text{air}} = 1$ in Equation 8–7.) Thus, for the general case,

$$\frac{n_2}{n_1} = \frac{\sin\phi_1}{\sin\phi_2} \quad \text{or} \quad n_1\sin\phi_1 = n_2\sin\phi_2 \qquad (8\text{--}8)$$

where the subscripts 1 and 2 refer to light going from medium 1 into medium 2. Equation 8–8 is known as *Snell's law*. Once the index of refraction of any medium has been determined with respect to vacuum or air, it can be used as a secondary standard to determine indices of refraction of other media.

The velocity and the index of refraction in any medium other than vacuum are functions of the temperature and the frequency. This last statement is extremely important because it indicates that light of different frequencies is refracted at different angles.

A prism utilizes the variation of refractive index as a function of wavelength to achieve *dispersion*. Dispersion is described qualitatively as the *angular separation* of the different wavelengths which make up a beam of radiation. When a beam of radiation passes from air into a denser medium, it will be refracted (bent) toward the perpendicular, and will undergo an opposite effect if and when it re-emerges into the air. When the two surfaces are parallel, as in a glass plate, the overall effect of the refraction is a small displacement of the beam with no net change in direction or angular deviation. However, the faces of a prism are not parallel and an angular deviation results. These effects are shown in Figure 8–7.

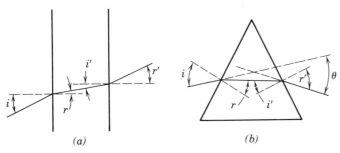

(a)   *(b)*

Figure 8–7 Deviation of monochromatic radiation by (a) a glass plate and (b) a prism.

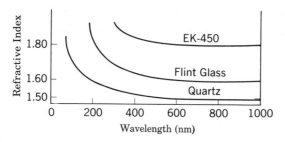

Figure 8–8 Refractive index as a function of wavelength for several materials.

The angle of incidence to the first surface is represented by i, the angle of refraction at the first surface by r, the angle of incidence to the second surface by i', and the angle of refraction by r'. The angle of deviation which the incident ray has undergone upon passing through the prism is denoted by θ. The angle of refraction and consequently the angle of deviation for any wavelength of radiation is determined by the refractive index of the prism material for that particular wavelength. The variation of refractive index with wavelength is shown for various substances in Figure 8–8. The refractive index decreases at longer wavelength; therefore, from the laws of refraction, the angle of deviation is larger at shorter wavelength. The fate of a ray of collimated polychromatic radiation is depicted in Figure 8–9.

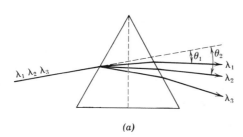

(a)

Figure 8–9 Dispersion of polychromatic radiation by a 60° prism.

In measuring and reporting values of the index of refraction, it is customary to report both the temperature and the frequency used, for example, n_D^{20} means the measurement was made at 20°C with the D line of a sodium lamp ($\lambda_D = 589.3$ nm). The dependence of the angle of refraction on the frequency is, of course, the explanation of how a prism disperses different wavelengths of radiation.

The determination of the refractive index involves only the determination of two angles. These angles may be measured quite accurately and thus n may be determined with a high degree of precision. Some instruments allow a precision of 10 parts per billion. Very careful control of temperature and wavelength of radiation is extremely important to high precision refractive index measurements. To illustrate the wide applicability of refractive indices, they are used to confirm the identity of substances, analyze mixtures, estimate molecular weights, molecular sizes and shapes, and to calculate properties such as reflectivity and optical dispersion. Some refractive index values of characteristic materials are presented in Table 8–3.

It should be emphasized that the index of refraction at a given temperature is very specific for a pure material. In organic qualitative analysis a conventional approach includes determining elemental composition, solubility, melting point, boiling point, and refractive index.

Table 8–3 Refractive Indices of Some Characteristic Materials

Substance	n_D^{20}	Substance	n_D^{20}
Methanol	1.3288	n-Hexane	1.3749
Water	1.3328	Cyclohexane	1.4266
Acetone	1.3588	Toluene	1.4929
Ethanol	1.3590	Benzene	1.4979
Acetic acid	1.3698	Pyridine	1.5095
Ethyl acetate	1.3701	Aniline	1.5863

8–6 INTERFERENCE AND DIFFRACTION

In the previous section we noted that a prism will produce a spectrum of colors because of the phenomenon of refraction. Another convenient way to produce a spectrum is to use a diffraction grating which also disperses polychromatic radiation into its component wavelengths.

Diffraction Gratings. Reflection diffraction gratings are now generally used in the monochromators of ultraviolet, visible, and infrared instruments. They consist of a highly reflective aluminized surface on which are etched a large number of equally spaced parallel grooves (sometimes called lines). Typical gratings have from 600 to 2000 lines per mm depending on the region of the spectrum for which they are intended.

In order to understand how a diffraction grating can disperse radiation, we must first consider the concepts of constructive and destructive interference of radiation. In Figure 8–10, radiant energy "waves" (1) and (2) are "in phase" (i.e., they cross the axis in the same direction at the same time). If (1) and (2) are superimposed, the two waves "reinforce" each other with an amplitude equal to the sum of the two component waves. Light waves (3) and (4) in Figure 8–10 are 180° out of phase and when they are superimposed they cancel each other. The amplitude of the resultant wave is zero—a phenomenon called *destructive interference.*

Polychromatic radiation emitted by a source of radiant energy and collimated by a lens or mirror is coherent, i.e., all radiations of the same wavelength are "in phase." Lines I and II in Figure 8–11 represent the boundaries of a collimated coherent ray which has a "wavefront" that is perpendicular to the direction of travel. Consider the situation when this ray strikes a reflection grating at an angle of incidence, i. In order for constructive reinforcement to occur in the reflected radiation, the difference in path lengths along lines I and II must be equal to an integral (i.e., 1, 2, 3, . . . , n) number of wavelengths. Application of the laws of diffraction to the many lines of a diffraction (or reflection) grating indicates that destructive interference is essentially complete at all angles other than those at which constructive reinforcement occurs. Constructive reinforcement of monochromatic radiation from a reflection grating is illustrated in Figure 8–11. Path II is longer than path I by the distance $CB - AD$. From the geometric construction of the two right triangles ACB and ADB, it is not difficult to see that $CB - AD = d(\sin i - \sin \theta)$. However, θ is considered a negative angle, so the difference between the two paths is $d(\sin i + \sin \theta)$. Therefore, for constructive

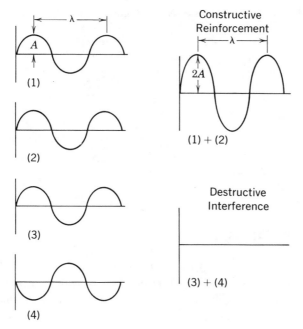

Figure 8-10   Constructive and destructive interference of monochromatic radiation.

reinforcement along the reflected path:

$$n\lambda = d(\sin i + \sin \theta) \tag{8-9}$$

where n is a small integer defined as the order of the radiation. From Equation 8–9, radiation of a particular wavelength striking a reflection grating at an angle of incidence, i, may undergo constructive reinforcement at several angles of θ, depending on the order. The reflection and dispersion of an incident beam of radiation extending from 1200 to 200 nm is illustrated in Figure 8–12. Equation 8–9 also indicates that overlapping of orders will occur; for instance, along the 600 nm first-order reflection angle in Figure 8–12, higher-order 300- and 200-nm radiation will be reinforced.

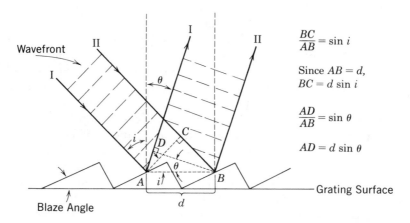

Figure 8-11   Constructive reinforcement of monochromatic radiation by a reflection grating.

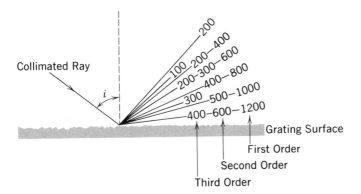

Collimated Ray

i

Grating Surface

First Order

Second Order

Third Order

Figure 8–12 Dispersion of polychromatic radiation by a reflection grating.

8–7   LIGHT SCATTERING

The term "scattering" suggests that a light beam is disrupted and deflected in random directions. This is, in fact, the observed phenomenon, but the details of the process can be related to molecular structure. Several mechanisms are involved depending on the wavelength of radiation relative to the dimensions of the particles of the medium.

If a beam of radiation encounters a particle which has dimensions no more than 10% of the wavelength, the particle will be subject to the oscillating electric field of the wave. If the particle can be "polarized" (partial separation of positive and negative charge), the electric field of the radiation will induce a dipole in the particle. As the wave passes, the induced dipole will oscillate at the same frequency as the radiation. The oscillating dipole thus sets up a secondary field of its own which acts as a source of radiation. This radiation is necessarily of the same frequency as the incident radiation but will be propagated in all directions, and appear to be scattered. All atoms and molecules can cause this type of scattering known as *Rayleigh scattering.* The intensity of Rayleigh scattering is proportional to the fourth power of the frequency, which is why we have blue skies and red sunsets.

Light is scattered from larger particles by a combination of processes. The angular distribution of the radiation is related to the size of the particles, and the intensity of the scattered radiation is related to the concentration of the particles. This type of scattering is the basis for *nephelometry,* which is used to measure the amount of suspended matter in liquids (colloidal precipitates) or gases (smoke and fog).

Still another type of scattering was discovered in 1928 by C. V. Raman, an Indian physicist. By using an intense light source and observing the scattered radiation at right angles, he found new wavelengths both longer and shorter than the wavelength of the incident radiation. This type of scattered radiation, resulting from the *Raman effect,* is only 0.01% as intense as the incident beam, thus the technique was not used much until powerful laser sources became available.

In ordinary (Rayleigh) scattering, the scattered photon is of the same frequency as in the incident beam, as shown in the center of Figure 8–13. Note that the upper ends of the energy arrows lie in a region between the highest vibration level of the ground electronic state and the lowest vibration level of the first excited electronic

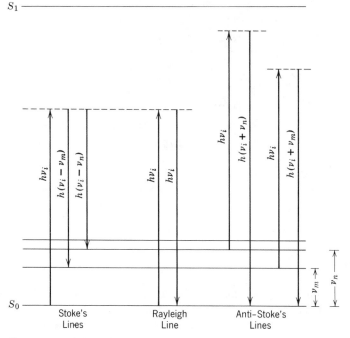

Figure 8–13 Schematic energy level diagram for scattering phenomena showing Rayleigh, Stokes, and anti-Stokes lines (Raman effect).

state. In Raman scattering, the molecule returns to a higher vibrational level than it started from (left arrows in Figure 8–13), therefore the re-emitted photon is of slightly lower energy (frequency), and *Raman lines* are shifted to lower frequencies by amounts equivalent to the vibrational energy differences. At the right side of Figure 8–13 is shown the opposite type of shift in frequency, resulting from a net transition to a lower vibrational level. The set of lower frequency lines are known as *Stokes lines*, and the higher frequency ones are known as *anti-Stokes lines*. Raman spectroscopy provides structural information from vibrational transitions; however, the expensive instrumentation has restricted its use up to the present.

8–8   ROTATION OF PLANE-POLARIZED LIGHT

When we examine the oscillation of a light wave along the direction of propagation from an end-on vantage point, the electric vector looks like Figure 8–14. An ordinary light beam, observed end-on, consists of waves oscillating in random directions like those shown in Figure 8–15a. Each of the vibrational directions can be resolved into two mutually perpendicular directions as shown in Figure 8–15b. The actual vibrational direction is the vector sum of the two

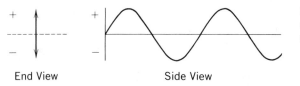

End View   Side View

Figure 8–14 Two views of oscillation of the electric component of radiation.

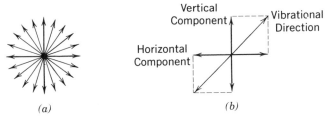

Figure 8–15 Vector diagram of (a) ordinary light, and (b) resolution of a random direction of oscillation into two perpendicular components.

components. Now if this ordinary light is collimated and passed through a crystal which allows only one vibrational orientation to be transmitted, the transmitted light is said to be *plane polarized*. Many natural crystals produce polarized light, but it is most conveniently obtained with commercially available Polaroid materials.

When plane polarized light is passed through substances which are asymmetric (have no plane of symmetry or center of symmetry), the direction of vibrational oscillation of the incident radiation undergoes rotation either to the right (dextrorotatory, +) or to the left (levorotatory, −). Dextrorotation is shown schematically in Figure 8–16.

What causes polarized light to undergo rotation? We are actually considering the rotation of an oscillating electric field vector as it is propagated through the medium. Even though absorption does not occur, it is not unreasonable to assume that there can be some interaction of the oscillating electric field of the radiation with the electric field generated by the electrons of the substance studied. If there is some preferential arrangement of electrons in the material, we might expect the interaction of radiation to be specific and not just random. We might also expect the resultant of the interaction to have specific characteristics, such as preferentially rotating light in one direction more than the other. A model has been proposed to explain preferential rotation in which polarized light is considered to be the vector sum of two oppositely rotating beams of circularly polarized light. In this model as the light is propagated through space, the electric component of each beam traces out a spiral trajectory which is either clockwise or counterclockwise. If the index of refraction is the same for both beams, they will rotate at the same angular velocity (but in opposite directions) as they travel through the medium. Thus, the vector sum of the rotating electric fields always lies along the same

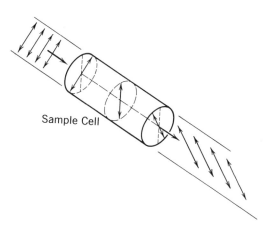

Figure 8–16 Dextrorotation of plane polarized light.

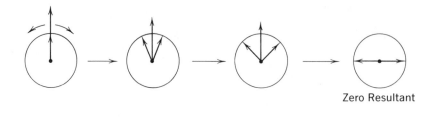

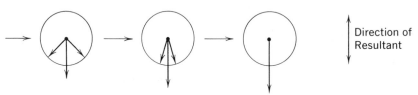

Figure 8–17   Combination of circular components to give plane polarization without rotation.

direction as shown in Figure 8–17. However, the behavior is more complex if the medium contains one of the forms of an asymmetric molecule. In this case the indices of refraction for the two circularly polarized beams are not the same. One of the beams interacts with the asymmetric molecule to a greater extent than the other. There is a difference in the angular velocities of the two beams and the resultant vector is no longer in the same plane as the incident radiation; this is illustrated in Figure 8–18. There is continual rotation of the resultant vector until the two beams emerge from the sample and reach a medium where the angular velocities are the same again. At this time, the resultant vector is again plane polarized and does not rotate. However, the resultant vector produced after the circularly polarized beams emerge from the sample does not have the same orientation as the one which entered the sample; it has undergone rotation. The extent of rotation clearly depends on the type and concentration of molecules present in the sample and the distance which the radiation travels through the sample. We have discussed how the index of refraction and thus the linear velocity of propagation varies with wavelength. The extent of rotation also depends on the wavelength of the polarized light. The sodium D radiation at 589.3 nm is used for most experimental work. The extent of rotation to some degree also depends on temperature and the nature of the solvent. The *specific rotation* is characteristic for an asymmetric substance and is defined as follows:

$$[\alpha]_\lambda^{t_o} = \frac{100\alpha}{l \times c} \qquad (8\text{--}10)$$

where $[\alpha]_\lambda^{t_o}$ = the specific rotation for a substance at temperature t_o, using plane polarized radiation of wavelength λ

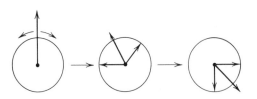

Figure 8–18   Result of interaction of two beams of circularly polarized light with asymmetric molecules.

α = the experimentally measured number of degrees of rotation undergone by the incident radiation

l = pathlength through the sample in decimeters

c = concentration of the sample in grams/100 ml of solution.

For a pure compound,

$$[\alpha]_\lambda^{t_c} = \frac{\alpha}{l \times d} \tag{8-11}$$

where d = density of the liquid in grams per milliliter. Another commonly used
expression in this field is *molecular rotation* which is defined as follows:

$$[M]_\lambda^{t_c} = \frac{[\alpha]_\lambda^{t_c} \times M}{100} \tag{8-12}$$

where M = the molecular weight of the optically active substance. Techniques
which are used to measure the angle of rotation are all classified under the general
title of polarimetry. Polarimetry is used commonly for kinetic and structural
studies involving optically active compounds.

QUESTIONS AND PROBLEMS

8-1. Calculate the following terms. Assume that the index of refraction is 1.000.
(a) The wavelength (in μm) of radiation with frequency = 8.58×10^{13} sec^{-1}.
Ans. 3.50 μm
(b) The wavelength (in cm) of radiation with energy = 7.95×10^{-12} erg/photon.
Ans. 2.50×10^{-5} cm
(c) The wavenumber of radiation with wavelength = 6.00 microns.
(d) The energy per photon (in ergs) of radiation with wavelength = 380 nm.
Ans. 5.23×10^{-12} erg/photon
(e) The frequency (in Hz) of radiation with wavelength = 700 nm.
(f) The wavenumber of radiation with energy = 4.41×10^{-13} erg/photon.
Ans. 2.22×10^3 cm^{-1}

8-2. The so-called sodium D line has a wavelength of 5890 Å in a vacuum. Calculate its
wavelength and frequency.
(a) in air ($n = 1.00027$). *Ans.* 5888 Å, 5.090×10^{14} Hz
(b) when transmitted through a solution with $n = 1.275$. *Ans.* 4620 Å, 5.090×10^{14} Hz

8-3. A meter is defined as 1,650,763.73 wavelengths in vacuum of the $2p_{10} - 5d_5$ transition
in ^{86}Kr. Calculate the following values for this radiation in vacuum ($c = 2.997925 \times 10^{10}$ cm/sec).
(a) wavenumber. *Ans.* 16,507.6373 cm^{-1}
(b) wavelength, in Å. *Ans.* 6057.8021 Å
(c) frequency. *Ans.* 4.9488659×10^{14} Hz

8-4. The angle of incidence is 0° for a certain transmission grating having 6000 lines/cm.
Calculate the wavelengths which would be diffracted at an angle of 30° for the first two
orders. *Ans.* 833.3 nm, 416.7 nm

8-5. At what angles would radiation of 500 nm appear in the first two orders of the grating
in Problem 8-4? *Ans.* 17.5°, 36.9°

8-6. The Raman spectrum of carbon tetrachloride is obtained with a helium-neon laser
(incident radiation at 6328 Å). Stokes lines are observed at 6416, 6456, and 6517 Å.
(a) Calculate the position of the Stokes lines in cm^{-1}.
(b) What are the Raman shifts of these lines?
(c) At what wavenumbers will the anti-Stokes lines appear?

8-7. The specific rotation of sucrose in an aqueous solution at 20°C is +66.5° using the sodium D line. What will be the rotation of a solution containing 50.0 g/l if a 20-cm sample cell is used? *Ans.* +6.65°

8–8. Define or explain the following terms: electromagnetic spectrum, absorption spectrum, line spectrum, continuous spectrum, intensity, wavenumber, photon, luminescence, fluorescence, phosphorescence, refraction, diffraction, dispersion, ground state, triplet state, scattering, specific rotation, plane polarization, refractive index, reciprocal centimeter.

REFERENCES

R. Bauman, *Absorption Spectroscopy*, Wiley, New York, 1962. Excellent advanced level treatment of the field.

N. Bauer and K. Fajans, *Techniques of Organic Chemistry*, A. Weissberger, Editor, Vol. I, Part II, 2nd ed., Wiley-Interscience, New York, 1949; Chapter 20. Excellent discussion of theory of refraction.

R. T. Conley, *Infrared Spectroscopy*, Allyn & Bacon, Boston, Mass., 1966.

W. Heller, *Techniques of Organic Chemistry*, A. Weissberger, Editor, Vol. I, Part II, 2nd ed., Wiley-Interscience, New York, 1949; Chapter 23. Comprehensive discussion of polarized light and its interaction with matter.

H. A. Strobel, *Chemical Instrumentation*, 2nd ed., Addison–Wesley, Reading, Mass., 1973. Very good discussion of optical methods of analysis.

J. D. Winefordner, S. G. Schulman, and T. C. O'Haver, *Luminescence Spectrometry in Analytical Chemistry*, Wiley-Interscience, New York, 1972.

9

QUANTITATIVE ANALYSIS BY ABSORPTION OF ELECTROMAGNETIC RADIATION

9–1   INTRODUCTION

In quantitative absorption studies, a beam of radiation is directed at a sample and the intensity of the radiation which is transmitted is measured. The radiation absorbed by the sample is determined by comparing the intensity of the transmitted beam when no absorbing species is present to the transmitted intensity when there is absorbing species present. The radiant power (i.e., intensity) of a collimated beam is proportional to the number of photons per second passing through a unit cross section. If the photons which strike the sample possess energy equal to that required to cause a quantized energy change, absorption *may* occur. The fraction of the radiant energy absorbed depends on the probability for the energy change involved. The absorption thus decreases the radiant power of the transmitted radiation. Scattering and reflection also lower the power of the radiation; however, for most systems which are studied, these losses are small compared to absorption. The instrumentation which is used to make absorption measurements will be discussed in Chapter 10.

9–2   QUANTITATIVE LAWS

Consider the changes in radiant power which occur as monochromatic radiation passes through the absorption cell in Figure 9–1. We first fill the cell with a "blank" solution, which normally consists of the solvent plus sample constituents *other than the principal absorbing species*. With this "blank" solution in the cell, the radiant power of the transmitted radiation represents the incident radiant power minus that lost by scattering, reflection, and any absorption by the other constituents (normally quite small). We denote this radiant power as I_0, because it serves as a "corrected" *incident* radiant power when the "blank" is replaced by the sample itself (this is a relatively good approximation).

Referring to Figure 9–1, let us consider what happens to the radiation as it passes through segment A of the sample. Using the differential notation of calculus, dI represents the decrease in radiant power in an infinitesimally small layer, db, that is, the amount of radiation absorbed in this layer. We will assume that the absorption of energy requires a physical interaction between a photon and an absorbing species. Therefore, the number of possible "collisions" occurring in this layer is proportional to both the number of absorbing species in the layer and the number of photons passing through. If the number of absorbing species is doubled, the number of collisions is doubled; likewise, doubling the number of photons also doubles the number of collisions. Thus the loss in radiant power, dI,

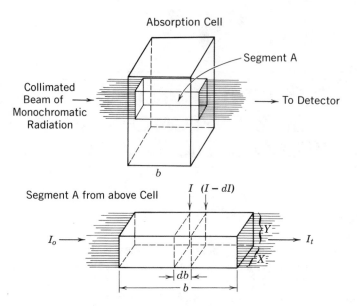

Figure 9–1   The absorption process.

is directly proportional to N (the number of absorbing species) and I (the number of photons per unit cross-sectional area per second). For layer db, the number of absorbing species is given by:

$$N = (6.02 \times 10^{20} \text{ species/mmole}) \, (c \text{ mmole/ml}) \, (db \times X \times Y \text{ ml})$$

where db, X and Y are the linear dimensions of the layer (assume 1 cc = 1 ml). Since X and Y are constant,

$$N = k'c \, db$$

where, $k' = (6.02 \times 10^{20})(X \times Y)$ species-cm^2/mmole.

The number of collisions is proportional to the product $N \times I$ or

$$dI \propto NI = k'Ic \, db$$

therefore,

$$dI = -kIc \, db \tag{9-1}$$

where k is a proportionality constant; the negative sign is introduced because radiant power decreases as db increases. Integration of Equation 9–1 over the entire cell length, b, gives the loss in radiant power due to absorption by the sample. Separating the variables in Equation 9–1 gives:

$$\int_{I_0}^{I_t} \frac{dI}{I} = -k \int_0^b c \, db$$

Solving,

$$\ln \frac{I_t}{I_0} = -kbc$$

and, converting from natural logarithms to base 10 logarithms (designated by "log"), we obtain:

$$2.303 \log \frac{I_t}{I_0} = -kbc$$

or

$$\log \frac{I_t}{I_0} = \frac{-k}{2.303} bc = -\epsilon bc \tag{9-2}$$

where ϵ is defined as the *molar absorptivity* (also called the molar extinction coefficient). If the concentration is given in grams/liter, ϵ is replaced by a, the *specific absorptivity*. The term I_t/I_0 is defined as the *transmittance* (symbol, T) which is the fraction of the incident radiant power transmitted by the sample. The percent transmittance is defined as $100 \times T$. Therefore, from Equation 9-2,

$$\log T = -\epsilon bc \qquad \text{or} \qquad -\log T = \epsilon bc$$

$-\log T$ is also defined as *absorbance* (symbol, A) or optical density; thus:

$$-\log T = A = \epsilon bc \qquad (9\text{-}3)$$

The value of ϵ is characteristic of the absorbing molecule or ion in a particular solvent and at a particular wavelength. The value of ϵ is independent of concentration and the path length of the radiation. Equation 9-3 has been alternately referred to as the Beer–Lambert law, the Bouguer–Beer law, or more simply, *Beer's law*. In the derivation of this law it is assumed that: (1) the incident radiation is monochromatic, (2) the absorbing species act independently of each other in the absorption process, (3) the absorption occurs in a volume of uniform cross section, (4) energy degradation is rapid (no fluorescence), and (5) the refractive index is independent of concentration (not true at high concentration). Equation 9-3 demonstrates that a determination of absorbance or transmittance will yield the concentration if ϵ and b are known.

Contemporary instruments are calibrated to readout in either percent transmittance, or absorbance, or both. The relationships between absorbance, transmittance, and concentration at a given wavelength are illustrated graphically in Figure 9-2. A typical spectrum is plotted in both coordinates in Figure 9-3. Both systems are in common use, and one must realize that absorption peaks appear as deep valleys in a transmittance plot.

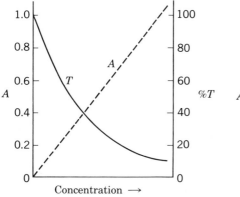

Figure 9-2 Absorbance and transmittance versus concentration at a given wavelength and cell pathlength.

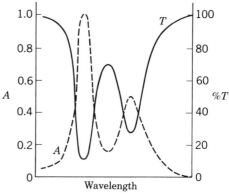

Figure 9-3 Typical absorption spectrum plotted in both absorbance and transmittance.

Example/Problem 9-1. Potassium chromate (K_2CrO_4) in basic solution exhibits an absorption maximum at 372 nm. A basic solution containing 3.00×10^{-5} M K_2CrO_4 transmits 71.6% of the incident radiation at 372 nm when it is placed in a 1.000-cm cell.

(a) What is the absorbance of this solution?

Solution.

$$\text{If } \% \; T = 71.6; \; T = 0.716 \text{ and from Beer's law, } \log \frac{1}{T} = A$$

$$\text{therefore, } A = \log \frac{1}{0.716} = \log 1.396 = 0.145.$$

(b) What is the molar absorptivity of potassium chromate at 372 nm?

Solution.

$$A = \epsilon bc \text{ or } 0.145 = \epsilon (1.000 \text{ cm})(3.00 \times 10^{-5} \text{ mole/liter})$$

$$\text{therefore, } \epsilon = 4.83 \times 10^3 \text{ liter/mole-cm.}$$

(c) What would be the percent transmittance if the cell length were 3.000 cm?

Solution.

$$\log \frac{1}{T} = \epsilon bc$$

$$\log \frac{1}{T} = (4.83 \times 10^3 \text{ liter/mole-cm})(3.000 \text{ cm})(3.00 \times 10^{-5} \text{ mole/liter})$$

$$\log \frac{1}{T} = 0.435$$

$$\text{therefore, } T = 10^{-0.435} = 10^{0.565} \times 10^{-1} = 0.367$$

$$\text{and } \% \; T = 36.7\%.$$

Example/Problem 9–2. Compound X exhibits a molar absorptivity of 2.45×10^3 liter/mole-cm at 450 nm. What concentration of X in a solution will cause a 25% decrease in radiant power for 450 nm radiation when the solution is placed in a 1.000 cm absorption cell?

Solution. If the solution of X causes a 25% decrease in radiant power, then the percent transmittance of the solution is 75%. From Beer's law:

$$\log \frac{1}{T} = \epsilon bc$$

$$\log \frac{1}{0.75} = \log 1.33 = (2.45 \times 10^3 \text{ liter/mole-cm})(1.000 \text{ cm})c$$

$$\text{or } 0.124 = (2.45 \times 10^3 \text{ liter/mole})c$$

$$\text{and } c = 5.06 \times 10^{-5} \text{ mole/liter.}$$

Multiple Component Systems. When systems which contain more than one absorbing component are studied, it is assumed that the species act independently of one another and that their absorbances are additive. Figure 9–4 shows the absorption spectra of components I and II and their mixture. At the absorbance maximum for component I, λ_1, component II also has an appreciable absorbance; at the absorbance maximum for component II, λ_2, component I also absorbs radiation. The absorption spectrum for a mixture of I and II is simply the sum of the two individual curves. We can set up equations for the total absorbance at each wavelength as follows:

At λ_1:
$$A_I^{\lambda_1} = \epsilon_I^{\lambda_1} bc_I \quad \text{and} \quad A_{II}^{\lambda_1} = \epsilon_{II}^{\lambda_1} bc_{II}$$

$$A^{\lambda_1} = A_I^{\lambda_1} + A_{II}^{\lambda_1} = \epsilon_I^{\lambda_1} bc_I + \epsilon_{II}^{\lambda_1} bc_{II} \tag{9-4}$$

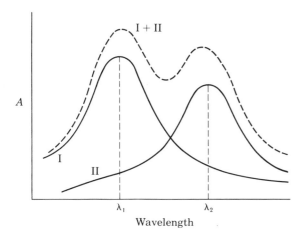

Figure 9-4 Overlapping spectra of two components and the spectrum of a mixture of the two components.

At λ_2:

$$A_I^{\lambda_2} = \epsilon_I^{\lambda_2} b c_I \quad \text{and} \quad A_{II}^{\lambda_2} = \epsilon_{II}^{\lambda_2} b c_{II}$$

$$A^{\lambda_2} = A_I^{\lambda_2} + A_{II}^{\lambda_2} = \epsilon_I^{\lambda_2} b c_I + \epsilon_{II}^{\lambda_2} b c_{II} \tag{9-5}$$

where A^{λ_1} and A^{λ_2} = observed absorbances of the mixture at wavelengths λ_1 and λ_2, respectively;

$A_I^{\lambda_1}$ and $A_I^{\lambda_2}$ = absorbances of component I in the mixture at λ_1 and λ_2;

$A_{II}^{\lambda_1}$ and $A_{II}^{\lambda_2}$ = absorbances of component II in the mixture at λ_1 and λ_2;

$\epsilon_I^{\lambda_1}$, $\epsilon_I^{\lambda_2}$, $\epsilon_{II}^{\lambda_1}$, and $\epsilon_{II}^{\lambda_2}$ = the molar absorptivities of components I and II at λ_1 and λ_2; and

c_I and c_{II} = the respective concentrations of components I and II in the mixture.

The molar absorptivities are determined by obtaining the absorption spectrum of each component separately in a solution of known concentration. Thus the two unknown concentrations are obtained by solving two simultaneous equations (9-4 and 9-5) obtained by measuring A at two different wavelengths. In general, if there are n components, the total absorbance expression at any wavelength, λ, takes the form:

$$A^{\lambda} = \sum_n A_n^{\lambda} = b \sum_n \epsilon_n^{\lambda} c_n \tag{9-6}$$

In principle, n absorbance measurements at n different wavelengths are required to determine the concentration of n components in a mixture. This provides n independent simultaneous equations in n unknowns. If possible, we select wavelengths so that all components except one do not absorb, thus reducing the number of terms in Equation 9-6. In general, select those wavelengths where the absorptivities vary the most.

Example/Problem 9-3. Titanium and vanadium form colored complexes with hydrogen peroxide. Separate solutions containing 5.00 mg of these metals were treated with perchloric acid and hydrogen peroxide and diluted to 100 ml. A third solution was prepared by dissolving 1.00 g of an alloy (containing Ti and V but no other interfering metals) and treating in the same manner as the standard solutions. The absorbances of the three solutions were measured at 410 and

460 nm in 1-cm cells. Calculate the % V and % Ti in the alloy.

Solution	A^{410}	A^{460}
Ti	0.760	0.513
V	0.185	0.250
Alloy	0.715	0.657

Solution.   From the standard solutions

$$A^{410}_{Ti} = a^{410}_{Ti} b c_{Ti}$$
$$a^{410}_{Ti} = 0.760/5 = 0.152$$

Similarly: $a^{460}_{Ti} = 0.103$, $a_V^{410} = 0.037$, $a_V^{460} = 0.050$.
For the alloy solution:

$$\text{At } 410 \text{ nm: } 0.715 = 0.152C_{Ti} + 0.037C_V$$
$$\text{At } 460 \text{ nm: } 0.657 = 0.103C_{Ti} + 0.050C_V$$

Solution of these simultaneous equations yields

$$C_{Ti} = 3.0 \text{ mg}/100 \text{ ml}, C_V = 6.9 \text{ mg}/100 \text{ ml}$$
$$\% \text{ Ti} = 0.30, \% \text{ V} = 0.69$$

9-3   DEVIATIONS FROM BEER'S LAW

Many absorbing systems in dilute solution follow Beer's law rather closely. For some systems, however, absorbance varies in a non-linear way with respect to concentration. Such behavior is called a "deviation from Beer's law." In order to treat these systems, a curved calibration plot may be prepared with samples of known concentration. Then the concentration of an "unknown" can be determined from its absorbance in the same cell following the same procedures used for the standard samples. A typical calibration curve is shown in Figure 9–5.

It is important to examine the reasons for deviations from Beer's law in order to provide some indication of the quantitative limitations of absorption measurements. True deviations from Beer's law occur only in systems where the concentration of the absorbing species is so high that the index of refraction for

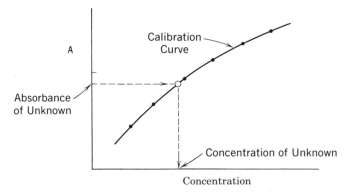

Figure 9–5   Nonlinear calibration curve of absorbance versus concentration at a given absorption maximum.

the absorbed radiation is changed. Consequently, Beer's law methods are used primarily in solutions with concentrations below 10^{-2} M. The lower limit of about 10^{-7} M is determined primarily by the ability of the instrument to detect small differences in radiant power. Apparent deviations from Beer's law may be due both to limitations of instrumentation and to effects of nonsymmetrical chemical equilibrium.

Instrumentation Limitations. Indeterminate instrumental variations which cause apparent deviations include: (1) stray radiation reaching the detector (reflected within the instrument), (2) sensitivity changes in the detector, and (3) power fluctuations of the radiation source and detector amplification system. Double beam operation (to be discussed in Chapter 10) tends to cancel out most of the random causes of deviation.

A more unavoidable instrumental cause of deviation is the necessity of working with a band of wavelengths rather than truly monochromatic radiation. Unless the molar absorptivity is invariant within the wavelength band used, the absorbance measured is not the proper average absorbance over the band. What the detector measures is the average intensity, whereas what is wanted is the average of the log intensity. The average of the logarithms of a set of values is not equal to the logarithm of the average. The greater the slope of the absorption curve through the wavelength band, the greater the deviation.

Figure 9–6 demonstrates that the shape of the calibration curve often depends on the band width. Two wavelength bands of equal width are designated λ_1 and λ_2. The best wavelength for quantitative analysis is λ_1, for two reasons. First, at the absorption maximum the change in absorbance with concentration is at a maximum; this yields greater sensitivity and higher accuracy. Second, within this band the molar absorptivity is relatively constant and a linear calibration curve is obtained as in Figure 9–6b. However, if a wavelength is selected on the side of an absorption peak, for example λ_2 in Figure 9–6a, the molar absorptivity varies across the band. At this wavelength the system does not follow Beer's law and a curved calibration is obtained, as in Figure 9–6c.

Narrow band widths are obviously desirable for best accuracy, but as the band is decreased, less energy reaches the detector. Consequently, there is always a compromise between accuracy, sensitivity, and detector requirements.

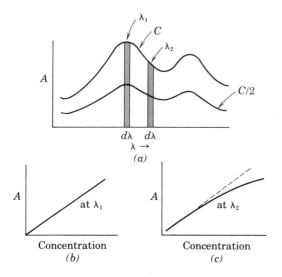

Figure 9–6 Effect of finite bandwidth on measured absorbance. (a) one component at two concentrations, (b) linear calibration curve at λ_1, (c) nonlinear calibration curve at λ_2.

We normally assume that all absorption cells used in a given analysis have the same transparency. To verify this assumption we may fill two cells with solvent; using a single beam instrument we then set $\% \ T = 100$ for one cell and measure the $\% \ T$ of the other cell. The second cell should also read 100% T at every wavelength studied. The difference between the two is the *transparency correction* (converted to units of absorbance). This correction is usually negligible above 300 nm wavelength.

Chemical Equilibrium. An absorbing species which is involved in a chemical equilibrium may exhibit "apparent deviations" from Beer's law. A few examples of such systems are presented below.

1. Let us consider an aqueous solution of a weak acid HB which has an absorption maximum at λ_1. The anion of the acid B^- is non-absorbing at λ_1. HB is involved in the equilibrium:

$$HB \rightleftharpoons H^+ + B^-$$

for which the mass action expression is:

$$K_a = \frac{[H^+][B^-]}{[HB]}$$

The analytical concentration, C_{HB}, includes both forms: $C_{HB} = [HB] + [B^-]$. As long as the ratio $[HB]/C_{HB}$ remains constant, Beer's law is valid when using either $[HB]$ or C_{HB}. But this ratio depends on the pH of the solution:

$$\text{fraction undissociated} = \frac{[HB]}{C_{HB}} = \frac{[HB]}{[HB] + [B^-]}$$

$$= \frac{[HB]}{[HB] + K_a[HB]/[H^+]} = \frac{1}{1 + K_a/[H^+]}$$

If $[H^+] \gg K_a$, the acid exists primarily as HB, and there is no problem. If $[H^+] \ll K_a$, the acid is highly dissociated with a correspondingly small absorbance. At a constant pH in the intermediate region (buffered solution), the fraction dissociated does not vary with total concentration and absorbance is proportional to C_{HB}. However, in unbuffered solutions the fraction dissociated changes with pH, which in turn is a function of total concentration, C_{HB}. Such a system will show an apparent deviation from Beer's law if C_{HB} is used as the concentration.

Example/Problem 9–4. An aqueous solution of a weak acid, HB ($K_a = 1.00 \times 10^{-5}$), absorbs ultraviolet radiation with a maximum at 280 nm, $\epsilon = 975$. B^- does not absorb. Assume that you start with a $2.000 \times 10^{-3} \ F$ solution of HB and then perform three successive one to one dilutions. The absorbance of each of these solutions is measured. Would the system show a positive or negative apparent deviation from Beer's law? A 1.000-cm absorption cell is used for all measurements.

Solution. First determine the equilibrium concentration of HB in each solution using K_a. For example, for the first solution:

$$\frac{[H^+][B^-]}{[HB]} = 1.00 \times 10^{-5}$$

$$C_{HB} = 2.000 \times 10^{-3} \ F \quad \text{and} \quad [HB] = C_{HB} - [B^-]$$

Since $[H^+] = [B^-]$,

$$\frac{[H^+]^2}{2.00 \times 10^{-3} - [H^+]} = 1.00 \times 10^{-5}$$

Solving quadratically,

$$[H^+] = 1.37 \times 10^{-4} \, M \quad \text{and} \quad [HB] = 1.86 \times 10^{-3} \, M$$

From Beer's law,

$$A = \epsilon b c_{HB} = (9.75 \times 10^2)(1.000)(1.86 \times 10^{-3})$$
$$A = 1.813$$

The other solutions are treated in a similar manner with the results summarized in Table 9–1. With increasing analytical concentration of HB, C_{HB}, absorbance shows a positive deviation from Beer's law. However, the absorbance is linear with the "true" equilibrium concentration; thus the deviation is an "apparent deviation." As the concentration is increased, a larger fraction of the acid remains undissociated.

Table 9–1 Data for Example/Problem 9–4

C_{HB}, M	[HB], M	[HB]/C_{HB}	pH	A	A/C_{HB}	$A/$[HB]
2.000×10^{-3}	1.86×10^{-3}	0.930	3.86	1.813	907	975
1.000×10^{-3}	9.05×10^{-4}	0.905	4.02	0.881	881	973
5.00×10^{-4}	4.34×10^{-4}	0.868	4.18	0.422	845	972
2.50×10^{-4}	2.05×10^{-4}	0.820	4.34	0.200	800	976

2. A similar approach to that used with the weak acids is applied in studying inorganic complexes. For example, in systems such as Cu(II)-ammonia or Co(II)-ethylenediamine there is a series of complexes formed. As an example, in the Co(II)-ethylenediamine system in aqueous solution, possible complexes might be $Co(en)(H_2O)_4^{+2}$, $Co(en)_2(H_2O)_2^{+2}$ and $Co(en)_3^{+2}$, where en represents an ethylenediamine molecule. Each of these complexes exhibits a different absorption maximum, and there is considerable overlap of the absorption spectra of some of the complexes. Therefore, in the absence of a large excess of ligand (en), the total analytical concentration of Co(II) is distributed (according to respective equilibrium constants) among more than one absorbing species. Consequently, the assumption that the total analytical concentration of Co(II) exists as a single complex ion will cause "apparent deviations" from Beer's law. However, if enough excess ligand is present the only complex in the system is the one of highest coordination (i.e., $Co(en)_3^{+2}$). In this case Beer's law could be applied to the spectra of this one complex and no deviation would be noted. This technique of having excess ligand present is commonly used to measure the concentration of inorganic complexes spectrophotometrically. It should also be pointed out that in most complexing media, metal ions are in competition with protons to bond with various ligands. Therefore, it is often necessary to adjust the pH to a rather high value to insure that the metal ion complexation reaction is essentially complete.

3. Aqueous solutions of potassium chromate offer another example of an "apparent deviation" from Beer's law which may be attributed to unsymmetrical chemical equilibria. In this system, CrO_4^{-2}, $HCrO_4^-$ and $Cr_2O_7^{-2}$ all absorb radiation. The relative amounts of the three species depend on the total concentration of Cr(VI) and the pH. This is illustrated by considering the equilibria:

$$2CrO_4^{-2} + 2H^+ \rightleftharpoons 2HCrO_4^- \rightleftharpoons Cr_2O_7^{-2} + H_2O$$

The overall mass action expression for this system is given by

$$\frac{[Cr_2O_7^{-2}]}{[CrO_4^{-2}]^2[H^+]^2} = K_{equilibrium}$$

When aqueous solutions of chromate or dichromate are diluted with water, there are apparent deviations from Beer's law. If the solutions are *strongly* acidic, essentially all of the Cr(VI) is in the form of $Cr_2O_7^{-2}$ and Beer's law is followed. Also, if the system is made *strongly* alkaline, the Cr(VI) is present as CrO_4^{-2} and Beer's law is again followed.

The three example systems discussed above illustrate the importance of understanding the chemical system under study and controlling conditions so that quantitative absorption measurements are meaningful.

9–4 PHOTOMETRIC ERROR

Under the best of conditions, the accuracy of photometric procedure is limited at both low and high values of absorbance. At low absorbances, the intensity of the incident and transmitted beams are nearly the same; thus there is a relatively large error in the absorbance which the instrument perceives as the *difference* in the intensities of the two beams. At high absorbances, so little energy is transmitted that it cannot be measured accurately. In most photoelectric instruments, the reading is directly proportional to the intensity of radiation striking the photocell (the scale is linear in T), and it can be assumed that there is a constant error in the reading, for example, the width of the meter needle or the pen line.

Let us assume that the error, dc, in the determination of concentration, c, is due entirely to the reading error, dT. From Beer's law:

$$c = -\frac{1}{ab} \log T \tag{9-7}$$

Differentiating Equation 9–7 yields:

$$dc = -\frac{1}{ab} \times \frac{\log e}{T} dT \tag{9-8}$$

Dividing Equation 9–8 by Equation 9–7 yields:

$$\frac{dc}{c} = \frac{\log e \, dT}{T \log T} \tag{9-9}$$

Equation 9–9 may be used to determine the relative error in concentration at various transmittances when the uncertainty in transmittance measurements is known. Figure 9–7 shows a plot of the relative error in concentration versus transmittance for potassium permanganate solutions at 523 nm where $\epsilon = 2400$. It is assumed that the uncertainty in reading the transmittance, dT, is $\pm1\%$ and that the path length is 1 cm. By inspection, the minimum relative error requires a transmittance of approximately 0.4.

A more accurate value of transmittance at which the relative error is minimized is obtained by differentiating Equation 9–9 and setting the derivative equal to zero:

$$\frac{d}{dT}\left[\frac{\log e \, dT}{T \log T}\right] = 0 = \log T + \log e \tag{9-10}$$

Therefore, $\log T = -\log e = 0.4343$ or $T = 0.368$

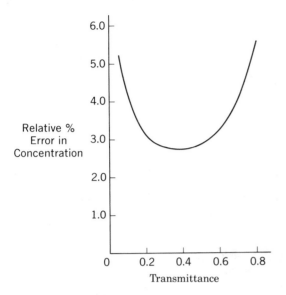

Figure 9–7 Photometric error versus transmittance for potassium permanganate solutions. Uncertainty in transmittance is ± 0.01.

Thus, the optimum $\% T$ for best accuracy is 36.8% (or $A = 0.43$), but the curve (Figure 9–7) shows that reasonable accuracy can be expected between $T = 15\%$ ($A = 0.8$) to $T = 65\%$ ($A = 0.2$). If the measured absorbance is outside this range, the solution should be diluted (or concentrated), a cell with a more appropriate light path used, or a more appropriate wavelength chosen.

Equation 9–10 must be regarded as a general approximation based on the assumption that the error, dT, is constant throughout the range. This is not necessarily true for all spectrophotometers, and the manufacturer may quote some other value for the optimum $\% T$.

9–5 INTENSITY OF LUMINESCENCE

The intensity of fluorescence (or phosphorescence) is proportional to the radiation which was absorbed by the species.

$$I_F = K(I_0 - I_t) \tag{9–11}$$

where I_F, I_0 and I_t are the intensities of the fluorescent, incident, and transmitted beams, respectively, and K is a constant which is a function of the quantum efficiency of the fluorescent process. Beer's law is given as:

$$I_t = I_0 10^{-abc} \tag{9–12}$$

Substitution of Equation 9–12 into Equation 9–11 yields

$$I_F = KI_0(1 - 10^{-abc}) \tag{9–13}$$

The exponential term in Equation 9–13 can be expanded as a Taylor series, giving:

$$I_F = KI_0 \left[2.3abc - \frac{(2.3abc)^2}{2!} + \frac{(2.3abc)^3}{3!} + \cdots \right] \tag{9–14}$$

Now if $abc < 0.05$, all the bracketed terms in Equation 9–14, except the first, become negligible, and we may write:

$$I_F = K'I_0abc \qquad (9\text{–}15)$$

Thus, the intensity of luminescence is proportional to the intensity of the incident beam, the absorptivity of the species, the path length in the cell, and the concentration of the species (within the limit that $abc < 0.05$).

QUESTIONS AND PROBLEMS

9–1. The nucleic acid adenine exhibits a molar absorptivity of 13,100 at 263 nm. What is the molar concentration of an adenine solution which exhibits 75.0% transmittance when 2.00 ml of the solution is placed in a 1.000-cm absorption cell?

Ans. 9.54×10^{-6} mole/liter

9–2. A 1.28×10^{-4} M solution of potassium permanganate has a transmittance of 0.500 at 525 nm in a 1.00-cm cell.
(a) What is the absorbance of this solution? *Ans.* 0.301
(b) If the concentration were doubled, what would be the absorbance and transmittance? *Ans.* $A = 0.602$; $T = 0.250$
(c) What concentration would have a transmittance of 0.750 in this cell?
Ans. 5.32×10^{-5} M
(d) What cell path would give the optimum accuracy for the concentration first quoted? *Ans.* 1.44 cm

9–3. A 2.00×10^{-4} M solution of a certain compound has an absorbance of 1.00 in a 1.00-cm cell at 320 nm.
(a) What is the molar extinction coefficient at this wavelength? *Ans.* 5.00×10^{3}
(b) What fraction of the incident light is transmitted through the cell? *Ans.* 0.100
(c) What fraction would be transmitted through a 4.00×10^{-4} M solution of this compound in the same cell? *Ans.* 0.010

9–4. A 500-mg sample containing a colored component X is dissolved and diluted to 500 ml. The absorbance of an aliquot of this solution, measured at 400 nm in a 1.00-cm cell, is 0.900. 10.0 mg of pure X is dissolved in 1 liter of the same solvent and the absorbance measured in a 0.100-cm cell at the same wavelength is 0.300. What is the % X in the first sample? *Ans.* 0.300

9–5. A sulfuric acid solution of cupric sulfate is analyzed for copper by transferring exactly 5.00 ml to a 1.000-cm absorption cell. The measured % transmittance is 75.3% at the wavelength of maximum absorption. A 1.000 ml portion of standard 0.01000 M cupric sulfate is added to the cell (original 5.00 ml of unknown still present) and the % transmittance changes to 62.5%. What is the concentration of the cupric ion (in moles/liter) in the original solution? *Ans.* 2.03×10^{-3} M

9–6. Standard solutions of Y were used to construct a calibration curve: 1.00 ppm, $A = 0.26$; 2.00 ppm, $A = 0.44$; 3.00 ppm, $A = 0.54$; 4.00 ppm, $A = 0.61$. If an unknown solution of Y had an absorbance of 0.50 under the same conditions, what is the concentration of Y?

9–7. The molar absorptivity of benzoic acid (Mol. Wt. = 122.1) in methanol at 275 nm is about 1950. If it is desired to use an absorbance not exceeding 1.25, what is the maximum allowable concentration in g/l that can be used in a 2.00-cm cell?

9–8. A weak acid with a dissociation constant of $K_a = 1.00 \times 10^{-4}$ shows an absorbance of 1.500 when a 0.100 F solution is measured at 500 nm in a 1.00-cm cell. The anion is colorless.
(a) What is the absorbance of a 1.00×10^{-3} F solution under the same conditions?
(b) Plot a Beer's law curve (A versus F) for this acid over the range from 0.100 F to 1.00×10^{-5} F.
(c) Plot a Beer's law curve (A versus F) for this acid in a buffered solvent (pH = 1.50) over the same range.

9–9. Estimate the K_a of a weak acid from the data below. One-gram samples are dissolved in equal quantities of the various buffers and all solutions are measured under the same conditions. The anion of the acid is the only substance which absorbs at the wavelength used.

pH:	4	5	6	7	8	9	10	11
A:	0.00	0.00	0.06	0.39	0.95	1.13	1.18	1.18

9–10. (a) Estimate the K_a of a weak acid from the data below. All of the various buffered solutions are 1.00 millimolar in sample, and all solutions were measured under the same conditions. The anion of the acid is the only substance which absorbs at the wavelength used. A is the absorbance.

pH:	4	5	6	7	8	9	10	11
A:	0.000	0.000	0.100	0.750	1.000	1.150	1.250	1.250

Ans. $K_a = 1.5 \times 10^{-7}$

(b) What is the transmittance of the sample at pH 4 using the conditions above?

Ans. $A = 0; T = 1.00$

(c) What is the value of the molar absorptivity for the anion at this wavelength if a 1.00-cm cell is used? *Ans.* 1.25×10^3

9–11. P and Q, which are colorless, form the colored compound PQ: $P + Q \rightleftharpoons PQ$. When 2.00×10^{-3} moles of P are mixed with a large excess of Q and diluted to 1 liter, the solution has an absorbance which is twice as large as when 2.00×10^{-3} moles of P are mixed with 2.00×10^{-3} moles of Q and treated similarly. What is the equilibrium constant for the formation of PQ? *Ans.* 1.00×10^3

9–12. Absorbances were measured for three solutions containing Y and Z separately and in a mixture, all in the same cell. Calculate the concentrations of Y and Z in the mixture.

Absorbance

	475 nm	670 nm
0.001 M Y	0.90	0.20
0.01 M Z	0.15	0.65
Mixture	1.65	1.65

Ans. $C_Y = 1.48 \times 10^{-3} M; C_Z = 2.08 \times 10^{-2} M$

9–13. The molar absorptivity of component A is 3070 at 520 nm and 2160 at 600 nm. The molar absorptivity of component B is 220 at 520 nm and 1470 at 600 nm. An unknown solution containing both A and B is analyzed spectrophotometrically in a 1.000-cm cell. The $\% T$ is 54.4% at 520 nm and 35.0% at 600 nm. What are the molar concentrations of A and B in the unknown?

Ans. $C_A = 7.16 \times 10^{-5} M; C_B = 2.06 \times 10^{-4} M$

9–14. A 2.00-ml portion of a Cu(II) solution containing $2F$ NH_3 is placed in a 1.00-cm absorption cell. The absorbance at a certain wavelength is 0.60. Then a 1.00-ml portion of 0.0100 F $CuSO_4$ is added to the first solution in the absorption cell. The absorbance now reads 0.80. What is the concentration of Cu(II) in the first solution?

Ans. 0.0050 M

9–15. Salicylic acid and ferric ion form a highly colored complex with maximum absorption at 525 nm. However, at pH 2.4, the EDTA complex of Fe(III) is much stronger than the salicylate complex

$$Fe-Sal + EDTA \rightarrow Fe-EDTA + Sal$$

A 10.0-ml aliquot of a ferric solution was diluted to 100 ml with a pH 2.4 buffer. After the addition of 1 ml of 6% salicylic acid, the solution was titrated with 0.100 F EDTA. During the titration, portions of the solution were transferred to an absorption cell, the absorbance at 525 nm measured, and the solution returned to the titration beaker. The following readings were obtained.

ml EDTA:	4.2	4.4	4.6	4.8	5.0	5.2	5.4
A:	1.13	0.83	0.53	0.23	0.10	0.08	0.08

How many mg of iron were in the 10-ml aliquot?   *Ans.* 27.4 mg

9–16. Consider the general titration reaction, carried out as described in Problem 9–15.

$$X + T \rightarrow P$$

where T is the titrant and P is the product. Sketch the titration curve (absorbance versus ml titrant) for each of the following situations:

	ϵ_X	ϵ_T	ϵ_P
(a)	2000	0	500
(b)	0	500	0
(c)	0	0	1000
(d)	0	800	200
(e)	2000	500	0
(f)	0	200	1000
(g)	200	1000	200

REFERENCES

R. P. Bauman, *Absorption Spectroscopy*, Wiley, New York, 1962. Good general discussion of quantitative methods.

L. Meites and H. C. Thomas, *Advanced Analytical Chemistry*, McGraw-Hill, New York, 1958; Chapter 8. Good treatment of quantitative laws and limitations.

10

INSTRUMENTATION
FOR
SPECTROMETRY

10–1   INTRODUCTION

We have discussed the theoretical principles of absorption of electromagnetic radiation in Chapters 8 and 9. We shall now consider how these principles are applied to the solution of chemical problems. In this chapter it is not our purpose to provide a detailed examination of all the electronic components that are used for absorption measurements. However, we do hope to take the student somewhat beyond the "black-box, twiddle-the-knobs" level of understanding of some typical instruments and how they function.

The instruments that are used to study the absorption or emission of electromagnetic radiation as a function of wavelength are called "spectrometers" or "spectrophotometers." The optical and electronic principles employed in these instruments are basically the same for all of the regions of the spectrum from the vacuum ultraviolet to the far infrared. There are, however, some important differences in the specific components used in the various regions.

The essential components of a spectrophotometer include: (1) a stable source of radiant energy, (2) a system of lenses, mirrors, and slits which define, collimate (make parallel), and focus the beam, (3) a monochromator to resolve the radiation into component wavelengths or "bands" of wavelengths, (4) a transparent container to hold the sample, and (5) a radiation detector with an associated readout system (meter or recorder). Commercial instruments may be very complex, but all spectrophotometers represent variations of the simple diagram in Figure 10–1.

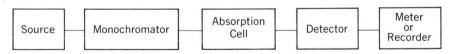

Figure 10–1   Block diagram of a spectrophotometer.

10–2   SOURCES OF RADIANT ENERGY

Sources of radiant energy consist of materials that are excited to high energy states by a high voltage electric discharge or by electrical heating. As the materials return to lower energy states or their ground states, they emit photons of characteristic energies corresponding to ΔE, the energy difference between the excited and lower quantum states. Some materials have numerous energy levels which are so close together that the wavelengths of emitted radiation take the form of a continuum of radiation extending over a rather broad region.

An ideal source of radiation for absorption measurements would emit a continuous spectrum of high, uniform intensity over the entire wavelength range of interest. Unfortunately, the intensity of a real source varies with wavelength as shown for a typical source in Figure 10–2. A change in the electrical power which provides energy for the source shifts the intensity-wavelength curve; consequently, a stable power supply is needed for sequential comparative absorption measurements when single-beam instruments are used. Otherwise, the intensity of the incident radiation (I_0 in Beer's law) may vary between the time of standardization of the instrument and the time of the measurement of the intensity transmitted by the sample (I_t), causing an error in the measured transmittance of the sample. Double-beam instruments are designed so that I_0 and I_t are measured and compared simultaneously; in these instruments, high source stability is not so critical. The source must also furnish sufficient intensity over the entire wavelength range to allow detection by the appropriate device.

Sources of Ultraviolet Radiation. The hydrogen lamp and deuterium lamp are the most common sources of ultraviolet radiation. They consist of a pair of electrodes which are enclosed in a glass tube provided with a quartz window and filled with hydrogen or deuterium gas at low pressure. When a stabilized high voltage is applied to the electrodes, an electron discharge occurs which excites other electrons in the gas molecules to high energy states. As the electrons return to their ground state they emit radiation which is continuous in the region roughly between 180 and 350 nm. Similarly, a xenon discharge lamp is used as a source of ultraviolet radiation. The xenon lamp produces higher intensity radiation, but it is not as stable as the hydrogen lamp. It also emits visible radiation which may appear as stray radiation in ultraviolet applications.

For atomic absorption spectrophotometry, a source that generates only sharp lines of the element in question is desired. The hollow cathode lamps which provide a characteristic emission spectrum for an element will be described in Section 13–7.

Sources of Visible Radiation. A tungsten filament lamp is the most satisfactory and inexpensive source of visible and near infrared radiation. The filament is heated by a stabilized d-c power supply, or by a storage battery. The tungsten filament emits continuous radiation in the region between 350 and 2500 nm. The carbon arc provides more intense visible radiation, however it is seldom used.

Sources of Infrared Radiation. The Globar and Nernst glower are the primary sources of infrared radiation. The Globar is a silicon carbide rod heated to approximately 1200°C. It emits continuous radiation in the 1- to 40-μm region. The Globar is an extremely stable source with an intensity profile similar to that shown in Figure 10–2. The Nernst glower is a hollow rod of zirconium and yttrium oxides heated to approximately 1500°C by electric current. It emits radiation in the range between 0.4 and 20 μm. It is not as stable a source as the Globar, but the Globar requires water cooling.

10–3   MONOCHROMATORS

As we have indicated, the sources of radiation commonly used emit *continuous* radiation over wide ranges of wavelengths. However, *narrow* band widths have

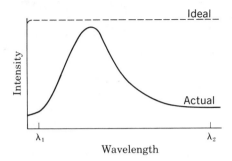

Figure 10–2 Source intensity versus wavelength at a given temperature.

many advantages: (1) narrow band radiation will allow the resolution of absorption bands which are quite close to each other, (2) with narrow band radiation a peak may be measured at its absorption maximum, thus increasing the sensitivity, and (3) the absorption of narrow band radiation will tend to show greater adherence to Beer's law because only that radiation which can be absorbed is measured.

Clearly, we must employ devices which resolve wide band polychromatic radiation from the source into narrow bands or, even better, monochromatic radiation. There are two types of resolving devices in use today, *filters* and *monochromators*. Filters prepared from special materials allow transmission of only limited wavelength regions while absorbing most of the radiation of other wavelengths. Filters typically transmit radiation with an effective band width of from 20 to 50 nm. The *effective band width* is defined as the range of wavelength over which the transmittance is at least one-half of its maximum value. This is illustrated in Figure 10–3. Monochromators, on the other hand, typically resolve polychromatic radiation into effective band widths varying from 35 nm to 0.1 nm. In years past, filter instruments were common because of the high cost of monochromators. However, at the present time, spectrophotometers which incorporate monochromators can be purchased for as little as $300.

As the name implies, a monochromator resolves polychromatic radiation into its individual wavelengths and isolates these wavelengths into very narrow bands. The components of a monochromator include: (1) an entrance slit which admits polychromatic radiation from the source; (2) a collimating device, either a lens or a mirror; (3) a dispersion device, either a prism or grating, which resolves the

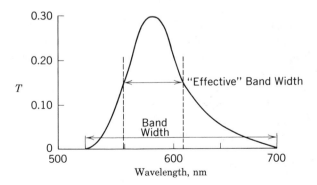

Figure 10–3 Effective band width of a band of radiation.

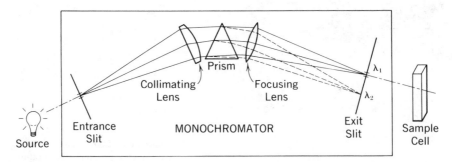

Figure 10–4   Schematic prism monochromator.

radiation into component wavelengths; (4) a focusing lens or mirror; and (5) an exit slit. All of the components in the monochromator must be transparent in the wavelength range to be studied, and the entire assembly is mounted in a light-tight box. A monochromator employing a prism for dispersion is shown schematically in Figure 10–4. The effective band width of radiation emerging from the monochromator depends on several factors, including the dispersing element and the slit widths of both the entrance and exits slits. Narrow slit widths isolate narrow bands; however, the slit width also limits the intensity which reaches the detector. Therefore, the minimum band width may be determined by the sensitivity of the detector. Since the effectiveness of the dispersing element is so important, let us consider the two kinds most widely used, namely, prisms and gratings.

Prisms.   The prism shown in Figure 10–4 resolves the polychromatic radiation into small bands of wavelengths each emerging from the prism at slightly different angles. To direct a particular wavelength of resolved radiation through the exit slit, the prism is rotated until the desired wavelength (or more correctly, a wavelength band centered about this wavelength) is focused on the exit slit.

We have already discussed the principles of refraction and defined refractive index in Chapter 8. High resolution of polychromatic radiation (e.g., good angular separation of the wavelengths) requires that the *dispersion* be as large as possible. Dispersion is quantitatively defined as $d\theta/d\lambda$, the change in angle of deviation with respect to the change in wavelength. The principal disadvantage of a prism is that the dispersion varies with wavelength (see Section 8–5). A prism's dispersion is greater for wavelengths close to its own absorption bands. Consequently, since most prism materials absorb low wavelength ultraviolet radiation, dispersion increases with decreasing wavelengths. For example, quartz prisms exhibit high dispersion and thus excellent resolution around 200 nm, however, dispersion and resolution are poor in the vicinity of 700 nm. In the infrared region, prisms typically absorb at the longer wavelengths and, correspondingly, dispersion and resolution increase with increasing wavelength.

In terms of resolution, the prism performs best at wavelengths near its own absorption band. However, as the prism absorption band is approached, the total transmitted radiant power decreases. Wider slits may be required in order to have enough radiant power to operate the detector. But wider slits pass broader wavelength bands with a significant loss of resolution. Consequently, high resolution requires the most favorable combination of dispersion characteristics of the prism material and the slit widths of the monochromator.

Prism Mountings.   The common 60° Cornu quartz prism, illustrated in Figure

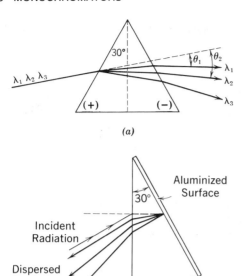

(a)

(b)

Figure 10–5 Dispersion of polychromatic radiation by (a) a Cornu prism and (b) a Littrow prism.

10–5a, is composed of one-half right-handed quartz (+ optical activity) and one-half left-handed quartz (− optical activity). Cementing the two forms of quartz together in this manner eliminates birefringence of radiation as it passes through the prism (i.e., one image is formed rather than two, which would be the case with normal anisotropic quartz). The Littrow prism is a 30° prism which passes radiation in both directions by reflection from an aluminized or silvered reflecting face. The Littrow prism is used in many commercial instruments and is illustrated in Figure 10–5b.

Prism Materials.   The prism material used in ultraviolet-visible and infrared monochromators must be selected wisely for best performance. Both transparency and dispersion should be considered. Silica (SiO_2) prisms of various types are used in the ultraviolet region. Quartz (a crystalline form of silica) and fused silica prisms transmit radiation down to approximately 200 nm even though a weak absorption band appears around 245 nm. High-grade silica transmits down to 185 nm. At the higher end, silica is transparent into the near infrared (3.3 μm), but exhibits very low dispersion in the visible region. Fluorite (CaF_2) is transparent to 125 nm and may be used in vacuum ultraviolet monochromators. Ordinary flint glass (containing lead) shows excellent transmission and dispersion in the visible and near infrared regions.

Ionic crystalline materials are used in the infrared region. The cut-offs, that is, regions of absorption, for these materials are predictable in terms of the masses of the atoms since absorption frequencies vary inversely with mass. Light atoms absorb at lower wavelengths and heavy atoms at higher wavelengths, thus LiF absorbs strongly around 5.6 μm and is used just below this, whereas CaF_2 absorbs strongly at 8.3 μm and is used for the region from 5.5 to 8.0 μm. Sodium chloride transmits throughout the visible and infrared regions to approximately 16 μm; however, it shows poor dispersion in the region from 1.0 to 5.0 μm. Sodium chloride has been the most commonly used prism material in infrared instruments. Unfortunately, like many salts, it is attacked by water vapor. Potassium bromide cannot be used for the near infrared; however it is quite satisfactory for the region from 14.3 to 23.5 μm. The mixed crystalline material, TlBr–TlI (commonly called KRS-5) is transparent from about 1 μm to approximately 35.7 μm; it is insoluble in water although its high reflectivity, tendency for plastic flow, toxicity, and high cost are distinct disadvantages. In the past, KRS-5 has been used in the region just above the KBr region; however, the lower cost CsBr is also used in this region. Cesium bromide is excellent for the region from 14.3 to 35.7 μm; unfortunately, it is highly susceptible to attack by water vapor and is quite expensive in comparison to KBr and NaCl. Cesium iodide extends the range of a prism monochromator to approximately 55.5 μm; however, grating monochromators are preferred in this region.

The approximate transmission limits of some commonly used materials are summarized in Table 10–1.

Table 10–1   Approximate Transmission Limits of Some Optical Materials

Material	Wavelength Range, μm		
Glass	0.4	to	2
Quartz	0.2		3.3
Sodium chloride	0.3		15
Lithium fluoride	0.2		5
Calcium fluoride	0.2		12
Silver chloride	0.4		25
Potassium bromide	0.3		30
KRS-5	1		30
Cesium bromide	0.3		50
Cesium iodide	0.3		70

Gratings.   The dispersion of radiation caused by a diffraction grating was discussed in Section 8–6. Reflection gratings exhibit a *constant* angular dispersion over the entire region of radiation dispersed. For instance, $d\theta/d\lambda$ is constant from 200 to 800 nm for the gratings commonly used in the visible and ultraviolet regions. The angular dispersion of a quartz prism changes ninetyfold over the same region. Constant angular dispersion gives gratings a distinct advantage over prisms. With a constant width exit slit, the grating monochromator supplies a constant band width over the entire range of application, whereas the band width of a prism instrument shows tremendous variation. With reflection gratings, there is no loss in intensity due to absorption by the optical material, as is the case with prisms. By optimizing the blaze angle, approximately 75% of the energy can be concentrated in the first-order spectrum. Another advantage is that gratings are less affected by water vapor.

Wavelength Calibration.   The accuracy of the wavelength dial of the instrument may be verified by running the spectrum of a substance which has known absorption maxima and comparing the known values with observed values. The mercury arc produces a sharp line at 546.1 nm and the hydrogen discharge tube provides a fairly intense line at 656.3 nm; either of these may be used to calibrate a spectrophotometer. However, it is desirable to have a number of calibration points. A didymium glass filter has nine accurately known, narrow absorption bands in the region 441 to 1067 nm. A polystyrene film is commonly used to calibrate infrared instruments; polystyrene has a large number of very precisely known absorption bands.

Lenses and Mirrors.   Radiation is collimated and focused by lenses and mirrors. Materials used for lenses must, of course, be transparent to the radiation being used. In the infrared region mirrors are used because most materials are not sufficiently transparent to infrared radiation and cause significant energy losses.

10–4   **SAMPLE HANDLING**

Sample Containers.   Samples to be studied in the ultraviolet or visible region are usually gases or solutions and are put in cells or cuvettes. Quartz or fused silica cells are used in the ultraviolet region; ordinary glass or more expensive quartz is used in the visible region. Gas cells vary in path length from 0.1 to 100 mm, whereas solution cells have typical path lengths of 1 to 10 cm. Microcells with a beam condenser are available for extremely small samples. The windows of the absorption cells must be kept scrupulously clean; fingerprint smudges and traces of previous samples may cause considerable error in quantitative measurements. Quartz and glass cells may be cleaned by rinsing with water or, if more drastic measures are required, with detergent solutions or hot nitric acid.

Samples for infrared analysis may be gases, liquids, or solids. Infrared gas cells consist of cylindrical glass tubes with NaCl, KBr, or CaF_2 windows. The path lengths vary from a few centimeters to several meters (by multiple reflections in the cell). Liquids are studied either as thin films of the pure compound (commonly referred to as "neat") or as solutions between NaCl, KBr, or CaF_2 salt plates. The plates are separated by 0.005 to 0.1 mm for "neat" liquids and 0.1 to 1 mm for solutions. Infrared cells clearly must *never* come in contact with water. They should be cleaned with organic solvents only. Solids are examined in the infrared as pressed KBr discs or as suspensions in high molecular weight liquids ("mulls"). Potassium bromide discs are prepared by thoroughly mixing approximately 1 mg of the solid sample and 100 mg of dry KBr, and then applying a pressure of 20,000 to 50,000 $lb/in.^2$.

Absorption cells are usually placed *after* the monochromator in ultraviolet and visible instruments in order to minimize possible decomposition or fluorescence, which might be caused by other high-energy wavelengths of the unresolved radiation. In infrared instruments, the sample is placed *before* the monochromator so that it will not hinder the focusing of the radiant beam on the detector. Wherever the cell is placed, it should be positioned so that the beam of incident radiation is perfectly normal (perpendicular) to the window or cell face; otherwise, there may be significant losses due to reflection and refraction. In addition, the container should be inserted so that the same cell face is presented to the radiation beam in consecutive measurements. Rectangular cells are preferable to cylindrical cells. However, if the more inexpensive cylindrical cells are used, they should somehow be marked to insure that the same position is used in each measurement.

Solvents.   The solvents used in spectrophotometric studies must (1) dissolve the sample and (2) transmit in the wavelength region under study. Common solvents which are used in the ultraviolet and visible regions are listed in Table 10–2 along with their lower transparency limits.

Common solvent absorption bands in the infrared range of 2 to 15 μm (5000 to 666 cm^{-1}) are given in Table 10–3. Carbon tetrachloride, carbon disulfide, and chloroform are the most commonly used.

Solution Preparation.   Components which are to be determined in the ultraviolet and visible regions often exhibit high molar absorptivities at the

Table 10–2 Approximate Lower Transparency Limits of Solvents Used in Ultraviolet-Visible Absorption Studies

Solvent (high purity)	Transparency Limit, nm
Acetone	330
Benzene	285
Carbon tetrachloride	265
Carbon disulfide	375
Chloroform	245
Cyclohexane	215
Dichloromethane	235
Dioxane	225
95% Ethanol	205
Ethyl ether	205
iso-Octane	215
Isopropanol	215
Methanol	215
Pyridine	305
Water	200
Xylene	295

absorbance maximum. High concentrations of these components will yield a very low percent transmittance. Recall that the percent transmittance of the solution should be in the range 20 to 65%, and that at *low* values of %T, the uncertainty is extremely high. Consequently, the sample may have to be diluted to give an absorbance in the optimum range.

The path length of the infrared solution cell is quite short, and the infrared absorption bands have rather low molar absorptivities. Consequently, fairly concentrated solutions of the absorbing components are required in order to obtain a measurable absorbance. Concentrations of the order of 0.5 to 10 weight percent are common for infrared studies. Quantitative accuracy in the infrared region is rather poor, and is limited primarily by the low energy of infrared radiation coupled with the difficulty of establishing a true 100%T reference line.

10–5 DETECTION DEVICES

Any detector absorbs the energy of the photons which strike it and converts this energy to a measurable quantity such as the darkening of a photographic plate, an electric current, or thermal changes. Most modern detectors generate an electrical signal which eventually activates some type of meter or recorder. Any detector must generate a signal which is quantitatively related to the radiant power striking it. The "noise" of a detector refers to the "background" signal generated when no radiant power from the sample reaches the detector. This noise may be caused either by random changes within the detector itself or by electrical "pick-up" of other signals in the vicinity of the detector unit. Important requirements for detectors include: (1) high sensitivity with a low noise level in order to allow the detection of low levels of radiant power, (2) short response time, (3) long term stability to insure quantitative response, and (4) an electronic signal which is

Table 10–3   Infrared Solvent Absorption Bands in Range 2 to 15 μm

Solvent	Absorption Bands	
	μm	cm^{-1}
Bromoform	3.2–3.5	3120–2860
	4.2–4.6	2380–2170
	7.4–7.8	1350–1280
	8.2–9.3	1220–1080
	11.3–11.8	880–850
	13.1–15.0	760–666
Carbon disulfide	4.2–4.8	2380–2080
	6.1–7.1	1640–1410
	11.4–11.8	877–847
Carbon tetrachloride	6.2–6.5	1610–1540
	7.8–8.3	1280–1200
	9.9–10.5	1010–950
	11.7–15.0	855–666
Chloroform	3.3–3.5	3030–2850
	4.1–4.3	2440–2330
	6.6–7.1	1520–1410
	10.6–15.0	943–666
Nujol oil (high molecular weight petroleum fraction for mulls)	3.4–3.8	2940–2630
	6.8–7.4	1470–1350
Fluorolube (perfluoro carbon for mulls)	7.7–15.0	1300–666
Tetrachloroethylene	7.2–7.5	1390–1330
	8.5–9.1	1180–1100
	9.9–15.0	1010–666

easily amplified for typical readout apparatus. Performance characteristics of common detectors are summarized in Table 10–4.

Ultraviolet and Visible Radiation Detectors.   Ultraviolet and visible photons possess enough energy to cause photoejection of electrons when they strike surfaces which have been treated with specific types of compounds. Their absorption may also cause bound, nonconducting electrons to move into conduction bands in certain semiconductors. Both processes generate an electric current which is directly proportional to the radiant power of the absorbed photons. Devices which employ these systems are called *photoelectric detectors* and are sub-classified as *phototubes* and *photovoltaic cells.*

Phototubes.   A phototube consists of: (1) an evacuated glass envelope (with a quartz window for use in the ultraviolet region); (2) a semi-cylindrical cathode which has an inner surface coated with a compound with relatively loosely bound electrons, such as an alkali or alkaline earth oxide; and (3) a central metal wire anode. A potential difference of approximately 90 volts is applied across the electrodes. The phototube and its associated circuitry are shown schematically in Figure 10–6. The radiation enters through the quartz window and strikes the photoemissive surface of the cathode. The photons are absorbed and transfer

Table 10–4   Characteristics of Detectors

Detector	Sensing Element	Sensitivity	Response Time $1 \mu sec = 10^{-6} sec$ $1 msec = 10^{-3} sec$	Recommended Wavelength Range of Application (μm)	Nature of Output
Photographic plate	Silver halide grains in emulsion	High, wavelength dependent	Slow	0.2–1.2	Density of silver metal deposit
Phototube	Alkali metal oxide	High, wavelength dependent	Fast $<1 \mu sec$	0.2–1	Electric current
Photo-multiplier tube	Alkali metal oxide, metals	Very high, wavelength dependent	Fast $<1 \mu sec$	0.16–0.7	Electric current
Photovoltaic cell	Semiconductor between two metals	Medium, wavelength dependent	Fast $<1 \mu sec$	0.4–0.8	Electric current
Photo-conductive cell	Lead sulfide or lead selenide	Very high, wavelength dependent	Moderate 100 to 1000 μsec	0.7–4.5	Resistance change
Thermocouple	Junction of two different metals connected to blackened metal leaf	High	50 to 100 msec	0.8–15.0	Electric potential difference at junction of two metals
Golay cell	Blackened membrane of gas chamber	High	3 to 30 msec	0.8–1000	Electric current

[After Strobel, *Chemical Instrumentation*, 2nd ed., Addison–Wesley, Reading, Mass., 1973, p. 306.]

their energy to the loosely bound electrons of the surface material. The electrons escape from the surface and are collected at the anode causing current to flow in the circuit. If the electron collection is essentially 100% efficient, the phototube current should be proportional to the radiant power of the incident radiation. However, the magnitude of the photocurrent also depends on the voltage applied to the electrodes and the wavelength of the incident radiation. The phototube

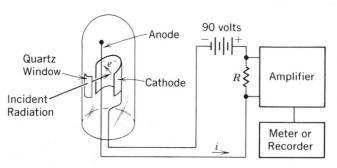

Figure 10–6   Schematic diagram of phototube circuit.

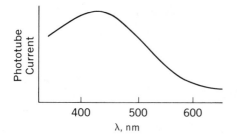

Figure 10–7 Phototube current versus wavelength of incident radiation at constant radiation intensity, I.

current at a given radiant power increases with applied voltage until a plateau is reached where it is no longer dependent on the voltage. The voltage at this point, called the saturation voltage, represents the point at which all the photoemissive electrons are being collected with 100% efficiency. Consequently, if a phototube is to respond linearly to incident radiant power, it must be operated above the saturation voltage. The variation in phototube sensitivity with wavelength demonstrated in Figure 10–7 clearly points out that narrow band radiation will produce a more linear phototube response than wide band radiation.

Phototube currents are quite small (as low as 10^{-11} ampere) and require amplification in order to operate any common type of readout device. This is accomplished by placing a high resistance (R in Figure 10–6) in the phototube circuit and applying the electrical potential difference (iR drop) across this resistor as the "input" to an amplification circuit. The "output" of the amplifier then is used to drive a meter or recorder.

A small phototube current, known as "dark current," is observed even when there is no incident radiation on the phototube. This is a result of the random thermal emission of electrons from the cathode surface. The magnitude of the dark current increases with increasing cathode surface area and increasing temperature.

Photomultiplier Tubes. In the discussion of phototubes we have indicated the mechanism by which an electron escapes from a photoemissive surface. If the ejected electron is accelerated by an electric field, it acquires more energy; and if it strikes another electron-active surface, it may transfer some of its energy, ejecting several more electrons. These electrons may in turn be accelerated to another surface and produce even more electrons, and so on. This is the principle of the photomultiplier tube; a cross section of this device is shown schematically in Figure 10–8. Each succeeding electron-active plate, or dynode, is at higher electrical potential and thus acts as an amplification stage for the original photon. After nine stages of amplification the original photon has been amplified by a

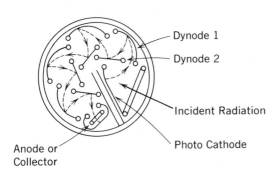

Figure 10–8 Schematic cross section of a photomultiplier tube.

Dynode 1

Dynode 2

Incident Radiation

Photo Cathode

Anode or Collector

factor of approximately 10^6. In practice, photomultiplier tubes are used only for low radiant power levels; otherwise, they exhibit great instability. The photocurrent may be further amplified by other means.

Near Infrared Detectors.   The photoconductive cell is commonly used for the detection of near infrared radiation (0.8 to 3 μm). The sensing element is a semiconductor (lead sulfide, lead telluride, or germanium with low percent impurity). Upon illumination with radiation of appropriate wavelength, the electrons of the semiconductor, most of which are nonconducting, are raised to conduction bands. The electrical resistance drops, and a large increase in current is noted if a small voltage is applied. The resistance of the system is such that the current may be amplified and finally indicated on a meter or a recorder.

Middle and Far Infrared Detectors.   When middle and far infrared photons are absorbed, their energies are converted to thermal energy (heat) and a corresponding temperature change is noted. Thus the detectors used in this region are various types of rapid response thermometers such as thermocouples, resistance thermometers (bolometer), and gas thermometers (pneumatic or Golay cell). Thermocouples used in infrared receivers typically consist of a blackened gold leaf-tellurium metal pin junction which develops a voltage that is temperature dependent. The thermocouple is often enclosed in a shielded, evacuated housing to avoid heat loss and unnecessary temperature fluctuations.

10–6   AMPLIFICATION AND READOUT OF DETECTOR SIGNALS

The electronic signal generated by any radiation detector must be translated into a form which the experimenter can interpret. The process is typically accomplished with amplifiers, ammeters, potentiometers, and potentiometric recorders.

Amplifiers.   In order to be measured, most detector signals must be amplified by several orders of magnitude. An *amplifier* takes an "input" signal from the circuit of the sensing component and, through a series of electronic operations, produces an "output" signal which is many times larger than the "input." The amplification factor (ratio of output to input) is called "the gain" of the amplifier. The amplifier input and output signals are normally electrical voltages. If the detector circuit generates an electric current, the input is taken from the voltage drop across a resistor inserted into the circuit (see Figure 10–6 or 10–9). Therefore,

$$E_{\text{input}} = i_{\text{detector circuit}} R_{\text{detector circuit}} \text{ (Ohm's law)}$$

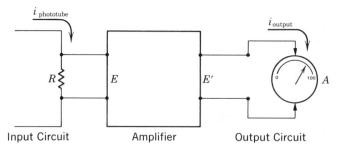

Input Circuit   Amplifier   Output Circuit

Figure 10–9   Schematic diagram of a meter readout system.

Meter Readout. Let us use a meter as the readout device in an ultraviolet-visible instrument. In Figure 10–9, A is an ammeter which measures the electric current in the output circuit. For convenience we will assume that the meter is calibrated in 100 divisions. Using Ohm's law, the output voltage, $E' = i_{output}R'$, where $R' =$ the resistance of the output circuit (assumed to be constant). Therefore, the following proportionalities should be correct:

$$\text{Meter reading} \propto i_{output} \propto E' \propto E \propto i_{phototube} \propto \text{transmitted intensity}$$

The validity of the overall proportionality requires linear amplification by the amplifier; that is, $E' = kE$, where k is the gain of the amplifier.

Quantitative Calibration. The circuit illustrated in Figure 10–9 can be easily modified to compensate for the dark current by introducing an additional equal but opposite signal into the input of the amplifier. The compensating signal is adjusted so that the meter reads 0 with no radiant energy striking the phototube (shutter closed). The instrument is then "standardized" by using a "blank" solution (normally the solvent) and adjusting the output signal so that the meter reads 100. This standardizing adjustment can be made by varying the gain of the amplifier, varying the slit width, varying the sensitivity of the meter (electrical shunt), or by partially blocking the light beam with a comb device. Thus the instrument is calibrated so that there are 100 units on the meter from $I_t = 0$ to $I_t = I_0$ and these units are linear with respect to I_t. When an absorbing sample is substituted for the "blank," the detector response will show between 0 and 100 units on the meter. For example, if the detector response for a particular sample produced a meter reading of 80 units (80% transmittance), $I_t = k(80)$ and $I_0 = k(100)$, where k is the proportionality constant of the detector and amplifier. Therefore, for this sample, $\log I_0/I_t = \log 100/80 = A = \epsilon bc$. If ϵ and b are known, the concentration of the absorbing sample may be determined.

A potentiometric recorder utilizes a "null-point" design by mechanically adjusting its output voltage to be equal and opposite to the output signal from the detector amplifier. (See Figure 10–10.) Any difference between the two outputs is used to actuate the balancing mechanism. A recording pen is attached to the balancing mechanism and this pen continuously plots the position (voltage) of the null point. If the amplifier output signal changes, the pen will automatically move

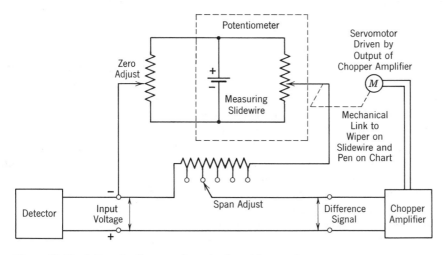

Figure 10–10 Schematic diagram of a potentiometric recorder.

to a new null point. The position of the pen is analogous to a meter reading, but it also leaves a permanent trace on a chart paper. The system is adjusted to read 100% transmittance when a "blank" is inserted in the sample radiation beam and 0% transmittance when an opaque object is placed in the radiation beam. Therefore, as the pen moves across the paper, it automatically records the percent transmittance of the samples as it changes with wavelength. The rate of movement of the chart paper (perpendicular to the pen travel) corresponds to the automatic rate of change of wavelength of the radiation from the monochromator. It makes little difference whether the pen travels in one direction and the chart paper moves under it in a perpendicular direction, or the paper is held fixed while the pen moves across the paper and its carriage moves lengthwise along the paper—both systems are used. In this manner, after the instrument is standardized, an absorption spectrum is recorded automatically, as shown in Figure 10–11. The potentiometric recorder is discussed in more detail in Section 25–2.

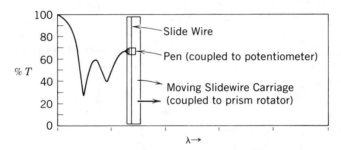

Figure 10–11   Automatic recording spectrophotometer.

10–7   SINGLE-BEAM VERSUS DOUBLE-BEAM OPERATION

Two basic instrument designs are employed in commercial spectrophotometers. One design uses only a single beam, whereas the other provides a double beam. We shall emphasize ultraviolet and visible instruments in the following discussion.

Single-Beam Operation.   A beam of radiation from the source enters the monochromator where it is dispersed by a prism or a grating. As the dispersing element is rotated, the various resolved bands of radiation are focused at the exit slit. The radiation then passes through the cell and on to a detector. The instrument is calibrated (0% and 100% T) and used by the methods just described.

The single-beam method requires stable, high quality components in the source, detector, and amplifier for high precision measurements. The instrumental parameters cannot fluctuate between the time of 100% T calibration with the "blank" and the determination of the transmittance of the sample. Direct-reading instruments using meters give immediate readout with an accuracy of ± 1 to 3% in transmittance. The null-point readout design is considerably more accurate ($\pm 0.2\%$ in transmittance), but more expensive. Unless high accuracy is required, the meter readout instruments are quite satisfactory. Single-beam instruments are simpler and less expensive than double-beam, but they are not readily adapted to recording because of the necessity of calibration at each wavelength.

Double-Beam Operation.   Double-beam instruments employ some type of beam splitter prior to the sample cells. One beam is directed through the "blank" cell (or reference cell) and the other beam through the sample cell. The two beams are then compared either continuously or alternately many times a second. Thus, in the double-beam design, fluctuations in the source intensity, the detector response and amplifier gain are compensated for by observing the ratio signal between "blank" and sample. As might be expected, double-beam instruments are more sophisticated electronically and mechanically than the single-beam designs and consequently are more expensive.

In ultraviolet-visible double-beam spectrophotometers, the beam splitting occurs after the monochromator; front surface mirrors and, more commonly, rotating sector mirrors are used. The rotating sector mirror alternately passes and reflects the beam several times a second and thus splits the beam and also chops it. This chopped radiation is used as the input source for a-c amplifiers, which provide amplification stability. An electronic null-readout system is employed in this type of instrument. One of these designs is shown schematically in Figure 10–12. The reference and sample beams alternately reach the detector at intervals which depend on the rotational frequencies of the choppers. The instrument records the ratio of the reference and sample signals. If the radiant powers of both of the beams are the same, the a-c amplifier generates no output signal. If the radiant powers of the two beams are different, the imbalance signal activates a servomotor which drives a potentiometric recorder such that the imbalance signal is "electronically nulled" by the recorder. The electrical bridge of the potentiometric recorder is calibrated in terms of percent transmittance of the sample and thus the position of the bridge balance point is used to determine percent transmittance.

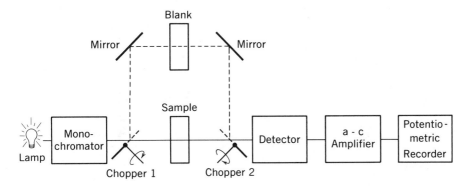

Figure 10–12   Double-beam spectrophotometer employing a chopper.

10–8 TYPICAL INSTRUMENTS

Bausch and Lomb Spectronic 20 Spectrophotometer.   This low-cost instrument ($\sim$ \$525) is particularly suitable for student use. It has a range from 340 to 625 nm, which can be extended to approximately 750 nm with simple modification. The Spectronic 20 is illustrated schematically in Figure 10–13. Radiation from a tungsten lamp is dispersed by a diffraction grating with 600 lines/mm. The slits allow a constant band width of 20 nm. The grating is turned by the calibrated wavelength control knob in order to select the desired radiation band. The instrument is standardized separately at each wavelength by (1) adjusting the amplifier control knob so that the meter reads 0% T when no radiation reaches the

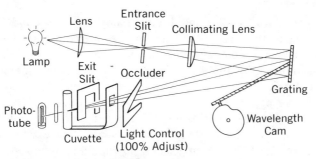

Figure 10–13   Schematic diagram of Bausch and Lomb Spectronic 20 spectrophotometer.

phototube ("dark current" adjust), and (2) adjusting the light control so that the meter reads 100% T with the "blank" cell in the beam. The sensitivity of the Spectronic 20 phototube is similar to that shown in Figure 10–7. It is clear that the calibration of the "blank" to readout 100% T requires more intensity at 600 nm than at 400 nm; the light control adjust performs the desired attenuations. The above calibrations should be made *each time* the wavelength is changed.

Beckman DU Spectrophotometer.   This was the first ultraviolet-visible instrument to be produced on a large scale in this country. A schematic optical diagram of this instrument is shown in Figure 10–14. The DU spans the wavelength region 190 nm to 1000 nm by employing a hydrogen discharge lamp for the ultraviolet region and a 6-volt tungsten lamp for the visible near-infrared region. The 30° quartz Littrow prism is rotated so that the proper radiation will pass through the exit slit. A calibrated dial which is attached to the prism indicates the wavelengths which are being used. The slit widths are adjusted manually. Two interchangeable phototubes are used; one is most sensitive in the ultraviolet and blue region and the other in the red range.

The DU employs a null balance readout design. The instrument is calibrated in much the same way as the Spectronic 20, except that the "blank" setting (100% T) can be adjusted by changing either the slit widths or the amplifier gain and, of course, one reads a potentiometer dial instead of a meter.

Perkin-Elmer Model 467 Spectrophotometer.   The Model 467 is a typical double-beam, medium priced, automatic recording infrared spectrophotometer. It operates in the range 2.5 to 15 μm. A schematic diagram is shown in Figure 10–15. The radiation from a heated ceramic source is split in half by a plane mirror and two

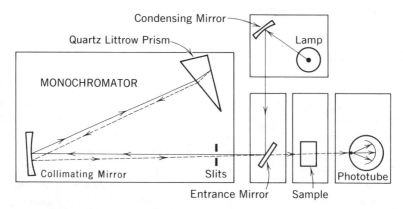

Figure 10–14   Schematic diagram of Beckman DU ultraviolet-visible spectrophotometer.

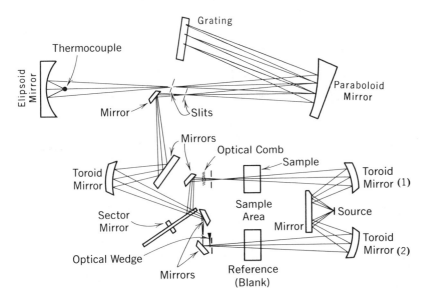

Figure 10–15 Schematic diagram of Perkin-Elmer Model 467 infrared spectrophotometer.

toroid mirrors. Toroid mirror 1 focuses the sample beam through the sampling area onto an optical comb (100% T adjust optical attenuator); toroid mirror 2 focuses the reference beam onto an optical wedge (reference beam attenuator). The sample cell is placed just in front of the optical comb and the reference ("blank") cell just in front of the optical wedge. After passing through the sample and "blank," the two beams are recombined by a semicircular sector mirror which rotates at 13 revolutions per second. This odd frequency avoids any possibility of interference from the 60 Hz power line. The single beam which emerges from the sector mirror then consists of alternate pulses of sample and reference radiation which eventually undergo dispersion by one of two gratings mounted back-to-back, 25 and 100 lines per mm, to cover the entire spectral range. The dispersed radiation is then focused on the exit slit as a band of radiation which finally reaches the thermocouple sensing element. If the two beams are of equal intensity, the thermocouple produces a d-c voltage which is ignored by the a-c amplifier. If the two beams are not equal, the detector produces a pulsating signal which is converted to an alternating voltage and amplified. The amplified signal is used to drive a servomotor which moves an optical wedge into or out of the reference beam to equalize (or null) the beam intensities. The recorder pen is coupled to the wedge and records its position. The position of the wedge is calibrated to correspond to percent transmittance and thus the pen records the percent transmittance of the sample at the wavelength band used. Since the wavelength band is changed by rotating the grating, the drive mechanism can also control the movement of chart paper on which the spectrum is to be recorded. The chart paper is precalibrated in units of wavelength (or wavenumbers) along one axis and in % T along the other.

For some routine infrared determinations, particularly in commercial and environmental analysis, a nondispersive infrared spectrometer serves the purpose. These instruments, which use unique filtering systems, are described in Section 27–8.

Spectrofluorimeters. Luminescent radiation is emitted from the sample in all directions, and it is convenient to observe it at right angles to the incident

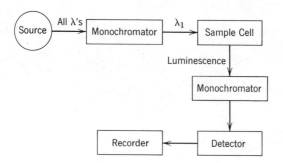

Figure 10–16 Block diagram of a spectrofluorimeter.

radiation from the source. A block diagram of a typical spectrofluorimeter is given in Figure 10–16. Since the intensity of fluorescence or phosphorescence is proportional to source intensity (and is at best very weak), the source is usually a powerful xenon arc lamp or laser. A filter or monochromator selects a suitable narrow band for excitation of the sample, and a second monochromator is needed in the luminescence beam.

REFERENCES

R. P. Bauman, *Absorption Spectroscopy*, Wiley, New York, 1962. Very comprehensive discussion of instrumentation.

G. W. Ewing, *Instrumental Methods of Chemical Analysis*, 4th ed., McGraw-Hill, New York, 1975.

H. A. Strobel, *Chemical Instrumentation*, 2nd ed., Addison-Wesley, Reading, Mass., 1973.

H. H. Willard, L. L. Merritt, and J. A. Dean, *Instrumental Methods of Analysis*, 5th ed., Van Nostrand Reinhold, New York, 1974.

11

INFRARED SPECTROSCOPY

11–1 INTRODUCTION

Infrared radiation was discovered in 1800 by Sir William Herschel who reported his findings on certain experiments in heat radiation to the Royal Society. At that time scientists did not have a clear understanding of the nature of radiation. Herschel's experiment consisted of resolving sunlight into its spectrum with a glass prism and placing thermometers at successive positions throughout the spectrum (Figure 11–1). The thermometer placed beyond the red end of the solar spectrum registered the most marked rise in temperature demonstrating that more heat existed outside and beyond the red region than within. This simple experiment was both the discovery of the infrared region (infra meaning beyond) and the construction of the first infrared spectrometer. In further experiments Herschel measured the absorption of this new radiation by numerous substances including sea water, distilled water, and, oddly enough, gin and brandy. Unfortunately he was not aware of the potential of his discovery, namely, that this absorption could reveal information about the molecular structure of organic compounds. Before the significance of infrared absorption was to be appreciated, the theory of the nature of radiation had to be better understood. Almost a century passed before the necessary theory, techniques, and instrumentation for infrared analysis were developed. The infrared region lies between the visible and radio portions of the electromagnetic spectrum (Figure 11–2). The most useful range lies between 4000 and 400 cm^{-1}.

Figure 11–1 The first infrared spectrometer improvised by Sir William Herschel in 1800.

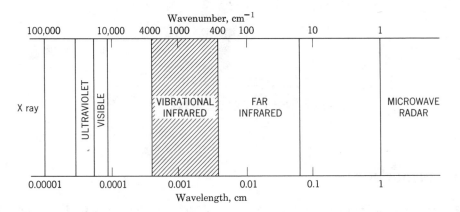

Figure 11-2 The position of the infrared region of the electromagnetic spectrum.

Beginning in 1903 William W. Coblentz, a young graduate student at Cornell University, improved the experimental techniques and set about at long last to measure the absorption spectra of pure substances. For two years he mapped spectra by a point-to-point plotting of values and finally in 1905 published the first collection of accurate infrared spectra of 131 substances. With more than half a century's progress since Coblentz, we now understand the basic theory of infrared spectroscopy that he so energetically tried to postulate. Organic chemists in the 1930's seriously began to consider infrared spectroscopy as a possible method to identify compounds through their functional groups. By 1935 a few of the larger chemical companies had invested in infrared spectrometers for organic quantitative analysis.

The major industrial impetus to the field came early in World War II when it was shown that infrared spectroscopy offered the most accurate and rapid method of analyzing the C_4 hydrocarbon fraction of interest in the production of synthetic rubber. The boom had started and commercial production of instruments was initiated during 1943. Availability of instrumentation stimulated infrared studies and the number of papers published rocketed sky high. Such basic research has nowadays enabled the ordinary bench chemist not only to record his own spectra but to interpret them in terms of structure. Infrared spectroscopy has gained worldwide acceptance by chemists to the extent that it is considered standard laboratory practice.

11-2   INFRARED THEORY

Vibrational Spectra.   Molecular vibrations can occur by two different mechanisms. Firstly, quanta of infrared radiation can excite atoms to vibrate directly—the absorption of infrared radiation gives rise to the infrared spectrum. Secondly, quanta of visible light can achieve the same result indirectly—the Raman Effect.

Most organic molecules are fairly large and their resultant vibrational spectra are complex. To introduce the basic concepts governing vibrational spectra a simple diatomic covalent bond will be considered as a spring with the atomic masses at either end (Figure 11-3). The stiffness of the spring is described by a *force constant, k.* If such a simple system is put into motion (by stretching and

$$\tilde{v} = \frac{1}{2\pi} \sqrt{\frac{k(m_1 + m_2)}{m_1 \cdot m_2}}$$

Mass   Force Constant   Mass

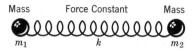

m_1 k m_2

Figure 11–3 Ball and spring representation of two atoms of a molecule vibrating in the direction of the bond.

releasing), the induced vibrations of the masses are adequately described by Hooke's law of simple harmonic motion.

Frequency of motion,

$$v = \frac{1}{2\pi} \sqrt{\frac{k}{\mu}} \qquad (11\text{–}1)$$

where μ is the *reduced mass*, that is, the harmonic mean of the individual masses

$$\frac{1}{\mu} = \frac{1}{m_1} + \frac{1}{m_2}$$

or

$$\mu = m_1 m_2 / (m_1 + m_2)$$

To a first approximation, this assumption of harmonic forces is in agreement with actual conditions in real molecules. However, quantum theory governs molecular motion and restricts the energy stored in the vibration, E_v, such that only certain energy transitions are allowed, as determined by a quantum number, v.

$$E_v = (v + 1/2)hv \qquad (11\text{–}2)$$

where $v = 0, 1, 2, 3, \ldots$ etc. For instance, if a molecule were to undergo a transition from the lowest level ($v = 0$) to the first level ($v = 1$) by absorption of infrared radiation, the frequency of that exciting radiation would be given by the *Bohr principle*, $hv = E_1 - E_0$.

Now Equation 11–2 gives

$$E_0 = 1/2hv \quad \text{and} \quad E_1 = 3/2hv$$

Therefore, by substitution:

$$(E_1 - E_0)/h = v$$

In summary, the absorption of infrared radiation causes the excitation of the molecule to higher vibrational levels and is quantized. The normal vibration has the same frequency as the electromagnetic radiation. The absorption process can occur only if there is a change in the magnitude and direction of the dipole moment of the bond.

In the same manner, a transition from the lowest level ($v = 0$) to the second level ($v = 2$) would occur at a vibrational frequency, $2v$. In musical terminology, v is the fundamental vibration frequency and $2v$ the overtone frequency.

According to Boltzmann's distribution law almost all molecules are in the lowest vibrational energy level ($v = 0$) at room temperature. Therefore, the vibrational transitions in a molecule are restricted to those occurring from the lowest level.

However, a selection rule forbids transitions which change the quantum number, v, by more than one unit. This rule limits the observable transitions to the fundamental frequencies, v, that is, vibrational spectra should exhibit only characteristic vibrational frequencies corresponding to the various bonds within the molecule. This selection rule is obeyed only for perfectly harmonic vibrations. In practice, molecular vibrations are not strictly harmonic and the selection rule fails as is evident by the appearance of weak overtone vibrations.

Evaluation of Equation 11–1 for the O—H bond (Table 11–1) indicates that the hydrogen atom bound to an oxygen atom vibrates back and forth approximately 10^{14} times per second in the direction of the linkage.

$$v = \frac{1}{2\pi} \sqrt{\frac{k}{\mu}} = \frac{1}{2\pi} \sqrt{\frac{7.7 \times 10^5}{0.941 \times 1.67 \times 10^{-24}}} = 1.11 \times 10^{14} \text{ Hz}$$

or

$$\bar{v} = \frac{v}{c} = \frac{1.11 \times 10^{14}}{3 \times 10^{10}} \text{ cm}^{-1} = 3700 \text{ cm}^{-1}$$

In summary, the O—H bond vibrates 1.11×10^{14} cycles per second and the electromagnetic radiation required to excite this particular vibration must have a frequency of approximately 1.11×10^{14} Hz or 3700 cm^{-1}. Experience has demonstrated that all alcohol spectra contain a characteristic absorption at approximately 3600 cm^{-1}. If the vibrations of the C—C linkage are calculated in a similar manner, the result is 3.5×10^{13} Hz corresponding to a frequency of 1100 cm^{-1}. Now in the case of the C=C linkage, the masses of the atoms involved are the same (i.e., $\mu = 6$), but the force constant, k, is greater, and the vibrational frequency is 4.9×10^{13} Hz, corresponding to a radiation of 1640 cm^{-1}. The force constant of the C≡C linkage is even greater and the absorption band is at 2100 cm^{-1}. This series illustrates the potential power of infrared to distinguish between combinations of the same two atoms linked differently.

Polyatomic Vibrational Spectra. In a polyatomic molecule, the atoms or covalent bonds are not rigidly linked together and are able to vibrate from their positions of

Table 11–1 Reduced Mass and Force Constants for Various Atom Pairs

Atom Pair	Force Constant[a]	$\mu = \dfrac{m_1 m_2}{m_1 + m_2}$[b]
C—C	4.5	6
C=C	9.6	6
C≡C	15.6	6
C—O	5.0	6.85
C=O	12.1	6.85
C—H	5.1	0.923
O—H	7.7	0.941
C—N	5.8	6.46
N—H	6.4	0.933
C≡N	17.7	6.46

[a] Expressed in 10^5 dynes/cm.
[b] Expressed in units of m_H, where m_H = mass of a hydrogen atom = 1.67339×10^{-24} g.

rest. In addition, there are the bond angles enclosed by the various individual diatomic bonds which result in a powerful qualitative method of describing the vibrations of polyatomic molecules.

Since each type of chemical bond in a molecule involves different values of force constants and reduced masses, absorption of radiation will occur over a range of frequencies. Thus, if infrared radiation of successive frequencies is passed through a substance, a series of absorption bands is recorded—the active fundamental modes of vibration. These can be subdivided into the following classes.

1. *Stretching vibrations* are those in which two bonded atoms continuously oscillate, changing the distance between them without altering the bond axis or bond angles. They are either isolated vibrations (e.g., the O—H bond) or coupled vibrations (e.g., the methylene group, $\overset{\diagup H}{\underset{\diagdown H}{\diagup C}}$). Coupled vibrations are symmetrical or unsymmetrical (asymmetric) as illustrated (Figure 11–4).

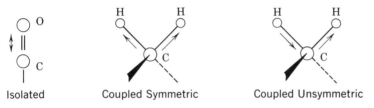

Isolated   Coupled Symmetric   Coupled Unsymmetric

Figure 11–4 Coupled stretching vibration for the methylene group.

In the symmetric case both hydrogen atoms move away from the carbon simultaneously while in the asymmetric case one hydrogen moves toward the carbon while the other moves away. Stretching vibrations generally require higher energies than bending and are denoted by the Greek symbol nu, ν, followed by the chemical group in parentheses afterwards, that is, $\nu(C{=}O) = 1600$ cm^{-1} indicates that the fundamental stretching vibration of the carbonyl group is observed at 1600 cm^{-1}.

2. *Bending vibrations* are characterized by a continuously changing angle between two bonds. Bending modes of aromatic C—H groups for instance, which take place in the plane of the phenyl nucleus, are denoted by the symbol delta, $\delta(C{-}H)$ while those which occur out of the plane are denoted by the symbol gamma, $\gamma(C{-}H)$. This nomenclature also applies to alkenes and alkynes.

3. *Wagging vibrations* result when a nonlinear three-atom structural unit oscillates back and forth in the equilibrium plane formed by the atoms and their two bonds (Figure 11–5). Such vibrations are denoted by the symbol omega, $\omega(CH_2)$.

4. *Rocking vibrations* occur where the same structural unit oscillates back and forth out of the equilibrium plane (Figure 11–5). The symbol to denote this particular mode of vibration is rho, $\rho(CH_2)$.

5. *Twisting vibrations* occur when the same structural unit rotates around the bond which joins it to the rest of the molecule (Figure 11–5). Such vibrations are recorded by the symbol tau, $\tau(CH_2)$.

6. *Scissoring vibrations* occur when two nonbonded atoms move back and forth toward each other (Figure 11–5). These are denoted by the symbol $s(CH_2)$.

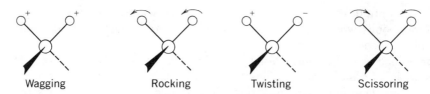

Figure 11–5 Bending vibrations (where + indicates movement out of the plane of the page and − denotes movement back of the plane of the page).

Apart from these fundamental modes of vibration, harmonic and combination vibrations may also occur. Harmonic vibrations possess frequencies which represent approximately integral multiples of the fundamental frequency, for example, 2ν or 2δ. Frequencies of combination vibrations are composed of the sum of $(\nu + \delta)$, or the difference between $(\nu - \delta)$ the frequencies of two or more fundamental or harmonic vibrations.

If the origin and theory of the infrared spectrum are kept in mind it is apparent that the spectrum represents a veritable wealth of information about the basic characteristics of the molecule, namely, the nature of the atoms, their spatial arrangement, and their chemical linkage forces. It is for this reason that the infrared spectrum gained recognition as being a "fingerprint of the molecule."

11–3   SAMPLE PREPARATION

In general the amount of sample necessary to obtain a good infrared spectrum is of the order of 1 to 5 mg. Techniques have been developed to handle the sample in any of its three possible phases—solid, liquid, and gas.

Solid Substances

As Solid State Spectra.   Solid state forces such as intermolecular hydrogen bonding render such spectra somewhat unreliable for diagnostic purposes. For successful examination of solids the following points must be clearly understood. First, the particle size should be less than the wavelength of the infrared radiation (1μm), otherwise pronounced scattering of the incident light occurs. This restriction can easily be overcome by very thorough grinding or even better by using a mechanical ball and mill technique. Second, these small particles must now be suspended in a medium of similar refractive index. In practice, grinding is continued in Nujol (medicinal paraffin) to obtain a "mull." The mull is then placed between two polished sodium chloride plates and placed in the spectrophotometer. Nujol has a relatively simple spectrum consisting only of ν(CH) at 2950 cm^{-1}, δ(CH) asy. for methylene and methyl groups at 1450 cm^{-1} and δ(CH) sym. for methyl groups only at 1380 cm^{-1}.

Another common method of suspending the solid particles is to intimately mix the substances with potassium bromide. The resulting mixture is compressed under high pressure to form a disc which can be placed directly into the spectrophotometer.

As Solution Spectra.   Separation of molecules can be achieved easily by dissolving them in a suitable solvent. In such an environment of solvent molecules we can deal with the molecule as an individual entity. Accurate quantitative work demands a dilute solution. By varying the concentration of alcohols and phenols in particular, a degree of control is exercised over the association or intermolecular hydrogen bonding between these molecules. In progressively more dilute solutions they will exhibit polymeric, trimeric, dimeric, and finally monomeric hydroxyl stretching frequencies, $\nu(OH)$.

Choice of solvent depends on the region of the spectrum of most interest. All solvents have some absorptions in the 4000 to 650 cm^{-1} region. By using "window areas", that is, transparent areas of the solvent, the whole spectrum may be covered. For instance, the most common use of carbon tetrachloride (CCl$_4$) is from 4000 to 1300 cm^{-1} and for carbon disulfide (CS$_2$), 1300 to 660 cm^{-1}. Double beam operation may cancel solvent absorption (assuming the same pathlengths are used), but only to a certain degree.

Sodium chloride cells are employed, the most useful thickness being 0.1 mm and 0.5 mm. Typical concentrations for oxygen-containing substances are 5 to 10% in the 0.1-mm cell and 1 to 2% in the 0.5-mm cell.

Finally, a word of warning regarding solvent effects. Shifts in the absorption frequencies may result on changing from one solvent to another. Measurements in nonpolar CCl$_4$ are usually regarded as the reference value. Water is not at all a suitable solvent due to its intense absorption spectrum even in thin films throughout the entire infrared region. In any case, sodium chloride cell windows would gradually dissolve if water were employed as solvent.

Liquid Substances

As Pure Liquid Spectra. These can be easily examined either as thin films pressed between two NaCl plates or in cells with known pathlengths ranging from 0.01 to 0.1 mm. Spectra of pure liquids often show strong intermolecular hydrogen bonding and association effects.

Gases

Gaseous samples are measured in cells with long pathlengths; 10 cm is common. For trace amounts, very long paths are needed which can be achieved only with mirrors mounted at the ends of the cell to give multiple reflections.

11-4 INFRARED ABSORPTION SPECTRA

It is not always convenient to reproduce spectra in a scientific paper or book, so parameters are quoted which allow the reader to visualize, reconstruct, or compare with his own data.

Position. This is normally quoted as the wave number of maximum absorption, $\nu(X—Y)$ cm^{-1}, where X and Y represent the two atoms in question.

Half-Band Width. The apparent half-band width, $\Delta\bar{\nu}_{1/2}$, is cited as the width in cm^{-1} at half-height.

Intensity. The apparent molar absorptivity measured at the peak maximum, ϵ^a, is given by Beer's law:

$$\epsilon^a = \frac{\text{absorbance} \times \text{molecular weight}}{\text{mg/ml} \times \text{cell path (cm)}} \qquad (11\text{-}3)$$

Integrated Intensity. A logical extension of the apparent molar absorptivity, ϵ^a,

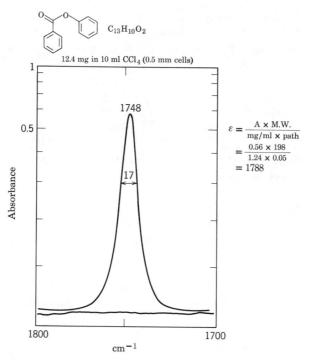

Figure 11–6   Measurement of parameters for a typical carbonyl absorption band.

is to measure the area under the peak. The integrated intensity, B, is given by

$$B = \int_{\bar{\nu}_1}^{\bar{\nu}_2} \epsilon \, d\bar{\nu} \text{ (in liter mole}^{-1} \text{ cm}^{-2}) \tag{11–4}$$

Figure 11–6 shows the calculations involved for a carbonyl absorption band. These parameters vary with the type of instrument employed and the conditions under which the recording is made. This variation is greatest with sharp peaks. Two important factors for such variations are the scanning speed and the slit width. Quantitative spectra must be scanned more slowly than is necessary for qualitative analyses. The error stems from the inability of the detecting system to respond to rapid changes in absorption.

   In the past the effect of slit width on prism infrared spectrophotometers has hindered the transfer of data from one laboratory to another. To a large extent the advent of grating spectrophotometers has removed this barrier. The $\nu(C{=}O)$ band of phenyl benzoate, shown in Figure 11–7, illustrates the effect of slit width. The areas under the peaks, however, are the same in each case. For quantitative analyses with prism instruments the same slit width should be used for recording the spectra of the unknown and calibration solutions. If possible, resolving power high enough to give no slit width error should be used, although, in general, only grating spectrophotometers will be capable of this.

   Two spectral regions are distinguished, the region of group frequencies and the fingerprint region. The group frequency region lies approximately between 4000 cm^{-1} and 1400 cm^{-1}. In this region the principal absorption bands may be assigned to vibration units consisting of only two atoms, and the frequency is characteristic of their masses and the force constant of their linkage. This simplification ignores the rest of the molecule. The functional groups of organic

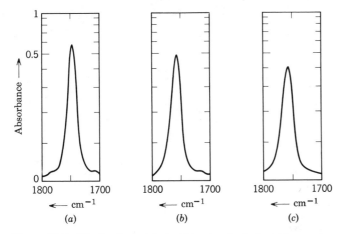

Figure 11–7   The fundamental stretching band of phenyl benzoate with (a) grating, (b) prism with narrow slits, and (c) prism with wide slits. (From A. J. Baker and T. Cairns, *Spectroscopic Techniques in Organic Chemistry*, Heyden & Son, London, 1967, pp. 13.)

molecules illustrate such vibrational units. To a first approximation the frequency of their fundamental stretching mode, $\nu(X—Y)$, is independent of the influence of the rest of the molecule (i.e., all alcohols have $\nu(O—H)$ at approximately 3600 cm^{-1}). Such influences do reveal themselves on careful study and can offer valuable evidence as to the nature of the neighboring atoms.

The fingerprint region extends from 1400 cm^{-1} to 400 cm^{-1}. Absorption bands found here are related to vibrations of the molecule as a whole, each atom exerting a mutual influence on the others (i.e., combination bands). The resulting bands are, of course, unique to a particular molecule and can be used for the unambiguous identification of that molecule provided you have a standard spectrum of that substance previously recorded on file. One exception to this rule is that a series of compounds such as long chain fatty acids can give almost identical spectra.

11–5   CORRELATION CHARTS

In the molecular diagnoses of vibrational frequencies or absorption bands it is extremely useful to refer to tabulated values of the various functional groups and their associated characteristic group frequency ranges (Colthup chart, Figure 11–8). This chart provides at a glance the expected range of frequencies for a particular functional group. To supplement this general data a series of more descriptive tables (Tables 11–2 through 11–26) has been prepared to deal with each major functional group in greater detail and the factors affecting the absorption band positions.

11–6   INTERPRETATION OF INFRARED SPECTRA

In almost every subject there are certain basic "alphabet-like" essential facts that must be mastered early and infrared spectroscopic identification is no exception. Although extensive correlation tables have systematized and rational-ized experimental findings, they have by no means removed the necessity of

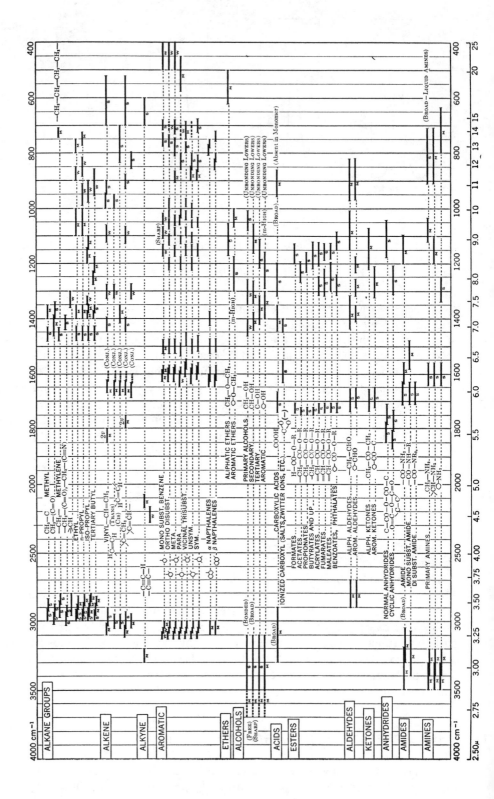

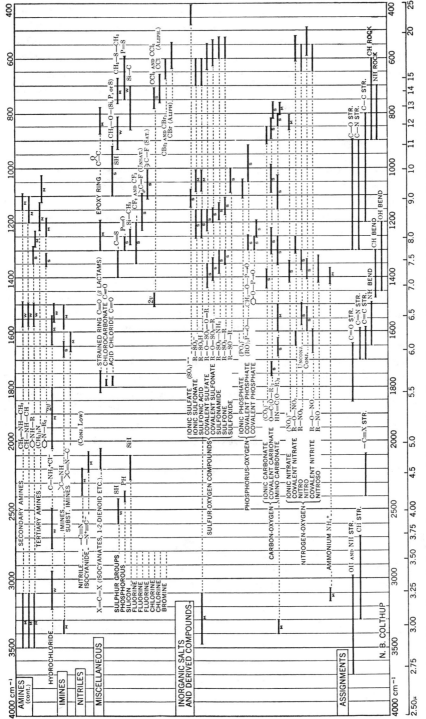

Figure 11–8 Colthup chart of characteristic group infrared absorptions. [From Colthup, *J. Opt. Soc. Am.,* **40,** 397 (1950).]

having almost reflex command of such elementary essentials as identification of functional groups.

To refresh the memory the infrared spectrum consists of two major portions— the group frequency region and the fingerprint region. The strength of the infrared method lies in the fact that each functional group has a characteristic absorption band in this so-called "group frequency region" while direct comparison of fingerprint regions between unknown and standard reference spectra give the true unambiguous identification. Initially, however, concentration is placed on "seeing" the various functional groups present in the molecule. This facility in turn demands the use of correlation tables listing such values until frequent usage commits them to memory. Many of the unknown spectra to be interpreted into molecular structure are not entirely devoid of some background information such as the exact molecular weight (from mass spectrometry) which can be translated into a molecular formula, $C_x H_y O_z$. This important piece of information might also have come from the more classical and still widely used combustion analysis. Today the chemist no longer concentrates on a solution to this problem from one particular method alone but by a joint attack by many associated techniques (UV, NMR, MS, m.p., b.p., etc.).

Double Bond Equivalent. For practical purposes let us assume that besides the unknown spectrum you are also given the molecular formula. Possession of this information alone allows the calculation of the "Double Bond Equivalent" (hereafter called D.B.E.) of the molecule; that is, the number of double bonds (or their equivalent) the molecule possesses relative to the appropriate saturated hydrocarbon. For example, what is the D.B.E. of C_6H_{10}? From the general formula, $C_n H_{2n+2}$ for the homologous series of saturated hydrocarbons, the C_6 member should have H_{14}. Therefore, the original formula, C_6H_{10}, is four hydrogens short of that prediction. Now, if it takes the removal of two hydrogen atoms from two adjacent carbon atoms to create a double bond, then a shortage of four hydrogens corresponds to two double bonds or two D.B.E. A triple bond corresponds to the shortage of four hydrogens, or 2 D.B.E. A D.B.E. of two can suggest two double bonds, or one double bond and one ring or one triple bond; three possible structures from this information alone are

If the molecular formula contains an oxygen atom or atoms, then simply substitute CH_2 for each one and calculate as before. For example,

$$\left. \begin{array}{l} C_6H_6O \rightarrow C_6H_6 + CH_2 = C_7H_8 \\[2mm] C_n H_{2n+2} \longrightarrow C_7H_{16} \end{array} \right\} \Delta = H_8 = 4 \text{ D.B.E.}$$

The number of D.B.E. that a molecule possesses enables you to search the spectrum to satisfy this criterion; for example, one carbonyl, two double bonds, and one ring might be a possible solution.

Use of Group Frequencies. The search for functional groups by looking for their characteristic stretching vibrations in the "group frequency region" then begins. It should be mentioned, however, that, if possible, confirmation of your findings should also be sought; for example, $\nu(CH) = 2980 \text{ cm}^{-1}$ could be either —CH_3 or —CH_2— but the appearance of the $\delta(CH)$ at 1370 cm^{-1} confirms the presence of the methyl group. In the case of a band at 3300 cm^{-1}, one might suspect a terminal acetylene. Confirmation can be easily provided by the appearance of the $\nu(C\equiv C)$ at 2300 cm^{-1}. Even if one had mistaken the 3300 cm^{-1} band to be $\nu(O—H)$ because of inexperience in recognizing band contours ($\nu(O—H)$ tends to be broad and dependent on dilution), the lack of any $\nu(C—O)$ at 1000 cm^{-1} would have made you re-think your proposed assignment. Only practice can provide that intuitive power of making correct assignments the first time.

Particular emphasis should be placed on the significance of the assignment of the presence of a carbonyl function, $\nu(C=O)$ at 1720 cm^{-1}. The presence of a carbonyl group can be attributed to any of the following chemical classes: (a) ketone, (b) aldehyde, (c) ester, (d) lactone, (e) anhydride, (f) carboxylic acid.

To differentiate between these is a matter of elimination. For instance, if the compound is an aldehyde, there should be two highly characteristic $\nu(C—H)$ bands at 2700 and 2800 cm^{-1}. Esters, in addition to the $\nu(C=O)$, should also exhibit a characteristic ester band at 1200 cm^{-1}, $\nu(C—O)$. Lactones show complex band patterns in the carbonyl stretching region, usually doublets. Anhydrides show much more complex band patterns in the $\nu(C=O)$ region and at higher values than lactones (molecular formula requires O_4). In the case of carboxylic acids the presence of a very broad $\nu(O—H)$ centered around 3000 cm^{-1} dominates the spectrum. In the event that none of the above applies, the compound is a ketone and the detailed search continues to discover its close environment (ring size, conjugation, etc.).

Use of Band Intensity. Spectra presented on the following pages are quite often recorded in solution in CCl_4. The calculation of apparent absorptivity is possible— yet another criterion in the identification of an absorption band. The first step is to convert the % transmittance values for the base line and peak maximum to absorbance by use of the following diagram and then apply the general equation

$$\epsilon^a = \frac{\text{Absorbance} \times \text{Molecular Weight}}{\text{Concentration (mg/ml)} \times \text{Cell path (cm)}} \qquad (11\text{--}3)$$

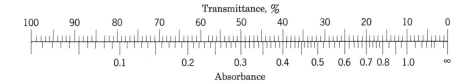

11-7 REPRESENTATIVE SPECTRA

Let us first examine the infrared spectra of some representative compounds and correlate absorption bands with molecular structure. We will begin with a simple hydrocarbon, *n*-hexane, and then modify the molecule in several ways and observe the effect on the spectrum.

The infrared spectrum of *n*-hexane (Figure 11–9) exhibits vibrational absorp-

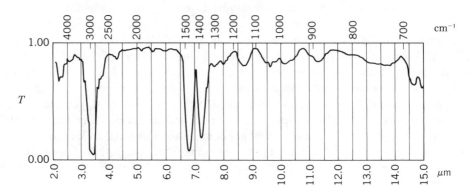

Figure 11–9   Infrared spectrum of *n*-hexane. [Courtesy of Sadtler Research Laboratories.]

tion bands which are characteristic of aliphatic hydrocarbons, that is, asymmetric and symmetric C—H stretching (averaged at 2899 cm^{-1}), —CH$_2$— scissoring (1471 cm^{-1}) and C—CH$_3$ bending (1381 cm^{-1}). The weak band near 725 cm^{-1} is caused by bending vibrations of the —(CH$_2$)$_4$— unit.

The *2-methylpentane* molecule represents a branched chain isomer of *n*-hexane. Let us consider how branching affects the absorption spectrum (Figure 11–10). The C—H stretching and —CH$_2$— scissoring vibrations are not changed significantly compared to the straight chain analogue. However, the C—CH$_3$ bending vibrations indicate a strong "doublet" at 1377 cm^{-1} and 1361 cm^{-1}. This doublet and a band at 1166 cm^{-1} are characteristic of the isopropyl group.

If the saturated hydrocarbon is rearranged into the form of an unstrained ring, the C—H stretching vibration is essentially unchanged and the —CH$_2$— scissoring vibration is displaced slightly to longer wavelength. For example, the spectrum of *cyclohexane* (Figure 11–11) exhibits C—H stretching vibrations at 2865 cm^{-1} and —CH$_2$— scissoring at 1451 cm^{-1} compared to the respective 2899 cm^{-1} and 1471 cm^{-1} values for *n*-hexane. When the ring compound is sterically strained, the C—H stretching vibrations move to higher frequency. For example, for bromocyclopropane the symmetric and asymmetric C—H stretching vibrations appear at 3077 cm^{-1} and 2985 cm^{-1}, respectively.

When two hydrogens are removed from *n*-hexane to give *hexene-1*, the spectral effect is shown in Figure 11–12. Two classes of C—H stretching vibrations are

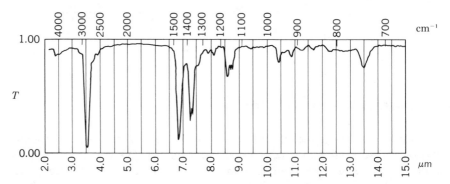

Figure 11–10   Infrared spectrum of 2-methylpentane. [Courtesy of Sadtler Research Laboratories.]

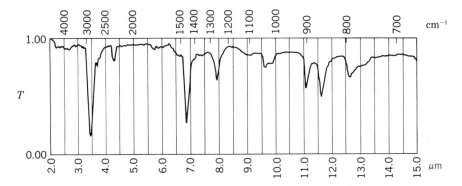

Figure 11–11 Infrared spectrum of cyclohexane. [Courtesy of Sadtler Research Laboratories.]

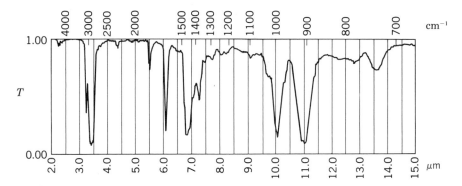

Figure 11–12 Infrared spectrum of hexene-1. [Courtesy of Sadtler Research Laboratories.]

apparent in this spectrum. The absorption band at 3067 cm^{-1} is characteristic of alkene carbons and, in particular, the vinyl group. Alkene C—H bonds which are not terminal normally absorb in the 3030 to 3012 cm^{-1} region. The C=C alkene stretching vibration is noted at 1637 cm^{-1}. Out of plane bending vibrations for alkene C—H groups occur at 992 cm^{-1} and 903 cm^{-1}. The typical aliphatic C—H stretching and bending vibrations are noted for the remaining groups of the molecule.

The spectrum for *benzene* is shown in Figure 11–13. Aromatic C—H stretching

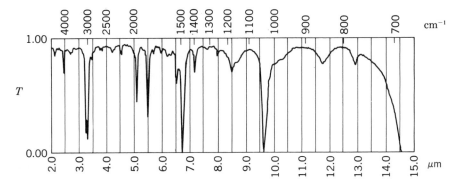

Figure 11–13 Infrared spectrum of benzene. [Courtesy of Sadtler Research Laboratories.]

vibrations appear at 3030 cm^{-1}. Ring skeletal vibrations (—C=C— vibrations) are indicated in the 1613 to 1471 cm^{-1} region with a strong maximum at 1481 cm^{-1}. C—H out of plane bending occurs in the broad absorption region below 769 cm^{-1}.

When an aliphatic hydrocarbon hydrogen atom is replaced by an —OH group, the spectrum changes to include absorptions due to the new O—H and C—O vibrations in addition to the carbon-hydrogen vibrations already mentioned. The spectrum of *n*-hexanol (Figure 11–14) shows an intense O—H stretching band at 3356 cm^{-1}. The broadness of this band is due to hydrogen bonding among the alcohol molecules. The broad band distributed around the 1062 cm^{-1} peak is characteristic of the C—O stretching vibration of primary alcohols. The C—H stretching and bending vibrations are essentially identical to those of the unsubstituted alkane.

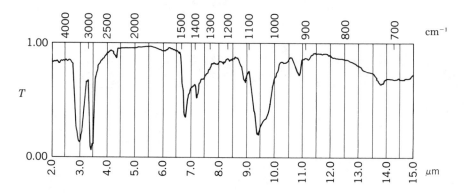

Figure 11–14   Infrared spectrum of *n*-hexanol. [Courtesy of Sadtler Research Laboratories.]

The effect of hydroxy substitution in benzene is illustrated in the spectrum of *phenol* (Figure 11–15). The O—H stretching vibration appears over a very broad band with a maximum at 3333 cm^{-1} and a "shoulder" at 3030 cm^{-1} (caused by aromatic C—H stretch). The bands at 1595 cm^{-1}, 1497 cm^{-1}, and 1468 cm^{-1} represent the common benzene skeletal vibrations. The band at 1359 cm^{-1} is due to O—H bending, and the intense broad band with a maximum at 1218 cm^{-1} is characteristic of the C—OH stretching vibration of phenol and its derivatives. The aromatic C—H out of plane bending vibrations at 750 cm^{-1} and 685 cm^{-1} indicate that the benzene ring is monosubstituted.

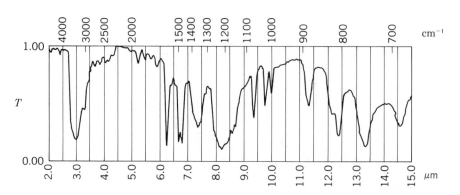

Figure 11–15   Infrared spectrum of phenol. [Courtesy of Sadtler Research Laboratories.]

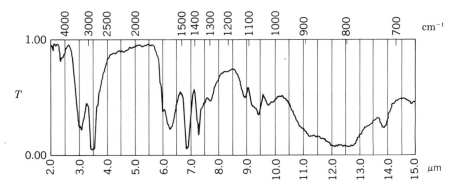

Figure 11-16   Infrared spectrum of *n*-hexylamine. [Courtesy of Sadtler Research Laboratories.]

If a terminal hydrogen on *n*-hexane is replaced by an amine group, C—N and N—H stretching and N—H bending vibrations appear. The spectrum of the resulting compound, *n-hexylamine*, is given in Figure 11-16. An N—H stretching vibration doublet appears at 3333 and 3247 cm^{-1}. The location of this doublet is somewhat variable because hydrogen bonding tends to shift it to lower wavenumber. N—H bending vibrations are indicated in both the broad band with the maximum at 1592 cm^{-1} and the very broad band from 909 to 714 cm^{-1}. The familiar aliphatic group vibrations also are present.

The replacement of a methylene group of *n*-hexane with an oxygen atom yields an ether, for example, *n-butyl-methyl ether*. The absorption spectrum for this compound (Figure 11-17) exhibits a strong C—O—C absorption band around 1126 cm^{-1} in addition to the other C—H vibrational bands.

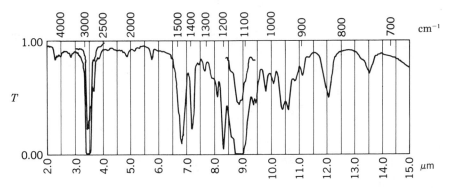

Figure 11-17   Infrared spectrum of *n*-butyl-methyl ether. [Courtesy of Sadtler Research Laboratories.]

The *carbonyl stretching absorption* band is an excellent barometer for molecular structure determination. The band intensity is very large and its wavelength maximum can be very precisely determined. The exact location of the carbonyl absorption band varies with (1) the physical state of the molecule, (2) electrostatic interactions with neighboring constituents, (3) conjugation, (4) hydrogen bonding, and (5) ring strain. The position of the band is essentially the same for aliphatic aldehydes, 1740 to 1720 cm^{-1}, and ketones, 1725 to 1705 cm^{-1}; however, it appears at considerably different locations for conjugated unsaturated systems, carboxylic acids, esters, anhydrides, and acyl halides. The locations of these various C=O stretching vibrations should be studied carefully.

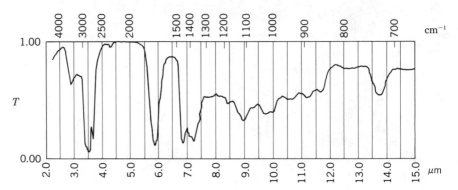

Figure 11–18 Infrared spectrum of *n*-hexanal. [Courtesy of Sadtler Research Laboratories.]

Hexanal represents the aldehyde oxidation product of 1-hexanol. Its spectrum (Figure 11–18) shows a strong carbonyl stretching vibration at $1727\ cm^{-1}$. In addition to the familiar methyl and methylene C—H stretching vibrations at 2924 and $2841\ cm^{-1}$, a sharp aldehyde C—H stretching vibration appears at $2710\ cm^{-1}$. This vibration is characteristic of the aldehyde functional groups and may be used to decide if the compound studied is an aldehyde or ketone; the C=O absorption band is not diagnostic for this purpose. The absorption peaks at 1464 and $1412\ cm^{-1}$ represent methylene bending vibrations and the one at $1393\ cm^{-1}$ indicates the typical C—CH₃ bending vibration.

The characteristic carbonyl stretching vibration is shifted to *lower* frequency in conjugated systems. This is indicated in the spectrum of *acetophenone* (Figure 11–19) in which the C=O absorption band appears at $1692\ cm^{-1}$.

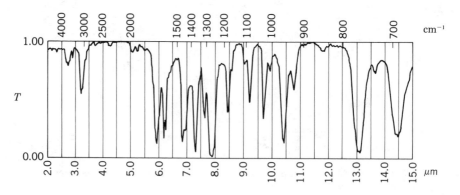

Figure 11–19 Infrared spectrum of acetophenone. [Courtesy of Sadtler Research Laboratories.]

The position of the C=O stretching vibration of esters also depends on conjugation and ring size. For example, compare *n-butyl acetate* (Figure 11–20) and *ethyl benzoate* (Figure 11–21). The C=O stretching vibration occurs at $1739\ cm^{-1}$ in the aliphatic ester, whereas in the aromatic ester it appears at $1724\ cm^{-1}$. In the *n*-butyl acetate spectrum, we also note a broad C—O stretching band centered at $1242\ cm^{-1}$, characteristic for the acetate group. Formates, propionates and *n*-butyrates exhibit similar broad bands centered around $1189\ cm^{-1}$. Esters of α,β-unsaturated and aromatic acids have two strong bands

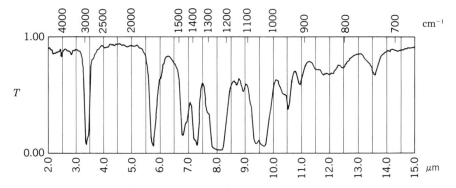

Figure 11–20   Infrared spectrum of *n*-butyl acetate. [Courtesy of Sadtler Research Laboratories.]

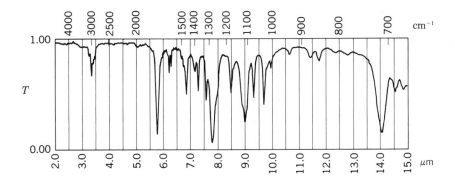

Figure 11–21   Infrared spectrum of ethyl benzoate. [Courtesy of Sadtler Research Laboratories.]

around 1310 to 1250 cm^{-1} and 1200 to 1100 cm^{-1}. These are evident as the 1277 cm^{-1} and 1109 cm^{-1} peaks in the ethyl benzoate spectrum.

Carboxylic acids commonly exist as dimers in the solid and liquid states. The two monomer units are held together by fairly strong hydrogen bonding between the two carboxyl groups. As a consequence, there are essentially no free O—H stretching vibrations in these molecules even in dilute solution, and the carboxylic acid O—H stretching absorption often consists of a continuous series of bands distributed over a very broad region (usually between 3330 and 2500 cm^{-1}). These absorption bands are illustrated in the spectrum of *n*-hexanoic acid (Figure 11–22). Even though the O—H stretching band is quite broad, the characteristic aliphatic C—H stretching vibration is evident at 2915 cm^{-1}. The carbonyl stretching band appears at 1698 cm^{-1}; it is more intense than the corresponding ester band. The doublet with maxima at 1460 and 1408 cm^{-1} is caused by —CH$_2$— scissoring and O—H bending. The broad absorption band between 1330 and 1190 cm^{-1} represents C—OH stretching vibrations (usually a doublet near 1250 cm^{-1}).

The spectrum of *benzoic acid* (Figure 11–23) also exhibits the very broad acid O—H stretching region between 3230 and 2500 cm^{-1} and a superimposed aromatic C—H stretching vibration at 3012 cm^{-1}. The intense band at 1678 cm^{-1} is characteristic of the conjugated carbonyl group, and the group of benzene skeletal vibration bands at 1605 cm^{-1}, 1587 cm^{-1}, and 1456 cm^{-1} indicate aromaticity. The

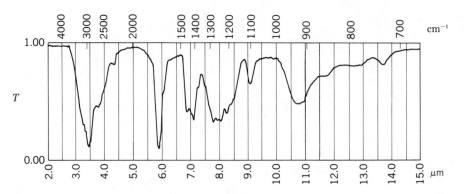

Figure 11–22   Infrared spectrum of *n*-hexanoic acid. [Courtesy of Sadtler Research Laboratories.]

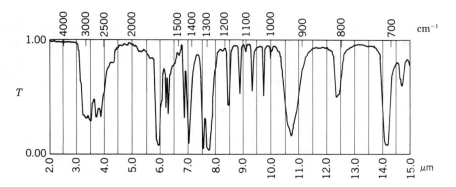

Figure 11–23   Infrared spectrum of benzoic acid. [Courtesy of Sadtler Research Laboratories.]

strong band at 1425 cm^{-1} is caused by O—H bending vibrations, and C—O stretching vibrations are evidenced in the maxima at 1325 and 1285 cm^{-1}. The strong band around 706 cm^{-1} is characteristic of a monosubstituted aromatic compound.

11–8   ELUCIDATION OF STRUCTURE

The following series of examples of interpreted spectra illustrate technique involved in structural elucidation by infrared spectra. In many of the cases discussed unambiguous structures result. In the case of ambiguous solutions the final deciding factor is the comparison with standard reference spectra if available.

Example 1   Volatile liquid, C_8H_{14}. Spectrum on page 185.

     A. *Calculations*
         (a) D.B.E.

$$C_n H_{2n+2} \longrightarrow \left. \begin{array}{l} C_8H_{14} \\ C_8H_{18} \end{array} \right\} \Delta = H_4 = 2 \text{D.B.E.}$$

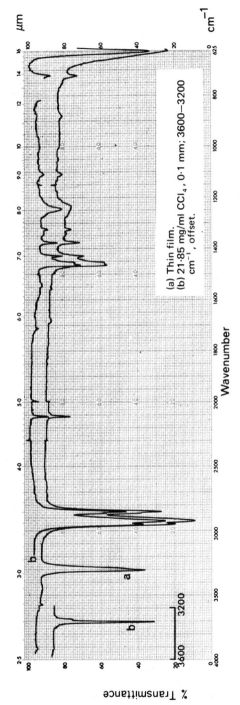

Infrared spectrum of compound discussed in Example 1. (Courtesy of Heyden and Son, Ltd.)

(b) Apparent absorptivities

3300 cm^{-1} band	2100 cm^{-1} band
% transmission for base line = 86 % transmission for peak max. = 32 Absorbance = 0.495 − 0.067 = 0.428	% transmission for base line = 98 % transmission for peak max. = 88 Absorbance = 0.055 − 0.010 = 0.045
$\epsilon^a = \dfrac{\text{absorbance} \times \text{mol. wt}}{\text{conc. (mg/ml)} \times \text{path (cm)}}$	$\epsilon^a = \dfrac{\text{absorbance} \times \text{mol. wt}}{\text{conc. (mg/ml)} \times \text{path (cm)}}$
$\epsilon^a = \dfrac{0.428 \times 110}{21.85 \times 0.01} = 215$	$\epsilon^a = \dfrac{0.045 \times 110}{21.85 \times 0.01} = 23$

B. Assignments

cm^{-1}	ϵ^a	Assignment	Structural Unit	D.B.E.	Formula Unit
3300 2100 625	215 23	$\nu(\equiv\!C\!-\!H)$ $\nu(C\!\equiv\!C)$ $\gamma(\equiv\!C\!-\!H)$	$-C\!\equiv\!C\!-\!H$	2	C_2H
2960–2850		$\nu(C\!-\!H)$ satd.			
1470 720		$\delta_a(C\!-\!H)$ $\rho(CH_2)$	$-(CH_2)_n-$ where $n \geqslant 5$		C_5H_{10} at least
1370		$\delta_s(C\!-\!H)$	$-CH_3$		CH_3

C. Comment

The existence of 2D.B.E. in this compound makes one search for functional groups to use up this quota. Absence of any strong sharp bands in the 1650 cm^{-1} region rules out any ethylenic linkages. However, the presence of a strong ($\epsilon^a = 215$) sharp band at 3300 cm^{-1} indicates a C≡C linkage which satisfies the 2D.B.E. requirement. Lack of experience might lead you to believe that this absorption is due to $\nu(O\!-\!H)$; however, it is not an oxygen containing compound and the $\epsilon^a = 215$ is far too high for alcohols (average value $\epsilon^a = 60$). Notice the weak $\nu(C\!\equiv\!C)$–a reflection of the dipole moment of this bond. In general, the greater the dipole moment, the more intense the absorption band. Since all $\nu(C\!-\!H)$ absorption bands in the spectrum lie below 3000 cm^{-1}, they belong to saturated entities. In practice, the recognition of unsaturated groups is aided by using this 3000 cm^{-1} marker—bands immediately above indicate unsaturated groups, while those immediately below belong to saturated groups. With reference to this particular spectrum, the presence of the 1370 cm^{-1} band indicates CH$_3$ group(s), while the 1470 cm^{-1} band indicates methylene groups. Notice the ratio of these two bands, weighted heavily in favor of many CH$_2$'s relative to CH$_3$'s (assuming both bands have equal intensity per hydrogen content). Confirmation for the existence of a number of methylene groups comes from the $\rho(CH_2)$ at 720 cm^{-1}. Usually this band appears only when there are at least five methylene groups in a chain. If this is so, we have only one methyl group since we have already accounted for seven of the eight carbons present. Our compound is therefore *oct-1-yne.*

Example 2 Liquid with molecular weight of 58 (C, H, and O only). Spectrum on page 188.

A. *Calculations*
   (a) Apparent absorptivities

3610 cm^{-1} band	1650 cm^{-1} band
% transmission for base line = 90	% transmission for base line = 95
% transmission for peak max. = 44	% transmission for peak max. = 85
Absorbance = 0.36 − 0.05 = 0.31	Absorbance = 0.07 − 0.02 = 0.05
$\epsilon^a = \dfrac{0.31 \times 58}{21.1/50 \times 0.5} = 85$	$\epsilon^a = \dfrac{0.05 \times 58}{21.1 \times 0.01} = 14$

B. *Assignments*

cm^{-1}	ϵ^a	Assignment	Structural Unit	Mol. Wt.
3620	85	ν(O—H) free		
(3350)		ν(O—H) bonded	$>$C—O—H	29
1030		ν(C—O) alcohol		
3100–3000		ν(C—H) unsat.		
1650	14	ν(C=C)	H H	27
995			C=C	
910		γ(C—H) vinyl	R H	
3000–2880		ν(C—H) satd.		

C. *Comment*
The first observation is that the broad band centered around 3350 cm^{-1} disappears on dilution, indicating that an intermolecular association had existed in the more concentrated solution. The appearance of a sharp band at 3620 cm^{-1} pinpoints the hydroxyl stretching frequency and $\epsilon^a = 85$ adds further support. Absorptions just above and below 3000 cm^{-1} indicate that both unsaturated and saturated C—H groups exist. Looking at the 1650 cm^{-1} region for a ν(C=C) absorption band, we find one—weak, but nevertheless there (once again the dipole phenomenon manifesting itself). The exact nature of this double bond is decided by the γ(C—H) pattern—in this case, vinyl. Combination of the hydroxyl group with the double bond accounts for 56 of the total 58 mass units the compound possesses, leaving two to be placed, that is, primary alcohol. Our compound is therefore *allyl alcohol*.

Example 3 Colorless volatile liquid, C_9H_{12}. Spectrum on page 189.

A. *Calculations*
   (a) D.B.E.

$$C_nH_{2n+2} \longrightarrow \left.\begin{matrix} C_9H_{12} \\ C_9H_{20} \end{matrix}\right\} \Delta = H_8 = 4 \text{ D.B.E.}$$

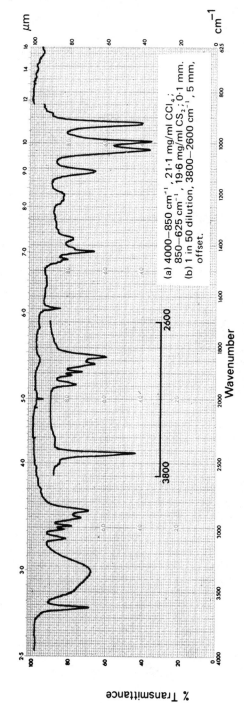

Infrared spectrum of compound discussed in Example 2. (Courtesy of Heyden and Son, Ltd.)

(a) 4000–850 cm⁻¹, 21·1 mg/ml CCl₄;
850–625 cm⁻¹, 19·6 mg/ml CS₂; 0·1 mm.
(b) 1 in 50 dilution, 3800–2600 cm⁻¹, 5 mm,
offset.

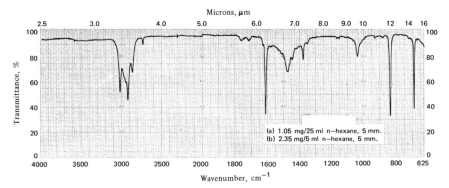

Infrared spectrum of compound discussed in Example 3. (Courtesy of Heyden and Son, Ltd.)

B. Assignments

cm^{-1}	Assignment	Structural Unit	Formula Unit	D.B.E.
3020–3000	ν(C—H) unsatd.			
1610	ν(C=C) aromatic		C_6H_3	4
840⎫ 690⎭	γ(C—H) 1,3,5-tri subst.			
3000–2880	ν(C—H) satd.	$(-CH_3)_n$		
1370	δ_s(C—H) for CH_3			

C. Comment

An aromatic nucleus is not always easy to spot. Very often the 1600-cm^{-1} band is weak and hardly visible. In this particular spectrum it is one of the most prominent features combined with the γ(C—H) absorptions at 840 and 690 cm^{-1}, indicating a 1,3,5-tri-substituted phenyl ring. The three double bonds and the ring take care of the 4 D.B.E. As a general rule always be suspicious that you have an aromatic nucleus when the D.B.E. equals or exceeds 4. The only other notable feature in the spectrum is the existence of methyl group(s). Since C_3H_9 of the molecular formula remains and we have to find three substituents for the phenyl ring, it is a good guess that our compound is *mesitylene*.

Example 4   Colorless crystalline hydrocarbon, $C_{14}H_{12}$. Spectrum on page 190.

A. Calculations
(a) D.B.E.

$$C_n H_{2n+2} \longrightarrow \left.\begin{matrix} C_{14}H_{12} \\ C_{14}H_{30} \end{matrix}\right\} \Delta = H_{18} = 9 \text{ D.B.E.}$$

(b) Apparent absorptivities

1620 cm^{-1} band	720 cm^{-1} band
$\epsilon^a = \dfrac{0.08 \times 180}{18.4 \times 0.01} = 78$	$\epsilon^a = \dfrac{0.59 \times 180}{18.4 \times 0.01} = 580$

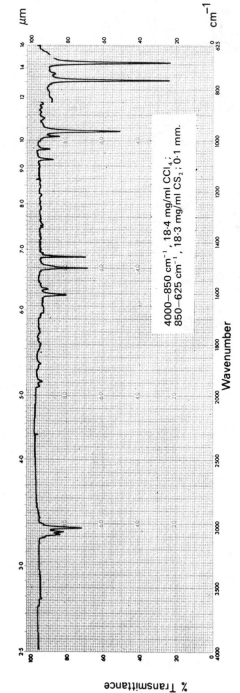

Infrared spectrum of compound discussed in Example 4. (Courtesy of Heyden and Son, Ltd.)

B. Assignments

cm^{-1}	ϵ^a	Assignment	Structural Unit	D.B.E.	Formula Unit
3100–3000		ν(C—H) unsatd.	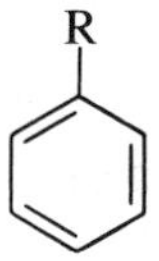 R		
1590⎫ 1500⎬ 1460⎭		ν(C=C) aromatic		4	C$_6$H$_5$
720⎫ 690⎭	580	γ(C—H) mono-subst. aromatic	R'		
1620	78	ν(C=C)	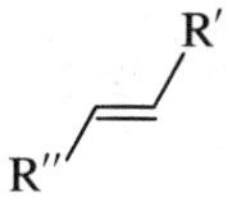	1	C$_2$H$_2$
990		γ(C—H) trans	R''		

C. Comment

The molecular formula suggests a highly unsaturated molecule of 9 D.B.E. with perhaps one or even two phenyl rings. There is strong evidence for a mono-substituted aromatic nucleus—in fact, this is a classic example. The only other dominant feature in the spectrum is the existence of a *trans*-disubstituted double bond. So far we have accounted for only 5 D.B.E. of the total of 9 and C$_8$H$_7$ of the molecular formula C$_{14}$H$_{12}$. This leaves a fragment of C$_6$H$_5$ with 4 D.B.E.—little or no doubt that we have a second mono-substituted aromatic nucleus. Our compound is *trans*-stilbene.

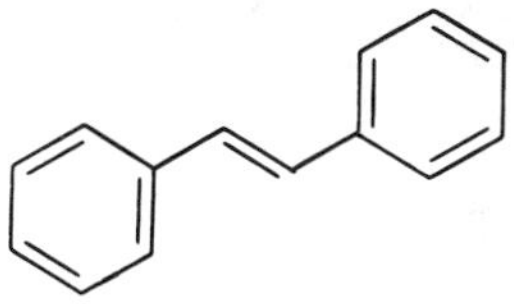

Example 5   Unstable liquid, C$_5$H$_6$O. Spectrum on page 192.

A. Calculations

(a) D.B.E.

$$\left.\begin{array}{l} C_5H_6O \longrightarrow C_5H_6 + CH_2 = C_6H_8 \\ C_nH_{2n+2} \longrightarrow C_6H_{14} \end{array}\right\} \Delta = H_6 = 3 \text{ D.B.E.}$$

(b) Apparent absorptivities

3300 cm^{-1} band	2100 cm^{-1} band
$\epsilon^a = \dfrac{0.78 \times 82}{26.7 \times 0.01} = 240$	$\epsilon^a = \dfrac{0.15 \times 82}{26.7 \times 0.01} = 46$
1640 cm^{-1} band	
$\epsilon^a = \dfrac{1 \times 82}{26.7 \times 0.01} = 310$	

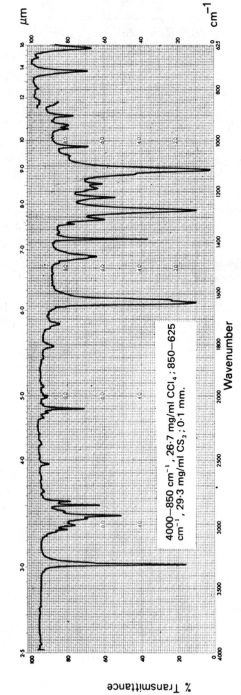

Infrared spectrum of compound discussed in Example 5. (Courtesy of Heyden and Son, Ltd.)

$4000-850$ cm^{-1}, $26 \cdot 7$ mg/ml CCl$_4$; $850-625$ cm^{-1}, $29 \cdot 3$ mg/ml CS$_2$; $0 \cdot 1$ mm.

B. Assignments

cm^{-1}	ϵ^a	Assignment	Structural Unit	D.B.E.	Formula Unit
3300	240	$\nu(\equiv$C—H$)$			
2100	46	ν(C$\equiv$C)	—C$\equiv$C—H	2	C$_2$H
630		γ(C—H)			
3060–3000		ν(C—H)	H H		
1640	310	ν(C=C)		1	C$_2$H$_2$
720		γ(C—H) cis-disubst.	R R'		
2850		ν(C—H) satd.			
2950		ν(C—H) suspect O—CH$_3$			
1110		ν(C—O)	—O—CH$_3$		CH$_3$O
1370		δ_s(C—H) for CH$_3$ enhanced intensity since O—CH$_3$ environment			

C. Comment

The evidence points to the presence of both a terminal acetylenic group and a *cis*-disubstituted double bond. Enhancement of intensity for the ν(C$\equiv$C) relative to that observed in Example 2 is an indication that the linkage is attached to something other than a methylene group—perhaps the *cis*-double bond (slight decrease in ν(C$\equiv$C) frequency also lends support to this idea). Enhancement of intensity in the 1370 cm^{-1} band (δ_s(C—H) for CH$_3$) is common when the methyl group is attached to O or N or a carbonyl group, for example, methyl ketones. Further evidence for this arrangement is the sharp ν(C—H) bands below 3000 cm^{-1}. So far we have accounted for all the D.B.E. and the entire molecular formula. Our compound is therefore *1-methoxybut-1-en-3-yne (cis)*.

Example 6   Volatile liquid, C$_7$H$_{12}$O. Spectrum on page 194.

A. Calculations
(a) D.B.E.

$$C_7H_{12}O \longrightarrow C_7H_{12} + CH_2 = C_8H_{14}$$
$$C_nH_{2n+2} \longrightarrow C_8H_{18}$$
$$\Delta = H_4 = 2 \text{ D.B.E.}$$

(b) Apparent absorptivity
1720 cm^{-1} band

$$\epsilon^a = \frac{0.47 \times 112}{9.8 \times 0.05} = 110$$

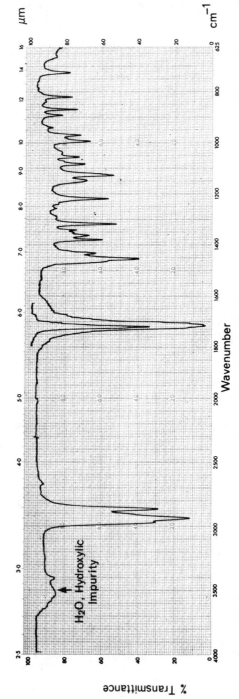

Infrared spectrum of compound discussed in Example 6. (Courtesy of Heyden and Son, Ltd.)

B. Assignments

cm^{-1}	ϵ^a	Assignment	Structural Unit	D.B.E.	Formula Unit
3000–2870		ν(C—H) satd.			
1720	110	ν(C=O) ketone		1	CO
1460		δ_a(C—H) for CH$_2$			
1370		δ_s(C—H) for CH$_3$ —(CH$_2$)$_n$— ring?			

C. Comment

This is a good example of a saturated ketone. As already explained one must eliminate all other possibilities first (i.e., aldehyde, ester, lactone, etc.). We have then a carbonyl, some CH$_2$ groups, and at least one CH$_3$ group (the ratio of 1470 to 1370 cm^{-1} bands is usually a good measure of the ratio of CH$_2$/CH$_3$ in the molecule). One methyl group would leave only C$_6$H$_9$O with one D.B.E. to account for with no other unsaturated centers present. The conclusion is that there is a ring. If we had a methyl ketone, enhancement of the 1370 cm^{-1} band would have occurred. Turning attention to the carbonyl band at 1720 cm^{-1}, we find that the only possible arrangement to satisfy this value is a substituted cyclohexanone (ring strain in a five-membered ring would have caused the frequency to increase). Three possible structures resulting from this information are as follows:

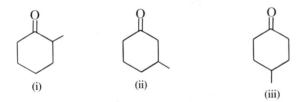

(i)   (ii)

(iii)

This is as far as one can go with the data given. Comparison of the fingerprint region to standard reference spectra would decide the matter. In any event the mere additional information that the compound contained three exchangeable hydrogens in alkaline D$_2$O pinpoints (i) as correct. Our compound is therefore *2-methylcyclohexanone.*

Example 7 Colorless liquid, C$_8$H$_8$O. Spectrum on page 196.

A. Calculations
(a) D.B.E.

$$\left.\begin{array}{l} C_8H_8O \longrightarrow C_8H_8 + CH_2 = C_9H_{10} \\ C_nH_{2n+2} \longrightarrow C_9H_{20} \end{array}\right\} \Delta = H_{10} = 5 \text{ D.B.E.}$$

(b) Apparent absorptivity
   1695 cm^{-1} band

$$\epsilon^a = \frac{0.7 \times 120}{2.5 \times 0.05} = 672$$

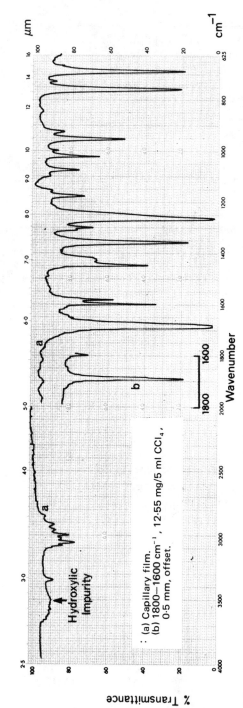

Infrared spectrum of compound discussed in Example 7. (Courtesy of Heyden and Son, Ltd.)

B. Assignments

cm^{-1}	ϵ^a	Assignment	Structural Unit	D.B.E.	Formula Unit
3100–3000		ν(C—H) unsatd.	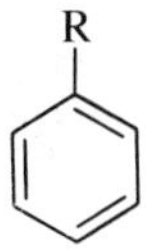		
1600			R		
1580		ν(C=C) aromatic		4	C_6H_5
1450					
760		γ(C—H) mono-subst.			
690					
1695	672	ν(C=O) ketone conj.?	R″ R′ C=O	1	CO
3000–2900		ν(C—H) satd.			
1360		δ(C—H) for CH$_3$ intensity enhanced by adjacent carbonyl	—CH$_3$	0	CH$_3$

C. Comments

Evidence for the identification of a mono-substituted aromatic nucleus is clear (cf. Example 4). Identification of the carbonyl function as a ketone can be done once again by elimination. However, the frequency observed is too low for a saturated carbonyl and suggests that the carbonyl function is adjacent to an aromatic nucleus. This takes care of the 5 D.B.E. and leaves only CH$_3$. Enhancement of the 1370 cm^{-1} band confirms our suspicions that we have a methyl ketone. Our compound is *acetophenone*.

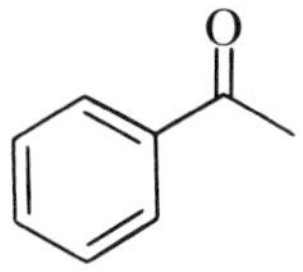

Example 8   Colorless liquid, $C_5H_{10}O_3$. Spectrum on page 198.

A. Calculations
(a) D.B.E.

$$C_5H_{10}O_3 \longrightarrow C_5H_{10} + C_3H_6 = C_8H_{16} \atop C_nH_{2n+2} \longrightarrow C_8H_{18}} \Delta = H_2 = 1 \text{ D.B.E.}$$

(b) Apparent absorptivity
1750 cm^{-1} band

$$\epsilon^a = \frac{0.46 \times 118}{5 \times 0.01} = 1085$$

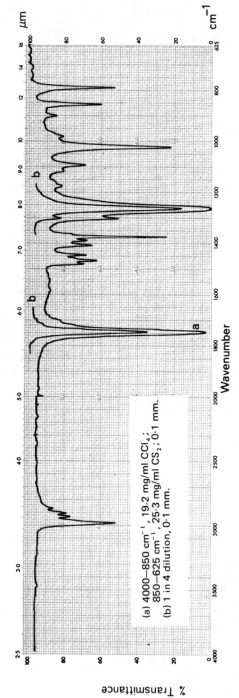

Infrared spectrum of compound discussed in Example 8. (Courtesy of Heyden and Son, Ltd.)

(a) 4000—850 cm⁻¹, 19.2 mg/ml CCl₄;
     850—625 cm⁻¹, 25.3 mg/ml CS₂; 0.1 mm.
(b) 1 in 4 dilution, 0.1 mm.

B. Assignments

cm^{-1}	ϵ^a	Assignment	Structural Unit	D.B.E.	Formula Unit
3000–2900		ν(C—H) satd.			
1470		δ_a(C—H) for CH$_2$	CH$_2$'s and CH$_3$'s		
1370		δ_s(C—H) for CH$_3$			
1750	1085	ν(C=O) Ester			
1280		ν(C—O)	R—C—O—R	1	CO$_2$
1030		ν(C—O)	CO—O—C		

(The structural unit for the 1750/1280/1030 rows shows R—C(=O)—O—R with O double-bonded to the central C.)

C. Comments

At first sight since the molecule contains O$_3$ one might suspect an acid. However, the spectrum contains no ν(O—H). Instead we have a high ν(C=O) together with a ν(C—O) at 1280 cm^{-1} (often referred to as the ester band), indicating an ester is appropriate. The ratio of the 1370/1470 cm^{-1} bands is almost unity and can be used as an intuitive clue to the fact that equal numbers of CH$_2$'s and CH$_3$'s are present. In addition the third oxygen must also be involved in an ester type linkage. Piecing together all these clues makes our compound *diethyl carbonate*.

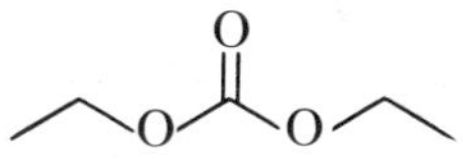

Example 9 Low boiling liquid, C$_3$H$_6$O$_2$. Spectrum on page 200.

A. Calculations
(a) D.B.E.

$$\left.\begin{array}{l} C_3H_6O_2 \longrightarrow C_3H_6 + C_2H_4 = C_6H_{10} \\ C_nH_{2n+2} \longrightarrow C_5H_{12} \end{array}\right\} \Delta = H_2 = 1 \text{ D.B.E.}$$

(b) Apparent absorptivity
1730 cm^{-1} band

$$\epsilon^a = \frac{0.7 \times 74}{1.2 \times 0.05} = 863$$

B. Assignments

cm^{-1}	ϵ^a	Assignment	Structural Unit	D.B.E.	Formula Unit
2900					
2800		ν(C—H) aldehydic			
1730	863	ν(C=O)	R—C—H	1	CHO
1150		ν(C—O) ester?			
1470		δ(C—H) for CH$_2$	C—O—?		
1370		δ_s(C—H) for CH$_3$	—CH$_2$— —CH$_3$		

(The structural unit for the 1730 row shows R—C(=O)—H; for the 1150 row shows C(=O)—O—?)

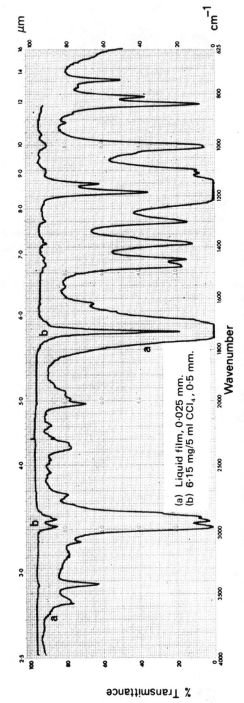

Infrared spectrum of compound discussed in Example 9. (Courtesy of Heyden and Son, Ltd.)

(a) Liquid film, 0·025 mm.
(b) 6·15 mg/5 ml CCl₄, 0·5 mm.

C. Comments

The presence of a carbonyl function together with the two very characteristic ν(C—H) bands leaves no doubt as to the identification of the aldehydic group. The intense 1190 cm^{-} band places the remaining oxygen atom in an ester type grouping. Evidence for CH$_2$'s and CH$_3$'s does exist and the formula can only accommodate one of each. Our compound is therefore *ethyl formate*

11-9 TABLES OF ABSORPTION CHARACTERISTICS

Alkanes	Table 11–2
Alkenes	Tables 11–3 and 11–4
Alkynes	Tables 11–5 and 11–6
Aromatics	Table 11–7
Alcohols and phenols	Tables 11–8 and 11–9
Ketones	Tables 11–10 through 11–14
Aldehydes	Tables 11–15 and 11–16
Ethers	Tables 11–17 and 11–18
Esters	Tables 11–19 and 11–20
Carboxylic acids	Tables 11–21 through 11–23
Anhydrides	Tables 11–24 and 11–25
Amines and amides	Table 11– 26

Hydrocarbons

Alkanes

Table 11–2 Absorption Characteristics for Alkanes[a], cm^{-1}

		ν(CH)$_a$	ν(CH)$_s$	δ(CH)$_a$	δ(CH)$_s$	Skeletal	ρ(CH$_2$)
Methine	—C—H	—	2890	—	1340		
Methylene	—CH$_2$	2926	2853	1465	—		720[b]
Methyl	—CH$_3$	2962	2872	1450	1365		
gem-Dimethyl							
—CH(CH$_3$)$_2$		Combination of —CH			1385[c]		
		and —CH$_3$			1370		
t-Butyl							
—C(CH$_3$)$_3$		As for —CH$_3$			1395[d]		
					1385		
					1365		
Cycloalkane							
(CH$_2$)$_n$		As for —CH$_2$				1030	

[a] Values quoted are for CCl$_4$ solution, ± 10 cm^{-1}. Subtract 10 cm^{-1} for solid state.
[b] Rocking mode characteristic only if five or more methylene groups in chain.
[c] Doublet bands of equal intensity.
[d] Triplet of approximately equal intensity.

Alkenes

Table 11–3 Absorption Characteristics for Alkenes[a], cm^{-1}

Group	Structure	ν(CH)	ν(C=C)	δ(CH)	γ(CH)
Vinyl	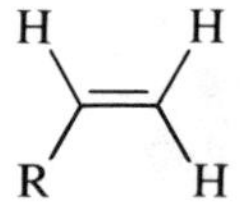	3085	1642	1415 1300	990 910
Vinylidine	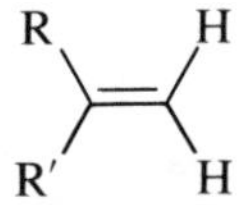	3080	1652	1415	890
Cis-disubst.	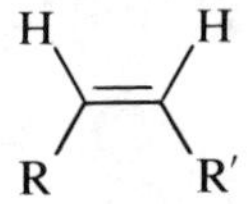	3020	1657	1300	690
Trans-disubst.	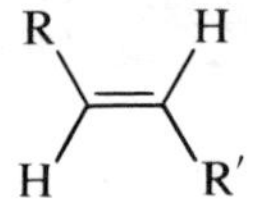	3020	1672	1410	980
Trisubst.	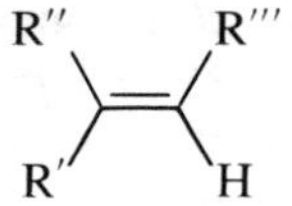	3030	1670	1345	840
Tetrasubst.	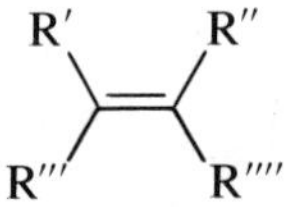	—	1670	—	—

[a] Mean positions quoted for CCl$_4$ solution, ± 10 cm^{-1}.

Table 11–4 Intramolecular Factors Affecting the ν(C=C) for Alkenes

Ring strain	As ring size diminishes, frequency is lowered due to change in bond angles. Cyclohexene is taken as the reference point. 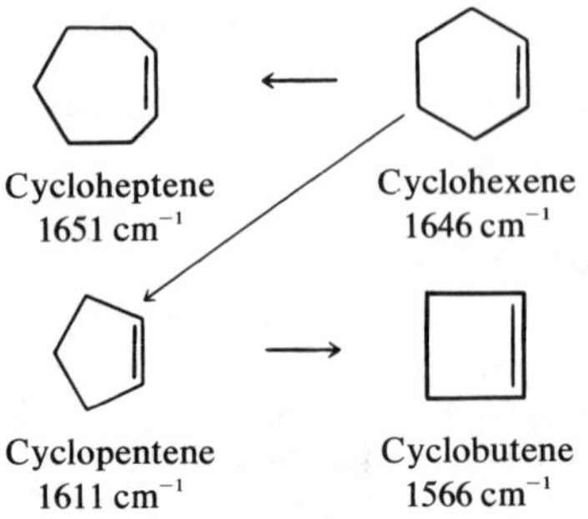

Cycloheptene
1651 cm^{-1}

Cyclohexene
1646 cm^{-1}

Cyclopentene
1611 cm^{-1}

Cyclobutene
1566 cm^{-1}

Table 11–4 (*continued*)

Methyl substitution	The presence of a methyl group on the double bond raises the frequency of $\nu(C{=}C)$ relative to the unsubstituted.

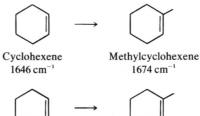

Cyclohexene	Methylcyclohexene
1646 cm^{-1}	1674 cm^{-1}

Cyclopentene   Methylcyclopentene
1611 cm^{-1} 1658 cm^{-1}

Ring condensation	Frequency is lowered due to bond angle change brought about by the formation of rigid con-formations.

Cyclohexene   Bicyclo(2,2,1)heptene
1646 cm^{-1} 1568 cm^{-1}

Conjugation	Frequency is lowered and two bands appear.

$$\overset{\longleftarrow \quad \longrightarrow}{-C{=}C{-}C{=}C-} \qquad \overset{\longrightarrow \quad \longrightarrow}{-C{=}C{-}C{=}C-}$$

$\nu(C{=}C)$ $\nu(C{=}C)$
sym   asy

Most marked reduction shown by aliphatic systems:

Butadiene   Vinyl acetylene   Styrene
1597 cm^{-1} 1600 cm^{-1} 1634 cm^{-1}

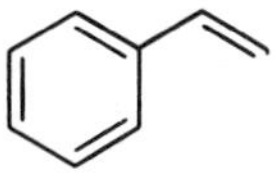

Allenes	Strong interaction (i.e., coupling) between the two adjacent double bonds results in a widely spread doublet.

$$CH_2{=}C{=}CH_2$$
Allene

$\nu(C{=}C)_{asy}$ 1980 cm^{-1}
$\nu(C{=}C)_{sym}$ 1070 cm^{-1}

Alkynes

Table 11–5 Absorption Characteristics for Alkynes, cm^{-1}

Group	Structure	$\nu({\equiv}C{-}H)$	$\nu(C{\equiv}C)$	$\gamma({\equiv}C{-}H)$
Alkynyl	$-C{\equiv}C{-}H$	3300	2140–2100	650

Table 11–6 Intramolecular Factors Affecting the $\nu(C\equiv C)$ in Alkynes

Conjugation	As for alkenes the effect is to lower the frequency. However, the $\nu(C\equiv C)$ suffers a much larger decrease than $\nu(C=C)$.

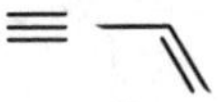

Vinyl acetylene	Diacetylene
2099 cm^{-1}	2024 cm^{-1}

Disubstitution	Intensity is reduced markedly since there is little change in dipole moment. Often a weak doublet is observed.

$$H_3C\text{—}\equiv\text{—}CH_3$$

Dimethylacetylene
2310 and 2233 cm^{-1}

Aromatics

Table 11–7 Absorption Characteristics for Aromatics, cm^{-1}

Group	$\nu(CH)$	$2\gamma(CH)$	$\nu(C=C)$	$\gamma(CH)$	$\gamma(CH)$
	3100–3000	2000–1650	1650–1450	1225–960	900–670

$\nu(CH)$	One to five bands possible. Usually three. Monosubstituted aromatics show most bands.
$\nu(C=C)$	Usually four bands arising from the combined stretching and deformation of the benzene ring.

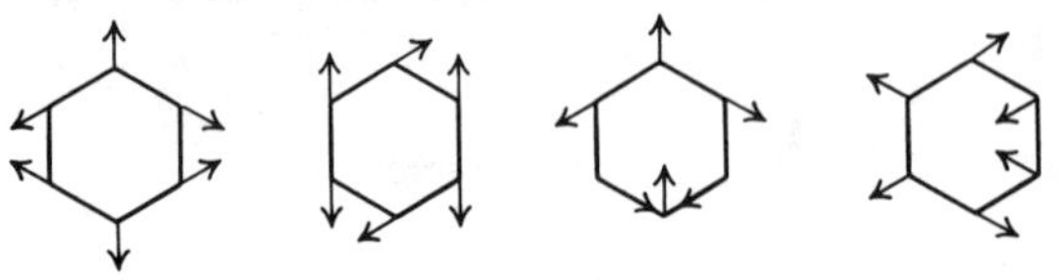

In most cases, however, the appearance of two strong bands are sufficient to determine the presence of an aromatic nucleus (i.e., 1650–1560 cm^{-1} and 1525–1475 cm^{-1}).

$\delta(CH)$	Up to six bands may be observed but since they occur within the fingerprint region they are of little diagnostic value.
$\gamma(CH)$	These bands originate from the motion of the hydrogen atoms out of the plane of the benzene ring. 

Table 11–7 (*continued*)

The number and position of the bands depend primarily on the number of adjacent hydrogen atoms remaining on the benzene ring. The band patterns are highly characteristic and provide a useful method of determining the substitution of the aromatic nucleus.

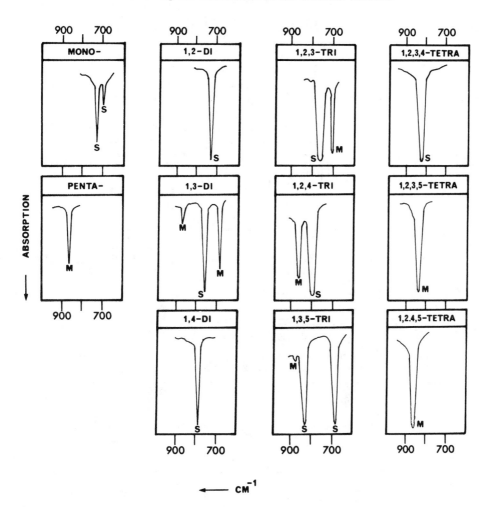

(Reproduced by kind permission from *Spectroscopic Techniques in Organic Chemistry*, A. J. Baker and T. Cairns, Heyden & Son Ltd., London, 1966.)

2γ(CH)	Characteristic patterns of bands arise out of the overtones of γ(CH). These can be used diagnostically to ascertain the substitution pattern of the phenyl nucleus. Sometimes they are too weak to be observed and one must use the parent γ(CH) pattern, also characteristic of substitution.

Table 11–7 (*continued*)

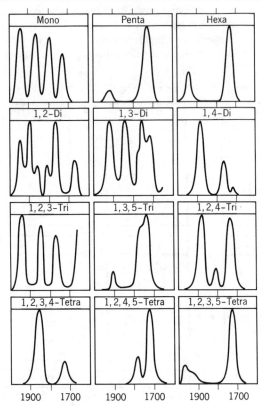

(Reproduced by kind permission from *Infrared Structural Correlation Tables and Data Cards*, Ed. R. G. J. Miller and H. A. Willis, Heyden & Son Ltd., London, 1969.)

Alcohols and Phenols

Table 11–8 Absorption Characteristics for Alcohols and Phenols[a], cm^{-1}

Group	Structure	ν(O—H)	δ(OH)	ν(C—O)
1° Alcohols	—CH₂OH	3640	1350–1250	1080 1010
2° Alcohols	>CHOH	3630	1350–1250	1120 1090
3° Alcohols	>C—OH	3620	1400–1300	1160 1140
Phenols	(phenol OH)	3610	1400–1300	1205 1170

[a] Values quoted for ν(OH) are for isolated or "free" hydroxyl groups in CCl₄ and are accurate to within 1 cm^{-1} recorded with a grating spectrophotometer. However, the hydroxyl group can participate in hydrogen bonding. Such a situation leads to a pronounced broadening of the hydroxyl stretching frequency, ν(OH), and a shift to lower frequencies. The various situations that can arise are summarized in Table 11–9 with certain criteria for identification.

Table 11–9 Hydrogen Bonding in Alcohols and Phenols

Type	ν(O—H) (cm^{-1})	Band Shape	Dilution Behavior
Free isolated R—O—H	3650–3600	Sharp	No change
Intermolecular (between molecules)			
(a) Dimeric 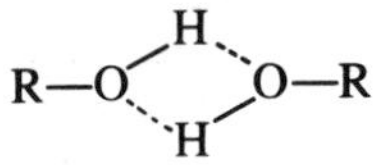	3550–3450	Fairly sharp	Decrease in intensity as dilution increases with appearance of free ν(OH).
(b) Polymeric 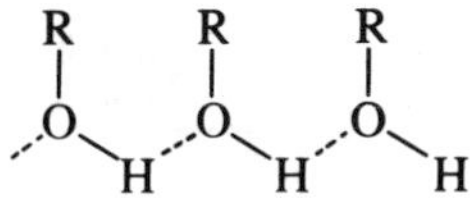	3400–3200	Broad	
Intramolecular (within the same molecule)			
(a) 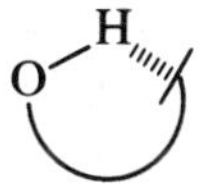	3570–3450	Fairly sharp	No change
(b) Chelate 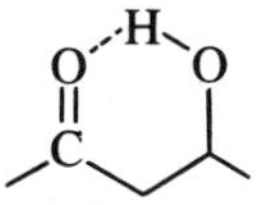	3200–2500	Very broad	

Ketones

Table 11–10 Absorption Characteristics for Ketones

Type	ν(C=O) (cm^{-1})	ϵ
Aliphatic sat. 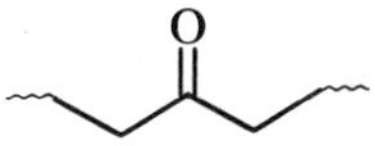	1725–1705	300–500

Table 11–11 Intramolecular Factors Affecting the ν(C=O) in Ketones

Conjugation	In general, conjugation lowers ν(C=O).		
	$\alpha\beta$-Unsatd.	Aryl	Di-aryl
	1685–1665 cm^{-1}	1700–1680 cm^{-1}	1670–1660 cm^{-1}

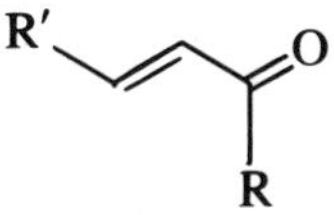

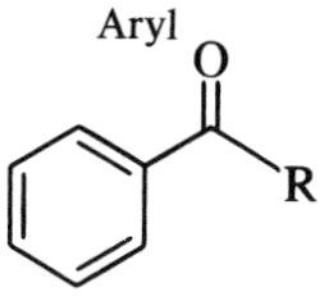

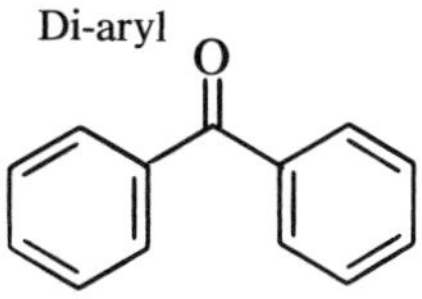

Table 11–12 Intermolecular Factors Affecting $\nu(C{=}O)$

Intermolecular hydrogen bonding	In most cases, values quoted for $\nu(C{=}O)$ are for measurements in CCl$_4$ (a nonpolar solvent). Measurements in a polar solvent such as chloroform (CHCl$_3$) will give a different $\nu(C{=}O)$ value since specific solute–solvent interactions exist—a form of hydrogen bonding in which the acidic proton of the solvent molecule hydrogen bonds to the electronegative oxygen atom of the carbonyl group, for example:

$$\overset{\delta_+}{>}C{=}\overset{\delta_-}{O}\cdots\cdots\cdots H{-}CCl_3$$

Table 11–13 Dependence of $\nu(C{=}O)$ on Solvent, cm^{-1}

| | | | Solvent Polarity $\longrightarrow$ | | | |
Solute	Vapor	Hexane	CCl$_4$	CS$_2$	CHCl$_3$	Ether
(acetophenone)	1709	1697	1692	1690	1683	1694

Comments	As can be seen, $\nu(C{=}O)$ in CHCl$_3$ is 10 cm^{-1} lower than in CCl$_4$ because of the existence of intermolecular association between the solute and solvent. This sort of shift to lower frequencies can often be used to prove the existence of a carbonyl function rather than a carbon–carbon double bond, especially if the carbonyl has a low $\nu(C{=}O)$, taking it into the double bond region. Also worthy of comment is the return to 1694 cm^{-1} in ether. The reason is simple. In this particular case the solvent offers no acidic hydrogen atom for intermolecular association.

Diketones

Table 11–14 Absorption Characteristics for Diketones[a], cm^{-1}

Type	Structural Unit	$\nu(C{=}O)$	$\nu(O{-}H)$
α-Diketones	$-\underset{\underset{O}{\|}}{C}-\underset{\underset{O}{\|}}{C}-$	1730–1710	—
β-Diketones	(a) Keto form $-\underset{\underset{O}{\|}}{C}-CH_2-\underset{\underset{O}{\|}}{C}-$	1730–1710	—
	(b) Enol form $-\underset{\underset{O\cdots H-O}{\|}}{C}-CH{=}C-$	1640–1535	2800–2600
γ-Diketones	$-\underset{\underset{O}{\|}}{C}-CH_2-CH_2-\underset{\underset{O}{\|}}{C}-$	1725–1705	—

[a] Values quoted are for measurement in CCl$_4$.

Aldehydes

Table 11–15 Absorption Characteristics for Aldehydes[a]

Structure	$\nu(CH)$ (cm^{-1})	$\nu(C=O)$ (cm^{-1})	ϵ
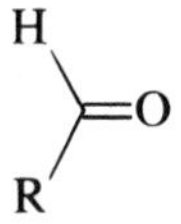	2870–2810[b] 2720–2660	1740–1720	300–500

[a] Values quoted are for measurement in CCl$_4$.
[b] Doublet and highly characteristic bands for an aldehyde.

Table 11–16 Intramolecular Factors Affecting the $\nu(C=O)$ for Aldehydes

Conjugation	As for ketones the observed shift is to lower frequencies.

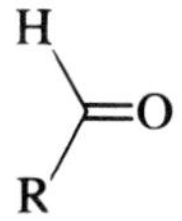

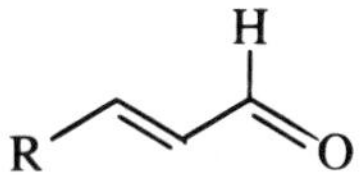

1740–1720 cm^{-1} 1705–1680 cm^{-1}

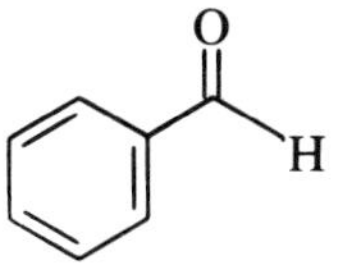

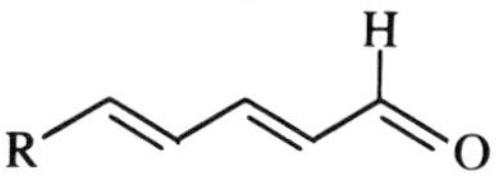

1715–1695 cm^{-1} 1680–1660 cm^{-1}

Ethers

Table 11–17 Absorption Characteristics for Ethers

Type	$\nu(C-O)$ (cm^{-1})	Comment
Saturated —CH$_2$—O—CH$_2$—	1150–1080	In spite of its position it is easily recognizable by shape and intensity

Table 11–18 Intramolecular Factors Affecting the ν(C—O) in Ethers

Conjugation	Unsaturation increases frequency of ν(C—O).

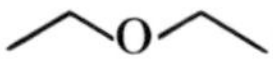

Diethylether
1145 cm^{-1}

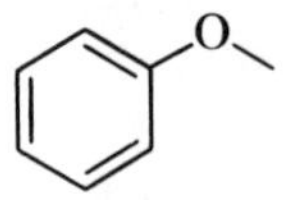

Anisole
1242 cm^{-1}

Ring strain	No correlation whatsoever.

Ethylene oxide
1270 cm^{-1}

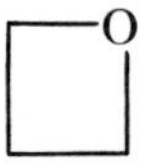

Oxacyclobutane
1010 cm^{-1}

Tetrahydrofuran
1174 cm^{-1}

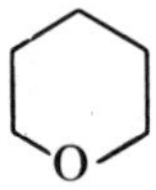

Pyran
1099 cm^{-1}

Furan
1170 cm^{-1}

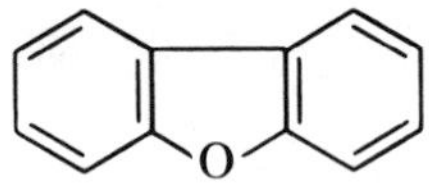

Diphenylene oxide
1198 cm^{-1}

Esters

Table 11–19 Absorption Characteristics for Esters[a]

Type	ν(C=O) (cm^{-1})	ϵ^{δ}
Saturated esters	1780–1720	500
Formates	1724–1720	—
Acetates	1755–1740	—
Propionates	1740–1730	—
n-Butyrates	1735–1725	—
Long chain esters	1735–1725	—
Triglycerides	1751–1748	—

[a] Liquid spectral values.

Table 11–20 Intramolecular Factors Affecting the $\nu(C{=}O)$ of Esters

Conjugation	Lowers frequency. Similar to both aldehydes and ketones.

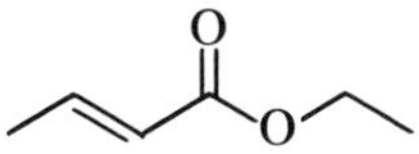

Ethyl crotonate
1712 cm^{-1}

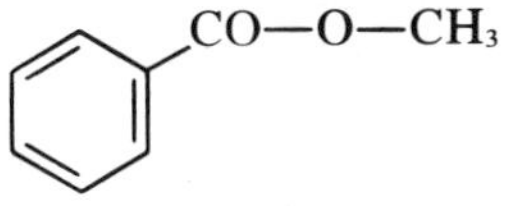

Methyl benzoate
1724 cm^{-1}

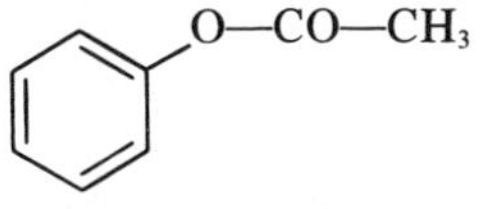

Phenyl acetate
1740 cm^{-1}

Intramolecular hydrogen bonding	As with ketones the involvement of the carbonyl in hydrogen bonding lowers its $\nu(C{=}O)$.

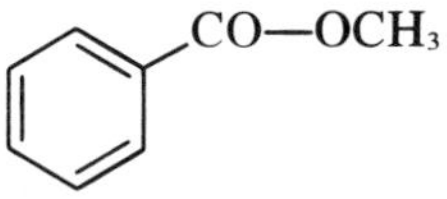

Methyl benzoate
1724 cm^{-1}

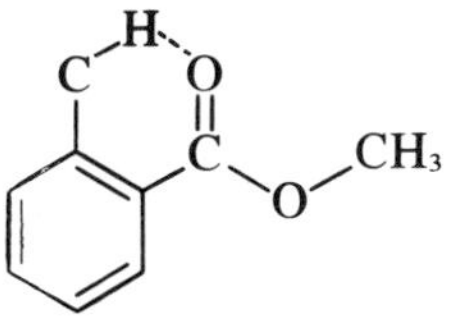

2-Hydroxymethyl benzoate
1683 cm^{-1}

α-Halogen subst. Approximate increases observed.

α-Fluorine $+15$ cm^{-1}
α-Chlorine $+12$ cm^{-1}
α-Bromine $+3$ cm^{-1}
α-Iodine   no shift detected

Examples:

CH$_3$COOEt	1740 cm^{-1}		
CH$_2$FCOOEt	1750 cm^{-1}	CH$_2$ClCOOEt	1760 cm^{-1}
CHF$_2$COOEt	1778 cm^{-1}		
CF$_3$COOEt	1790 cm^{-1}	CCl$_3$COOEt	1775 cm^{-1}

Carboxylic Acids

Table 11–21 Absorption Characteristics for Carboxylic Acids, cm^{-1}

Type	$\nu(OH)$ free	$\nu(OH)$ bonded	$\nu(C{=}O)$	$\nu(C{-}O)$
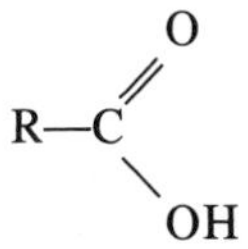	3560–3500	3000–2500	1740–1650	1320–1210

Table 11–22 Intramolecular Factors Affecting the $\nu(C{=}O)$ of Carboxylic Acids

Increasing chain length	Increasing the chain length causes the $\nu(C{=}O)$ to move to lower frequencies.

		$\nu(C{=}O)$ cm^{-1}
Acetic acid	CH$_3$COOH	1712
Propionic acid	EtCOOH	1715
Caproic acid	C$_5$H$_{11}$COOH	1710
Lauric acid	C$_{12}$H$_{25}$COOH	1700

Conjugation As for other carbonyl containing compounds, the observed trend is that the $\nu(C{=}O)$ moves to lower frequencies.

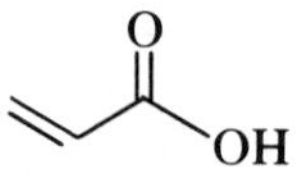

Acrylic acid	Benzoic acid	Cinnamic acid
1705 cm^{-1}	1685 cm^{-1}	1680 cm^{-1}

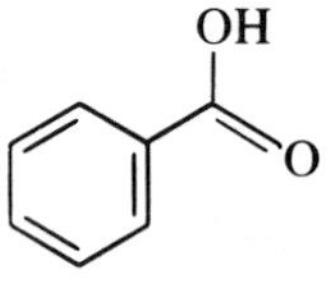

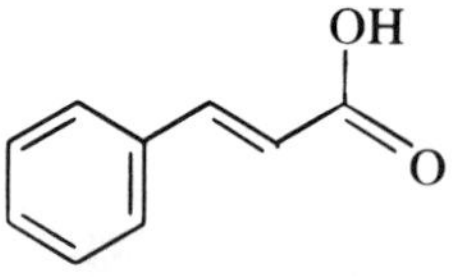

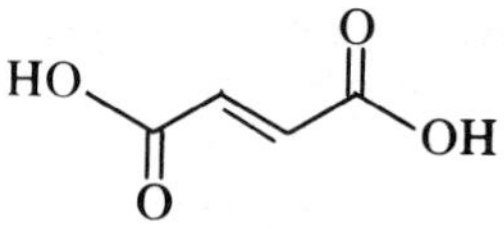

Fumaric acid	Maleic acid
1680 cm^{-1}	1705 cm^{-1}

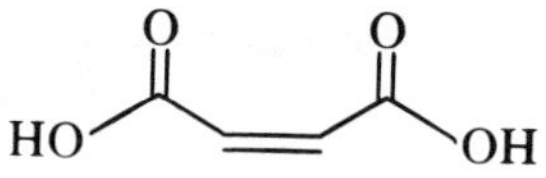

α-Halogen	Raises the frequency of $\nu(C{=}O)$

CH$_3$COOH	CH$_2$ClCOOH
1712 cm^{-1}	1735 cm^{-1}

Table 11–23 Hydrogen Bonding in Carboxylic Acids

Intermolecular hydrogen bonding	Carboxylic acids possess the correct geometry to form dimers which can exist to low concentrations in CCl$_4$. Consequently, it is sometimes difficult to observe the free $\nu(OH)$ value.

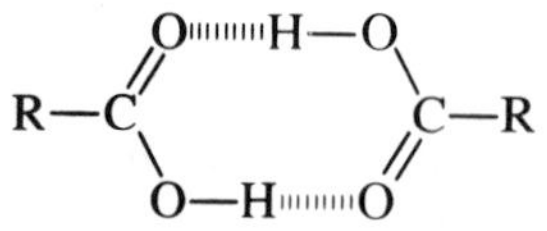

The bonded hydroxyl appears as a very broad band centered around 2800 cm^{-1} and in many cases obscures the $\nu(C{-}H)$ absorptions around 3000 cm^{-1}. Similarly the $\nu(C{=}O)$ is lowered and broadened.

Table 11–24 Absorption Characteristics for Anhydrides, cm^{-1}

Type	ν(C=O)	ν(C—O)
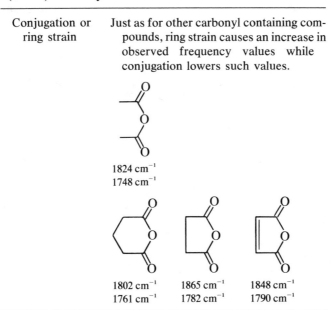	1880–1725[a]	1140–980

[a] Doublet arises from the in-phase and out-of-phase stretching of the two carbonyl groups attached to a common oxygen atom.

Table 11–25 Intramolecular Factors Affecting the ν(C=O) of Anhydrides

Conjugation or ring strain	Just as for other carbonyl containing compounds, ring strain causes an increase in observed frequency values while conjugation lowers such values.

1824 cm^{-1}
1748 cm^{-1}

| 1802 cm^{-1} | 1865 cm^{-1} | 1848 cm^{-1} |
| 1761 cm^{-1} | 1782 cm^{-1} | 1790 cm^{-1} |

Table 11–26 Absorption Characteristics for Amines and Amides[a], cm^{-1}

Group	Structure	ν(N—H)	δ(N—H)	γ(C=O)
1° Amine	RNH$_2$	3500 / 3400	1600	—
2° Amine	R$_2$NH	3350	1600	—
1° Amide	—CONH$_2$—	3500 / 3400	1600	1690
2° Amide	—CONH—	3430	1600	1680

[a] Values recorded are in dilute solution for free groups.

QUESTIONS AND PROBLEMS

11-1. The absorption band maximum of 2-pentanone occurs at shorter wavelength in 95% ethanol solvent than in *n*-hexane solvent. Why?

11-2. What infrared band or bands (give *all* the useful absorptions, if more than one) would you expect if you thought an unknown compound might be one of the following: (a) an ester, (b) an acetylenic alcohol, (c) an aromatic ether, (d) a primary amide, and (e) an olefinic ketone (nonconjugated). Give the *type of vibration* (e.g., "C—C stretch") and the *approximate region* (to the nearest $100 \, cm^{-1}$).

11-3. Compound A, C_5H_8O, has the following infrared bands (among others): 3020, 2900, 1690, and $1620 \, cm^{-1}$; in the ultraviolet it absorbs at 227 nm ($\epsilon = 10^4$). Propose a structure and say whether it is the only possible one. The compound is *not* an aldehyde.

11-4. The IR spectrum of an unknown compound has strong absorption bands at 2924, 2841, 2710, 1721, 1464, and $1393 \, cm^{-1}$. Elemental analysis of the unknown indicates that it has the following composition *by weight*: 66.6% C; 22.2% O; and 11.2% H. Determine the molecular formula for this compound. Show all reasoning and calculations.

11-5. Deduce the molecular structure of the following compounds whose infrared spectra are given below. If ambiguous structures are possible, list them and try to eliminate as many as possible suggesting how you would be able to differentiate between them on the basis of additional information from an associated technique.

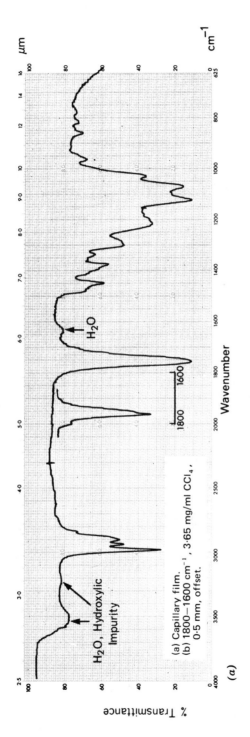

(a) Colorless liquid, $C_8H_{16}O_4$. (Spectrum courtesy of Heyden and Son, Ltd.)

215

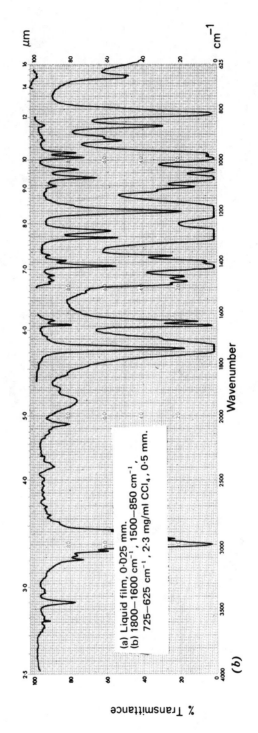

(b) Easily polymerized volatile liquid, $C_5H_8O_2$. (Spectrum courtesy of Heyden and Son, Ltd.)

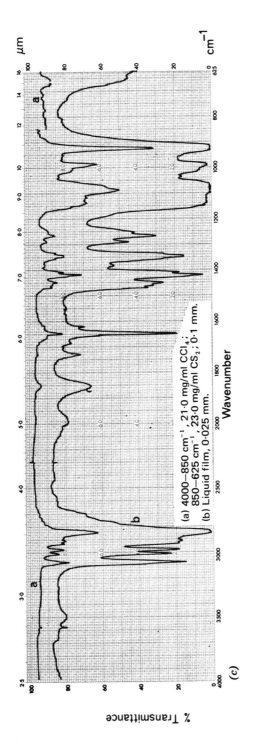

% Transmittance

Wavenumber

μm

cm⁻¹

a

b

(a) 4000—850 cm⁻¹ , 21·0 mg/ml CCl₄ ;
    850—625 cm⁻¹ , 23·0 mg/ml CS₂ ; 0·1 mm.
(b) Liquid film, 0·025 mm.

(c) Colorless liquid, $C_6H_{10}O$. (Spectrum courtesy of Heyden and Son, Ltd.)

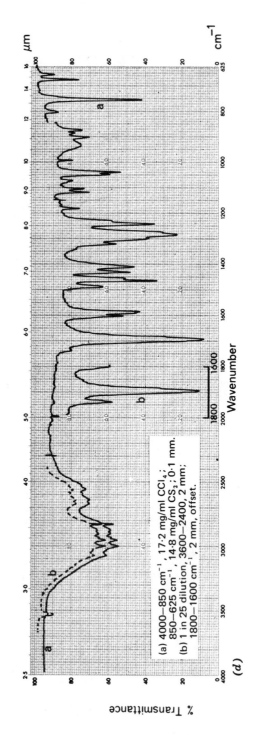

(d) Colorless crystalline solid, $C_8H_8O_3$. (Spectrum courtesy of Heyden and Son, Ltd.)

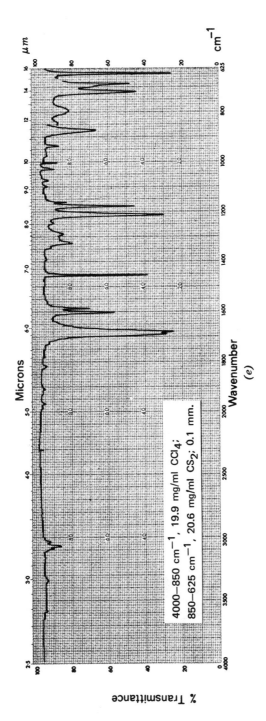

(e) Pale yellow compound, $C_{14}H_{10}O_2$. (Spectrum courtesy of Heyden and Son, Ltd.)

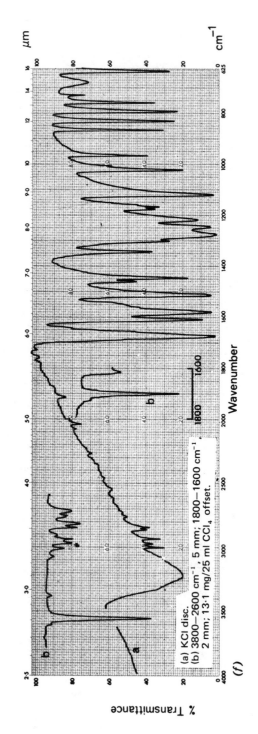

(f) Colorless solid, $C_8H_8O_3$. (Spectrum courtesy of Heyden and Son, Ltd.)

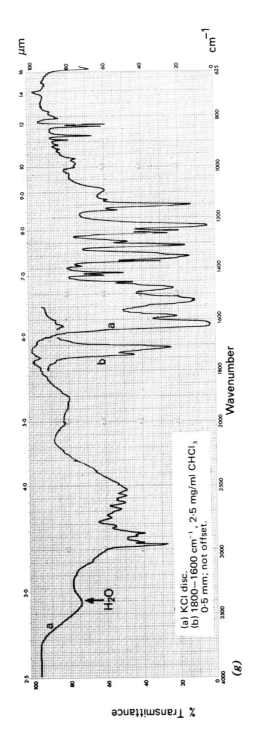

(g) Crystalline compound, $C_8H_{12}O_2$. The solution spectra of this compound show marked solvent and concentration dependence. (Spectrum courtesy of Heyden and Son, Ltd.)

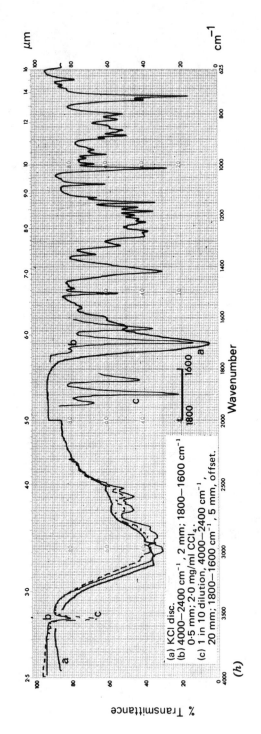

% Transmittance

Wavenumber

(a) KCl disc.
(b) 4000—2400 cm⁻¹, 2 mm; 1800—1600 cm⁻¹
    0·5 mm; 2·0 mg/ml CCl₄.
(c) 1 in 10 dilution, 4000—2400 cm⁻¹,
    20 mm; 1800—1600 cm⁻¹, 5 mm, offset.

(h)

(h) Crystalline solid, $C_7H_6O_3$. (Spectrum courtesy of Heyden and Son, Ltd.)

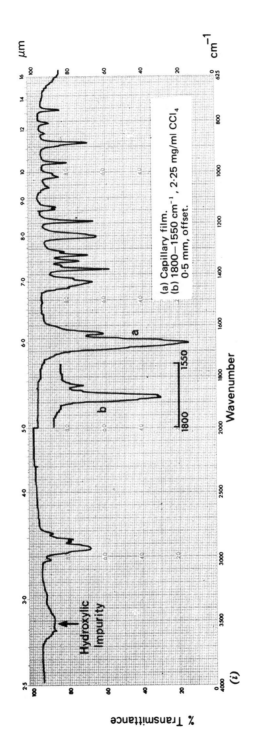

(a) Capillary film.
(b) 1800—1550 cm^{-1} , 2·25 mg/ml CCl$_4$ 0·5 mm, offset.

μm

cm^{-1}

Wavenumber

% Transmittance

Hydroxylic Impurity

1800 1550

a

b

(i) Liquid, C$_7$H$_{10}$O. (Spectrum courtesy of Heyden and Son, Ltd.)

(i)

223

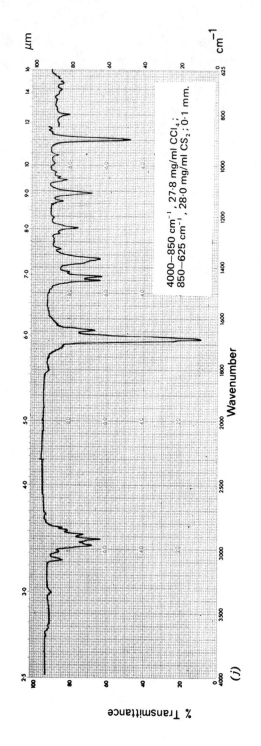

(i) Naturally occurring terpenoid, $C_{10}H_{14}O$. (Spectrum courtesy of Heyden and Son, Ltd.)

4000–850 cm^{-1}, 27·8 mg/ml CCl$_4$;
850–625 cm^{-1}, 28·0 mg/ml CS$_2$; 0·1 mm.

REFERENCES

R. P. Bauman, *Absorption Spectroscopy*, Wiley, New York, 1962. Good general discussion of applications.

J. R. Dyer, *Applications of Absorption Spectroscopy of Organic Compounds.* Prentice-Hall, Englewood Cliffs, N.J., 1965. Very lucid short paperback which includes discussions of ultraviolet and infrared spectrophotometry.

The Sadtler Standard Spectra, Sadtler Research Laboratories, Philadelphia, Pa., 1965. Comprehensive collection of known compounds. Index available.

R. M. Silverstein, G. C. Bassler and T. C. Morrill, *Spectrometric Identification of Organic Compounds.* 3rd ed., Wiley, New York, 1974; Chapter 3. Thorough presentation of applications of ultraviolet and infrared absorption spectrophotometry in organic chemistry.

B. W. Cook and K. Jones, *A Programmed Introduction to Infrared Spectroscopy*, Heyden & Son, London, 1975. Good self-teaching guide.

12

ULTRAVIOLET SPECTROSCOPY

12-1 INTRODUCTION

The visible and ultraviolet spectra of molecules and ions are associated only with transitions between electronic energy levels of certain types or groups of atoms within the molecule and do not characterize the molecule as a whole. In contrast, as we have already seen, absorption of radiation in the infrared region (energy content ~5 kcal/mole) is sufficient to stimulate the whole molecule under investigation to undergo vibrational and rotational changes which do reflect the whole structural entity. The much larger energies involved in the absorption of radiation in the ultraviolet region (energy content ~100 kcal/mole) cause electronic as well as associated vibrational and rotational changes. Since complete resolution of the electronic and closely spaced vibrational and rotational bands is not possible, the overall shape of the absorption band is broad. Two distinct regions exist in the ultraviolet, (a) the near-ultraviolet (400 to 190 nm) and (b) the far-ultraviolet (190 to 100 nm). Routine instrumentation, however, concentrates on the near-ultraviolet since the absorption by silica (used for cells and optics) and by atmospheric oxygen below 190 nm deter measurement in the far-ultraviolet.

A group of atoms that gives rise to absorption in the near-ultraviolet is commonly known as a *chromophore*. Most unsaturated groups and heteroatoms carrying lone pair electrons are potential chromophores and are the main subject of structural elucidation in the near-ultraviolet.

12-2 THEORY

Molecular Orbital Theory. When two atoms form a chemical bond, electrons from both atoms participate in the bond and occupy a new orbital, a molecular orbital. The bonding electrons are associated with the molecule as a whole and not with any particular nucleus. Two atomic orbitals from the two bonding atoms combine to form one "bonding" molecular orbital of low energy and one "antibonding" molecular orbital of very high energy. Recall that covalent bonds consisting of electron pairs may be σ (sigma) or π (pi) bonds. Sigma bonds are formed when there is "head-on" atomic orbital overlap and pi bonds result when there is "parallel" atomic orbital overlap, for example, parallel overlap of two atomic p orbitals. Thus, according to molecular orbital theory, each bonding σ orbital must have a corresponding σ^* antibonding orbital; and each π bonding orbital must have a corresponding π^* antibonding orbital. Electron-cloud probability distributions for these four types of molecular orbitals are illustrated in Figure 12-1. Valence electrons which are not participating in chemical bonding in molecules are referred to as nonbonding or "n" electrons. In organic molecules, n electrons are located principally in the atomic orbitals of nitrogen, oxygen, sulfur, and the halogens. Electronic transitions for organic molecules involve the

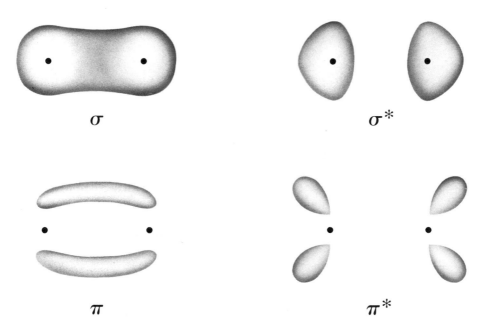

Figure 12–1 Bonding and antibonding molecular orbitals.

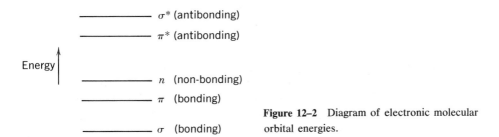

—————— σ^* (antibonding)

—————— π^* (antibonding)

Energy

—————— n (non-bonding)

—————— π (bonding)

—————— σ (bonding)

Figure 12–2 Diagram of electronic molecular orbital energies.

absorption of ultraviolet or visible radiation by electrons in n, σ or π orbitals and result in their promotion to some higher energy antibonding orbital (excited state). The order of energy levels of the various molecular orbitals is indicated in Figure 12–2. Ultraviolet and visible radiation absorption promotes the electronic transitions $\sigma \to \sigma^*$, $n \to \sigma^*$, $n \to \pi^*$, and $\pi \to \pi^*$. The energy level diagram in Figure 12–2 indicates that the ΔE values for transitions are in the order: $n \to \pi^* < \pi \to \pi^* < n \to \sigma^* \ll \sigma \to \sigma^*$. The energy required for the $\sigma \to \sigma^*$ transition is very large; consequently, compounds such as saturated hydrocarbons with no n electrons and only σ bonds do not absorb near-ultraviolet radiation. For example, propane exhibits a $\sigma \to \sigma^*$ absorption maximum at 135 nm which is well into the far-ultraviolet region and inaccessible for study except under extraordinary conditions. As indicated above, compounds which contain nonbonding electrons display absorptions due to the $n \to \pi^*$ and $n \to \sigma^*$ transitions. Compounds which have characteristic $n \to \sigma^*$ transitions include methyl chloride, H_3CCl (λ_{max} = 173 nm), methyl alcohol, H_3COH (λ_{max} = 183 nm), and trimethylamine $(H_3C)_3CNH_2$ (λ_{max} = 227 nm). The lowest energy transition is represented by $n \to \pi^*$ and thus requires the longest wavelength for electronic absorption. For example, saturated ketones and aldehydes show a low intensity $n \to \pi^*$ absorption at about 285 nm. Molecular electronic transitions may be represented in the following manner.

$$\underset{R}{\overset{R}{>}}C{=}\ddot{O} \ + \ h\nu \ \xrightarrow{\ \pi\to\pi^*\ } \ \underset{R}{\overset{R}{>}}C{\because}\ddot{O}$$

$$\underset{R}{\overset{R}{>}}C{=}\ddot{O} \ + \ h\nu \ \xrightarrow{\ n\to\pi^*\ } \ \underset{R}{\overset{R}{>}}C{=}\overset{.}{O}$$

Some representative molecular electronic transitions are given in Table 12–1. Remember that the electronic transitions we have discussed necessarily include possible superimposed, associated molecular vibrational and rotational changes.

   Electronic Spectra.   Common terms which are used in discussions of electronic spectra include *chromophore, auxochrome, bathochromic effect*, and *hypso - chromic effect*. The first two terms apply to the wavelength region near and above 200 nm and were introduced before the relatively recent interest in far-ultraviolet studies ($<$200 nm). For our purposes, the definitions are useful because we will discuss electronic transitions only in the wavelength region near and above 200 nm, which is the lower wavelength limit of common commercial ultraviolet-visible spectrophotometers. *Chromophores* are defined as functional groups which absorb near-ultraviolet or visible radiation when they are bonded to a nonabsorbing, saturated residue which possesses no unshared, nonbonding valence electrons (e.g., a hydrocarbon chain). Most chromophores have unsaturated bonds. *Auxochromes* are functional groups such as —OH, —NH$_2$, and —Cl which have nonbonding valence electrons and do not absorb radiation at wavelengths $>$200 nm. They do, however, exhibit intense absorption in the far ultraviolet ($n \to \sigma^*$ transitions). When an auxochrome is attached to a chromophore, the chromophore absorption band typically shifts to longer wavelength (*batho - chromic effect*) and increases in intensity (i.e., the molar absorptivity, ϵ_{max}, at the wavelength of maximum absorbance, λ_{max}, increases). The *hypsochromic effect* is a shift of the absorption band to shorter wavelength; this effect is often noted when positive charge is introduced into the molecule and when changing from nonpolar to polar solvents.

12–3   THE CHROMOPHORES

   Wavelengths of maximum absorption for some common chromophores are presented in Table 12–1. In this table, the high intensity absorption bands represent higher probability $\pi \to \pi^*$ transitions, and the low intensity bands are due to the lower probability $n \to \pi^*$ transitions. The location of a chromophore absorption band will be essentially the same for all of its compounds in which the R constituents are saturated groups which cannot conjugate with the chromophore.

   For simplicity, the chromophores have been organized according to their chemical class. The following chemical functional groups will be discussed in some detail.

   1. Ethylene—the isolated double bond
   2. Dienes and Polyenes
   3. Poly-ynes and En-ynes
   4. Aldehydes and Ketones
   5. α,β-Unsaturated Acids, Lactones, and Esters
   6. Aromatics and Heterocyclics.

Table 12–1 Characteristic Absorption Bands and Electronic Transitions for Common Chromophores

Chromophore	Compound Types	Example	Solvent	Absorption Band		Electronic Transition
				λ_{max}(nm)	ϵ_{max}	
Alkene	RCH=CH—R	Ethylene	Vapor	$\begin{cases}165\\193\end{cases}$	15,000 / 10,000	$\pi \rightarrow \pi^*$
Alkyne	RC≡CR	2-Octyne	Heptane	$\begin{cases}195\\223\end{cases}$	2,100 / 160	$\pi \rightarrow \pi^*$
Carbonyl (ketone)	R—C(=O)—R	Acetone	Hexane	$\begin{cases}188\\279\end{cases}$	900 / 15	$\pi \rightarrow \pi^*$ / $n \rightarrow \pi^*$
Carbonyl (aldehyde)	R—C(=O)—H	Acetaldehyde	Vapor / Hexane	180 / 290	10,000 / 17	$\pi \rightarrow \pi^*$ / $n \rightarrow \pi^*$
Carboxyl	R—C(=O)—OH	Acetic acid	95% Ethanol	208	32	—
Amido	R—C(=O)—NH$_2$	Acetamide	Water	220	63	—
Nitro	R—NO$_2$	Nitromethane	Methanol	201	5,000	—
Nitrate	R—ONO$_2$	n-Butyl nitrate	95% Ethanol	270	17	—
Nitroso	R—N=O	Nitroso butane	Ethyl ether	300 / 665	100 / 20	— / —
Azo	R—N=N—R	Azomethane	95% Ethanol	338	4	—

Ethylenes—the Single Isolated Bond. It is unfortunate that the $\pi \to \pi^*$ transition for isolated double bonds occurs in the far-ultraviolet below 200 nm. Measurements in this region are extremely difficult and usually omitted. In fact the only useful diagnostic piece of information that can be obtained with the conventional ultraviolet spectrophotometer is the presence of the tetra-substituted double bond by observing the so-called end absorption curve at 200 to 220 nm, that is, the tail end of the absorption peak. This fact in itself can be extremely useful since neither infrared nor proton magnetic resonance techniques can spot this group although these techniques are supreme in other double bond systems.

Dienes and Polyenes. In the case of dienes and trienes the absorption maximum moves to longer wavelengths indicating that conjugation reduces the energy required for the $\pi \to \pi^*$ transition. Table 12–2 illustrates the various parameters for some dienes and trienes.

Table 12–2   Absorption Characteristics for Some Dienes and Trienes

Compound	Structure	Transition	λ_{max}, nm	ϵ
Butadiene		$\pi \to \pi^*$	217	21,000
Hexatriene		$\pi \to \pi^*$	256	22,400
2,4-Dimethyl-2,4-pentadiene		$\pi \to \pi^*$	228	8,500
1,3-Cyclohexadiene		$\pi \to \pi^*$	256	8,000

Woodward was the first to recognize that diene absorption is influenced in a regular fashion by structure. His set of rules to calculate the expected λ_{max} is given in Table 12–3. Examples of those rules worked out are shown in Table 12–4.

Table 12–3   Woodward's Rules for Diene Absorption

Parent heteroannular diene[a]	214
Parent homoannular diene[b]	253
Add for each substituent:	
Double bond extending conjugation	30
Alkyl substituent, ring residue, etc.	5
Exocyclic double bond	5
N(alkyl)$_2$	60
S(alkyl)	30
O(alkyl)	6
OAc	0
$\lambda_{max}^{calc} = $ TOTAL, nm	

[a] Diene system not involved in a ring structure.
[b] Diene system within ring structure.

Table 12–4 Worked Examples of Woodward's Rules

Parent (hetero)	$= 214$
Alkyl subst. (4×5)	$= 20$
Exo C=C (2×5)	$= 10$
λ_{max}^{calc}	$= 244$ nm (obs. 247 nm)

Parent (homo)	$= 253$
Ring residues (5×5)	$= 25$
Exo C=C (3×5)	$= 15$
C=C ext. conj.	$= 30$
λ_{max}^{calc}	$= 323$ nm (obs. 320 nm)

These rules, however, cannot be extrapolated to calculate the λ_{max} for polyenes containing greater than five double bonds in conjugation. Certain naturally occurring polyenes such as the carotenoids are therefore excluded from the above rules.

Poly-ynes and En-ynes. As with the isolated double bond the simple triple bond exhibits a $\pi \to \pi^*$ transition at 175 nm ($\epsilon = 6000$). A remarkable transformation results, however, when we have two acetylenic groups in conjugation—a series of medium intensity peaks ($\epsilon = 300$) at intervals of 2300 cm^{-1} (Table 12–5). Extension of this conjugated system to four and six triple bonds gives rise to a set of high intensity bands ($\epsilon = 10^5$) thereby providing instant recognition of such polyacetylenic systems—an important group of naturally occurring compounds. The conjugation of an acetylenic linkage with an ethylenic one (the en-yne system) shows absorption at 223 nm with an inflection at 229 nm thereby providing a distinguishing feature to differentiate the en-yne from the diene.

Aldehydes and Ketones. Simple nonconjugated aldehydes and ketones show a weak intensity $n \to \pi^*$ absorption band about 280 nm ($\epsilon = 14$) and a medium intensity $\pi \to \pi^*$ absorption near 190 nm ($\epsilon = 1000$), for example, acetone (Table 12–6). As with dienes, α,β-unsaturated ketones and aldehydes have an intense, solvent-variable absorption band in the far-ultraviolet above 220 nm. The $n \to \pi^*$ absorption band now appears at 310 to 350 nm ($\epsilon = 50$ to 100).

α,β-Unsaturated Acids, Lactones, and Esters. Compounds belonging to this class absorb at much shorter wavelengths than the corresponding enones because the electron affinity of the carbonyl group is reduced by resonance thereby increasing the energy of the transition (Table 12–7).

Table 12–7 Absorption Characteristics of α,β-Unsaturated Acids, Lactones, and Esters

Compound	Structure	Transition	λ_{max}, nm	ϵ
Propenoic acid	CH_2=CH—CO_2H	$\pi \to \pi^*$	200	10,000
2-Butenoic acid	CH_3—CH=CH—CO_2H	$\pi \to \pi^*$	205	14,000
trans-2-Butenoate	CH_3—CH=CH—CO_2CH_3	$\pi \to \pi^*$	205	16,500
5-Methyl-δ-butyrolactone		$\pi \to \pi^*$	205	10,700

Table 12–5 Absorption Characteristics for Diynes, Poly-ynes, and En-ynes

Type	Structure	Transition	λ_{max}, nm	ϵ	λ_{max}, nm	ϵ	λ_{max}, nm	ϵ
Diynes	$CH_3-C\equiv C-C\equiv C-CH_3$	$\pi \to \pi^*$	227	370	236.5	390	249	210
	$Et-C\equiv C-C\equiv C-Et$	$\pi \to \pi^*$	227.5	360	238.5	340	253	230
Poly-ynes	$CH_3-(C\equiv C)_4-CH_3$	$\pi \to \pi^*$	234	281,000	354	105	—	—
	$CH_3-(C\equiv C)_6-CH_3$	$\pi \to \pi^*$	384	445,000	—	—	—	—
En-ynes	$CH_2=CH-C\equiv CH$	$\pi \to \pi^*$	219	6,400	228	4,000	—	—
	$HOCH_2-CH=CH-C\equiv CH$	$\pi \to \pi^*$	223	14,000	229	10,000	—	—

Table 12–6 Absorption Characteristics of Aldehydes and Ketones

Compound	Structure	Transition	λ_{max}, nm	ϵ	Transition	λ_{max}, nm	ϵ
Acetone	$CH_3-CO-CH_3$	$\pi \to \pi^*$	188	1,860	$n \to \pi^*$	276	13
Cyclohexanone	(cyclohexanone structure)	—	—	—	$n \to \pi^*$	285	14
Acetaldehyde	CH_3-CHO	$\pi \to \pi^*$	194	—	$n \to \pi^*$	292	14
Propenal	$CH_2=CH-CHO$	$\pi \to \pi^*$	208	10,000	$n \to \pi^*$	328	13
2-Methylpent-2-en-4-one	$CH_3\!-\!\underset{CH_3}{C}\!=\!CH-CO-CH_3$	$\pi \to \pi^*$	235	14,000	$n \to \pi^*$	314	6
Cyclohexenone	(cyclohexenone structure)	$\pi \to \pi^*$	224	10,300	—	—	—

Nielson has established parameters to predict with reasonable accuracy the position of the λ_{max} for α,β-unsaturated acids (Table 12–8).

Table 12–8 Characteristic Absorptions for α,β-Unsaturated Acids[a]

Type	Transition	λ_{max}^{EtOH}, nm
Mono-substituted α or β	$\pi \to \pi^*$	208 ± 5
Di-substituted $\alpha\beta$ or $\beta\beta$	$\pi \to \pi^*$	217 ± 5
Tri-substituted $\alpha\beta\beta$	$\pi \to \pi^*$	225 ± 5

[a] For each exocyclic double bond or double bond endocyclic to a 5- or 7-membered ring add 5 nm.

Aromatics and Heterocyclics. Comparison of the spectrum of benzene with that of 1,3,5-hexatriene shows no similarities. Benzene exhibits three bands at 255 nm ($\epsilon = 230$), 200 nm ($\epsilon = 8,000$) and 185 nm ($\epsilon = 60,000$). The effect of substitution on the aromatic nucleus has been studied extensively. Both alkyl and alkenyl substitution cause a bathochromic shift. With polar solvents (e.g., —OH, —NR₂, etc.) that allow for the $p \to \pi$ conjugation and $>$C=O and —NO₂ where polarizability is of importance, absorption due to electron transfer transitions is apparent. These two types of transfer can be illustrated as shown below.

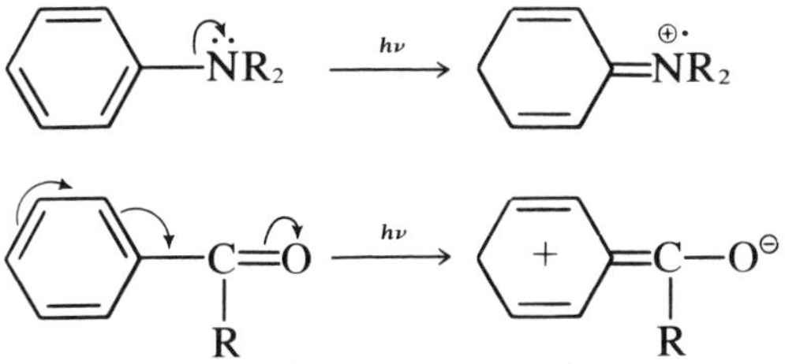

In the case of heterocyclic compounds it is essential to have the spectrum of a model compound for comparison if interpretation is attempted. Table 12–9 lists the absorption characteristics for a number of prominent aromatic and heterocyclic compounds for such reference purposes.

Table 12–9 Absorption Characteristics of Some Aromatic and Heterocyclic Compounds

	λ_{max}, nm	Approx. ϵ	Solvent
Benzene	200	8,000	H
	255	230	
Toluene	261	300	H
Xylenes	266	400–800	H
Diphenylmethane	262	490	A
Diphenyl	250	18,000	A
Styrene	244	12,000	A
	282	450	

continued on page 234

Table 12–9 (*continued*

	λ_{max}, nm	Approx. ϵ	Solvent
Stilbene (*trans*)	295	27,000	A
(*cis*)	280	13,500	
Acetophenone	240	13,000	A
	278	1,100	
	319	50	
Benzaldehyde	244	15,000	A
	280	1,500	
	328	20	
Benzoic acid	230	10,000	W
	270	800	
Cinnamic acid	273	20,000	A
Benzophenone	252	20,000	A
	325	180	
Nitrobenzene	252	10,000	H
	280	1,000	
	330	125	
Phenol	210.5	6,200	W
	270	1,450	
Phenol (anion)	235	9,400	aq. OH⁻
	287	2,600	
Phenyl acetate	263	500	
Aniline	230	8,600	W
	280	1,430	
Aniline (cation)	203	7,500	aq. H⁺
	254	160	
Acetanilide	242	14,400	A
	280	500	
Furan	200	10,000	H
Pyrrole	210	16,000	H
	340	300	
Pyridine	195	7,500	H
	250	2,000	
Quinoline	275	4,500	H
	311	6,300	
Isoquinoline	262	3,700	H
	317	3,500	
Acridine	252	10,000	A
	347	8,000	
		log ϵ	
Naphthalene	220	5.05	A
	275	3.75	
	314	2.50	
Anthracene	250	5.20	A
	380	3.90	
Pentacene (blue)	580	4.10	A
Tropolone	247	4.82	H
	331	4.51	
	380	4.14	

W = water; A = alcohol; H = *n*-hexane.

(Reproduced by kind permission from *More Spectroscopic Problems in Organic Chemistry*, A. J. Baker, T. Cairns, G. Eglinton and F. J. Preston, Heyden & Son Ltd., London, 1975.)

Nucleic Acids. An important group of heterocyclic compounds are the bases which form the nucleic acids. Because of their function as building blocks of the living cell, nucleic acids are extremely important. There are two types of nucleic acids, deoxyribonucleic acid (DNA) and ribonucleic acid (RNA). Both types consist of two long chains of alternating sugar and phosphate residues. The chains are linked at the sugar units by nitrogenous bases which are either substituted purines or substituted pyrimidines. Purine and pyrimidine are heteroaromatic compounds with the structures:

Purine   Pyrimidine

The substituted purines and pyrimidines are the units which absorb ultraviolet radiation in the nucleic acids and their many derivatives. The sugar and phosphate residues are transparent above 200 nm.

DNA and RNA contain the two substituted purines, adenine and guanine, and the substituted pyrimidine, cytosine. RNA also contains the pyrimidine, uracil, whereas DNA contains thymine. The absorption bands of these various bases in water are listed in Table 12–10.

Table 12–10 Absorption Data for Nitrogenous Bases of Nucleic Acids

	pH	λ_{max} (nm)	ϵ
Adenine	2	263	13,100
	12	269	12,300
Guanine	2	248	11,400
		275	7,350
	11	246	6,350
		273	8,000
Cytosine	2	210	9,700
		274	10,200
Uracil	2	259	8,200
	12	284	6,500
Thymine	2	207	9,500
		264	7,900
	12	291	5,440

12-4 SOLVATION AND SUBSTITUTION

As already explained, the energy required to bring about an electronic transition in the ultraviolet region also induces associated vibrational and rotational changes. The benzene molecule has a high degree of symmetry and in the vapor state

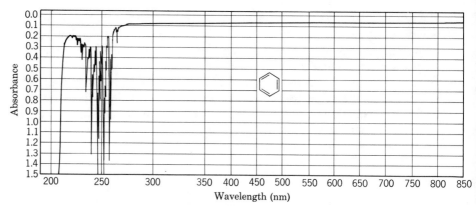

Figure 12–3   Ultraviolet spectrum of benzene in the vapor state. (Spectrum courtesy of Perkin-Elmer Corp.)

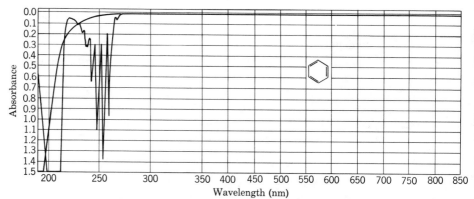

Figure 12–4   Ultraviolet spectrum of benzene in *iso*-octane. (Spectrum courtesy of Perkin-Elmer Corp.)

(Figure 12–3) a great deal of vibrational fine structure is clearly evident. Benzene in solution (Figure 12–4), however, illustrates the effect of solvation. Most of the fine structure it exhibited in the vapor state has disappeared. Notice also the appearance of the spectrum of the solvent itself. Figure 12–5 illustrates the spectrum of fluorobenzene in iso-octane. Its spectrum is somewhat similar to that of benzene in the same solvent with the exception that a small wavelength shift of

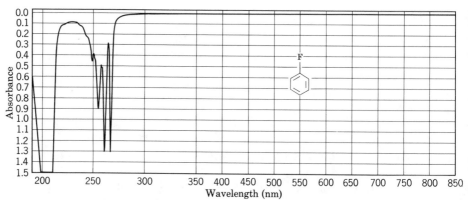

Figure 12–5   Ultraviolet spectrum of fluorobenzene in *iso*-octane. (Spectrum courtesy of Perkin-Elmer Corp.)

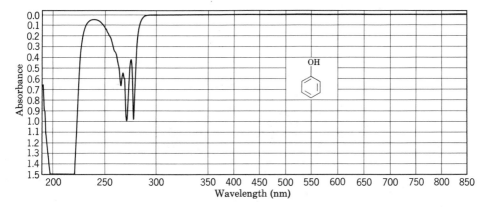

Figure 12–6 Ultraviolet spectrum of phenol in *iso*-octane. (Spectrum courtesy of Perkin-Elmer Corp.)

approximately 6 nm is observed and the intensity has increased fifty-fold. The spectrum of phenol (Figure 12–6) shows a similar pattern to benzene but a wavelength shift of 17 nm is observed. Phenol dissolved in water (Figure 12–7) illustrates an even greater effect of solvation since the remaining fine structure has disappeared completely and only one main absorption band remains.

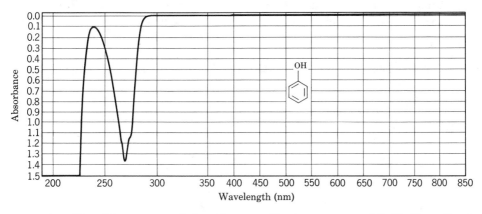

Figure 12–7 Ultraviolet spectrum of phenol in water. (Spectrum courtesy of Perkin-Elmer Corp.)

12–5 INTERPRETATION AND USES OF ULTRAVIOLET SPECTRA

In contemporary practice the ultraviolet spectrum of a particular compound is recorded in conjunction with other spectral data such as infrared in an attempt to deduce its molecular structure. Ultraviolet spectra, however, do not furnish prime information as such but tend to act as complimentary or even supplementary evidence to infrared. In discussing the following spectra it has been assumed, therefore, that functional groups have been successfully detected and we shall concern ourselves with ascertaining substitution patterns.

Example 1 An unsaturated aldehyde or ketone. (Spectrum courtesy of Heyden and Son, Ltd.)

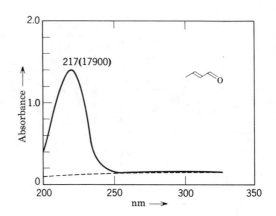

This spectrum shows one absorption band at 217 nm (ϵ = 17,900) which can be assigned to an $\alpha\beta$-unsaturated aldehyde or ketone. A diene can be ruled out since only butadiene itself (λ_{max} = 217 nm, ϵ = 21,000) absorbs below 220 nm. If an $\alpha\beta$-unsaturated ketone is considered only a vinyl ketone such as methyl vinyl ketone (λ_{max} = 219 nm, ϵ = 3600) can be considered, but the ϵ value for this type of compound is about 5000. Thus we are led to assign this spectrum to a mono-substituted $\alpha\beta$-unsaturated aldehyde, either α- or β-substituted, for which the calculated values of λ_{max} are 217 nm and 219 nm.

Example 2 A poly-yne. (Spectrum courtesy of Heyden and Son, Ltd.)

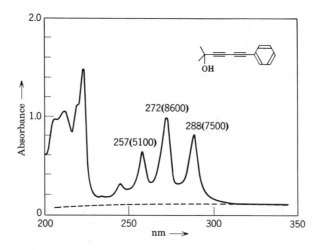

This spectrum illustrates the typical band pattern associated with poly-ynes. Note the peak separation of *ca.* 2300 cm^{-1}. The strong absorption in the 220 nm region is due to benzenoid absorption. Conjugation of the diacetylene with the benzene ring causes a bathochromic shift of all the poly-yne bands together with an enhancement of the intensity of the absorptions.

Example 3 A poly-yne. (Spectrum courtesy of Heyden and Son, Ltd.)

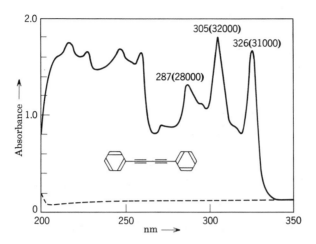

This spectrum, like the preceding example, shows the characteristic poly-yne bands and regular peak separation. Conjugation with the two benzene rings causes a further bathochromic shift and a substantial increase in intensity.

12–6 APPLICATIONS OF ULTRAVIOLET SPECTRA

Kinetic Studies. The hydrolysis of methyl salicylate may be studied by repetitively scanning the spectrum from 270 to 390 nm. The rate of decrease of the ester absorption at 331 nm and the increase in the anion absorption at 298 nm can easily be calculated (see Figure 12–8).

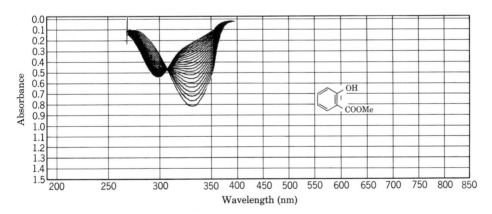

Figure 12–8 Ultraviolet spectra produced by repetitive scanning of methyl salicylate while undergoing hydrolysis with base in cell. (Spectrum courtesy of Perkin-Elmer Corp.)

Structural Differentiation. The nitrophenol isomers can be differentiated by ultraviolet spectroscopy as can be seen with reference to Figure 12–9.

Decomposition Products. An assessment of the quality of fish may be made by the detection and quantitation of its trimethyl amine content which increases as

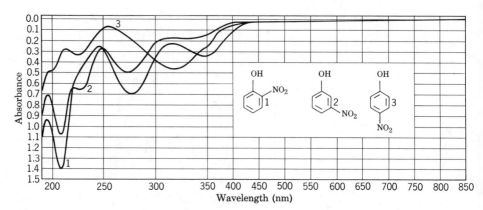

Figure 12-9 Ultraviolet spectra of nitrophenol isomers. (Spectrum courtesy of Perkin-Elmer Corp.)

decay proceeds. The amine itself absorbs at 199 nm so the picrate is usually monitored at 358 nm (Figure 12–10.

Drugs. Barbiturates, for example, are easily recognized by their behavior under varying conditions of pH (Figure 12–11).

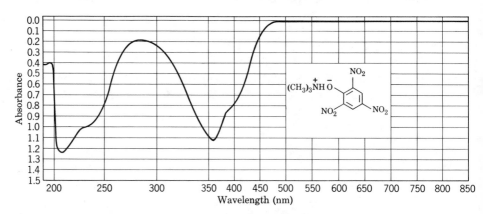

Figure 12–10 Ultraviolet spectrum of trimethyl amine picrate. (Spectrum courtesy of Perkin-Elmer Corp.)

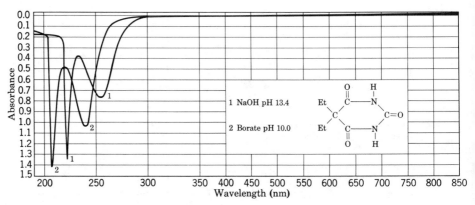

Figure 12–11 Ultraviolet spectra of diethyl barbituric acid at pH 10.0 and 13.4. (Spectrum courtesy of Perkin-Elmer Corp.)

QUESTIONS AND PROBLEMS

12–1. Using the data in Table 12–1 calculate the transition energies (in electron volts and kcal) for the $n \to \pi^*$ and $\pi \to \pi^*$ transitions for acetone.

Ans. 4.4 eV, 102 kcal; 6.59 eV, 152 kcal

12–2. A volatile constituent, cosmene ($C_{10}H_{14}$), is isolated from the *Compositae* species and found to have an ultraviolet spectrum of four distinct peaks, the most intense at 296 nm ($\epsilon = 47,000$). Establish the chromophore present and suggest possible structures.

12–3. Calculate the absorption maxima of the diene,

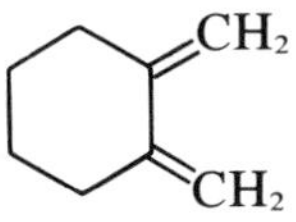

and explain the deviation from the observed value of 220 nm ($\epsilon = 10,000$).

12–4. Calculate the absorption maxima of
(a) 1,3-cyclopentadione enol ether:

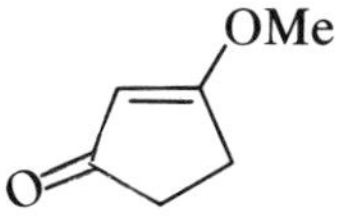

Ans. 249 nm

(b) citral:

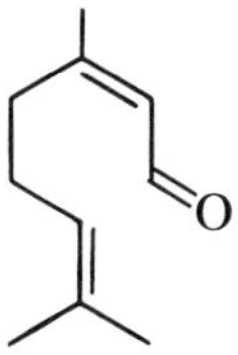

Ans. 239 nm

(c) 3-methylcyclohexa-1,2-dione:

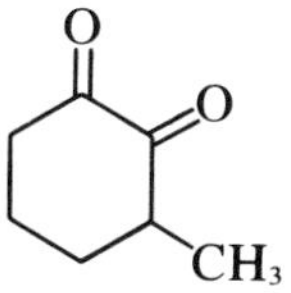

Ans. 274 nm

12–5. Show how the ultraviolet spectrum can be used to distinguish between the two following isomeric compounds.

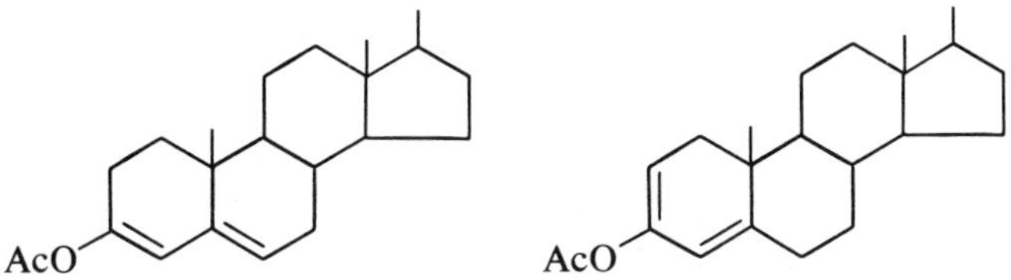

(A)   (B)

REFERENCES

J. R. Dyer, *Applications of Absorption Spectroscopy of Organic Compounds*, Prentice-Hall, Englewood Cliffs, N.J., 1965.

C. R. Hare, *Visible and Ultraviolet Spectroscopy* in *Guide to Modern Methods of Instrumental Analysis*, T. H. Gouw, Ed., Wiley-Interscience, New York, 1972; Chapter V.

F. Grum, *Visible and Ultraviolet Spectrophotometry* in *Techniques of Chemistry*, A. Weissberger, Editor, Vol. 1, Part II, Wiley-Interscience, New York, 1972.

H. H. Willard, L. L. Merritt, and J. A. Dean, *Instrumental Methods of Analysis*, 5th Ed., Van Nostrand Reinhold, New York, 1974.

H. H. Jaffé and M. Orchin, *Theory and Application of Ultraviolet Spectroscopy*, Wiley, New York, 1962.

13

FLAME EMISSION
AND
ATOMIC ABSORPTION
SPECTROSCOPY

13-1   INTRODUCTION

The use of flame emission spectra, or *flame photometry*, for quantitative analysis dates back to about 1930. Absorption techniques have been developed more recently, and this has been mainly due to the work of Walsh and his colleagues in Australia. However, the first reported application of *atomic absorption spectroscopy* was due to Kirchoff who, in 1860, deduced the presence of various elements in the solar atmosphere from the Fraunhofer absorption lines. Flame emission and absorption are complementary techniques in that some elements can be detected at lower concentrations by emission measurements and some by absorption, but absorption measurements are less prone to inter-element interference effects. In both techniques, sample molecules are dissociated into their atoms and we observe *atomic* spectra only.

Atomic fluorescence spectroscopy is a related, but even more recent technique. It is still in the development stage but offers the selectivity of atomic absorption and the possibility of a further improvement in detection limits.

13-2   EMISSION AND ABSORPTION IN FLAMES

In emission spectroscopy radiation is emitted by atoms which are in excited energy states, whereas in absorption spectroscopy the atoms which can absorb energy are in the ground state. For an equilibrium distribution, the ratio of the number of atoms in the excited state, N_j, to the number in the ground state, N_0, is given by

$$N_j/N_0 = (g_j/g_0) \exp(-E_j/kT) \tag{13-1}$$

where g_j and g_0 are the statistical weights for the excited and ground states, respectively, and E_j is the excitation energy. Values for several elements are given in Table 13-1.

Flame emission and absorption measurements are generally made with flame temperatures below 3000°C, and consequently most atoms are in the ground state. It has therefore been argued that absorption should be virtually independent of temperature, whereas the number of excited atoms and hence the emission intensity varies exponentially with temperature. However, the dissociation processes which result in the production of atoms from the sample molecules are also temperature-dependent. Because of this the number of ground state atoms, and

Table 13–1 Percentage of Atoms in Excited States at Various Temperatures

Element	Line, nm	g_i/g_0	Temperature, °K		
			2000	3000	4000
Cesium	852.1	2	0.04	0.72	2.98
Sodium	589.0	2	1×10^{-3}	0.06	0.44
Calcium	422.7	3	1×10^{-5}	4×10^{-3}	0.06
Zinc	213.9	3	7×10^{-13}	6×10^{-8}	1×10^{-5}

[Data from A. Walsh, *Spectrochim. Acta*, **7**, 108 (1955).]

hence the amount of absorption, will change with temperature although not to the same extent as emission.

In principle, the same equipment can be used for both emission and absorption, although a separate radiation source is required for absorption measurements, as shown in Figure 13–1. There have also been differences in burner design, which will be discussed later, and for absorption spectroscopy the source radiation is chopped to distinguish it from the radiation emitted by the flame. Fluorescence radiation is normally viewed at right angles to the radiation source in order to separate the low intensity re-emitted, or fluorescence, radiation from that due to the incident radiation beam.

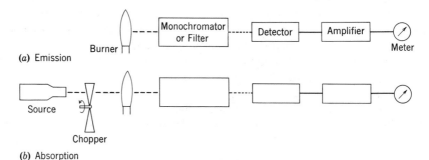

(a) Emission

(b) Absorption

Figure 13–1 Flame photometer (a), and atomic absorption spectrometer (b).

Another important difference between emission and absorption is the relationship between the instrument response and the concentration of the measured species. Within certain limits, discussed in Section 13–8, emission (and fluorescence) varies linearly with concentration whereas absorption follows Beer's law. The linear relationship is more convenient for quantitative analysis but the logarithmic relationship provides a much greater concentration range over which measurements can be made without altering the instrument settings.

In atomic absorption spectroscopy, radiation sources are used that are specific for each element to be measured. This overcomes the problem of overlap of spectral bands that frequently occurs in flame photometry. Partly for this reason and partly because inter-element effects have been a major problem in flame photometry, atomic absorption is generally preferred to flame photometry. This is in spite of the fact that the use of a separate radiation source for each element is somewhat inconvenient. Furthermore, with atomic absorption it is not normally possible to scan through the entire wavelength range to obtain a qualitative analysis of a sample as is sometimes possible with a flame photometer.

13-3 ATOMIZATION AND IONIZATION

Emission of radiation in flame photometry and absorption of radiation in atomic absorption spectroscopy is mainly a process involving atoms. At high sample concentrations some nonatomic absorption can occur due to the presence of molecules in the optical path, but this is generally a relatively minor effect.

The major problem with these techniques, and the factor which is mainly responsible for the present limitations in sensitivity and stability, appears to be the atomization process. Until recently flames have been used to dissociate the sample, but the desire to measure trace metals in small samples has resulted in the development of a number of small high-temperature furnaces for use in atomic absorption.

When a solution of the sample to be analyzed is introduced into a flame as an aerosol, a sequence of processes occurs. The initial step is evaporation of the solvent, leaving fine salt particles suspended in the flame. These particles are vaporized and some or all of the resulting vapor is dissociated into atoms, that is, it is *atomized*. This process may be due partly to the direct effect of the heat generated by the flame and partly to chemical reduction of the species present in the flame.

A complication arises due to ionization of atoms in the flame. Alkali metals have low ionization potentials and are readily ionized in flames. Since ionic spectra differ from atomic spectra, it is important to maintain the degree of ionization at a constant level, preferably so low that it can be neglected. We can consider the thermal ionization of an atom to be a dissociation process:

$$M \rightleftharpoons M^+ + e^-$$

If x is the degree of ionization in the flame and P is the partial pressure of both ionized and unionized metal atoms in the flame, the partial pressure of M^+ (or of e^-) in the flame will be given by:

$$\frac{x}{(1-x)+x+x} P = \frac{x}{1+x} P$$

The equilibrium constant is therefore given by:

$$K = \frac{[M^+][e^-]}{[M]} = \frac{\left(\dfrac{x}{1+x} P\right)^2}{\left(\dfrac{1-x}{1+x}\right)P} = \left(\frac{x^2}{1-x^2}\right)P \tag{13-2}$$

If x is small, x^2 is negligible compared with unity so that

$$x \simeq (K/P)^{1/2} \tag{13-3}$$

This shows that the degree of ionization is concentration-dependent, increasing with decreased sample concentration. Because of its dependence on K, it also increases as the flame temperature increases and as the ionization potential of the metal decreases. This is shown in Table 13–2 for which the partial pressure of metal in the flame is 10^{-6} atmosphere. The calculated figures are based on the assumption that equilibrium conditions exist in the flame. This will be approximately true in the upper regions of the flame but not in the reaction zone. The figures also assume a single flame temperature, whereas in reality the flame temperature varies across the flame front. Excessive ionization can be overcome either by using a lower temperature flame or by adding an excess of a more easily ionized metal. The latter

Table 13-2   Percent Ionization of Alkali and Alkaline Earth Metals in Various Flames

Metal	Ionization potential (eV)	Propane-air (2200°K)	Acetylene-air (2450°K)	Acetylene-nitrous oxide (3200°K)
Lithium	5.36	0.6	3	63
Sodium	5.12	1.1	5	78
Potassium	4.32	9	33	98.3
Rubidium	4.16	14	45	99.0
Cesium	3.87	30	71	99.7
Magnesium	7.61	0.003	0.03	3
Calcium	6.09	0.2	1.0	39
Strontium	5.67	0.5	3	68
Barium	5.19	2	9	91

Equilibrium figures calculated from the Saha equation—see, for example, A. G. Gaydon and H. G. Wolfard, *Flames: Their Structure, Radiation and Temperature,* Chapman and Hall, London, 3rd ed., 1970, p. 302.

produces a high concentration of electrons in the flame and thus suppresses the ionization of the original metal in accordance with Equation 13-2.

13-4   FLAMES

The choice of fuel and oxidant depends largely on the flame temperature required to atomize the sample, although chemical factors which reduce the formation of metal oxides may also be important. It is also desirable that the background emission should not interfere with the analysis. Of the commonly used flames, acetylene-nitrous oxide exhibits the most intense background emission. Its emission spectrum is shown in Figure 13-2. The effect of flame emission can be eliminated in atomic absorption by the use of a chopped radiation beam, as shown in Figure 13-1, provided that the steady background signal from the flame does not saturate the photomultiplier. The chopper blocks the source radiation at a fixed frequency so that the detector sees alternately the flame background alone and the flame background plus the source radiation (partially absorbed by the sample). The steady background signal (dc) is filtered out and the amplifier sees only the alternating difference signal (ac). There is no similar technique to eliminate the flame background in emission measurements.

The recommended flames for the determination of various elements by atomic absorption spectroscopy are shown in Figure 13-3. Maximum flame temperatures and burning velocities are given in Table 13-3. For flame emission, acetylene-nitrous oxide is generally recommended except for the alkali metals, iron and ruthenium, for which the lower temperature acetylene-air flame is preferred.

In the normal premixed flame, that is, one in which the fuel and oxidant are mixed before combustion, there are at least two zones: a primary reaction zone and a secondary reaction or diffusion zone. In the primary zone of the acetylene-air flame the main reaction products are carbon monoxide, hydrogen, and water, produced by a chain of reaction steps which involve H, O, and OH radicals. In the secondary zone, which surrounds the primary zone, the carbon monoxide and

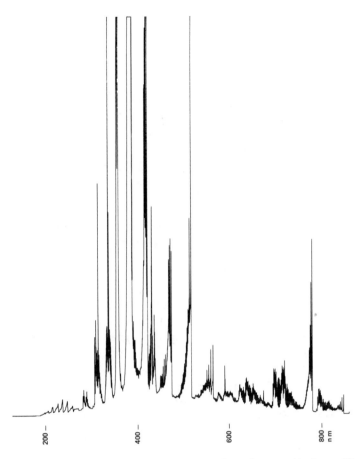

Figure 13–2 Emission spectrum of an acetylene-nitrous oxide flame. [Courtesy of Varian-Techtron.]

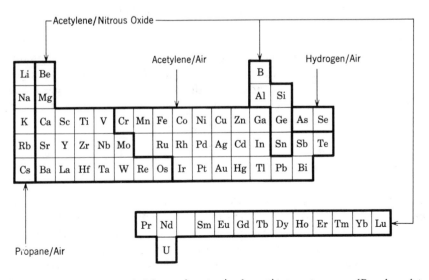

Figure 13–3 Recommended flames for atomic absorption spectroscopy. [Based on data from Varian-Techtron.]

Table 13–3 Maximum Temperatures
(and Burning Velocities) of Various
Flames, °K (cm/sec)

Fuel	Air	Nitrous Oxide
Acetylene	2450 (160)	3200 (220)
Propane	2200 (45)	2900 (250)
Hydrogen	2300 (320)	2900 (380)

[From W. J. Price, *Analytical Atomic Absorption
Spectrometry*, Heyden & Son Ltd., London, 1972, p.
20.]

hydrogen are burnt in atmospheric oxygen, and thereby produce a strong
background signal. In strongly reducing flames, which are fed with an excess of
fuel, C_2 and CH radicals also exist in the lower part of the flame.

In the slightly fuel-rich acetylene-nitrous oxide flame, the primary reaction
zone—whitish blue in color—is separated from the blue secondary reaction zone
by a red interconal zone. The red zone gives strong CN and NH band emission,
and the high degree of atomization of some oxide-forming elements in this flame is
believed to be due at least partly to the reducing atmosphere provided by the CN
and NH radicals. These species also exist in the yellow flame of the very fuel-rich
acetylene-nitrous oxide flame.

The most commonly used flame is a mixture of acetylene as fuel and air as
oxidant. A propane-air flame is preferred for the alkali metals because the lower
flame temperature reduces the amount of ionization, but this flame is very slow
burning, which makes it susceptible to drafts and leads to instability. The
hydrogen-air flame is more transparent than the acetylene-air flame in the UV
region (below 220 nm) and this, together with its reducing properties, make it
suitable for the determination of arsenic and selenium by atomic absorption.
However, the low flame temperature can result in severe chemical interference
when other elements are present in the sample, and in those circumstances the
higher temperature acetylene-nitrous oxide flame would be preferred. (Maximum
sensitivity for arsenic and selenium is obtained by converting them into the
corresponding hydrides before passing them into the flame.) For those elements
which tend to form stable refractory oxides (e.g., aluminum, silicon, titanium, and
the lanthanides), a high temperature and strongly reducing environment is
required. This is achieved by the use of the acetylene-nitrous oxide flame, but
ionizable metals such as calcium, strontium, and barium, and most of the rare
earths, require an ionization suppressant.

13–5 BURNERS AND NEBULIZERS

Introduction of the sample into the flame in a constant and uniform manner
requires a device which disperses the sample uniformly throughout the flame. The
only practical method is to transform the liquid sample into fine droplets, hence
the term *nebulizer*. The nebulizer and burner form an integral unit.

Total-Consumption Burner. Burners can be broadly classified as *total-
consumption* burners or *premix* burners. The construction of a total-consumption
burner is shown in Figure 13–4. The sample is aspirated by the Venturi effect

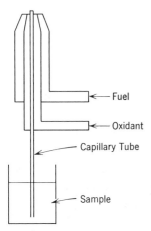

Figure 13-4 Total consumption burner.

directly into the flame and, as the name implies, the entire sample supposedly is "consumed" by the burner. However, the droplet size varies widely. The mean diameter is typically about 20 μm and a significant number of particles may have diameters in excess of 40 μm. The larger droplets pass completely through the flame without even being totally evaporated and, as a result, the overall efficiency is poor. The incomplete vaporization of metal from the larger aerosol particles also makes this burner more prone to chemical interference effects than the premix burner. Nevertheless, the flame is concentrated in a small area and this type of burner has been used in instruments which are designed solely for flame emission work.

Premix (Laminar-Flow) Burners. In a premix burner, Figure 13-5, the sample is mixed with the fuel and oxidant in a mixing chamber before being burnt. The larger droplets, which would otherwise pass through the flame unexcited, are removed and the remaining sample (approximately 10%) enters the burner. Some increase in sensitivity can be obtained by placing an impact bead directly in the path of the droplets, thereby further breaking up the larger droplets. The average drop size entering the flame in a premix burner is less than 10 μm. The premix system allows the use of a long narrow burner which provides a longer path length for absorption and therefore an increased signal-to-noise ratio. Furthermore, the removal of the larger droplets and the laminar flow characteristics of this type of burner make the flame more stable than the turbulent flame of the total-consumption burner. In emission measurements the longer path length results in self-absorption, that is, absorption of some of the emitted radiation by the flame itself. This causes a decrease in the emission intensity at higher sample concentrations and hence curvature of the emission intensity versus sample concentration plot. It can be avoided by making emission measurements at right angles to the burner slot, but with a consequent reduction in sensitivity (a factor of 2 to 3 at low concentrations).

A problem with the premix burner is that the flame may burn back into the mixing chamber and cause an explosion. This can be prevented by using narrow burner slots (0.3 mm for acetylene-nitrous oxide and 0.5 mm for acetylene-air) but these create problems because of deposition of salts which eventually block the orifice. In practice, somewhat wider slots are used and the flash-back problem is avoided by turning the gases on and off in the correct sequence, and by using the correct fuel-to-oxidant ratio. The problem is more pronounced with a gas mixture that has a high burning velocity, and for this reason an acetylene-nitrous oxide

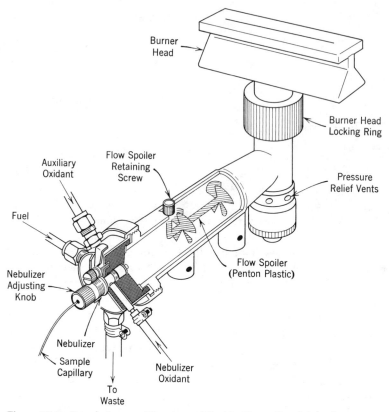

Figure 13–5 Premix burner. [Courtesy of Perkin-Elmer Corporation.]

mixture (burning velocity ~ 220 cm/sec) is preferred to acetylene-oxygen (1130 cm/sec) which gives a similar flame temperature. A lower burning velocity also means that the atoms are in the reaction zone for a longer time, and therefore a more intense signal should be obtained.

The fuel, air, and sample flow rates required depend on the burner dimensions and on the sample components. Typical values for the acetylene-air flame are 2 to 5 l/min of acetylene, 5 to 10 l/min of air, and 3 to 6 ml/min of sample solution. Gas flow rates for the acetylene-nitrous oxide flame are typically 4 to 5 l/min of acetylene and 5 to 10 l/min of nitrous oxide. The burner head is made of stainless steel, or of titanium to provide better corrosion resistance. Grooves cut in the head alongside the slot improve the air entrainment pattern so that a steadier flame is obtained, and better air access with the grooved burner makes it relatively free from carbon deposition.

Recently there has been some interest in the use of separated flames. The primary and secondary zones can be separated by surrounding the bottom of the flame by nitrogen or argon, which restricts the flow of atmospheric air into the flame. An intermediate flame region is thus produced that is relatively free from the normal strong emission background due to the secondary zone. This reduction in background signal is of value in both emission and fluorescence, the latter using relatively wide monochromator slits which allow high levels of background radiation to fall on the photomultiplier. The separated primary zone is also more reducing and therefore more free from metal oxide and hydroxide formation.

Ultrasonic Nebulizers. An alternative method of nebulization is to break up a fine stream of the sample solution by using high-frequency vibrations. This produces a very uniform and controllable droplet size and it is even possible to study the behavior of single droplets in a flame by this method. However, the equipment required is considerably more expensive than the simpler pneumatic nebulizers, the droplet size is larger than normal, and there are problems with the efficient changing of samples.

Emission and Absorption Flame Profiles.   The regions of the flame in which maximum emission and maximum absorption occur depend on both the flame operating parameters and on the particular element being determined. The flame absorption profiles for several elements in a hydrogen-air flame are shown in Figure 13–6. Silver shows a continuous increase in absorbance with increased exposure to the heat of the flame, whereas the other elements exhibit maxima at different heights above the burner due to the formation of stable oxides. Emission and absorption maxima occur at different points in the flame. It is obviously important to work near the maximum, and this may vary from day to day because of slightly different flame operating conditions.

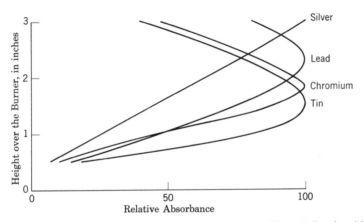

Figure 13–6   Absorption profiles in a turbulent flame. [From J. Ramirez-Munoz, *Atomic-Absorption Spectroscopy*, Elsevier Publishing Co., New York, 1968, p. 111.]

13–6   NONFLAME ATOMIZATION

Conventional flame techniques are very inefficient. Much of the sample is not atomized, either because it never reaches the flame or because the droplets are too large, and the fraction that is atomized stays in the optical beam for only a fraction of a second. The desire to analyze small samples containing low levels of the components of interest has resulted in the development of a number of flameless atomization devices.

Most commercial flameless atomizers are based on the Massmann furnace, shown in Figure 13–7. A measured amount of sample, which may be only a few μl of a liquid or a few mg of a solid, is placed in a graphite tube or cup. This is heated by passing an electric current through the graphite. Heating normally proceeds in three stages which can be made completely automatic. A relatively low temperature is used to remove the solvent, a higher temperature to ash the sample (i.e., to remove the chemical matrix by volatilization and pyrolysis), and the tube is then heated to incandescence to atomize the sample. Oxidation of the graphite is prevented by surrounding it with a nitrogen or argon atmosphere. (Argon is better for some elements.) The result is a relatively sharp absorption peak, as the

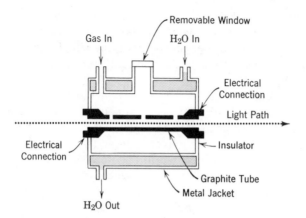

Figure 13–7   Massmann furnace. [Courtesy of Perkin-Elmer Corporation.]

gaseous sample atoms cross the radiation path, and this is usually recorded on a pen recorder. Alternatively, it can be measured by a peak-picking device which determines the peak maximum, or by integrating the area under the peak. The concentration is obtained by comparison with a standard sample.

The effect of the three separate heating stages is shown in Figure 13–8a, in

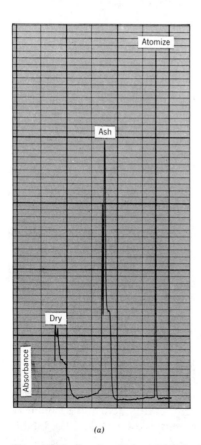

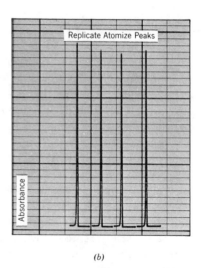

(a)   (b)

Figure 13–8   Results obtained with the Varian-Techtron carbon rod atomizer for $5 \, \mu l \times 0.1 \, \mu g/g$ cadmium in tomato soup (a), and $2 \, \mu l \times 0.5 \, \mu g/l$ zinc in seawater (b). [Courtesy of Varian-Techtron.]

which the first two peaks are due to molecular absorption during drying and ashing and the third sharp line is the true atomic absorption peak for the metal. The reproducibility obtainable by this technique is shown in Figure 13–8b. Both these sets of results were obtained with a Varian-Techtron carbon rod atomizer which is, in effect, a mini-Massmann furnace. The detection limit obtainable depends on the particular element and operating parameters, but is typically about 1 to 10 picograms (10^{-12} g).

Cathodic sputtering has been used as a source of atomic vapor for the direct determination of metals and alloys. A flat water-cooled metal specimen is used which forms the cathode in a glow-discharge tube operating at a pressure of about 5 torr. With some metals, such as aluminum and zinc, oxygen must be rigorously excluded because the formation of oxide layers on the specimen surface can lead to poor reproducibility. However, for iron-base alloys a reproducibility of $\pm 1\%$ has been reported.

Mercury can be determined in the vapor phase without a flame because of its high volatility. Inorganic mercury compounds can be reduced to elemental mercury by treating an acidified solution of the sample with stannous chloride, and then sweeping out the released mercury by passing air through the reaction vessel. This method is much more sensitive for mercury than the conventional flame technique, the detection limit being 0.00003 μg/ml compared with 0.1 μg/ml for mercury in a flame.

13-7 RADIATION SOURCES AND OPTICAL SYSTEMS

Hollow Cathode Lamps. The most commonly used radiation source for atomic absorption spectroscopy is the hollow cathode lamp. The structure of a typical lamp is shown in Figure 13–9. The cathode is made of the element to be determined or is lined or coated with it, and the anode is of tungsten, nickel, or zirconium. The latter is used as a "getter" and therefore prolongs the life of the lamp. The window is made of Pyrex glass or quartz, depending on the wavelength of the emitted radiation. The lamp is filled with neon or sometimes argon at a pressure of a few torr. These gases emit sharp line spectra, neon producing a red glow near the cathode and argon a blue glow, and the filler gas is selected to provide least spectral interference. The discharge is confined to the interior of the cathode by a shield and by selection of the appropriate gas pressure and operating voltage. The recommended operating current depends on the particular element, and varies between 3 and 25 mA. Normally these lamps are constructed for individual elements but multi-element lamps are also available, some for up to six elements. To distinguish between the lamp and flame emission the lamp radiation

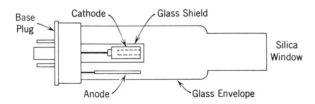

Figure 13–9 Varian-Techtron hollow-cathode lamp.

is chopped, either by means of a mechanical chopper or by pulsing the current through the lamp. This is then used with a suitable a.c. amplifier.

Electrodeless Discharge Tubes. Electrodeless microwave-excited discharge tubes have recently become available for antimony, arsenic, cadmium, iodine, lead, mercury, phosphorus, selenium, tellurium, and tin. For some elements these lamps are claimed to be more intense and more stable than the corresponding hollow cathode lamps. For other elements, a level of stability and reliability comparable to the hollow cathode lamp has not yet been achieved. The development of more intense sources will be of little benefit in atomic absorption spectroscopy because the source intensity and stability is usually not a limiting factor. However, more intense sources will be of value in atomic fluorescence where the output signal is proportional to the intensity of the light source.

Continuous Radiation Sources. In addition to the line emission sources, hydrogen and deuterium lamps are available which provide continuous radiation sources over the range from 190 to 320 nm. These lamps can be used to correct for nonatomic absorption which can be a problem when trace amounts of material are determined in complex samples.

Atomic absorption spectrometers are generally single-beam instruments, but correction for nonatomic absorption is most conveniently accomplished by using a dual-beam instrument in which the radiation from the element hollow cathode lamp and a deuterium lamp pass alternately through the same section of the flame. When only a single source is used in a dual-beam instrument, a reference beam external to the flame can be provided which will compensate for changes in the lamp intensity, amplifier gain, and detector performance (but not for variations which occur within the flame).

The hydrogen lamp is a hollow cathode lamp that is interchangeable with other hollow cathode lamps. It can be used to manually compensate for nonatomic absorption by taking separate readings with the element lamp and the hydrogen lamp. The deuterium arc requires a separate power supply, but is much more intense than the hydrogen lamp and for this reason is used in dual-beam systems.

It might be thought that a continuous source would be preferable to the use of discrete line sources for individual elements. However, flame absorption lines are very narrow, of the order of 0.01 to 0.001 nm (0.1 to 0.01 Å), so that an extremely high resolution monochromator would be required. Furthermore, the total energy emitted over such a small spectral range is small. This problem is largely overcome by the use of a line source. The flame absorption line width is determined mainly by Doppler and pressure broadening. The former is due to the random directional distribution of atomic velocities, and the latter to collisional interactions which reduce the lifetimes in the upper energy states. The emission line from a hollow-cathode lamp is narrower because of its lower operating pressure and temperature, but not so narrow that we should expect absorption of radiation to strictly follow Beer's law. This will be evident in comparing Figure 9–6 with Figure 13–10, which shows the profiles of two hollow-cathode emission lines and the corresponding flame absorption lines. The emission line profiles vary widely, depending on the hyperfine structure, and pressure broadening causes a slight shift of the flame profiles toward longer wavelengths.

Monochromator and Detector. A hollow-cathode lamp generally emits more than one composite line for each element, as shown for example in Figure 13–11, but the required spectral line can be isolated by means of a relatively low dispersion monochromator. Most of the lines in Figure 13–11 are nonabsorbing lines because they involve transitions other than from the ground state. The most

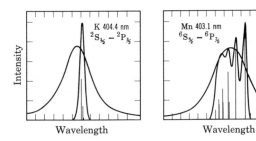

Figure 13–10 Atomic line profiles for the potassium 404.4 nm and manganese 403.1 nm transitions in a hollow-cathode lamp and in an acetylene-nitrous oxide flame. Each baseline division represents 0.1 cm^{-1} (0.0016 nm). [From H. C. Wagenaar and L. deGalan, *Spectrochimica Acta,* **28B**, 168, 169 (1973).]

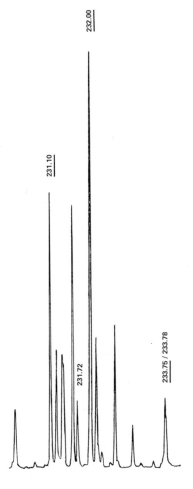

Figure 13–11 Spectral output of a nickel hollow-cathode lamp: ground state transitions are underlined. [Courtesy of Varian-Techtron.]

intense absorption line is chosen to provide maximum sensitivity, but weaker lines involving a ground-state transition can be used for the analysis of more concentrated samples. Alternatively, the burner can be rotated so that a shorter length of flame is placed in the path of the radiation beam.

The inlet and exit slit widths of the monochromator need only be sufficiently narrow to isolate the particular line being used. The requirements depend on the focal length and grating ruling of the monochromator and on the complexity of the spectrum of the element being determined; the acceptable spectral bandwidth varies between 0.1 and 1.0 nm. An unnecessarily large value will not increase the amount of atomic absorption but may well result in increased unwanted molecular absorption.

Filters provide adequate isolation of the required spectral line if the element has a very simple line spectrum. Thus filter photometers are frequently used for the determination of sodium, potassium, calcium, and magnesium in clinical samples. Iron and nickel have more complex spectra and for these elements a monochromator with a bandpass of about 0.2 nm is required to isolate individual lines.

A photomultiplier is used as the detector, and for some elements the photomultiplier needs to have a sensitive response in the UV region. The useful wavelength range is from 194 nm for arsenic to 852 nm for cesium. The overall gain of the system is controlled by varying the voltage applied to the multiplier dynodes.

13-8   FLAME SPECTRA AND INTERFERENCES

Atomic Line Overlap.   Overlap of emission lines may occur because of the finite bandwidth of the wavelength selection system. Filter photometers are particularly prone to this form of interference, because of the relatively wide band of wavelengths that they pass. The spectral bandwidth for a glass filter is typically about 80 nm, and for an interference filter 15 nm. However, the effect also occurs with monochromators. For example the magnesium line at 285.21 nm cannot be isolated from the sodium line at 285.28 nm.

Until recently it was believed that this form of interference did not occur in atomic absorption spectroscopy because a specific radiation source is used and the absorption lines are very narrow. Several examples have now been reported but they are relatively rare in comparison with spectral interference effects in flame emission spectroscopy.

Molecular Emission and Absorption.   Flame emission spectra are superimposed on a continuous background radiation which arises from the flame itself (see Figure 13-2) or from molecular species present in the flame. Allowance must be made for this in emission spectroscopy when the emission intensity from the element of interest is relatively low. For example, the presence of a large amount of calcium interferes with the determination of small amounts of sodium and barium because of the emission bands of calcium hydroxide in the vicinity of the 589-nm sodium and 554-nm barium lines.

The effect of flame emission on absorption measurements is virtually eliminated by modulating the radiation source. However, molecular absorption by species such as strontium oxide and calcium hydroxide may interfere in atomic absorption spectroscopy. The effect is generally more pronounced at low wavelengths and high concentrations, and is reduced by the use of higher flame temperatures. With a hollow-cathode line source, the wavelengths in the vicinity of the resonance line cannot be scanned to obtain a background correction, but corrections can generally be made by measuring the effect of the molecular absorption on an otherwise nonabsorbing line close to the resonance line. An example is the 231.7-nm nickel nonabsorbing line which can be used to correct for nonatomic absorption at the 232.0-nm resonance line (see Figure 13-11). A more convenient method in the 190 to 320 nm range is to use a hydrogen or deuterium continuum lamp, in which case the line absorption and molecular absorption can both be measured at the same nominal wavelength. This is possible because the spectral slit width is much wider than the absorption line and the line absorption due to the element of interest will have a negligible effect on the hydrogen or deuterium continuum.

Physical Effects and Solution Properties.   Changes in the physical properties of the aspirated solution change the absorption or emission intensity. The viscosity of the solution affects the rate at which it is aspirated into the flame and the surface tension, density, and viscosity of the solution, together with the gas velocity, determine the droplet size. It is therefore important that the physical properties of the sample and standard solutions should be matched as closely as possible. The use of an organic solvent frequently increases the sensitivity by a factor of 3 to 5. The amount of sample carried into the flame is increased because of the lower viscosity compared with water. A reduction in droplet size also occurs because of the lower surface tension and this increases the volatilization rate as does a reduction in the solvent boiling point. Furthermore, with aqueous solutions the large excess of water tends to interfere by forming nonemitting radicals or even oxides with some metals, and its molecular emission spectrum may interfere with some emission measurements. Organic solvents reduce these effects, although some water is still formed in the flame. They also increase the reducing conditions in the flame.

Self-Absorption.   Part of the energy emitted from the interior of the flame may be absorbed by atoms in the ground state in the outer regions of the flame. At low sample concentrations this effect is negligible, but at high concentrations it can be so pronounced that the emission intensity becomes nearly proportional to the square root of the sample concentration. The effect is greatest when a long path burner is used, as shown in Figure 13–12. Self absorption may also occur with hollow-cathode lamps if they are operated at too high a current. Unexcited metal atoms in front of the cathode absorb radiation from the center of the resonance line, and this distorts the line profile. The distortion leads to reduced sensitivity and more pronounced curvature of the calibration plots, as shown in Figure 13–13. However, the increased lamp emission may provide a better signal-to-noise ratio with the acetylene-nitrous oxide flame in cases where a high flame background emission produces an appreciable noise level in the photomultiplier.

Chemical Effects and Condensed Phase Interference.   Chemical interference

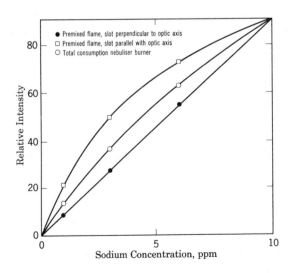

Figure 13–12 Effect of the flame length on self-absorption of the sodium 589.0 nm line. [From E. E. Pickett and S. R. Koirtyohann, *Spectrochimica Acta*, **23B**, 238 (1968).]

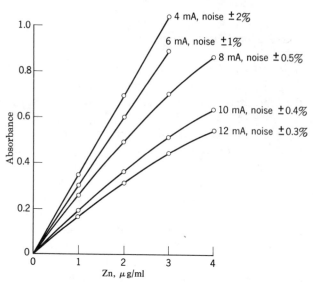

Figure 13–13   Effect of the hollow-cathode lamp current on calibration curves for zinc. [From J. B. Willis, in R. Mavrodineanu (ed.), *Analytical Flame Spectroscopy*, Centrex Publishing Co., Eindhoven, Netherlands, 1970, p. 559.]

occurs as the result of formation in the flame of compounds that are incompletely volatilized or dissociated. It is believed that most chemical interference effects take place during the formation of solid particles in the flame prior to their evaporation, and this is therefore often referred to as condensed phase interference. An example is phosphorus interference in the determination of calcium, which is due to the formation of calcium phosphate.

The effect of interferents can be established by measuring the emission or absorption of a series of samples containing different concentrations of interferent. In some cases the effect can be eliminated by changing the flame conditions; for example, increasing the fuel flow rate to provide more reducing conditions minimizes the formation of stable oxides. In others, a higher flame temperature may be required and it is for this reason that the acetylene-nitrous oxide flame is often preferred to the acetylene-air flame. However, it appears that the flame chemistry is more important than flame temperature where the formation of refractory oxides is concerned. Thus it has been found that refractory compounds are poorly atomized in the hydrogen-nitrous oxide flame, although its temperature is similar to that of the acetylene-nitrous oxide flame. This indicates the need for carbon or carbon-containing radicals in the flame to minimize the formation of metal oxides.

Alternatively, the interferent can be removed by selective extraction, or an excess of a releasing agent can be used that will combine preferentially with the interferent. An example of the latter is the addition of lanthanum or strontium nitrate to calcium solutions to prevent the formation of calcium phosphate in the air-acetylene flame. A similar result can be achieved by protecting an element by chelating it with EDTA. This forms a complex which protects the element from unwanted reactions but is readily decomposed in the flame.

Flame Temperature.   The presence of other chemical compounds in the flame affects both emission and absorption by changing the flame temperature. It is obvious that any temperature change will affect the emission intensity, because of the change in population of the upper energy state in accordance with Equation

13–1. The absorption will also change because most compounds are incompletely dissociated, even in the acetylene-nitrous oxide flame, and the degree of dissociation is therefore temperature dependent. Differentiating Equation 13–1 with respect to temperature, and neglecting dissociation for the moment, we have

$$\frac{d}{dT}\left(\frac{N_j}{N_0}\right) = \frac{g_j}{g_0}\frac{E_j}{kT^2}\exp\left(-E_j/kT\right) = (N_j/N_0)E_j/kT^2$$

Since N_0 is virtually independent of temperature (see Table 13–1) we can rewrite this as

$$dN_j/dT \simeq N_jE_j/kT^2 \tag{13–4}$$

Consequently, a 1% change in the number of atoms in the upper state (dN_j/N_j) is equivalent to a temperature change ΔT given by:

$$\Delta T \simeq 0.01\,kT^2/E_j \tag{13–5}$$

or

$$\Delta T \simeq 7.0 \times 10^{-10}\lambda\,(nm)\,T^2. \tag{13–6}$$

For a flame temperature of 3000°K, and a transition which produces a line at 300 nm (equivalent to 4 eV), this corresponds to a temperature change of about 2°K. If the dissociation energy is also of the order of 4 eV, we can expect a temperature change of 2°K to produce a change in absorption of 1% or in emission intensity of 2%.

Ionization. As has already been discussed, pronounced ionization of elements with low ionization potentials occurs in high-temperature flames. Because the extent of ionization decreases with increased metal concentration in the flame, an upward curvature of the calibration plot results. As shown in Figure 13–14, the effect can be overcome by adding an excess of a more readily ionized element. This creates a high electron concentration and suppresses the ionization of the less easily ionized element. A relatively high concentration of potassium, in the order of 2000 μg/ml, is generally used for this purpose.

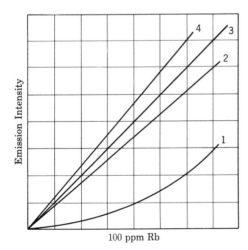

Emission Intensity

100 ppm Rb

Figure 13–14 Suppression of rubidium ionization (1) in an acetylene-air flame by the addition of 500 (2), 1000 (3) and 2500 (4) ppm of potassium. [Reprinted from I. Rubeska, in J. A. Dean and T. C. Rains (eds.), *Flame Emission and Atomic Absorption Spectrometry*, Vol. 1, Dekker, New York, 1969, p. 336. By courtesy of Marcel Dekker, Inc.]

13–9 QUANTITATIVE ANALYSIS

Sample Preparation. The sample pretreatment required depends on the element to be determined, the substrate material, and the method of atomization.

With the graphite furnace most samples require no chemical pretreatment because the chemical matrix is removed by ashing prior to atomization. For flame atomization most liquid samples can be sprayed directly into the flame after dilution with a suitable solvent. Solid samples are generally dissolved in acid but some may require alkali fusion before acid treatment. Hydrochloric, nitric, and sulfuric acids are commonly used for the dissolution of metals and alloys. Nitric acid gives compounds that are relatively easily decomposed but not highly volatile, and is therefore used for ashing in preference to hydrochloric acid. A 3 : 1 : 1 mixture of nitric, sulfuric, and perchloric acids is useful for the wet oxidation of organic matter. (*Caution*: Perchloric acid and perchlorates may be unstable, and, when mixed with carbonaceous materials, form explosive mixtures. See Christian and Feldman, p. 193, concerning the use of this oxidizing mixture.) Note that reagent grade acids can contain significant amounts of trace metal impurities; chromium in nitric acid, lead in hydrochloric acid and cadmium in sulfuric acid are particularly common.

An organic solvent may be used to selectively extract metals after complexing from an aqueous solution, and the solvent can then be aspirated directly into the flame. Methyl isobutyl ketone and ethyl acetate are preferred for this purpose because of their suitability as extractive agents and for use in premix burner systems. Detailed method sheets for the analysis of a wide variety of samples by atomic absorption spectroscopy are available from a number of instrument manufacturers.

Standards. Standards and sample solutions should be as similar as possible, both in chemical and physical properties. Reagents need to be checked to ensure that they do not introduce significant amounts of the element to be determined. Standards and samples should be stored only in polyethylene bottles because some metals adsorb onto glass surfaces. Standards containing low concentrations (less than 1 mg/l) should be made up daily from more concentrated stock solutions to avoid adsorption problems.

Calibration Techniques. The calibration technique which is most appropriate for a particular analysis depends on the number of samples to be analyzed, the linearity of the calibration graph, and the extent to which other components in the sample interfere with the analysis.

Working Curves. If a number of samples are to be analyzed, the simplest procedure is often to prepare a series of standards covering the concentration range of interest and from these obtain a "working curve" as shown in Figure 13–15. This tends to be nonlinear, but devices are commercially available that will electronically compensate for the curvature and thereby enable a direct readout linear in concentration to be obtained. Sometimes an alternative wavelength can be selected that yields a more linear calibration curve. For example, the 235.4 or 286.3-nm absorption line for tin provides a more linear response than the 224.6-nm line which is very close to an interfering line.

Neglecting interference effects, flame emission provides a signal which is directly proportional to concentration. Atomic absorption, however, follows Beer's law because an absorption process is involved. In the latter case the amplifier gain is set to provide a signal corresponding to 100% transmission when a blank solution is aspirated into the flame. The corresponding absorbance for each standard is then given by

$$A = -\log T = 2.000 - \log(\%T) \qquad (13\text{--}7)$$

which follows directly from the use of Beer's law. Alternatively, the %T

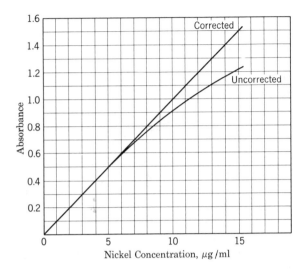

Figure 13-15 Typical atomic absorption calibration curve, before and after electronic curve correction. [Courtesy of Varian-Techtron.]

calibration data can be plotted directly on semi-logarithmic paper. With most modern instruments the absorbance is determined electronically by means of a logarithmic function generator, and can be read out directly.

The working curve must be checked from time to time, since the slope of the curve may change due to minor fluctuations in the fuel and oxidant pressures or in the sample flow rate. Appropriate adjustments can then be made to comply with the previously determined calibration data. The measurements will almost always be made at the wavelength corresponding to the emission or absorption maximum, because this is least affected by a small shift in the wavelength setting such as may result from a change in temperature of the monochromator. It should also be remembered that emission and absorption varies with the position of the burner, and for any one element the emission and absorption maxima (see Figure 13-6) rarely occur at the same height in the flame.

In emission spectroscopy there is generally a significant contribution to the signal due to the flame background and this may need to be subtracted from the measured signal intensity. This can be done by measuring the background signal close to the emission line used for determination of the test element. If the background varies with wavelength, an average value for the background at the line center must be obtained from measurements on either side of the line. Any variation will be obvious with a recording instrument which scans across the line but with a nonrecording instrument it may be necessary to take three separate measurements, one at the center of the line and one on either side of it. It is obviously necessary to ensure that the same three wavelengths are used for each separate determination. In atomic absorption spectroscopy, a correction may occasionally be required for nonatomic absorption, particularly at wavelengths below 220 nm. In that case the background absorption is determined at the line center by using a hydrogen or deuterium continuum lamp.

Internal Standardization. Minor variations in the fuel supply or in the physical properties of the sample may be compensated for by adding a fixed amount of an internal standard to each sample. The calibration graph is then a plot of the ratio of the emission intensity, or the absorbance, of the element being determined to that of the added standard. The standard, which must not already be present in the sample, should be chosen such that its spectral line is close to that of the test element and arises from a similar transition. Its ionization potential should also be similar to that of the test element.

Method of Standard Additions. If the calibration graph is linear, the sample concentration can be determined by adding known amounts of the test element to the sample. This gives a section of the calibration graph above the unknown sample concentration, and the resulting straight line can be extrapolated back to zero signal intensity. The concentration scale is determined by the standard additions, and the unknown concentration is given by the point at which the extrapolated line crosses the concentration axis.

Sample Bracketing. This is another method suitable for use with a small number of samples; it can also be used when the calibration graph is nonlinear. After an initial run on the unknown, two standards are chosen, one with a concentration higher than the unknown and one lower. When the calibration graph is curved, the standards can be chosen so that they are very close to the unknown concentration. The segment between the two standards can then be considered linear, and an accurate analysis obtained.

Detection of Interference. Curvature of the calibration graph indicates interference in one form or another. A quick check of linearity can be obtained by measuring the emission or absorption signal at two sample concentrations, preferably in the ratio of 2:1. However, the fact that this check indicates a reasonable degree of linearity may not necessarily mean that there are no interference effects present, as shown by curve 2 in Figure 13–14. Curvature may be due to chemical interference, self-absorption, or spectral interference. In the latter case the calibration plot may be linear but in emission spectroscopy it will not pass through the zero point because of the presence of background emission.

Chemical interference effects are frequently small at low levels of interferent, but increase markedly as the concentration of interferent increases. At still higher levels the change in signal with increase in the amount of additive tends to level out so that a buffering effect is achieved. This means that an excess of added interferent will have a similar effect on both the sample and the standards, and the initial concentration of contaminant is no longer significant. An example is provided by the addition of an excess of phosphate ion to calcium samples. However, except for the suppression of ionization, the net effect of an excess of contaminant will be a decrease in the signal intensity.

13–10 TYPICAL APPLICATIONS

In recent years flame photometry has been largely overshadowed by atomic absorption spectroscopy. This is partly because the latter is somewhat less prone to inter-element interference effects and partly because the detection limits obtained by atomic absorption spectroscopy are generally comparable to, or better than, those for flame emission. The major application for flame photometry is the measurement of sodium and potassium, particularly for clinical purposes. Because these elements have simple emission spectra, filter photometers generally provide adequate spectral resolution and simple instruments can be used.

Atomic absorption spectroscopy has been used for the determination of approximately 70 elements. Applications include clinical and biological samples; forensic materials; foods and beverages; water and effluents; soils, plants, and

Table 13–4 Detection Limits (μg/ml) for Atomic Absorption, Flame Emission, and Atomic Fluorescence

Element	Atomic Absorption		Flame Emission		Atomic Fluorescence	
	λ, nm	μg/ml	λ, nm	μg/ml	λ, nm	μg/ml
Aluminum	309.3	0.03	396.2	0.03	—	—
Antimony	217.6	0.03	—	—	217.6	0.03
Arsenic	193.7	0.1(0.003)[a]	193.7	10	193.7	0.1
Barium	553.6	0.01	455.4	0.01	—	—
Beryllium	234.9	0.001	234.9	1	234.9	0.01
Bismuth	223.1	0.03	306.8	30	302.5	0.03
Boron	249.7	3	518.0[b]	0.03	—	—
Cadmium	228.8	0.001	326.1	10	228.8	0.000001
Calcium	422.7	0.001	422.7	0.0003	422.7	0.03
Cerium	—	—	494.0[b]	1	—	—
Cesium	852.1	0.01	455.6	1	—	—
Chromium	357.9	0.003	425.4	0.003	357.2	0.03
Cobalt	240.7	0.01	345.4	0.1	240.7	0.003
Copper	324.7	0.003	324.8	0.03	324.7	0.001
Dysprosium	421.2	0.03	404.6	0.03	—	—
Erbium	400.8	0.03	400.8	0.1	—	—
Europium	459.4	0.01	459.4	0.1	—	—
Gadolinium	368.4	1	622.0	3	—	—
Gallium	294.4	0.1	417.2	0.3	417.2	0.3
Germanium	265.1	0.1	265.1	1	265.2	10
Gold	242.8	0.01	267.6	3	267.6	0.3
Hafnium	307.3	1	531.2	30	286.6	3
Holmium	410.4	0.03	410.4	0.1	—	—
Indium	303.9	0.03	451.1	0.01	410.5	0.1
Iridium	208.9	0.3	550.0	0.3	254.4	> 1000
Iron	248.3	0.01	372.0	0.03	248.3	0.01
Lanthanum	550.1	3	442.0	0.1	—	—
Lead	217.0	0.01	405.8	0.3	405.8	0.03
Lithium	670.8	0.001	670.8	0.0003	—	—
Lutetium	336.0	0.3	451.9	1	—	—
Magnesium	285.2	0.0003	285.2	0.1	285.2	0.001
Manganese	279.5	0.003	403.3	0.01	279.4	0.01
Mercury	253.7	0.1(0.00003)[a]	253.7	10	253.7	0.1
Molybdenum	313.3	0.03	390.3	0.3	379.8	3
Neodymium	492.5	1	492.5	1	—	—
Nickel	232.0	0.01	352.5	0.03	232.0	0.003
Niobium	334.9	3	405.9	1	—	—
Osmium	290.9	0.1	442.1	1	—	—
Palladium	244.8	0.01	363.5	0.1	340.4	3
Platinum	266.0	0.1	265.9	10	265.9	30
Potassium	766.5	0.003	766.5	0.00003	—	—
Praseodymium	495.1	10	495.1	0.1	—	—
Rhenium	346.1	1	346.1	0.1	—	—
Rhodium	343.5	0.003	343.5	0.0003	369.2	3
Rubidium	780.0	0.003	794.8	3	—	—
Ruthenium	349.9	0.1	372.8	0.3	372.8	300
Samarium	429.7	1	476.0	0.3	—	—
Scandium	391.2	0.03	402.4	0.1	390.7	10
Selenium	196.0	0.3(0.003)[a]	—	—	196.1	0.1
Silicon	251.6	0.3	251.6	3	251.6	1
Silver	328.1	0.003	328.1	0.01	328.1	0.0001
Sodium	589.0	0.0003	589.0	0.001	589.2	100
Strontium	460.7	0.003	460.7	0.0003	460.7	0.03
Tantalum	271.5	3	474.0	3	—	—

Table 13–4   *(Continued)*

Element	Atomic Absorption λ, nm	μg/ml	Flame Emission λ, nm	μg/ml	Atomic Fluorescence λ, nm	μg/ml
Tellurium	214.3	0.01	486.6	3	214.3	0.03
Terbium	432.7	0.3	534.0[b]	0.03	—	—
Thallium	276.8	0.01	377.6	0.01	377.6	0.01
Thorium	—	—	492.0	10	—	—
Thulium	371.8	0.01	—	—	—	—
Tin	286.3	0.03	317.5	0.3	303.4	0.1
Titanium	364.3	0.03	334.9	0.3	394.9	10
Tungsten	255.1	1	430.2	1	—	—
Uranium	358.5	30	544.8	10	381.2	3
Vanadium	318.5	0.03	437.9	0.1	—	—
Ytterbium	398.8	0.003	398.8	0.01	—	—
Yttrium	410.2	0.1	597.2	0.03	—	—
Zinc	213.9	0.001	636.2	10	213.9	0.00003
Zirconium	360.1	1	360.1	30	352.0	3

[a] vapor generation—as hydrides for As and Se.
[b] band emission.
[Atomic absorption data from Varian-Techtron "Analytical Methods for Flame Spectroscopy," 1971–1972, and Model AA-6 brochure, 1973; Flame emission data from G. D. Christian and F. J. Feldman, *Atomic Absorption Spectroscopy*, Wiley-Interscience, New York, 1970, pp. 174–75; Atomic fluorescence data from A. Syty in J. A. Dean and T. C. Rains (Eds.), *Flame Emission and Atomic Absorption Spectrometry*, Dekker, New York, Vol. 2, 1971, p. 225.]

fertilizers; iron, steel, and various other alloys; minerals; petroleum products; pharmaceuticals; and cosmetics.

Detection limits for atomic absorption, flame emission, and atomic fluorescence are given in Table 13–4. The atomic absorption figures on which this table is based were obtained with Varian-Techtron AA-5 and AA-6 spectrometers, using the appropriate flames shown in Figure 13–3, and with digital readout of the data to avoid any operator bias. The emission figures were obtained with an Instrumentation Laboratory model 153 spectrometer, and an acetylene-nitrous oxide flame. The fluorescence figures are from various sources. In each case the original figures have been rounded to the nearest factor of 3 to simplify the data. Thus values originally quoted as 0.2 to 0.5 are given as 0.3, whereas a value of 0.6 is given in this table as 1.0.

The detection limit is defined as "That concentration in solution of an element which can be detected with a 95% certainty. This is that quantity of the element that gives a reading equal to twice the standard deviation of a series of at least ten determinations at or near blank level." This should not be confused with sensitivity, which is defined for atomic absorption as "That concentration of an element in aqueous solution (expressed in μg/ml) which absorbs 1% of the incident radiation passing through a cloud of the atoms being determined." This value corresponds to the concentration at which the absorbance equals 0.0044. The detection limit is generally about one tenth of the sensitivity.

The data given in Table 13–4 show that the detection limit obtainable by flame emission is between 10 and 100 times better than atomic absorption for 8 elements (boron, lanthanum, potassium, praseodymium, rhenium, rhodium, strontium, and terbium). In comparison, atomic absorption is better than emission by a factor of 10 or more for some 25 elements, and by a factor of 10,000 for cadmium and zinc.

Sulfur and phosphorus can be determined by their molecular band emission in low-temperature flames, but chemical interferences are severe. Alternatively, sulfur can be determined indirectly by atomic absorption by precipitation as barium sulfate, phosphate (and silicate) by formation of a molybdenum complex, and chloride by precipitation as silver chloride.

Atomic fluorescence spectroscopy has provided several examples of detection limits lower than those obtained by either direct flame emission or atomic absorption, and the development of more intense radiation sources offers the possibility of further improvements. However, the technique is susceptible to self-absorption effects in the flame and scattering of the exciting radiation by solvent droplets and unvaporized solute particles. The latter effect depends on the sample matrix, which can make the application of this technique to practical problems rather difficult.

QUESTIONS AND PROBLEMS

13–1. Show diagrammatically the construction of an atomic fluorescence spectrometer.

13–2. Explain why an increase in lamp intensity is of direct benefit in atomic fluorescence but of little benefit in atomic absorption.

13–3. Why is integration of the peak produced by a flameless atomizer not necessarily any better than measurement of the peak maximum?

13–4. Show diagrammatically the construction of a dual-beam atomic absorption spectrophotometer which incorporates a deuterium arc lamp for automatic background correction.

13–5. What monochromator bandwidth is required to isolate the main lines in the nickel spectrum shown in Figure 13–11?   *Ans.* 0.2 nm

13–6. What effects result in:
(a) Upward curvature of calibration graphs
(b) Downward curvature of these graphs?

13–7. Boron has been determined by flame photometry by measuring the intensity of the oxide molecular emission band at 518 nm. The background emission is measured at 505 nm, which corresponds to the minimum (or trough) between the 518 and 492 nm emission bands. The following data have been reported [J. A. Dean and C. Thompson, *Anal. Chem.*, **27**, 43 (1955)]:

Standard (ppm boron)	Emission Intensity (Scale Divisions)	
	518 nm	505 nm
0	27	29
50	42.5	35
100	55	40
150	69	45
200	82	49
300	107	58

What concentration of boron would a differential reading (518-nm reading less 505-nm reading) of 37 divisions correspond to?   *Ans.* 225 ppm

13–8. The following calibration data have been obtained for the determination of nickel by atomic absorption spectroscopy:

Standard (ppm nickel)	2	4	6	8	10
%T	62.4	39.8	26.0	17.6	12.3

Plot the working curve for this analysis (cf. Figure 13–15), and calculate the nickel concentration for a reading of 20.4% T. *Ans.* 7.2 μg/l

13–9. A 1.013-g sample of a copper-beryllium alloy is dissolved in 5 ml of dilute nitric acid, and the solution is diluted to 1 liter with distilled water. The atomic absorption working curve is nonlinear but the sample is known to contain approximately 2% of beryllium. Standards have therefore been prepared containing 20 and 25 μg/ml of beryllium and each standard also contains 1000 μg/ml of copper.

(a) Why include this amount of copper in the standards?

(b) If the percent transmittance readings for the standards are 44.7 and 38.0, and for the unknown 40.9, what is the concentration of beryllium in the sample?

Ans. 2.24%

13–10. Duplicate 0.5-g samples of finely powdered gold-bearing ore are treated in the following manner. Digest in 5 ml of hydrochloric acid, add 3 ml of nitric acid, and evaporate to dryness. Add 15 ml of hydrochloric acid and 40 ml of hot water, and allow to cool. To one sample 1 ml of a 5.0 μg/ml gold standard is added. Each sample is then treated with 5 ml of hydrochloric acid and the resulting gold bromide complex is extracted into 5 ml of methylisobutyl ketone. If the atomic absorption calibration graph is linear, what is the original gold concentration in the sample in μg/g if the two absorbance readings are 0.22 and 0.37, respectively? *Ans.* 14.7 μg/g

REFERENCES

G. D. Christian and F. J. Feldman, *Atomic Absorption Spectroscopy—Applications in Agriculture, Biology and Medicine*, Wiley-Interscience, New York, 1970.

J. A. Dean and T. C. Rains, eds., *Flame Emission and Atomic Absorption Spectrometry*, Dekker, New York, Vol. 1, 1969, Vol. 2, 1971.

W. J. Price, *Analytical Atomic Absorption Spectrometry*, Heyden & Son, London, 1972.

J. D. Winefordner, ed., "Spectrochemical Methods of Analysis", in *Advances in Analytical Chemistry and Instrumentation*, Vol. 9, Wiley-Interscience, New York, 1971.

14

X-RAY METHODS

14–1 INTRODUCTION

If a beam of high energy electrons is directed at a metal target in an evacuated tube, some of the kinetic energy may be converted into a very penetrating form of electromagnetic radiation. These "mysterious rays" were first observed by William Röntgen, a German physicist, in 1895. Because he did not understand their origin, he called them "X rays." Some twenty years later, Max von Laue, another German physicist, discovered that these X rays are diffracted by the ordered arrangement of atoms in a crystal in the same way that ordinary light is diffracted by a line grating. Shortly thereafter, Henry Moseley, an English physicist, studied X rays emitted by a number of elements and found that the wavelengths are characteristic of the element. In fact, if the elements are arranged in the order in which they appear in the periodic table, there is a regular decrease in wavelength (increase in frequency), now known as the *Moseley equation* (see Equation 14–2 below). This fundamental property of an element helped to explain a few atomic weight inversions in the periodic table (Ar–K and Cu–Zn), and gave rise to the term *atomic number.*

X rays are, of course, similar to other forms of electromagnetic radiation and are generally defined as having wavelengths from 0.1 to 25 Å. In this chapter we discuss their properties and uses under three broad categories: emission, absorption, and diffraction. Much of our knowledge of atomic and crystal structure derives from extended studies of the interaction of X rays with matter.

14–2 EMISSION, DISPERSION, AND DETECTION OF X RAYS

Three types of sources are commonly used to produce X rays for analytical purposes: (1) electron bombardment of a metal target, (2) irradiation of a target (the sample) with a primary beam of high energy X rays to produce a secondary beam of fluorescent X rays, and (3) exposure of a sample to a radioactive source which generates X rays. The X-ray emission spectra produced may be continuous or discontinuous, or a combination of both.

Continuous Spectra Produced by an Electron Beam. The *Coolidge tube*, illustrated schematically in Figure 14–1, is the most frequently used X-ray source. A heated tungsten cathode generates electrons that are accelerated to high velocities in an electric field of 50 to 70 kV (up to 100 kV for special purposes). When the electrons strike the anode (target), 1 to 2% of their kinetic energy is converted to X rays, as a result of their rapid deceleration. The remaining 98 to 99% of the energy is released as heat, requiring special water cooling of the target. The X rays produced are observed at a glancing angle and may be detected and analyzed as described below.

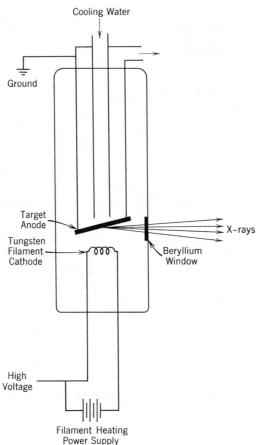

Figure 14–1 Schematic diagram of a Coolidge tube.

Any solid material can be used as a target, but copper, molybdenum, and tungsten are most frequently used for general purposes. Several spectra produced by a tungsten target at various applied voltages are shown in Figure 14–2. Continuous spectra from other target metals would look the same although the integrated intensity increases with atomic number. Thus, continuous spectra are of no value as a means of identification. The most interesting feature of these spectra is the sharp cut-off at the short wavelength limit. The minimum wavelength depends on the accelerating voltage of the tube, but is independent of the nature of the target. This behavior can be explained if we assume that the highest energy (lowest wavelength) X rays that can be produced result from complete conversion of the kinetic energy of the bombarding electron (Ve) into radiant energy ($h\nu$):

$$Ve = h\nu = hc/\lambda$$

$$\lambda_{min} = \frac{hc}{Ve} = \frac{12,400}{V}\, \text{Å} \tag{14–1}$$

X rays of longer wavelength are produced when a bombarding electron loses only a portion of its kinetic energy on impact and then undergoes a series of collisions with other atoms in the target.

Characteristic Line Spectra Produced by an Electron Beam. As the accelerating voltage is increased above a critical value (characteristic for each element),

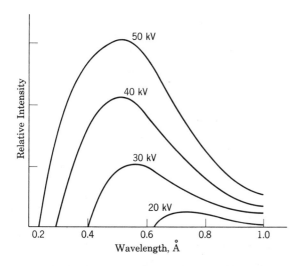

Figure 14–2 Continuous X-ray spectrum of tungsten at various voltages.

intense emission lines appear as spikes superimposed on the continuous back-ground. The emission spectrum of molybdenum at 35 kV is shown in Figure 14–3 in which the K_α and K_β lines are identified. Additional lines (the L series, not shown) would appear at wavelengths >4 Å. These sharp lines occur when the bombarding electrons have sufficient energy to "knock out" an electron from the K or L shell of the target atom. X-ray emission follows when an outer electron falls into the vacancy in the lower energy, inner shell. As shown in Figure 14–4, a

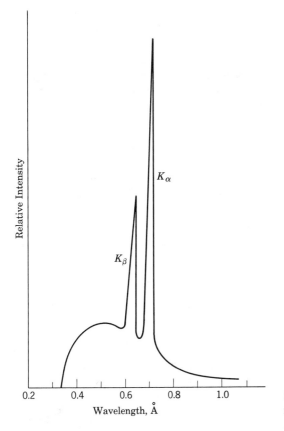

Figure 14–3 Line spectrum of molybdenum at 35 kV.

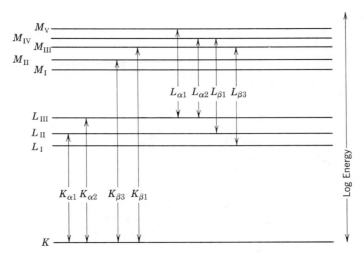

Figure 14–4 Abbreviated energy level diagram of a typical element identifying energy transitions with X-ray lines.

K_α X ray is emitted when an L electron falls into the K shell, and a K_β X ray is emitted when an M electron falls into the K shell. Elements before vanadium ($Z = 23$) in the periodic table produced only the K series of X rays, while those with higher atomic numbers produce both K and L X rays. As noted earlier the frequency of a given K or L line increases in a regular manner with atomic number, Z, (the Moseley equation):

$$\frac{c}{\lambda} = \nu = a(Z - \sigma)^2 \qquad (14\text{–}2)$$

where a and σ are constants (for the K_α line, $a = 2.48 \times 10^{15}$ and $\sigma = 1$). X-ray line spectra are related to the energy levels of the innermost atomic orbitals and are independent of the physical state or chemical combination in which the atom is found. Thus, the line spectrum is a positive identification of the element. Some characteristic emission lines (along with absorption edge values to be discussed later) are given in Table 14–1.

Table 14–1 Wavelengths of Emission Lines and Absorption Edges of Selected Elements

Element	Z	Emission Lines		Absorption Edge,[a]
		K_α, Å	K_β, Å	Å
Mn	25	2.106	1.910	1.895
Ni	28	1.658	1.500	1.487
Cu	29	1.542	1.392	1.380
Br	35	1.040	0.933	0.920
Zr	40	0.790	0.702	0.688
Mo	42	0.713	0.632	0.620
Ag	47	0.560	0.496	0.484
Pb	82	0.170	0.146	0.95

[a] K absorption edge is given for all elements except lead (L_{III} edge).

X-Ray Fluorescence Spectra. An incident beam of high energy X rays (the primary beam) may be used to excite the target atoms sufficiently to knock out *K* electrons. After a short time the excited ion returns to the ground state, producing a fluorescence spectrum similar to the emission spectrum. However, only the line spectrum appears without the continuous background. Thus, fluorescence spectra show a much greater signal-to-background level and are preferred for analytical work.

X-Ray Emission Spectrometers. Although the sample could be made the target in an X-ray tube, it is more convenient to generate its fluorescence spectrum with high energy X rays from a Coolidge tube, as shown in Figure 14–5. No suitable transparent materials are available for the fabrication of lenses; therefore X rays are collimated by passage through a series of slits or a collection of long narrow tubes. Likewise no prisms are available, but fortunately crystals of many salts are able to disperse X rays by diffraction (see Section 14–4 below) and serve as excellent monochromators. Properties of a few of these crystals are given in Table 14–2.

Table 14–2 Properties of Crystals Used for Diffracting X rays

Crystal	d, Å	λ_{min}, Å	λ_{max}, Å
Topaz	1.356	0.25	2.65
LiF	2.014	0.35	4.00
NaCl	2.820	0.50	5.55
EDDT[a]	4.404	0.75	8.65
ADP[b]	5.325	0.95	10.50

[a] EDDT is ethylenediamine *d*-tartrate.
[b] ADP is ammonium dihydrogen phosphate.

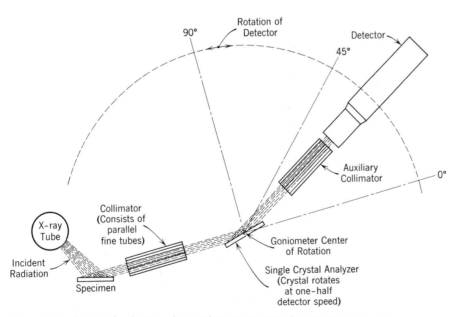

Figure 14–5 Schematic diagram of essential components of an X-ray spectrometer.

From Table 14–2 it is clear that no single material is suitable for the entire wavelength range. The large single crystal analyzer (monochromator) is rotated on its axis to obtain the spectrum.

X-Ray Detectors. Three types of detectors are used to measure the intensity of an X-ray beam. (1) The simplest detector is a photographic film or plate which will darken when exposed to X rays. The developed film is then scanned with a densitometer to record the spectrum. (2) Gas ionization detectors, such as ionization chambers, proportional counters, and Geiger tubes used to measure radioactivity, are also suitable for X rays. (3) Some crystals fluoresce in the ultraviolet or visible region when exposed to X rays. These crystals may be incorporated in a scintillation counter that also utilizes a photomultiplier tube to measure the intensity of the fluorescence. All of these detectors are described more fully in Section 23–4.

Nondispersive Instruments. Most detectors respond equally to photons of all energies within a certain range—they do not discriminate. However, the proportional counter generates a pulse of electrons for each X-ray photon detected, and the number of electrons in the pulse is proportional to the energy (frequency) of the X-ray photon. Thus, a 1.0-Å photon produces a pulse height that is twice that of a 2.0-Å photon. With suitable electronic filtering circuits (pulse height analyzers), the photons in narrow energy bands can be counted separately. The proportional counter thus replaces a monochromator. Furthermore, there is no need for a collimator so that the path length from source to sample to detector can be made very short, as in Figure 14–6. The vastly increased efficiency means that a relatively low intensity source can be used, such as a small amount of ^{241}Am. Nondispersive spectrometers are compact, relatively inexpensive, and give performance comparable to crystal monochromator instruments, except for somewhat poorer resolution of closely spaced lines.

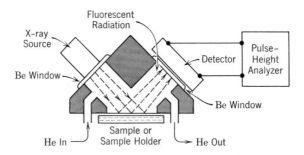

Figure 14–6 Schematic diagram of a nondispersive X-ray spectrometer to obtain fluorescence spectra.

Applications. All types of solids and liquids are easily handled; if necessary the sample may be placed in a thin walled cell or deposited as a film under cellophane. All elements above calcium ($Z = 20$) are readily detected, those between sodium ($Z = 11$) and calcium with difficulty, and the lighter elements not at all. X-ray emission spectroscopy is widely used for the analysis of steels and other alloys, and for the determination of heavy elements in organic samples (e.g. lead and bromine in aviation fuels). It is highly specific with limits of detectability as low as a few parts per million.

14-3   ABSORPTION OF X RAYS

Absorption Process.   The absorption of X rays is analogous to the absorption of other electromagnetic energy in the ultraviolet, visible, or infrared regions. The only significant difference is the energies involved. As with the emission process, we are concerned with the innermost electrons where the energy levels are independent of the chemical or physical state. The absorption spectrum identifies the element regardless of its environment. In the absorption process the energy brought by the incident photon, $h\nu$, is parceled between the kinetic energy of the ejected electron and the potential energy acquired by the excited ion. Absorption of the X-ray photon is most probable if the energy of the incident photon just equals the energy required to eject the electron. That is, the electron leaves with essentially zero kinetic energy. For example, the absorption spectrum of lead, shown in Figure 14-7, consists of a few broad peaks with sharp discontinuities, called *absorption edges*. Each absorption edge corresponds to the energy required to eject a K or L electron. The wavelength of an absorption edge is slightly less than the wavelength of the corresponding emission line because the energy required to eject the electron completely from the atom is greater than the energy associated with an outer electron (already in the atom) falling into the vacancy. Table 14-1 gives absorption edge data for several elements.

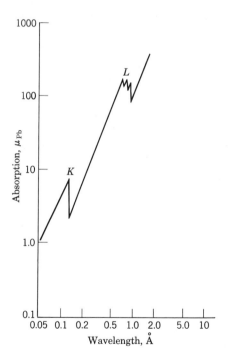

Figure 14-7   The absorption spectrum of lead.

Absorption Coefficients.   Beer's law is valid for the absorption of X rays, and is usually written as:

$$2.303 \log (I_0/I) = \mu x \qquad (14-3)$$

where I_0 is the incident intensity and I is the intensity transmitted through a sample thickness of x cm. The proportionality constant, μ, in Equation 14-3 is called the *linear absorption coefficient*. The numerical value of μ depends on the nature of the element and the number of its atoms per cm² of cross-sectional area through which the beam is traveling. It is customary to express the coefficient as a *mass absorption coefficient, μ_m*:

$$\mu_m = \mu\rho \tag{14-4}$$

where ρ is the density of the absorbing substance. An empirical relationship defines μ_m in terms of the atomic weight, A, of element number Z, Avogadro's number, N, and the wavelength of the X rays:

$$\mu_m = \frac{CNZ^4\lambda^3}{A} \tag{14-5}$$

The same constant, C, applies to all elements within limited regions of wavelength. Mass absorption coefficients are additive in multi-component samples. For a given wavelength, a mean absorption coefficient, μ_S is obtained as follows:

$$\mu_S = W_A\mu_A + W_B\mu_B + W_C\mu_C + \cdots \tag{14-6}$$

where μ_A, μ_B, μ_C are the mass absorption coefficients of elements A, B, and C, each present at weight fractions of W_A, W_B, and W_C.

X-Ray Absorption Methods. The broad absorption bands, as exemplified by Figure 14–7, lead to interferences among neighboring heavy elements. For the most part, X-ray absorption methods are limited to samples containing a single heavy element in an organic matrix (e.g., lead in gasoline or chlorine in chloro-compounds).

Absorption Edge Analysis. The discontinuity observed at an absorption edge provides a unique and selective means of identifying an element. A plot of the mass absorption coefficient of lead versus wavelength in the vicinity of its L_{II} edge is shown in Figure 14–8. Only two measurements are needed for an actual determination, one on either side of the edge. The K_α and K_β emission lines of bromine at 1.04 and 0.93 Å neatly bracket the lead L_{III} edge at 0.95 Å; thus, the fluorescence induced in NaBr is a convenient source for this determination of lead. The difference in the absorption coefficient of the sample between the two wavelengths is a measure of the amount of lead it contains. Suitable bracketing

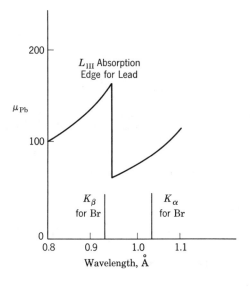

Figure 14-8 The absorption spectrum of lead superimposed on the emission lines of bromine.

lines are known for about 40 elements. This method is less subject to interference from other elements in the matrix than emission spectroscopy, but it is more cumbersome and not as widely used as emission analysis.

Beta Filters. One of the most important uses of the absorption edge effect is the so-called *beta filter*. In general, the K_α and K_β lines in an emission spectrum are so close together that they are not easily resolved by low resolution instruments. A beta filter is a simple mechanism for achieving a monochromatic source. It consists of a thin foil (10 to 100 mg/cm^2) of an element with atomic number of one or two less than the target element generating the X rays. The object is to select the filter so that its absorption edge falls between the wavelengths of the K_α and K_β radiation, as for example the nickel filter used to isolate the K_α line of copper in Figure 14–9. Several other target–filter combinations are available, for example, the data in Table 14–1 show that a zirconium filter ($Z = 40$) will isolate the K_α line of molybdenum ($Z = 42$). Beta filters are commonly used in X-ray diffractometers where a monochromatic source is essential.

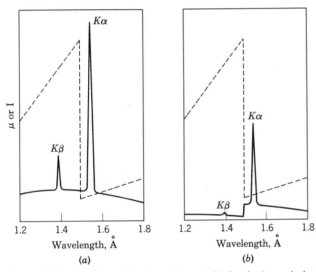

Figure 14–9 Principle of the beta filter. Solid line is the emission spectrum of copper observed (a) without and (b) with a nickel filter. Dashed line is the mass absorption coefficient of nickel.

Bulk Analysis. By far the most familiar use of X-ray absorption techniques is in the field of medicine to observe the physical state of bones, teeth, and various vital organs in living bodies. X rays are also indispensable in checking large solid objects for interior holes or other defects without having to destroy the sample; and, of course, for checking luggage for concealed metal objects.

X-ray methods are said to be nondestructive; however, their indiscriminate use is certainly a health hazard which should not be ignored.

14–4 DIFFRACTION OF X RAYS

The wavelengths of X rays are of the same order as the distances between atoms in crystalline materials (Table 14–2); therefore crystals will act as diffraction gratings for X rays. In Figure 14–10 the beam of X rays striking the crystal

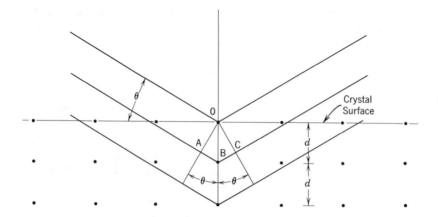

Figure 14–10 Diffraction of X rays by a crystal.

surface is partly scattered by the atoms in the first layer. Another portion is scattered by the second layer, and so on. Constructive and destructive interference operate in the same way as with optical gratings (Section 8–6). The requirement for constructive interference was first expressed by W. L. Bragg:

$$AB + BC = n\lambda$$
$$AB = BC = d \sin \theta$$

therefore

$$n\lambda = 2d \sin \theta \tag{14–7}$$

which is known as *Bragg's equation.* In their early studies of X rays, W. H. and W. L. Bragg used large crystals of sodium or potassium chloride. From the densities of the crystals and Avogadro's number, they calculated a value of d for these crystals. Through Equation 14–7 they obtained accurate values of λ for a number of X-ray lines. The Braggs' studies provided the basis for *X-ray crystallography.*

Diffraction phenomena are used in three different ways. (1) Large, single crystals of known geometry provide the monochromator element in X-ray spectrometers. (2) The diffraction pattern produced by a crystalline substance is perhaps the most positive means of identification. (3) For crystallography, diffraction patterns obtained from a large number of directions may be interpreted to yield a precise and detailed three-dimensional map of the atoms composing the crystal.

Measurement of X-ray Diffraction. Large single crystals of the sample are not always available. Fortunately, for routine identifications, this is not necessary, and it is more convenient to grind the sample to a fine powder (200 to 300 mesh). In such a sample, the large number of microcrystals will be randomly oriented so that there will be many particles oriented to satisfy the Bragg condition for constructive interference from all possible interplanar spacings. The sample need not be carefully oriented as is necessary with single crystals, nor is it necessary to reposition it to obtain the entire set of diffraction patterns.

Diffraction patterns are usually recorded with a *Debye-Scherrer powder camera,* shown schematically in Figure 14–11. The powdered sample is held in a thin glass tube or spread with a binder on a plastic film. X rays from a Coolidge tube are filtered to produce a nearly monochromatic beam. The undiffracted portion of the

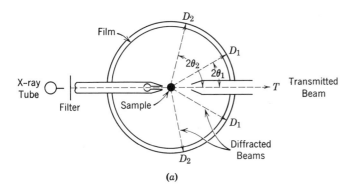

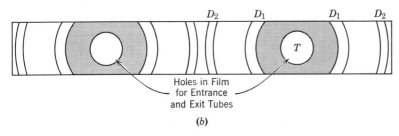

Figure 14–11 Schematic diagram of (a) Debye-Scherrer powder camera and (b) film strip after exposure and development. D_1, D_2, and T identify locations on film mounted in the camera.

beam is trapped and absorbed, while the diffracted rays take a conical shape emanating from the sample. Portions of these beams are intercepted by a photographic film mounted in a circular fashion on the inside wall of the camera. Thus, the diffraction pattern appears as a series of concentric arcs whose radii are determined by the diffraction angles, θ, and the interplanar distances, d. Since both λ and θ are known, values of d are calculated from Equation 14–7. For most routine work, the identification of crystals is based on a comparison of d-spacings and relative intensities with files of known compounds.

Modern X-ray spectrometers are highly automated, incorporating electronic scanning detectors. A computer may control the entire operation including the comparison to known compounds for which the relevant data are stored in its memory. In contrast to emission and absorption methods, diffraction leads to the identification of *compounds*.

Determination of Crystal Structure. Single crystals (at least a few tenths of a millimeter) are required for studies of crystal structure because of the need to know the orientation of the crystal planes with respect to the direction of the X-ray beam. The crystal is mounted in a *goniometer*, a mechanical device for rotating and tilting the crystal in every direction at accurately measured angular displacements. Diffraction patterns are obtained sequentially from hundreds (or even thousands) of orientations. From all these data the crystallographer generates maps of electron density through various cross-sections of the crystal. The details of the procedure, which is greatly facilitated by computers, are beyond the scope of this discussion, but the method is unique in the type of information it can provide; for example, the classic determination of the structure of Vitamin B_{12} whose molecule contains 200 atoms.

14–5 ELECTRON MICROPROBE ANALYSIS

The focusing of electron beams by specially shaped electromagnets has been perfected to a degree that makes it possible to examine a portion of the sample as small as $1 \ \mu m^3$ ($10^{-12} \ cm^3$). The X rays emitted are analyzed with a crystal spectrometer. The complete instrument includes an optical microscope so that the operator can locate the exact spot to be analyzed. Such an instrument, though complex and expensive, is invaluable in studying imperfections in metals and ceramics, and in examining the diffusion of dopants (deliberately introduced impurities) in semiconductors. With appropriate computer interfacing it is possible within a few hours to scan automatically $1 \ cm^2$ of surface of a sample and determine 3 elements in each $1 \ \mu m^3$ of the sample surface, detecting each element at a level of about $10^{-14} \ g$.

REFERENCES

H. A. Liebhafsky, H. G. Pfeiffer, E. H. Winslow, and P. D. Zemany, *X-Rays, Electrons and Analytical Chemistry*, Wiley, New York, 1972.

L. S. Birks, *X-Ray Spectrochemical Analysis*, 2nd ed., Wiley, New York, 1969.

H. P. King and L. E. Alexander, *X-Ray Diffraction Procedures*, Wiley, New York, 1954.

D. B. Wittry, "X-Ray Microanalysis by Means of Electron Probes," in I. M. Kolthoff and P. J. Elving, eds., *Treatise on Analytical Chemistry*, Wiley-Interscience Publishers, New York, 1964, pt. 1, vol. 5, chap. 61.

C. W. Bunn, *Chemical Crystallography*, 2nd ed., Oxford University Press, New York, 1961.

15

NUCLEAR MAGNETIC RESONANCE SPECTROSCOPY

15–1 INTRODUCTION

In 1946 two physicists, Purcell at Harvard University and Bloch at Stanford University, independently announced a far-reaching discovery about the behavior of the atomic nucleus. Most nuclei, including the proton and the electron, possess inherent magnetic fields, though the effects of the nuclear fields are too small to be observed in the ambient magnetic field of the earth. However, in an intense magnetic field the nuclei can assume specific orientations with corresponding potential energy levels. Bloch and Purcell invented techniques to detect the minute amount of energy absorbed or emitted as the nuclei jump from one energy level to another. Thus was born nuclear magnetic resonance, or n.m.r., a new spectroscopic tool. There are few methods that have developed so fast from a highly theoretical study to practical applications as varied as the determination of the detailed structure of complex biochemical compounds to the prospecting for minerals.

15–2 MAGNETIC PROPERTIES OF THE NUCLEUS

The magnetic properties of certain nuclei are conveniently explained if we assume that the nuclear charge is spinning around an axis. Such a nucleus possesses angular momentum given in terms of a spin number, I, which is assigned half-integral values $0, \frac{1}{2}, 1, \frac{3}{2}, \ldots, \frac{9}{2}$ depending on the particular nucleus. Some representative nuclei and their spin quantum numbers are listed in Table 15–1.

Nuclei which have either an *odd* number of protons *or* an *odd* number of neutrons, *but not both odd*, exhibit half-integral spin quantum numbers; for example, ^1H, ^{11}B, ^{19}F, and ^{31}P. Figure 15–1 portrays one of these nuclei as a spinning sphere. The circulating charge generates a magnetic field, much like an electric current does in a coil of wire. There is an associated nuclear magnetic moment, μ, along the axis of spin.

For nuclei in which the numbers of neutrons and protons are *both odd*, the charge is distributed nonsymmetrically. For example, ^2H and ^{14}N have an integral spin number, $I = 1$.

Nuclei in which the number of protons and the number of neutrons are *both even* have no angular momentum ($I = 0$) and exhibit no magnetic properties. Thus ^{12}C, ^{16}O, ^{32}S, etc., are magnetically inert and are not detected in n.m.r. experiments.

Magnetic nuclei ($I > 0$) interact with an external magnetic field by assuming discrete orientations with corresponding energy levels. The number of quantized

Table 15–1 N.M.R. Data for Selected Nuclei

Nucleus	Natural Abundance, %	Spin Number, I	Magnetic Moment, μ Nuclear Magnetons[a]	Magnetogyric Ratio, γ gauss^{-1} sec^{-1}	Resonance Frequency, ν(MHz) at $H_0 = 14{,}092$ gauss
^{1}H	99.98	1/2	2.79268	2.675	60
^{2}H	0.0156	1	0.85741	0.4102	9.2
^{11}B	81.17	3/2	2.688	0.8583	19.25
^{12}C	98.9	0	0	0	No resonance
^{13}C	1.1	1/2	0.7023	0.6721	15.08
^{14}N	99.62	1	0.4037	0.1931	4.33
^{16}O	99.76	0	0	0	No resonance
^{17}O	0.039	5/2	−1.893	−0.3625	8.13
^{19}F	100	1/2	2.628	2.5236	56.6
^{31}P	100	1/2	1.1305	1.083	24.29

[a] 1 nuclear magneton $= 5.05 \times 10^{-24}$ erg gauss^{-1}.

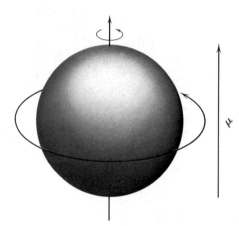

Figure 15–1 Spinning charge in nucleus generates a magnetic field with a magnetic moment, μ, directed along the axis of spin.

energy levels available depends on I, and is given by the series:

$$I, I-1, I-2, \ldots, -I$$

For a given nucleus, there are $2I + 1$ possible levels or orientations. For the proton in particular, with $I = \frac{1}{2}$, there are only two available orientations, described as aligned with the applied field (lower energy) or against the applied field (higher energy). Figure 15–2 shows this separation into two levels, each with an energy of

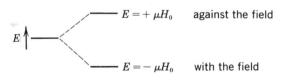

$E = +\mu H_0$ against the field

$E = -\mu H_0$ with the field

No Applied Magnetic Field   Applied Magnetic Field, H_0 Alignment

Figure 15–2 Proton spin states under the influence of an applied magnetic field.

$+\mu H_0$ or $-\mu H_0$, where H_0 is the intensity of the applied magnetic field. The difference in energy, ΔE, is equal to $2\mu H_0$. This is a special case of the general relationship:

$$\Delta E = \mu H_0 / I \qquad (15-1)$$

15-3 NUCLEAR RESONANCE

Equation 15-1 embodies a wealth of information. ΔE is determined in part by the value of the magnetic moment, μ, a characteristic property of each kind of nucleus. Typical values of μ are included in Table 15-1. ΔE is also directly proportional to the intensity of the applied magnetic field. In n.m.r. experiments, a field of about 14,000 gauss is common. For protons in this field, we find from Equation 15-1 and data in Table 15-1 that ΔE is 5.7×10^{-3} cal/mole, a very small energy indeed. At room temperature, the thermal energy is considerably larger, and is sufficient to maintain nearly equal populations in the two levels. In fact, at 25°C the Boltzmann distribution tells us that for every million protons, there is an excess of only three protons in the lower level. What kind of radiant energy corresponds to this small energy difference. We have that

$$\Delta E = h\nu = 2\mu H_0 \qquad (I = \tfrac{1}{2}) \qquad (15-2)$$

from which $\nu = 60 \times 10^6$ Hz or 60 MHz. This is in the radio-frequency range with $\lambda = 5$ m.

Now suppose that we are able to supply energy of just this frequency. A radio-frequency transmitter is an appropriate energy source. A proton in the lower level may absorb this energy and jump to the upper level. This absorption process is called "magnetic resonance," or we may say that the nucleus "resonates" at the proper "resonance frequency." Equations 15-1 and 15-2 are easily combined:

$$2\pi\nu = \frac{2\pi\mu H_0}{hI} = \gamma H_0 \qquad (15-3)$$

where 2π is introduced to convert linear frequency to angular units of frequency and a new parameter, γ, is introduced. Obviously, $\gamma = 2\pi\mu/hI$ and is a characteristic property of the nucleus, called *the magnetogyric ratio*. Another definition of γ is easily derived from Equation 15-3:

$$\text{Magnetogyric ratio} = \gamma = \frac{2\pi\nu}{H_0} \qquad (15-4)$$

Equation 15-4 is a fundamental equation for n.m.r. It defines the "resonance condition"—a function of the *ratio* of frequency to field. With a fixed field, H_0 (for example, 14,092 gauss), we may vary the frequency until we locate the resonance condition. Each type of nucleus will resonate at a different frequency as required by its magnetogyric ratio. Typical resonance data are given in Table 15-1 and Figure 15-3.

Some instruments are built to maintain a fixed frequency and vary the magnetic field. This is an entirely analogous procedure, and gives us the same information. In this type of experiment it is common to convert values in gauss to values in hertz, using Equation 15-4. For the proton, 1 gauss is equivalent to 4260 Hz.

Let us summarize the information we have given thus far. A nucleus $(I > 0)$ in an intense magnetic field may find itself in one of $2I + 1$ equally spaced energy levels. The energy differences between levels are small, so that although the

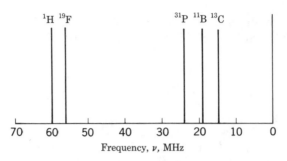

Figure 15–3 Resonance frequencies for typical nuclei at an applied field of 14,092 gauss. (*Note:* n.m.r. spectra are customarily plotted with frequency increasing from right to left.)

nucleus would prefer to be in the lowest energy level, thermal motions nearly equalize the populations. The small excess number of nuclei in the lower level can be promoted to higher levels by absorption of radiant energy, in this case in the radio-frequency region. If we can detect this absorption, we obtain a spectrum, for example, Figure 15–3. We must now ask, "What processes allow the nuclei to return to the lower level?" Unless the nuclei can fall back to lower levels, absorption of energy will cease as soon as the populations in the two states become equal. This question brings us to the study of nuclear relaxation—or, "How does an excited nucleus 'relax' to a lower energy condition?"

15–4   NUCLEAR RELAXATION

A sophisticated discussion of the theory of radiation is beyond our scope, but it can be shown that the emission of radiation may occur either spontaneously or by a process stimulated by an electromagnetic field. At radio frequencies, the probability of spontaneous emission is negligible. However, the probability of the stimulated process of emission is exactly the same as the probability of absorption of energy from the field. If the populations in the lower and upper levels are equal, absorption by nuclei in the lower state exactly balances induced emission by nuclei in the upper state. Thus, in the n.m.r. experiment, the absorption signal may dwindle to zero as soon as the populations are equalized, unless there are other mechanisms to maintain an excess population in the lower level. We know of two kinds of mechanisms, called spin–spin relaxation (or transverse relaxation) and spin–lattice relaxation (or longitudinal relaxation).

Spin–Lattice Relaxation. The excited nucleus can transfer energy to other nuclei in the surrounding molecular framework (lattice). Such energy is conserved within the system, but appears as extra translational or rotational energy distributed around the lattice. The same sort of process establishes the Boltzmann distribution in the first place and serves to maintain a constant slight excess of population in the lower level. Spin–lattice relaxation is characterized by a relaxation time, T_1, which is a measure of the half-life required for the system in the excited state to return to an equal population state.

Spin–Spin Relaxation. In this process, a nucleus in the upper level can transfer its energy to a neighboring nucleus by an exchange of spin. Because of this mutual interchange, there is no net effect on the population distribution; however, it does limit the average time a given nucleus spends in a given spin state. There is an associated spin–spin relaxation time, T_2, which affects the natural line width of an n.m.r. absorption peak.

Natural Width of an Absorption Line.   Heisenberg's uncertainty principle tells us that the uncertainty in energy of an excited state, ΔE, is inversely proportional to the average time a system spends in the excited state. Since $\Delta E = h \Delta \nu$, where $\Delta \nu$ is the width of the spectral line, narrow lines are associated with long relaxation times and broad lines with short relaxation times. Both T_1 and T_2 enter into the relaxation efficiency. Solids and viscous liquids have relatively rigid lattices, so spin–spin relaxation is efficient, T_2 is small, and the lines are relatively broad. In nonviscous liquids and dilute solutions, relaxation times are long and the lines are relatively sharp. Apparently, an examination of line widths in the spectrum can tell us much about the nature of the lattice, but we will not consider this aspect of n.m.r. here. Additional relaxation mechanisms operate in the case of nuclei with $I > \frac{1}{2}$. Therefore, their lines tend to be broader than those associated with nuclei where $I = \frac{1}{2}$. We must also point out that magnetic field inhomogeneity contributes more to line width than T_1 or T_2 in most high resolution work.

15–5   N.M.R. INSTRUMENTATION AND TECHNIQUES

An n.m.r. spectrometer contains a massive and intricate collection of electronics of which we can consider only the basic elements. We must remember that we are dealing with intense magnetic fields requiring enormous, precisely controlled power supplies, and precisely controlled frequencies. Unfortunately, most of the kilowatts required by the instrument are dissipated as heat and the few microwatts of signal we obtain from the sample must be amplified by another intricate electronic system. The cost ($5,000 to $500,000) and complexity of operation of the instrument represent a barrier to its widespread use, but even so it is rapidly becoming nearly as available as an infrared spectrometer.

The basic elements of a typical n.m.r. spectrometer, shown in Figure 15–4, include:

1. A magnet with a strong, stable, homogeneous field. The field must be constant over the area of the sample and over the period of time of the experiment to better than 1 part in 10^8.
2. A sweep generator which supplies a variable d-c current to a secondary magnet so that the total applied magnetic field can be varied (swept) over a limited range.
3. A radio-frequency oscillator (transmitter) connected to a coil which transmits energy to the sample in a direction perpendicular to the magnetic field.
4. A radio-frequency receiver connected to a coil encircling the sample. The two coils are perpendicular to each other and to the magnetic field.
5. A readout system consisting of an amplifier, recorder, and possibly additional components for increased sensitivity, accuracy, or convenience.
6. A sample container, usually a glass tube spun by an air-driven turbine to average the magnetic field over the sample dimensions.

We hasten to point out that this simple description does not adequately or accurately represent an actual commercial spectrometer. The requirements for high resolution are so severe that it is not practical to construct an instrument for wide range use. Thus each instrument is normally used to examine only one kind of

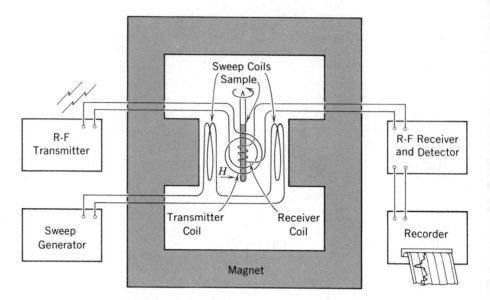

Figure 15–4 Schematic diagram of an n.m.r. spectrometer. [Courtesy of Varian Associates.]

nucleus, the proton for example. Some high resolution instruments are provided with interchangeable oscillators, so that several nuclei can be studied, but not simultaneously. For the study of solids, where the lines are inherently broad, high resolution is of little value, therefore some instruments are built for low resolution but greater flexibility.

Sample Handling.   A dilute solution (~2 to 10%) is normally used. If we are to study protons in the sample, the ideal solvent should contain no additional protons. Carbon tetrachloride meets this requirement, but deuterated chloroform, $CDCl_3$, or deuterated benzene are often preferred. Deuterated solvents may give a small additional peak for a residual proton impurity. Deuterium oxide (deuterated water) is available for samples soluble only in aqueous solutions.

A 5-mm OD glass tube serves as a sample container. It is held by a propeller arrangement so that it can be spun by a jet of compressed air.

Integration of Peak Areas.   The n.m.r. signal is directly proportional to the number of nuclei that are responsible for the resonance. In effect, the spectrometer "counts hydrogen nuclei (protons)" (or the relevant nuclei causing the absorption peak). As an aid in measuring peak areas, most recorders are equipped with automatic integrators which give peak areas directly. The integrator trace is often superimposed on the spectrum, and the height of a step on the integration trace measures the area in arbitrary units.

Nuclear magnetic resonance signals may be very weak and lost in the "noise" (random signals introduced by the instrument or its surroundings). By summing repetitive scans, the random fluctuations are averaged to zero, revealing the desired signal. A device known as a "computer of average transients", or CAT, performs this function, and is available as an accessory for this and other applications. (See Section 25–7.)

By far the most work has been done with spectrometers built specifically to examine the n.m.r. behavior of protons. In the remainder of this chapter we will consider primarily proton magnetic resonance.

15–6 THE CHEMICAL SHIFT

The behavior of an isolated nucleus is really of no concern to a chemist. An actual sample contains an enormous number of charged species (nuclei and electrons). Each charged particle is subject to the influence of the surrounding magnetic fields. Electrons in covalent bonds normally have paired spins and have no net magnetic field. But an applied magnetic field induces additional modes of circulation for these paired electrons which generate a small local magnetic field proportional to but opposing the applied field. This phenomenon is called "diamagnetic shielding" because it shields the nucleus to some degree from the effects of the applied field, as shown in Figure 15–5. The nucleus thus finds itself in an effective field which is somewhat smaller than the applied field H_0, that is

$$H_{eff} = H_0 - \sigma H_0 \qquad (15\text{–}5)$$

where σ is the shielding parameter. The value of σ depends on the electron density around the proton, and this in turn is a function of the structure of the compound.

Methanol, for example, has two kinds of protons, CH_3 and OH. The oxygen atom is more electronegative than the carbon atom, and therefore the electron density around the methyl protons is higher than that around the hydroxyl proton. Stated another way, the shielding parameter is greater for methyl protons, $\sigma_{CH_3} > \sigma_{OH}$. *At a given H_0, H_{eff} for CH_3 protons must be lower than H_{eff} for the OH proton.* Thus the resonance equation (15–4) must be modified by changing H_0 to H_{eff} *at the nucleus.* To bring each kind of proton into resonance *at a fixed frequency,* we must change H_0 as shown in Figure 15–6a. As a result, the n.m.r. spectrum of methanol has two peaks as in Figure 15–6b.

We can repeat the same argument in terms of frequency. *At a fixed applied field, H_0,* the hydroxyl proton finds itself in a greater effective field than do the methyl protons. Therefore, by Equation 15–4 a higher frequency is required to bring the OH proton into resonance. The variation of the resonance line with chemical structure is called *the chemical shift,* defined as the change in position of a resonance peak from that of an arbitrary reference line.

The most common reference compound used is tetramethyl silane, $(CH_3)_4Si$, hereafter abbreviated TMS. This compound has a symmetrical structure; each proton is identical to all others and is found in an identical electronic environment which provides very high shielding. As a result, TMS exhibits a single sharp resonance line at a high applied field, well beyond the protons in most other organic compounds. In addition TMS is chemically inert, relatively volatile (b.p. = 27°C), and soluble in most organic solvents. TMS is not soluble in aqueous

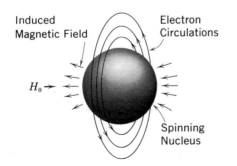

Induced Magnetic Field

Electron Circulations

$H_0 \rightarrow$

Spinning Nucleus

Figure 15–5 Induced electronic circulations generate a local magnetic field opposing H_0. (Diamagnetic shielding.)

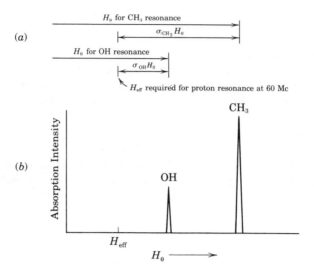

Figure 15–6 (a) Different values of H_0 are required to bring CH_3 and OH protons into resonance at 60 MHz. (b) The n.m.r. spectrum of methanol, CH_3OH.

solutions; for these media sodium 2,2-dimethyl-2-silapentane-5-sulfonate (DSS)

$$(CH_3)_3SiCH_2CH_2CH_2SO_3Na$$

may be used as a reference. The protons in the methyl groups of DSS give a strong sharp line. The methylene protons of DSS give a series of small peaks which can be ignored.

Several cautions are in order before we consider how chemical shift is related to chemical structure. First, the absolute magnitude of the shift is extremely small. For example, in a field of 14,092 gauss, methyl protons resonate at 60,000,054 hertz and methylene protons resonate at 60,000,075 hertz, a difference of only 21 hertz out of 6×10^7. Even so, accurate measurement of the *difference* is commonplace. Second, the magnitude of the shift is directly proportional to H_0. Therefore, the absolute value is of little use. We avoid this dilemma by using a reference compound, and employing a relative value, δ, for the chemical shift.

$$\delta = \frac{H_0(\text{reference}) - H_0(\text{sample})}{H_0} \times 10^6 \text{ ppm} \qquad (15\text{–}6)$$

H_0 for the reference is usually higher than H_0 for the sample, so subtraction in the direction indicated gives a positive δ. In terms of frequency units, δ takes the form:

$$\delta = \frac{\nu(\text{sample}) - \nu(\text{reference})}{\nu(\text{reference})} \times 10^6 \text{ ppm} \qquad (15\text{–}7)$$

For the common 60-MHz, 14,092-gauss spectrometer, and with TMS as a reference, a convenient equation is

$$\delta = \frac{\Delta \nu}{60 \times 10^6} \times 10^6 \qquad \text{or} \qquad \Delta \nu / 60 \text{ ppm} \qquad (15\text{–}8)$$

Note that δ as defined in Equation 15–6 or 15–7 is independent of the value of H_0, although the simplified Equation 15–8 is restricted to $H_0 = 14,092$ gauss and a specific reference, TMS. δ is dimensionless, but is given in parts per million (ppm). Typical values range from 0 to 10. Another common expression for

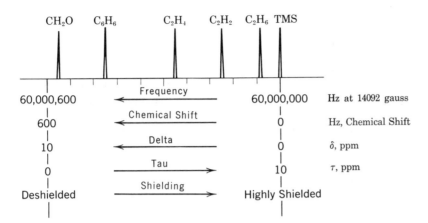

CH₂O	C₆H₆	C₂H₄	C₂H₂	C₂H₆	TMS

60,000,600   ← Frequency →   60,000,000   Hz at 14092 gauss

600   ← Chemical Shift →   0   Hz, Chemical Shift

10   ← Delta →   0 δ, ppm

0   Tau →   10 τ, ppm

Deshielded   Shielding →   Highly Shielded

Figure 15–7 Resonance frequencies at 14,092 gauss and chemical shift values for typical protons, with typical terms used to describe the shift.

chemical shift is called tau, τ:

$$\tau = 10 - \delta \qquad (15\text{–}9)$$

Let us mark the various ways of expressing the chemical shift and associated terminology on a single chart. Figure 15-7 provides an orientation to the various directional terms for n.m.r. spectra observed at a constant applied field. A corresponding chart showing chemical shifts at a constant frequency would appear identical if plotted with the field increasing from left to right; for example, deshielding or "negative" shielding results in lines found at a low field or "down field" at constant frequency, or at a higher frequency at constant field. The resonance lines for protons in several simple compounds are included in the figure.

Next, we shall attempt to explain why lines are shifted as they are. Our approach is necessarily qualitative and empirical—it is much easier to explain the behavior after we have observed it than to predict it in advance. We shall consider two effects resulting from (a) local diamagnetic currents and (b) magnetic anisotropy.

Local Diamagnetic Effects. The extent of shielding resulting from local magnetic fields is related to the electron density in the immediate vicinity of the proton, as we have demonstrated above. Thus the order of magnitude of δ is predictable from our general knowledge of electronegativity. For example, we could predict at least the order of the chemical shifts in the following sequences:

	CH₃—H	CH₃—I	CH₃—Br	CH₃—Cl	CH₃—F
δ:	0.2	2.2	2.7	3.0	4.3

	CH₃—CH₃	CH₃—N<	CH₃—O—	CH₃—F
δ:	0.9	2.2	2.5	4.3

Acidic protons have a very low electron density, and experience low diamagnetic shielding. Their resonance lines are found far downfield. For example, the proton in RCOOH resonates at $\delta = 10.8$ ppm.

Magnetic Anisotropy. Additional shielding and *deshielding* effects arise from electronic circulations induced within molecules. Consider this series of compounds:

$$CH_3—CH_3 \qquad CH_2{=}CH_2 \qquad CH{\equiv}CH$$

δ: 0.9   5.8   2.9

The order of δ values would be predicted erroneously from a consideration of local diagmagnetic effects alone. We have similar troubles with the abnormally large shift of an aldehydic proton, $—C\!\!\begin{smallmatrix}\nearrow H \\ \searrow O\end{smallmatrix}$, $\delta \sim 10$, and the aromatic protons of benzene, $\delta \sim 7.3$. All these examples contain double or triple bonds, and we must consider the behavior of the π electrons in the unsaturated bonds.

Let us consider first the linear molecule of acetylene, where the electron distribution in the triple bond is symmetric about the axis. If the axis is parallel with the magnetic field, an electronic circulation perpendicular to the axis will be induced. Figure 15–8 shows the orientation of this circulation and the resultant induced magnetic field which opposes the applied field. The lines of force of the induced field extend beyond the ends of the molecule. Thus the protons of acetylene find themselves in a lower effective field. This shielding effect of the triple bond overshadows the deshielding effect of electron withdrawal by the carbon atom as it operates in the saturated ethane molecule. In a triple bond the electronic circulations are restricted to a plane normal to the bond axis, so that if the bond axis is normal to the magnetic field there can be no induced circulation. In a real sample of acetylene the molecules are all rotating rapidly, and one observes only a single line for the average orientation, intermediate between the two extremes.

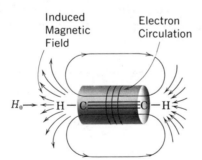

Induced Magnetic Field   Electron Circulation

Figure 15–8   Anisotropic shielding in the acetylene molecule.

In the double bond, the restrictions on electronic circulations are different. Consider an ethylene molecule oriented as in Figure 15–9. The π-electron clouds sit to the left and right of the bond axis in a plane normal to the paper. Again, induced circulations within the planes generate a magnetic field which opposes the applied field, but here the lines of force are perpendicular rather than parallel to the bond axis. The protons of ethylene find themselves in an effective field which is *enhanced* by the induced field. This corresponds to a greater deshielding than we expected from diamagnetic effects alone, and a larger chemical shift. Both effects combine forces in the case of an aldehydic proton, which accounts for its abnormally large chemical shift.

Aromatic rings likewise have π-electron clouds situated to the right and left of the ring pictured in Figure 15–10. The explanation of the so-called "ring current" is analogous to that given for ordinary double bonds. As in the first two cases, all

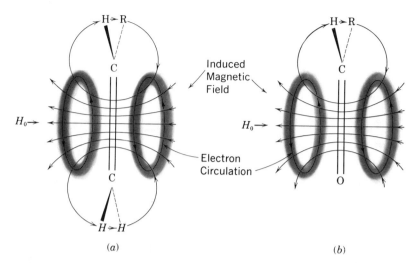

Figure 15–9   Anisotropic shielding in (a) olefins and (b) aldehydes.

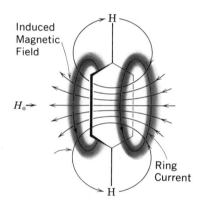

Figure 15–10   Anisotropic shielding of aromatic protons caused by ring current effect.

orientations of the ring with respect to the field are averaged because of the rapid rotation of molecules in the sample.

These effects are summarized in Figure 15–11, in which are depicted zones (cones) of (positive) shielding and (negative) deshielding. Obviously the effects vary continuously from point to point with maximum effects along the cone axes and in regions equidistant from the cones.

The anisotropic effects of the σ electrons in a C—C single bond are much smaller than those we have been discussing and will not be considered here.

Correlation of Chemical Shift with Structure.   In a complex molecule, a number of effects combine to establish the value of the chemical shift of each type of proton. The exact value may also depend on the concentration of the species and the nature of the solvent. The last two effects are normally small, unless the proton is subject to exchange with other protons in the system or hydrogen bonding occurs, as for example with alcohols or amines.

Although in most cases a detailed calculation of δ is not possible, there is already a wealth of data available for known compounds. These data are conveniently summarized in Tables 15–2, 15–3, 15–4, and 15–5.

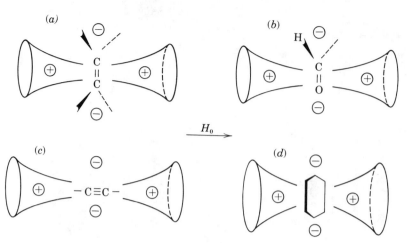

Figure 15–11 Cones of positive shielding for (a) double bonds, (b) aldehydes, (c) acetylenes, and (d) aromatic rings. Negative space represents deshielding.

Table 15–2 Chemical Shift Values (δ) for Protons of Methyl, Methylene, and Methine Attached to Various R Groups

(Reproduced by kind permission from *Spectroscopic Techniques in Organic Chemistry*, A. J. Baker and T. Cairns, Heyden & Son Ltd., London, 1967.)

Attached Functional Group	Methyl	Methylene	Methine
R	$\boxed{CH_3}$ R	$R''\cdot\boxed{CH_2}\cdot$ R	$R''\cdot\boxed{CH}\cdot$ R
$-\overset{\cdot}{C}\cdot$	0.9	1.25	1.5
⤢ (double bond)	1.7	1.9	
⤢–R'	1.9		
R' branched alkene	2.0		
$-CO_2R'$	2.0	2.1	
$-CN$	2.0		
$-CO_2H$	2.07	2.34	2.57
$-CONR'_2$	2.02	2.05	
$-COR'$	2.1	2.4	2.48
$-CHO$	2.17	2.2	2.39
R' (cyclopropyl)	2.1		
$-SR'$	2.1	2.4	
$-NR'_2$	2.15	2.5	2.87
$-I$	2.15	3.15	4.2
(phenyl)	2.34	2.62	2.87
$-NHCOR'$	2.85	3.2	
$-Br$	2.7	3.3	4.1
$-Cl$	3.05	3.4	4.0
$-OR'$	3.3	3.35	3.8
$-OH$	3.38	3.56	3.85

290

continued on p. 291

Table 15-2 (*continued*)

Attached Functional Group	Methyl	Methylene	Methine
R	$\boxed{CH_3}\,R$	$R'' \cdot \boxed{CH_2} \cdot R$	$R'' \cdot \boxed{CH} \cdot R$
—OCOR'	3.65	4.15	5.01
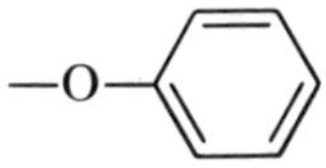 —O—⬡	6.3	3.9	
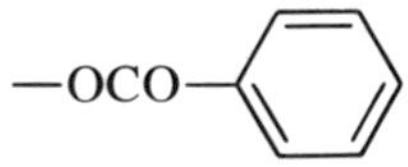 —OCO—⬡	3.9	4.23	5.12

Table 15-3 Chemical Shift Values (δ) for Protons on Unsaturated Groups

(Reproduced by kind permission from *Spectroscopic Techniques in Organic Chemistry*, A. J. Baker and T. Cairns, Heyden & Son Ltd., London, 1967.)

Group	δ ppm
—C≡CH	2.35
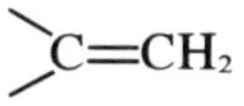 >C=CH$_2$	4.65
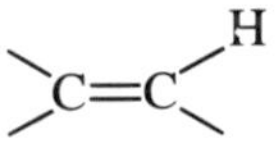 >C=C< (with H)	acyclic 5.4 cyclic 5.6
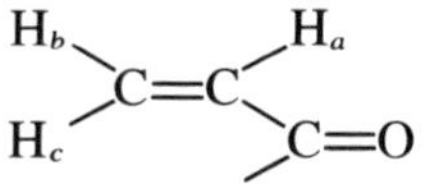 H$_b$, H$_a$ / C=C / H$_c$, C=O	H$_a$ 5.8 H$_b$ 6.0 H$_c$ 6.2
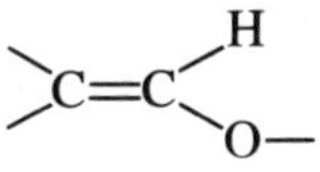 >C=C< (H, O—)	6.8
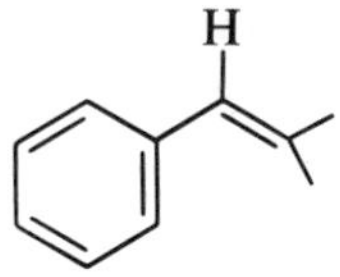 ⬡ =CH— (phenyl vinyl)	7.0
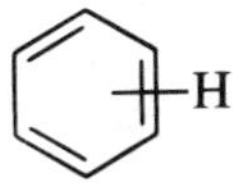 ⬡—H	7.27
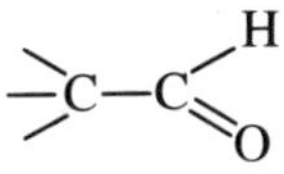 >C=C—C(H)=O	9.65
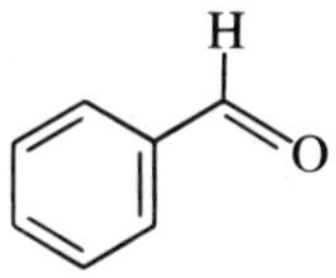 ⬡—C(H)=O	9.9
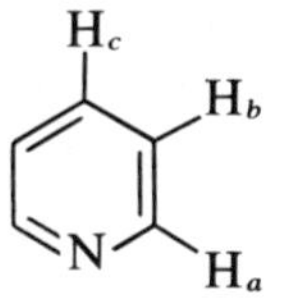 pyridine (H$_c$, H$_b$, H$_a$)	H$_a$ 8.5 H$_b$ 6.99 H$_c$ 7.35

Table 15–4 Chemical Shift Deviations in the Position of Aromatic Protons ($\delta = 7.27$) Caused by Various Ring Substituents

Substituent	ortho	meta	para
—CH₃	−0.15	−0.1	−0.1
—≡	0.2	0.2	0.2
—COOH, —COOR	0.8	0.15	0.2
—CN	0.3	0.3	0.3
—CONH₂	0.5	0.2	0.2
—COR	0.6	0.3	0.3
—SR	0.1	−0.1	−0.2
—NH₂	−0.8	−0.15	−0.4
—N(CH₃)₂	−0.5	−0.2	−0.5
—I	0.3	−0.2	−0.1
—CHO	0.7	0.2	0.4
—Br	0	0	0
—NHCOR	0.4	−0.2	−0.3
—Cl	0	0	0
—NH₃⁺	0.4	0.2	0.2
—OR	−0.2	−0.2	−0.2
—OH	−0.4	−0.4	−0.4
—OCOR	0.2	−0.1	−0.2
—NO₂	1.0	0.3	0.4

15–7   SPIN–SPIN SPLITTING

We have seen that n.m.r. lines are shifted to higher or lower field by the electronic environment of the proton. So far we have neglected the effects of other protons within the molecule. Recall that each proton is considered to be a spinning magnet which in an applied field is oriented with or against the field. We might expect that the resonance position of a proton would be affected by the spin of nearby protons. With high resolution instruments we can observe this effect as a "splitting" of the lines, or a "coupling" of the protons. Coupling occurs through bonds (not space) by means of a slight unpairing of the bonding electrons.

Let us consider two protons, A and B, on neighboring carbon atoms. The effective field at A is either decreased or enhanced by the local field generated by B, depending on whether B is oriented with or against the applied field. The resonance line for A is shifted to the left by one orientation of B and to the right by the other. In any sample, there are an enormous number of A and B protons, and the distribution between the two orientations is essentially equal. Both the downshift and the upshift will be observed. Figure 15–12 shows the result as a doublet. In a similar fashion, proton B is split into a doublet by A. We would expect intuitively that the separation of lines within the doublets should be the same for both. The spacing between the lines is called the *coupling constant, J,* and is given in hertz. We shall return to the use of coupling constants later, but first let us examine the splitting patterns of an actual molecule.

The high resolution n.m.r. spectrum of ethyl bromide is given in Figure 15–13, where the methyl protons are labeled (*a*) and the methylene protons (*b*). Each of the two (*b*) protons is randomly oriented with cr against the field, represented by

Table 15-5 General Regions of Chemical Shifts

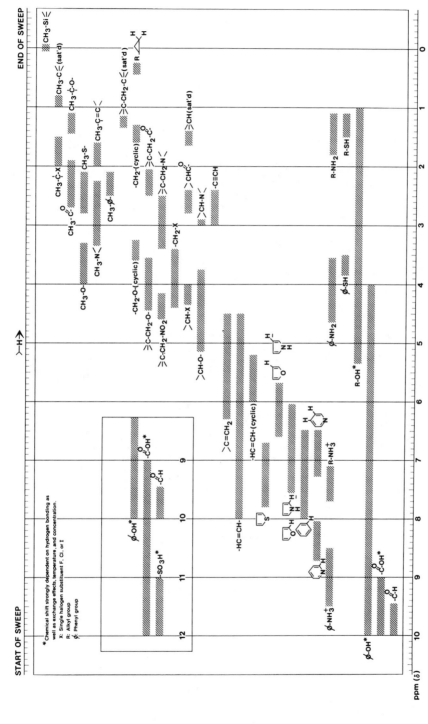

[Courtesy of Varian Associates.]

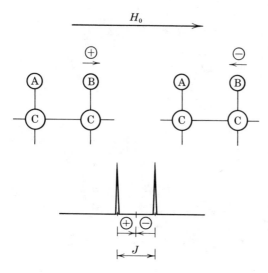

Figure 15–12 The effect of the alignment of proton B on the resonance of proton A. Alignment with the field, $\oplus$, shifts the line for A downfield. Alignment against the field, $\ominus$, shifts the line for A upfield.

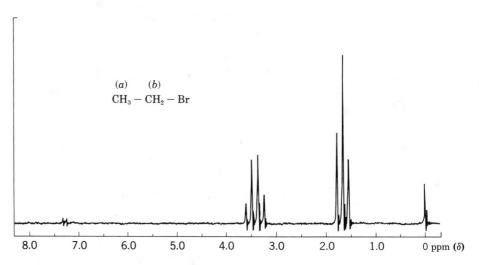

Figure 15–13   n.m.r. spectrum of ethyl bromide. [Courtesy of Varian Associates.]

the following possible combinations, all equally probable

where b_1 and b_2 represent the two (b) protons. The two combinations in the center are indistinguishable and represent no net effect. In a sample, the (a) protons experience three different effective fields due to the splitting effect, and the resonance appears as a triplet with sub-areas in the ratio $1:2:1$. In an analogous fashion, we can sort out the various combinations of orientations of the three (a) protons:

a_1	$\longrightarrow$	$\longrightarrow$	$\longleftarrow$	$\longleftarrow$
a_2	$\longrightarrow$	$\longrightarrow$	$\longleftarrow$	$\longleftarrow$
a_3	$\longrightarrow$	$\longleftarrow$	$\longrightarrow$	$\longleftarrow$

$\xrightarrow{\;\;H_0\;\;}$

a_1	$\longrightarrow$	$\longleftarrow$
a_2	$\longleftarrow$	$\longrightarrow$
a_3	$\longrightarrow$	$\longleftarrow$

a_1	$\longleftarrow$	$\longrightarrow$
a_2	$\longrightarrow$	$\longleftarrow$
a_3	$\longrightarrow$	$\longleftarrow$

All of the eight combinations are equally probable, but the center groups of three are equivalent. Thus the (b) protons appear as a quartet with sub-areas in the ratio $1:3:3:1$. The total area of each multiplet remains in the ratio of $2:3$ as it must for two (b) protons and three (a) protons.

We can generalize these observations by stating that a proton with n equivalent protons on the neighboring carbon atoms will be split by the n protons into $n+1$ lines (a multiplet) with relative sub-areas given by the coefficients of the binomial expansion $(a+b)^n$. But what if the given proton finds neighboring protons on more than one neighboring carbon atom?

In propane, $CH_3CH_2CH_3$, the protons in the end methyl groups are equivalent; the six methyl protons split the methylene protons into a 7-membered multiplet, with sub-areas in the ratio of $1:6:15:20:15:6:1$. Nitropropane presents a more interesting problem:

$$H-\underset{\underset{H}{|}}{\overset{\overset{H}{|}}{C}}-\underset{\underset{H}{|}}{\overset{\overset{H}{|}}{C}}-\underset{\underset{H}{|}}{\overset{\overset{H}{|}}{C}}-N\overset{\displaystyle{\nearrow O}}{\underset{\displaystyle{\searrow O}}{}}$$

$$(a)\ (b)\ (c)$$

Coupling through more than three bonds is very inefficient, and is not normally observed, therefore the interpretation of the peaks for the (a) and (c) protons is straightforward. Both are split into triplets by the two (b) protons. However, the (b) protons are split two-fold, once by the (a) protons and again by the (c) protons. Figure 15–14 is a diagram of the development of the splitting pattern; it makes no difference which splitting is considered first. (Try splitting into a triplet first!) The final pattern is a multiplet of 12 lines, but some of them are too weak to be seen in the actual spectrum given in Figure 15–15. Note that the triplets for the (a) and (c) protons appear to be distorted; the sub-areas are not precisely in the ratio $1:2:1$. This behavior is a general phenomenon. As the chemical shifts of two kinds of protons become more nearly the same, the peak areas become distorted with the inner peaks increasing at the expense of the outer peaks. When the chemical shifts become identical, both multiplets merge into a singlet as shown in Figure 15–16. In other words, *equivalent protons do not split each other.*

Let us recapitulate with a few rules and additional observations.

1. Spin–spin interactions are independent of the strength of the applied field. If a set of peaks cannot be identified as a multiplet or as a collection of several individual peaks, change the field. This does not change the coupling constant, but does change the chemical shift values (all measured in hertz).

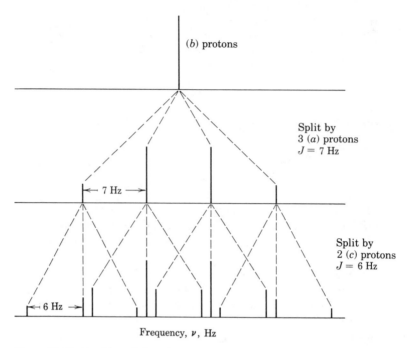

Figure 15–14   Double splitting of the methylene (b) protons in 1-nitropropane.

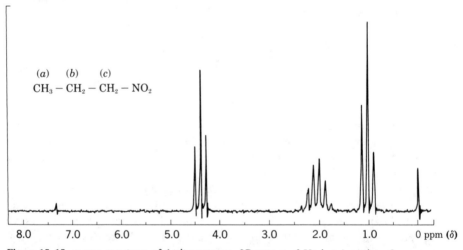

Figure 15–15   n.m.r. spectrum of 1-nitropropane. [Courtesy of Varian Associates.]

2. Equivalent nuclei do not interact with each other to cause observable splitting.

3. Multiplicity is determined by the number of neighboring groups of equivalent nuclei. For protons, the multiplicity equals $n + 1$, provided $\Delta\nu \gg J$.

4. Intensities of multiplets are symmetric (unless $\Delta\nu \approx J$) with relative intensities given by the coefficients of the terms in the expansion of $(a + b)^n$, where n is the number of equivalent neighbors causing the splitting.

5. Coupling constants decrease with distance; coupling is rarely observed at a distance greater than three bond lengths unless there is a charge delocalization as in conjugated or aromatic systems, or other special effects.

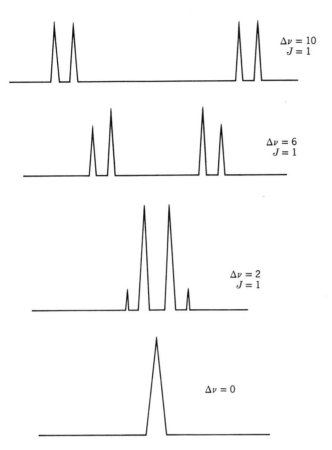

$\Delta\nu = 10$
$J = 1$

$\Delta\nu = 6$
$J = 1$

$\Delta\nu = 2$
$J = 1$

$\Delta\nu = 0$

Figure 15-16 Behavior of peaks for two protons as the difference in chemical shift, $\Delta\nu$, approaches the coupling constant, J.

6. Coupling constants are the same in both multiplets of a pair which are interacting.
7. Coupling constants rarely exceed 20 hertz, whereas chemical shifts vary over 1000 hertz. For two coupled protons, if $\Delta\nu/J > 7$, the simple pattern of two doublets appears, but as $\Delta\nu \to J$ the inner peaks increase and the outer peaks decrease.
8. In undistorted multiplets, the chemical shift is measured at the center of the system of peaks. In distorted multiplets, the shift is measured at the "center of gravity."
9. Spin–spin coupling systems are usually denoted by assigning letters of the alphabet (A, B, C, ... M, N, ... X, Y, ... etc.) to the different types of protons they contain. If the difference in chemical shift values is much greater (60 Hz or more) than the coupling constant (< 10 Hz), then letters at opposite ends of the alphabet are chosen (e.g., AX) to convey the fact that first-order splitting patterns are observed. Sometimes, more than two groups of protons are involved (e.g., $CH_3CH_2CH_2Cl$) with first-order splitting still operative—in these cases the nomenclature, AMX, is adopted. Furthermore, to indicate the relative numbers of protons of each chemical shift value, subscript numbers are assigned denoting identical coupled nuclei (e.g., $CH_3CH_2CH_2Cl$ is represented by $A_3M_2X_2$). In assigning letters one may elect to start either from downfield (10 δ) toward upfield (0 δ), or vice versa. In the

case of nuclei which are similarly shielded and have chemical shift values close to each other (less than 60 Hz), adjacent letters of the alphabet are chosen, for example, *AB*, thereby indicating a departure from strict first-order splitting patterns. The most common *AB* system encountered in proton magnetic resonance studies is the aromatic ring. Here again, subscript numbers indicate the relative numbers of protons of each chemical shift value (e.g., A_2B_2 for *p*-nitrophenol). If a third group is present (an *ABC* system), it is usually very difficult to predict the spectrum.

The interpretation of spectra in which multiplets overlap requires a mathematical analysis and will not be considered here. However, with the information at hand, we make use of the splitting pattern and the coupling constants to help unravel the structure of an unknown compound. A number of coupling constants for pairs of protons, (*a*) and (*b*), are given in Table 15–6.

Table 15–6   Proton Spin–Spin Coupling Constants

Type	J_{ab}, hertz	*Type*	J_{ab}, hertz
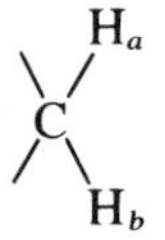 H_a, H_b on same carbon	10–15	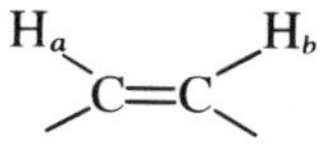 H_a, H_b cis/trans on $C=C$	6–12
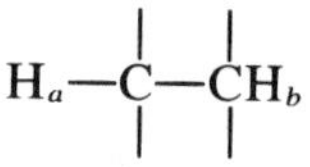 H_a-C-CH_b	6–8	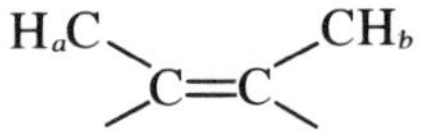 H_aC, CH_b on $C=C$	1–2
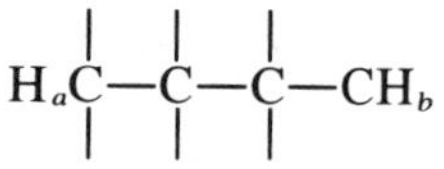 $H_aC-C-C-CH_b$	0	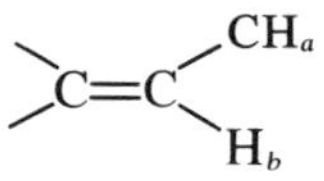 CH_a on $C=C-H_b$	4–10
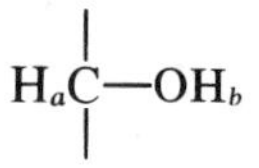 H_aC-OH_b (no exchange)	4–6	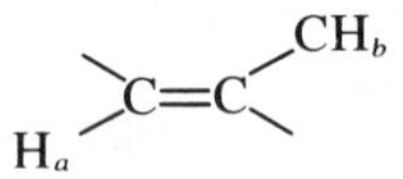 CH_b on $C=C-H_a$	0–2
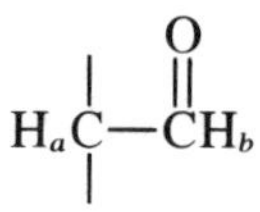 H_aC-CH_b with $C=O$	2–3	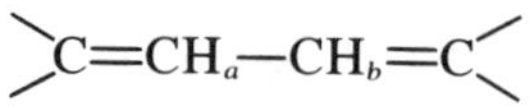 $C=CH_a-CH_b=C$	9–12
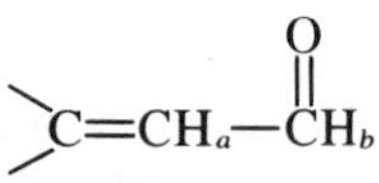 $C=CH_a-CH_b$ with $C=O$	5–7	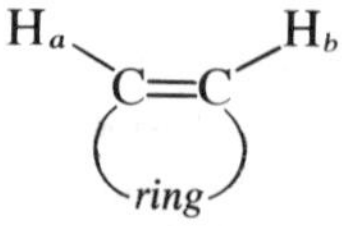 H_a, H_b on $C=C$ ring	5 mem. 3–4 6 mem. 6–9 7 mem. 10–13
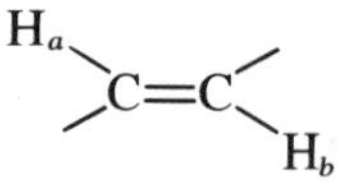 H_a, H_b on $C=C$	15–18		

Table 15–6 (*continued*)

Type	J$_{ab}$, hertz	Type	J$_{ab}$, hertz
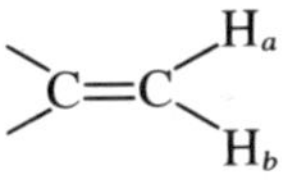 [structure: C=C with H$_a$, H$_b$]	0–2	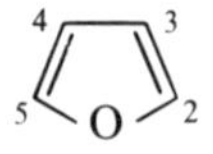 [structure: furan, O at position 1]	J$_{2\text{-}3}$ 1.5–2 J$_{3\text{-}4}$ 3–4 J$_{2\text{-}4}$ 0–1 J$_{2\text{-}5}$ 1–2
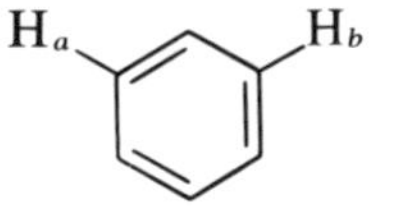 [structure: benzene ring with H$_a$, H$_b$]	ortho 6–10 meta 1–3 para 0–1	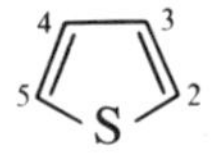 [structure: thiophene, S at position 1]	J$_{2\text{-}3}$ 5–6 J$_{3\text{-}4}$ 3.5–5 J$_{2\text{-}4}$ 1.5 J$_{2\text{-}5}$ 3.5
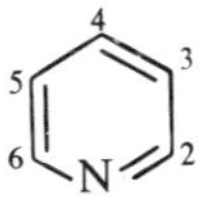 [structure: pyridine ring]	J$_{2\text{-}3}$ 5–6 J$_{3\text{-}4}$ 7–9 J$_{2\text{-}4}$ 1–2 J$_{3\text{-}5}$ 1–2 J$_{2\text{-}5}$ 0–1 J$_{2\text{-}6}$ 0–1	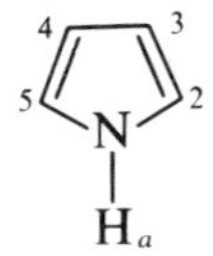 [structure: pyrrole, N–H$_a$]	J$_{a\text{-}2}$ 2–3 J$_{a\text{-}3}$ 2–3 J$_{2\text{-}3}$ 2–3 J$_{3\text{-}4}$ 3–4 J$_{2\text{-}4}$ 1–2 J$_{2\text{-}5}$ 2

15–8 INTERPRETATION OF N.M.R. SPECTRA

We are now ready to interpret some simple first-order n.m.r. spectra. In this context the term "first-order" means that $\Delta\nu \gg J$. The spectrum provides three kinds of information.

1. *Chemical shifts*, which identify the types of protons based on their electronic environment.
2. *Spin–spin splitting patterns*, which help to identify neighboring protons.
3. *Peak areas* (or intensities), which are proportional to the number of protons causing a given resonance line.

Example/Spectrum 15–1. CH_3—$\overset{\displaystyle O}{\overset{\displaystyle \|}{C}}$—$CH_2^A CH_3^X$

The spectrum of methylethylketone (Figure 15–17) illustrates a typical A_2X_3 system (the ethyl group). The methylene and methyl protons of the ethyl group have different chemical shift values (2.5 and 1.0 ppm respectively). Two equivalent methylene protons, denoted by A, are split into a quartet (1:3:3:1) by the neighboring equivalent methyl protons. In turn, the three equivalent methyl protons are split into a triplet (1:2:1) by the two equivalent methylene protons. Integration of the respective signal areas affords valuable information on the relative numbers of protons represented by these signal groups. Where the molecular formula is known (i.e., exact number of hydrogens), these relative numbers become absolute. Usually this kind of evidence can give the first clue as to the nature of the group, that is, methyl 3H, methylene 2H, etc. The coupling constant, J_{AX}, is given by the distance between the central line and the outer lines of the triplet signal of the methyl protons. When two groups interact and couple, both signal groups should display the same coupling constant value—yet another indicator to relate signal groups within a spectrum.

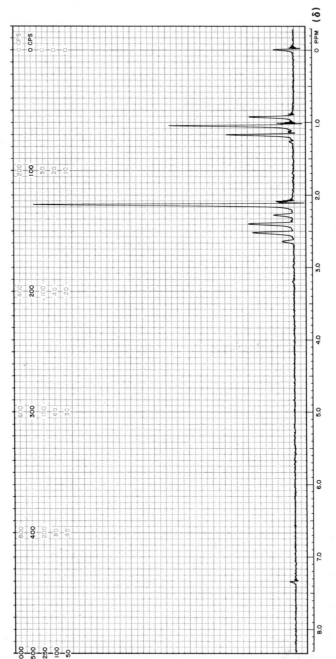

Figure 15–17 n.m.r. spectrum of methyl ethyl ketone. [Courtesy of Varian Associates.]

The chemical shift value observed for the methyl group (of the ethyl group) is the expected value when adjacent to a methylene group. The methylene signal at 2.5 ppm, however, is evidence that this group is deshielded by the carbonyl group; that is, the chemical shift value is larger than anticipated for a hydrocarbon environment. A further example of deshielding is that of the methyl group attached directly to the carbonyl function. The single intense signal at 2.1 ppm represents three chemically equivalent protons—no coupling is possible since the neighboring carbon atom carries no hydrogens to couple with. The shift downfield experienced from the normal hydrocarbon position (from 1.0 to 2.1 ppm) is a direct result of the electronegativity of the adjacent carbonyl function.

Example/Spectrum 15–2.

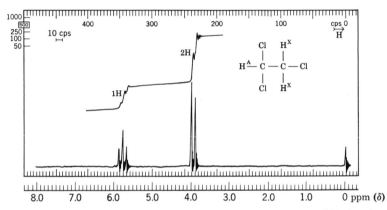

Figure 15–18   n.m.r. spectrum of 1,1,2-trichloroethane. [Courtesy of Varian Associates.]

The spectrum of 1,1,2-trichloroethane (Figure 15–18) illustrates a typical AX_2 system. The methine and methylene protons have different chemical shifts. The single methine proton, denoted by A, is split into a triplet by the neighboring equivalent methylene protons with intensity ratio $1:2:1$. Conversely, the two methylene protons, X_2, are in turn split into an equal intensity doublet. The coupling constant, J_{AX}, is given by the distance between the two lines of the methylene signal, or the distance between the central line and the outer lines of the triplet signal of the methine proton. This spectrum illustrates the ability of the chlorine nuclei to deshield protons as seen by the chemical shift values which are larger than those observed for similar protons in hydrocarbons.

Example/Spectrum 15–3.

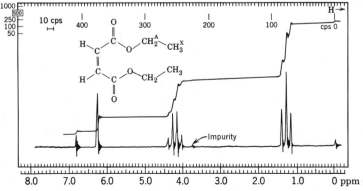

Figure 15–19   n.m.r. spectrum of diethyl maleate. [Courtesy of Varian Associates.]

The spectrum of diethyl maleate (Figure 15–19) illustrates a typical A_2X_3 system. Two equivalent methylene protons, A_2, are split to give a quartet (1:3:3:1) while the three equivalent methyl protons, X_3, are split to give a triplet (1:2:1). The methylene protons are moved to a lower field (large chemical shift) by the adjacent electronegative oxygen. Two uncoupled *cis*-olefinic protons give the single line at 6.2 ppm. Since no splitting is apparent, they must be chemically equivalent, that is, they experience the same magnetic field and have the same chemical shift value. The small signal at 6.8 ppm is probably due to the presence of a small amount of the isomeric compound, diethyl fumarate, where the olefinic protons are trans to each other. Interestingly, the trans system also results in chemical shift equivalence.

Example/Spectrum 15–4.

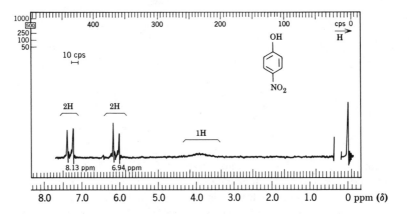

Figure 15–20   n.m.r. spectrum of *p*-nitrophenol. [Courtesy of Varian Associates.]

The spectrum of *p*-nitrophenol (Figure 15–20) is a good example of an A_2B_2 system. The two protons ortho to the nitro group have a chemical shift value of 6.94 ppm. The normal position for aromatic protons is 7.27 ppm due to the unusual shielding afforded by the ring current (anisotropic shielding). However, the presence of an electron-withdrawing nitro group perturbs the ring current around these ortho protons. Thus, the shielding is reduced and the two ortho protons appear at 6.94 ppm, as predicted by Table 15–4. On the other hand, the two protons ortho to the hydroxyl group are further deshielded and are displaced from the mean position of 7.27 to 8.13 ppm. Since the two groups of protons have chemical shifts close to each other the spin system A_2B_2 is selected. The resulting four-line spectrum illustrates the deviation from true intensity distribution as predicted by first-order rules. As two chemical shift values approach each other (see Figure 15–16), the two inner lines of the four-line spectrum gain in intensity at the expense of the outer lines, but the overall integration of the group of lines is unaffected. Conversely, as the difference in chemical shift values increases, the lines gradually become equally intense.

Example/Spectrum 15–5.

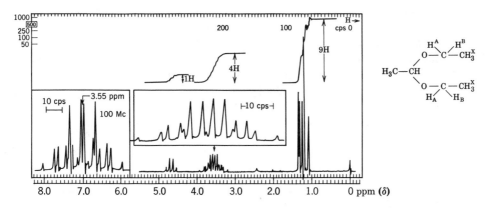

Figure 15–21 n.m.r. spectrum of acetaldehyde diethylacetal. [Courtesy of Varian Associates.]

The spectrum of acetaldehyde diethylacetal (Figure 15–21) illustrates an ABX_3 system. Until now the methylene protons of an ethoxy group have been considered equivalent. In this particular configuration they are not chemically equivalent and constitute an AB system. From the AB system alone, a four-line spectrum would be expected as in the previous example. However, each of these four signals should be further split into quartets by the three equivalent methyl protons. In summary, the AB system should consist of a total of 16 lines. At 60 MHz it is difficult to count that many, but the general intensity pattern is as predicted by theory. With increased resolution at 100 MHz a 16-line spectrum is clearly visible. In other cases of such spin systems it is not always easy to count the total number of lines expected. More often than not the lines overlap.

The quartet signal at 4.7 ppm is the methine proton adjacent to the methyl group; that is, an AX_3 system. The corresponding methyl doublet expected from this coupling is superimposed upon the methyl signals from the $O-CH_2CH_3$ groups.

Example/Spectrum 15–6.

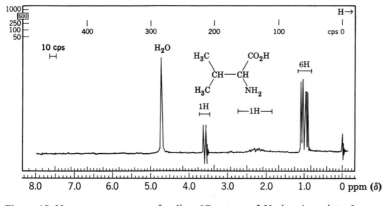

Figure 15–22 n.m.r. spectrum of valine. [Courtesy of Varian Associates.]

The spectrum of valine (Figure 15–22) illustrates another example of two similar groups which at first sight might be taken as chemically equivalent. In this particular case the two methyl groups differ slightly in chemical shift, 0.95 and 1.1 ppm. Both signals are split into equal intensity doublets by the adjacent methine proton at 2.2 ppm. The methine proton, being adjacent to six methyl protons, is by first-order splitting a seven-line pattern of signals, which in turn is split even further by the methine proton carrying both the amino and carboxyl groups. The absence of both the OH and NH_2 protons is explained by the fact that the spectrum was recorded in D_2O where exchange took place and deutero derivatives were formed.

Example/Spectrum 15–7.

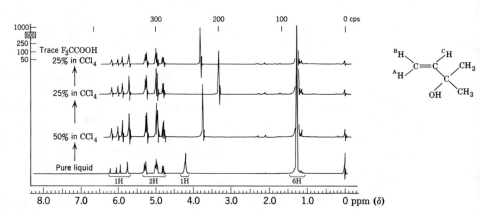

Figure 15–23   n.m.r. spectrum of 2-methylbut-3-en-1-ol. [Courtesy of Varian Associates.]

The spectrum of 2-methylbut-3-en-1-ol (Figure 15–23) includes two very interesting features. Firstly, the three olefinic protons of the vinyl group represent an *ABC* system. Such spin systems are extremely difficult to analyze by first-order splitting rules. Secondly, the chemical shift of the hydroxylic proton varies, depending on the sample preparation. Such protons may undergo proton transfer with a neighboring molecule of either the same or different type. This means that the proton can be in two different electronic environments. The time frame of the n.m.r. technique is such that it may see either of these pictures. In general, however, an average of such transfers is observed (i.e., the mean position of the proton between the two sites involved). As dilution increases, the signal moves upfield. Thus, the OH signal of an alcohol does not appear at a fixed value but its presence can be detected by dilution techniques.

Example/Spectrum 15–8.

The spectrum of 1-*p*-nitrobenzoyloxy-3-methylallene (Figure 15–24) illustrates one of the few cases where coupling extends through four or five bonds instead of the normal limit of coupling through three bonds. The two allenic protons, since their chemical shifts differ by more than 1 ppm, are denoted by *A* and *M*. The entire side chain constitutes an *AMX₃* system. The methyl signal is split by the *M*

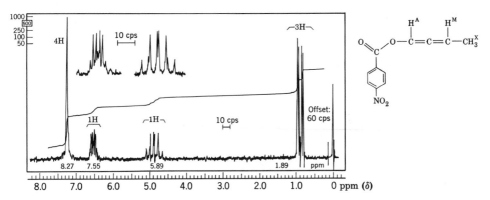

Figure 15–24 n.m.r. spectrum of 1-*p*-nitrobenzoyloxy-3-methylallene. [Courtesy of Varian Associates.]

proton into an equal intensity doublet with a relatively large coupling constant. However, the *A* proton is also capable of interacting with this methyl doublet via the allenic system. Hence the methyl doublet is further split into a pair of equal intensity doublets in which the coupling constant, J_{AX}, is much smaller than J_{MX}. Long-range coupling nearly always results in very small coupling constants. By first-order splitting rules both the *A* and the *M* protons should be 8-line patterns. Long-range coupling is normally restricted to allenes and acetylenes.

15–9   ADDITIONAL TOPICS

Proton Exchange.   As noted in Figure 15–23, the chemical shift of a hydroxylic proton (OH) is variable, depending on the solvent and concentration. At room temperature, proton transfer or hydrogen bonding occurs at such high exchange rates relative to the time frame of the n.m.r. observation that a single peak is recorded corresponding to the mean position of the proton between its two terminal sites. As the exchange rate is decreased by reducing the temperature of the sample to 0°C, a broad band develops indicating that the proton is observed in positions other than the mean. In cases where the exchange rate is very slow (at −100°C) two distinct signals can be observed equidistant from the mean, indicating the location of the proton on its two possible sites.

[13]C N.M.R. Spectroscopy.   Until recently, those nuclei of low natural abundance (Table 15–1) have not been studied extensively due to the poor sensitivity of the instruments available. To the organic chemist, the n.m.r. spectrum of carbon would be of enormous help in structural determination. The low natural abundance of the stable, magnetically active nuclide of carbon (^{13}C, 1.1% natural abundance) drastically reduces the probability of finding two [13]C's in the same molecule. Several technological advances, such as time averaging techniques (see Section 25–7), have at long last allowed observation of the otherwise weakly coupled spectra. Since the proton is a spin-1/2 nucleus, it is capable of coupling with [13]C (also a spin-1/2 nucleus). For instance, a ^{13}C of a methyl group would give a quartet signal (1:3:3:1) by coupling with its associated protons. Additional information of this kind can assist in structure determination of organic molecules.

QUESTIONS AND PROBLEMS

15–1. Predict what the high-resolution n.m.r. spectra of the following compounds would look like. Indicate the approximate δ values for the different groups of equivalent protons in these compounds.

(a) acetone   (e) ethyl benzene
(b) acetic acid   (f) n-propionamide
(c) ethyl chloride   (g) 2-iodobutane
(d) 1-bromo-3-chloropropane

15–2. Interpret and discuss the n.m.r. spectra of the following compounds and suggest a structure for the compound. Indicate the nature of any additional spectroscopic information which would help to verify your prediction. In each case, interpret the spectrum in terms of molecular structure, commenting on any unusual values.

(a) Compound A has the molecular formula $C_3H_6O_2$, Figure 15–25.

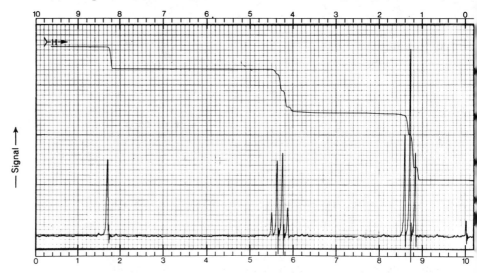

Figure 15–25   n.m.r. spectrum of compound A, $C_3H_6O_2$. [Courtesy of Heyden and Son Ltd.]

(b) Compound B, molecular formula $C_5H_{10}O_3$, Figure 15–26.

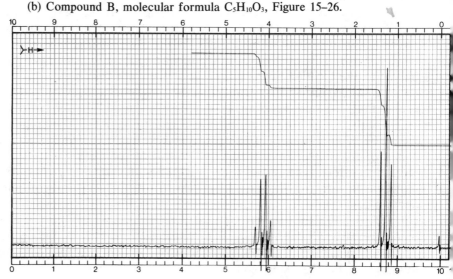

Figure 15–26   n.m.r. spectrum of compound B, $C_5H_{10}O_3$. [Courtesy of Heyden and Son Ltd.]

(c) Compound C, molecular formula $C_5H_8O_4$, Figure 15–27.

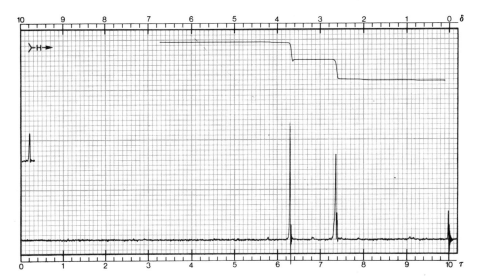

Figure 15–27 n.m.r. spectrum of compound C, $C_5H_8O_4$. [Courtesy of Heyden and Son Ltd.]

(d) Compound D, molecular formula $C_{14}H_{12}$, Figure 15–28.

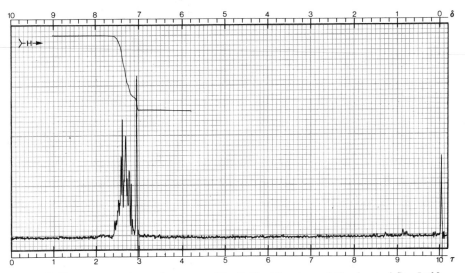

Figure 15–28 n.m.r. spectrum of compound D, $C_{14}H_{12}$. [Courtesy of Heyden and Son Ltd.]

(e) Compound E, molecular formula C_9H_{12}, Figure 15–29.

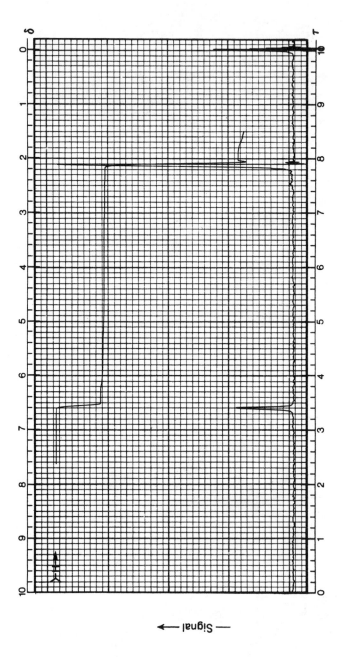

Figure 15–29 n.m.r. spectrum of compound E, C_9H_{12}. [Courtesy of Heyden and Son Ltd.]

(f) Compound F, molecular formula $C_{11}H_{12}O_2$, Figure 15-30.

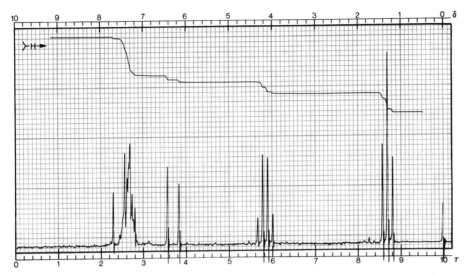

Figure 15-30 n.m.r. spectrum of compound F, $C_{11}H_{12}O_2$. [Courtesy of Heyden and Son Ltd.]

(g) Compound G, molecular formula $C_8H_8O_3$, Figure 15-31.

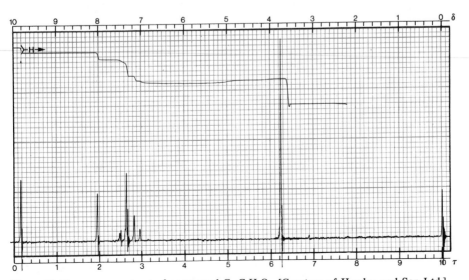

Figure 15-31 n.m.r. spectrum of compound G, $C_8H_8O_3$. [Courtesy of Heyden and Son Ltd.]

(h) Compound H, molecular formula $C_6H_{12}O_3$, Figure 15.32.

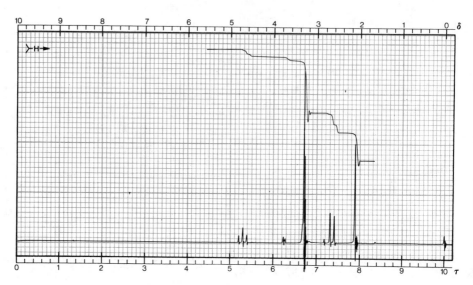

Figure 15–32   n.m.r. spectrum of compound H, $C_6H_{12}O_3$. [Courtesy of Heyden and Son Ltd.]

(i)   Compound I, molecular formula C_4H_8O, Figure 15–33.

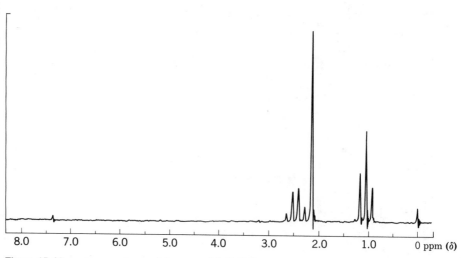

Figure 15–33   n.m.r. spectrum of compound I, C_4H_8O.

(j) Compound J, molecular formula C₇H₇OCl, Figure 15-34.

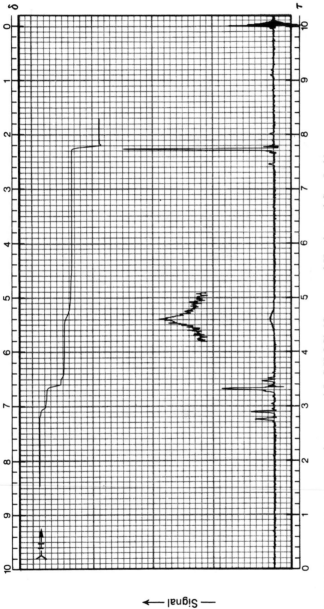

Figure 15-34 n.m.r. spectrum of compound J, C₇H₇OCl. [Courtesy of Heyden and Son Ltd.]

(k) Compound K, molecular formula C$_3$H$_8$O, Figure 15–35.

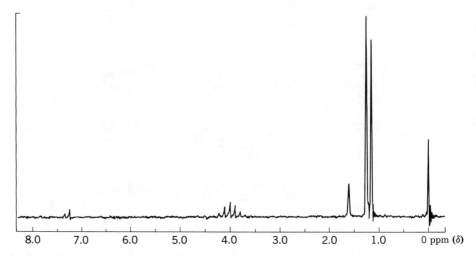

Figure 15–35 n.m.r. spectrum of compound K, C$_3$H$_8$O.

(l) Compound L, molecular formula C$_4$H$_8$O, Figure 15–36.

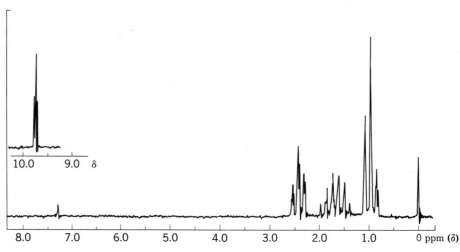

Figure 15–36 n.m.r. spectrum of compound L, C$_4$H$_8$O.

(m) Compound M, molecular formula C_7H_8, Figure 15–37.

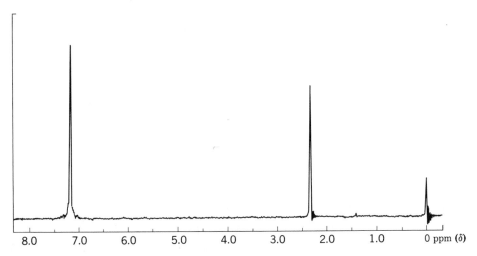

Figure 15–37   n.m.r. spectrum of compound M, C_7H_8.

15–3. The n.m.r. spectrum of a dichloro compound (Mol. Wt. 113) has a quintet with $\delta = 2.20$ and a triplet with $\delta = 3.70$. Assign the absorption peaks and identify the compound.

15–4. The n.m.r. spectrum of $C_9H_{10}O_2$ shows three single peaks at $\delta = 7.29$ (area 84), $\delta = 5.00$ (area 34) and $\delta = 1.98$ (area 50). What is the structural formula of the compound?

15–5. In very pure ethanol, $CH_3—CH_2—OH$, the (a) proton is coupled with the (b)
(c)   (b)   (a)
proton, $J_{ab} = 5.0$ Hz and the (b) protons also couple with the (c) protons $J_{bc} = 7.2$ Hz. Work out the splitting pattern for the (b) protons, similar to Figure 15–14.

15–6. Proton magnetic resonance spectra have been recorded with high resolution instruments employing magnetic field strengths such that the observed resonance is at 220 MHz. What is the value of the applied magnetic field? Are such fields stable? What advantages does recording proton spectra at 220 MHz afford?

15–7. Deduce the molecular structure of the compound represented in Figure 15–38.

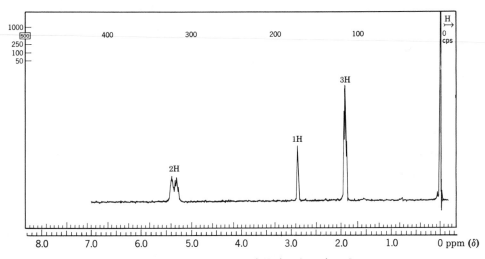

Figure 15–38   n.m.r. spectrum of C_5H_6. [Courtesy of Varian Associates.]

15–8. Deduce the structure of the furan derivative whose n.m.r. spectrum is given in Figure 15–39.

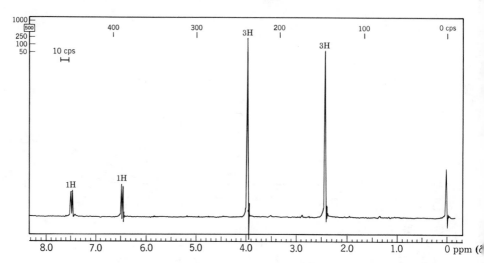

Figure 15–39 n.m.r. spectrum of furan derivative, $C_7H_8O_3$. [Courtesy of Varian Associates.]

15–9. Deduce the structure of the compound whose n.m.r. spectrum is given in Figure 15–40. This compound is the reaction product of formaldehyde with 2 moles of 5,5-dimethylcyclohexadi-1,3-one.

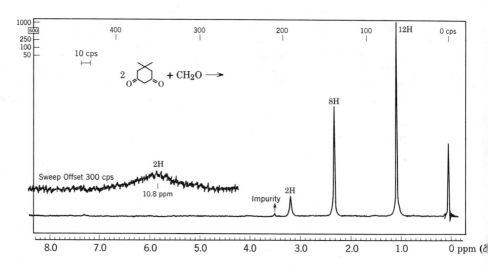

Figure 15–40 n.m.r. spectrum of a reaction product. [Courtesy of Varian Associates.]

15–10. Determine the molar ratios of naphthalene, tetralin, and decalin from the spectrum of a mixture of all three as shown in Figure 15–41.

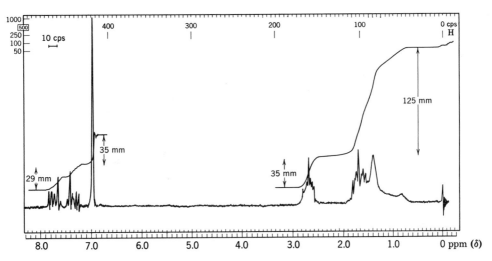

Figure 15–41   n.m.r. spectrum of a mixture of naphthalene, tetralin, and decalin. [Courtesy of Varian Associates.]

REFERENCES

R. M. Silverstein, G. C. Bassler, and T. C. Morrill, *Spectrometric Identification of Organic Compounds*, 3rd ed., Wiley, New York, 1974; Chapter 4. Elementary theoretical discussion with comprehensive data (e.g., chemical shift charts, coupling constants, shielding constants, etc.).

J. R. Dyer, *Applications of Absorption Spectroscopy of Organic Compounds*, Prentice-Hall, Englewood Cliffs, N.J., 1965. Paperback with good intermediate discussion of n.m.r.

L. M. Jackman, *Applications of Nuclear Magnetic Resonance Spectroscopy in Organic Chemistry*, Pergamon Press, New York, 1959. Short monograph with excellent discussions on the correlation of chemical shifts with molecular structure.

J. A. Pople, W. G. Schneider, and H. J. Bernstein, *High-Resolution Nuclear Magnetic Resonance*, McGraw-Hill, New York, 1959. Definitive detailed treatment. Considered by many to be the "bible" of n.m.r.

F. A. Bovey, "Nuclear Magnetic Resonance", *Chem. Eng. News*, August 30, 1965, p. 98. A review of principles and selected applications with examples.

R. J. Myers, *Molecular Magnetism and Magnetic Resonance Spectroscopy*, Prentice-Hall, Englewood Cliffs, N.J., 1973. Basic monograph on proton magnetic resonance.

G. A. Gray, "Carbon-13 Nuclear Magnetic Resonance Spectroscopy", *Anal. Chem.*, May, 1975, p. 546A. An excellent review of the state of the art.

MASS SPECTROMETRY

16

MASS SPECTROMETRY
OF
ORGANIC COMPOUNDS

16–1   INTRODUCTION

Mass spectrometry is a method involving the production of gas-phase ions from a sample and their resulting separation according to their mass to charge ratio (m/e)—a process somewhat analogous to the dispersion of light by a prism according to the wavelengths present. Recent advances have placed this particular spectroscopic technique in the forefront of research capability giving more information from a microgram of sample than any other single technique.

The first reported demonstration that positive ions could be deflected using electrical and magnetic fields was by Wien in 1898. Fourteen years later Thompson proved the existence of two neon isotopes using simple magnetic deflection. By 1918 Aston and Dempster had successfully constructed research instruments capable not only of the separation of the various isotopes but also the accurate measurement of their relative abundances. Commercial instruments became available about 1940 and were used primarily in the petroleum industry. Today the design of instruments has so many ramifications that almost any type of material may be successfully examined. Computerized data acquisition techniques have added a new dimension to mass spectrometry by eliminating the tedious and time-consuming calculations sometimes required.

Three distinct operations must be performed by the mass spectrometer system: namely, the production, separation, and detection of ions. A schematic diagram is given in Figure 16–1.

16–2   METHODS OF ION PRODUCTION

Electron Ionization (EI).   A simple electron impact ion source is illustrated in Figure 16–2. The sample molecules stream in because of the difference in pressure across the inlet leak, from about 10^{-2} to 10^{-5} torr. An electrically heated filament A produces electrons that are accelerated by anode B, thus creating a beam of electrons that intersects the flow of sample molecules. Positively charged ions produced by electron-molecule collisions are withdrawn by the electric field existing between the positively charged repeller plate C and the first accelerator

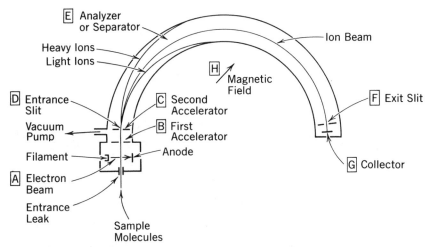

Figure 16-1 Schematic diagram of a single deflection 180° mass spectrometer. Magnetic field is perpendicular to plane of the paper.

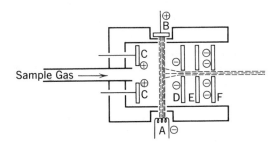

Figure 16-2 Schematic diagram of an electron-impact source.

plate D. The intermediate plate E serves to focus the ion beam and the second accelerator plate F imparts a final acceleration to the ions.

The energy of the electron beam is controlled by the potential on anode B. The ionization potential of most organic compounds is about 10 eV, and with this minimum energy the primary process is the production of singly charged molecular ions:

$$M + e^{\ominus} = M^{\oplus} + 2e^{\ominus}$$

giving a mass spectrum consisting almost entirely of a single peak corresponding to the mass of the original molecule (but see isotope effects to be discussed later). Increasing the energy of the electron beam will yield a more highly excited ion that will fragment if it is complex, or a second electron may be knocked out as shown in Figure 16-3 for argon atoms. For most applications the electron beam is given an energy of 50 to 70 eV, which yields the most reproducible spectra. Doubly charged ions are rare even at this potential.

The difference in potential between C and D (Figure 16-2) is only a few volts but is sufficient to remove the positive ions from the electron beam. Thus the positive ions reach electrode D with variable but relatively small kinetic energy.

Essentially all of the acceleration is accomplished between electrodes D and F which differ in potential from a few hundred to a few thousand volts. At a given accelerating voltage, all singly charged ions are given the same kinetic energy

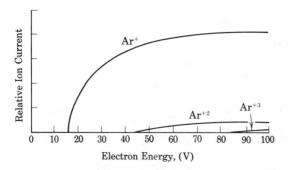

Figure 16–3 Ion production efficiency for argon as a function of electron beam energy.

defined by the relation:

$$\text{Kinetic Energy} = \tfrac{1}{2}mv^2 = eV \tag{16–1}$$

where m is the mass of the ion, v is its velocity, e is the electronic charge and V is the accelerating potential. A monoenergetic beam of ions is essential for the proper separation of the ions, and is a prime consideration in obtaining very accurate mass measurements.

Chemical Ionization (CI). With the advent of the study of high pressure mass spectrometry by Munson and Fields, a novel approach to ion production has emerged. To illustrate this technique, methane gas is chosen to describe the salient features. Normally, the pressure maintained during the operation of a mass spectrometer is extremely low (10^{-5} to 10^{-7} torr). Under these conditions, methane molecules entering the source will be bombarded with a stream of electrons (EI) causing ionization and fragmentation (Figure 16–4).

$$CH_4 + e^\ominus \longrightarrow CH_4^\oplus + CH_3^\oplus + CH_2^\oplus + CH^\oplus + C^\oplus$$
$$m/e\ 16 \qquad 15 \qquad 14 \qquad 13 \qquad 12$$

Now, if the source pressure is preferentially maintained at a much higher pressure (0.5 to 1 torr) than the rest of the mass spectrometer, ion-molecule reactions will

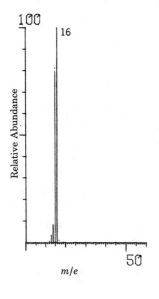

Figure 16–4 Normal mass spectrum of methane at low pressure, 10^{-5} to 10^{-7} torr.

occur by collision.

$$CH_4^\oplus + CH_4 \longrightarrow CH_5^\oplus + CH_3$$
$$m/e \ 17$$

$$CH_4 + CH_3^\oplus \longrightarrow C_2H_5^\oplus + H_2$$
$$m/e \ 29$$

$$CH_4 + C_2H_5^\oplus \longrightarrow C_3H_5^\oplus + H_2$$
$$m/e \ 41$$

The resultant mass spectrum is illustrated in Figure 16–5 showing the relative abundances of these higher molecular weight positive ions. It is this beam of positive ions that is now employed to ionize sample molecules introduced into this beam. With a sample-to-methane ratio of 10^{-3}, the sample molecules undergo ionization via ion-molecule collisions. Such collisions usually involve a neutral molecule of sample with one of the positive ions from methane (m/e 17, 29, and 41). Most sample molecules collide with the abundant CH_5^+ (m/e 17) ion to produce a protonated molecular ion, that is,

$$M + CH_5^\oplus \longrightarrow (M + 1)^\oplus + CH_4$$

Interaction with the other positive ion species present (m/e 29 and 41) usually results in alkyl transfer processes, with production of adjunct ions; namely,

$$M + C_2H_5^\oplus \longrightarrow (M + 29)^\oplus$$
$$M + C_3H_5^\oplus \longrightarrow (M + 41)^\oplus$$

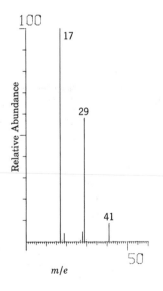

Figure 16–5 Mass spectrum of methane at "high" pressure, 1 torr.

The mass spectrum by CI methods (Figure 16–6) is always a simpler profile than those produced by EI techniques (Figure 16–7). The CI spectrum displays a clearly visible protonated molecular ion $(M + 1)^\oplus$. The presence of this quasi-molecular ion aids greatly in identifying the molecular weight of the compound under investigation, particularly where EI techniques do not indicate any $M^\oplus$ ions. Other reactions can occur in CI leading to proton abstraction $(M - 1)^\oplus$ and charge transfer $(M + 1 - 18)^\oplus$ (loss of water).

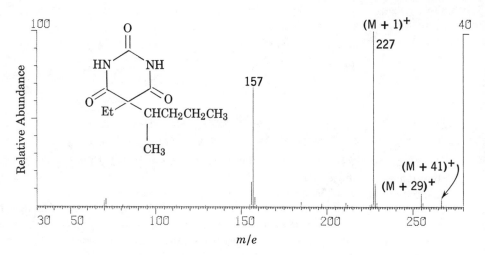

Figure 16–6   Methane chemical ionization mass spectrum of pentobarbital.

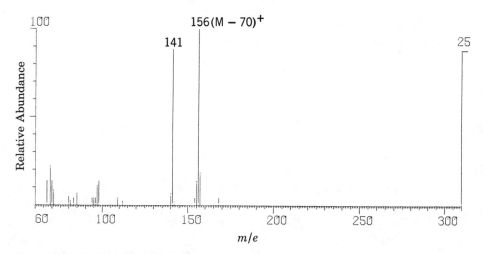

Figure 16–7   Electron impact mass spectrum of pentobarbital.

Field Ionization (FI).   If a molecule possessing a large dipole moment and high polarizability is placed in an intense electric field (10^8 V/cm), an electron may be withdrawn producing a positive ion. Such an ion source consists of an anode with a sharp point or edge and a cathode placed about 1 mm away. An impressed voltage of 5 to 20 kV creates an intense field in the vicinity of the anode. The design and materials used to construct the sharp point are critical factors—in fact, commercial razor blades have been used very successfully. Recently, multipoint sources have increased the efficiency considerably.

The greatest advantage of field ionization sources is their ability to produce stable molecular ions with little or no excess energy, and therefore almost no fragmentation. This is particularly useful in the study of natural products and other large molecules where the parent ion is difficult to find with electron impact sources. Unfortunately, the sharp edges are fragile and subject to damage from arcing but the technology is improving rapidly.

16-3   ION SEPARATORS

This part of the apparatus, sometimes called the *analyzer*, must separate ions according to their masses, with a very high differentiation between masses. However, the data from high resolution studies are not always strictly necessary since fragmentation under low resolution conditions (nominal mass numbers) usually yields sufficient evidence to elucidate a molecular structure. High resolution studies of molecular ions, $M^{\oplus}$, are beneficial in establishing the molecular formulae. For example, a good high resolution instrument can distinguish between $C_{16}H_{22}O_2$, Mol. Wt. 246.1620, and $C_{17}H_{26}O$, Mol. Wt. 246.1984; or between ^{12}CH, mass 13.0078 and ^{13}C, mass 13.0034. A second requirement is a high transmission of ions. Unfortunately, the two requirements are incompatible; the narrower the slits are made, the better the resolution but the smaller the ion current. As usual the designer must reach a compromise best suited for the purpose. The different kinds of mass spectrometers differ mainly in the way they sort the ions.

Single-Focusing Magnetic Deflection.   This is the type of separation most often described, and is the one shown in Figure 16-1. In the magnetic field H the charged particles experience a magnetic force F:

$$F = Hev \qquad (16\text{-}2)$$

and a counterbalancing centrifugal force,

$$Hev = mv^2/r \qquad (16\text{-}3)$$

where r is the radius of the circular path. A combination and rearrangement of Equations 16-1, 16-2, and 16-3 yields

$$m/e = H^2r^2/2V \qquad (16\text{-}4)$$

Equation 16-4 pertains to all ions, but only those with a particular path radius will be collected. For most spectrometers, H and r are fixed, so that the mass that is collected is inversely proportional to V. Some magnetic instruments employ a 60° or 90° sector rather than the 180° bend depicted in Figure 16-1. If the geometry is correctly arranged, both of these types of magnets focus an imperfectly collimated ion beam on the exit slit and the collector placed immediately behind it.

Double-Focusing.   In deriving Equation 16-4, it was assumed that all ions enter the magnetic field with the same kinetic energy; therefore, all ions with the same m/e ratio should have the same velocity. This is not strictly true because the ions vary in initial energy (because of the Boltzmann kinetic energy distribution and the potential gradient across the filament) and therefore leave the electron beam with variable energies. A much better focusing can be achieved if this energy spread is reduced before the ions enter the magnetic field. Mattauch and Herzog designed a "double-focusing" instrument in which a radial electrostatic field selects only those ions possessing a certain velocity, or kinetic energy. The geometry is shown in Figure 16-8. Since it can achieve such high resolution, the ion currents are extremely low. Double-focusing instruments are used whenever highest resolution is required, for example, in the determination of precise molecular weights.

Cycloidal Focusing.   Ions passing through crossed magnetic and electric fields will generate a cycloidal path as shown in Figure 16-9. The small radius of curvature permits the use of a smaller magnet without sacrificing range or resolution.

Time-of-Flight.   We have noted that all ions leave the acceleration field with

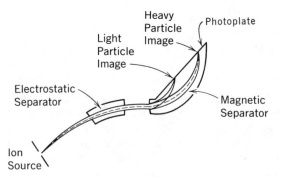

Figure 16–8 Schematic diagram of Mattauch-Herzog double-focusing mass spectrometer.

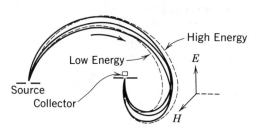

Figure 16–9 Schematic diagram of a cycloidal mass spectrometer.

the same kinetic energies but different velocities depending on their masses. With magnetic focusing, the ions are separated by changing their directions. However, if they are left to float in a straight line through a field-free region they will take different times to travel a given distance. The measurement of this "time-of-flight" is the basis for the nonmagnetic separator shown in Figure 16–10. If the ions were allowed to enter the drift tube continuously, there would be no possibility of measuring the time required for a given ion. This problem is solved by pulsing a control grid placed in the electron beam so that ions are produced in pulses lasting only about 0.25 μsec at a frequency of 10,000 times per second. The accelerating grid and collector device must also be pulsed in sequence. Equation 16–1 gives the velocity of the ions as they leave the accelerator

$$eV = \tfrac{1}{2}mv^2 \quad \text{or} \quad v = \sqrt{\frac{2eV}{m}}$$

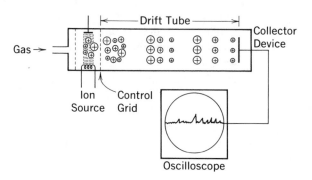

Figure 16–10 Schematic diagram of a time-of-flight mass spectrometer.

Therefore to travel the distance, d, to the detector, the time required is

$$t = \frac{d}{v} = d\sqrt{\frac{m}{2eV}} = k\sqrt{\frac{m}{e}} \qquad (16\text{–}5)$$

where k is a proportionality constant which depends on the length of the flight path. If the distance traveled is 40 cm, e is measured in electronic charge units, m in atomic mass units, and t in μsec, then k is nearly unity. Thus, for $N_2^\oplus$, $t = \sqrt{28} = 5.30$ μsec and for $O_2^\oplus$, $t = \sqrt{32} = 5.66$ μsec. Obviously, rather elaborate electronics are required to measure accurately these very short times. Oscilloscopes are commonly used to display the spectrum.

Quadrupole Mass Filters. As the name suggests, the separating mechanism in this technique utilizes four parallel rods arranged so that their axes are at the corners of a square (Figure 16–11). The ion beam is accelerated into the center of this rod arrangement along the z-axis. One pair of diagonally opposed electrodes is held at $+V_{dc}$ volts and the other pair at $-V_{dc}$ volts. An alternating radio frequency (rf) voltage is superimposed on the first pair of electrodes, and a second rf voltage (180° out-of-phase with the first) is applied to the second pair. Most of the ions introduced into the quadrupole will oscillate with an increasing amplitude and strike one of the electrodes. However, one particular m/e ratio (determined by the rf potential and frequency) can pass completely through the analyzer and be detected (Table 16–1).

Table 16–1   Basic Equations Governing Mass Spectrometer Performance

Type	Equation	Definitions
Magnetic sector	$m/e = \text{Const.} \times H^2 r^2/V$	V = accelerating voltage H = magnetic field r = radius of curvature of the trajectory
Time-of-flight	$m/e = 1.916 \times Vt^2/d^2$	t = time of flight in microseconds d = flight distance in cm
Quadrupole	$m = 0.136 \times V/r_0^2 \cdot f^2$	r_0 = rod separation in cm f = rf frequency in MHz (for singly charged species only)

Quadrupole spectrometers require no magnet, so are more compact and less expensive than magnetic focusing instruments. By scanning the rf frequency or dc potential a mass spectrum can be obtained. Such systems can only be operated under low resolution conditions, but produce linear mass scans as compared with the exponential mass scans of the magnetic scanning instruments.

Other Types of Ion Separators. There are several other kinds of instruments available, but space does not permit detailed descriptions. The "Omegatron" is based on the cyclotron principle of accelerating ions in a spiral path. The "radio-frequency" spectrometer, like the time-of-flight, separates on the basis of velocity, but the ions must make their way through a series of grids pulsed with an rf signal set to allow ions of only one velocity to get through.

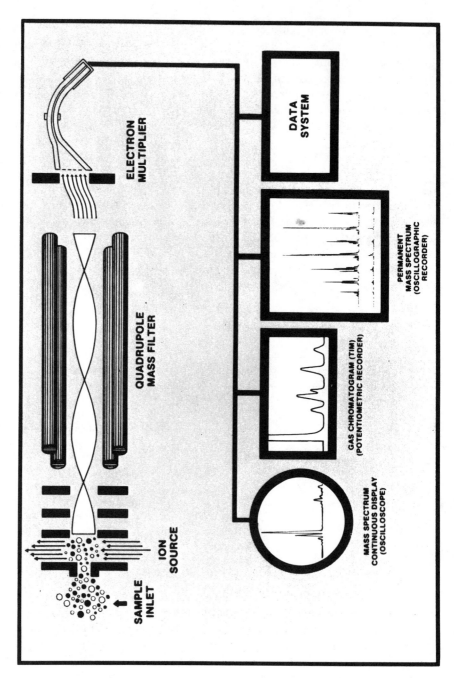

Figure 16–11 Schematic diagram of a quadrupole mass analyzer and auxiliary equipment for various data displays.

16–4   ION COLLECTION AND RECORDING

Electronic.   The separated ion beams are successively collected by an electron multiplier, whose signal is amplified and recorded by a variety of instruments from photographic chart recorders to small computers. The latter are now preferred where extremely fast scans of the mass range 1 to 1000 are employed. Such data acquisition systems are capable of storing information at very fast rates for later retrieval and plotting. (See Section 25–6.)

Photographic.   In photographic recording, the sensitive emulsion of a photo-plate (Figure 16–13) becomes exposed at each point of contact of each beam of ions at a number of different time exposures. The resultant densities of the silver deposits created on this plate correspond to the number of ions that struck the plate and are counted or measured by conversion to peak profiles or areas by the use of a densitometer or measuring device that allows calibration of the blackening on the plate to absolute values of optical density (Figure 16–12).

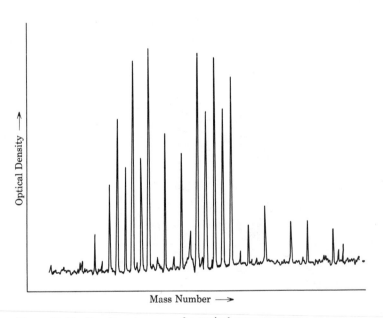

Figure 16–12   Densitometer scan of a typical mass spectrum.

16–5   SAMPLE HANDLING TECHNIQUES

The techniques most suitable for introducing a sample depend on its physical state, its chemical properties, and how it is to be ionized. The following systems can produce vaporized samples from gases to volatile solids which are to be ionized by electron bombardment.

Cold Inlet System.   A cold inlet system allows introduction of both gases and volatile liquids into a reservoir at room temperature. Mercury dosers allow reproducible sampling by vapor pressure control. A sintered glass leak maintains a constant rate of flow of sample vapor into the ion chamber (i.e., controlled leak across a vast pressure difference).

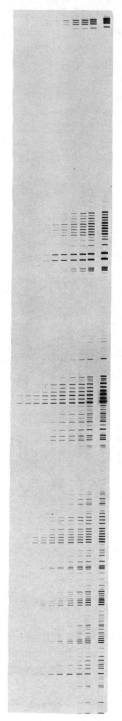

Figure 16–13 Photographic recording of a typical mass spectrum—10 different graded exposures are illustrated.

Heated Inlet System.   Less volatile compounds are introduced into the reservoir using a heated inlet system. It is similar to the cold inlet system, but can be heated to 350°C.

Direct Insertion Probe.   To avoid possible thermal decomposition, heat-sensitive compounds are introduced directly into the ion chamber on the tip of a ceramic tube. Often the sample is coated on the tip of the ceramic rod by evaporation of a solution containing the compound under investigation. An ingenious vacuum lock system allows facile insertion without seriously disturbing the high vacuum maintained in the ion chamber of the instrument.

Gas–Liquid Chromatographic Inlet Systems.   The separated components in the effluent stream of the typical gas liquid chromatograph can be fed directly into the ion chamber. With capillary columns the carrier gas is no problem since the normal rate of flow is only about 2 ml/min, which the pumping system of the mass spectrometer can easily handle. However, with packed columns where the flow rates usually exceed 20 ml/min, a molecular separator must be employed to remove most of the carrier gas. In one type of separator, illustrated in Figure 16–14a, the column effluent is passed through a porous tube located in a vacuum chamber. Helium readily diffuses through porous glass, and hydrogen readily diffuses through porous palladium. The heavier sample molecules continue through to the mass spectrometer inlet. Another type of separator utilizes a glass jet aimed at the outlet of an evacuated chamber, as shown in Figure 16–14b. The lighter carrier gas molecules diffuse from the main stream and are pumped away. Even the smaller peaks of a gas chromatogram produce a sufficient amount of sample for analysis by mass spectrometry, as shown in Table 16–2.

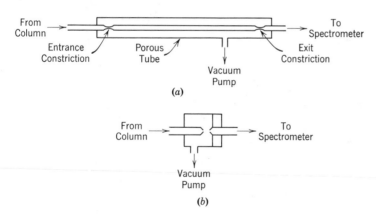

Figure 16–14   Schematic diagrams of molecular separators for interfacing a gas chromatograph to a mass spectrometer. (*a*) porous tube; (*b*) glass jet.

Table 16–2   Typical Sample Sizes Involved in Mass Spectrometry

Inlet System	Minimum Sample Size, μg
Cold-Inlet System	10
Heated Inlet System	1
Direct Insertion Probe	0.001
Gas Chromatographic Inlet	0.01

16-6   RESOLUTION

Most mass spectrometers can differentiate $NH_3^\oplus$ (mass 17) from $CH_4^\oplus$ (mass 16), but it would take a rather good one to differentiate $^{12}CH_3D^\oplus$ (mass 17.0376) from $^{13}CH_4^\oplus$ (mass 17.0347). The ability of a spectrometer to distinguish between two ions of nearly equal masses is called the "resolution" of the instrument.

Ideally, the spectrum consists of a series of narrow (zero width) peaks. This would represent infinite resolution. As seems to be inevitable in nearly every technique, there are a number of nonideal processes causing peak broadening with the probability that peaks with small mass differences will overlap (Figure 16–15):

1. Distribution of kinetic energies in the electron beam.
2. Variations in the accelerating voltage.
3. Poorly collimated ion beam.
4. Variations in the magnetic field.
5. Width of the ion beam as determined by the slits.
6. Space charge of the ion beam. (The positive charges repel each other, thus broadening the beam. This effect is particularly noticeable at high concentrations.)

Resolution is expressed as $M/\Delta M$, where M is the nominal mass of a pair of closely spaced peaks of equal height, one at M and the other at $M + \Delta M$, with the stipulation that the signal at the valley minimum between the peaks must not exceed 10% of the height of the peak maximum. (Some authors use a 2% figure rather than 10%.) A resolution in excess of 10,000 is required to separate fragments having the same nominal mass, for example $N_2^\oplus$ (mass 28.0061) and $CO^\oplus$ (mass 27.9949).

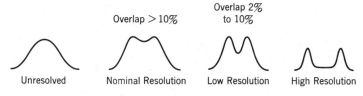

Figure 16–15   Typical resolution patterns.

16-7   THE MASS SPECTRUM

Part of a typical mass spectrum is reproduced in Figure 16–16. Notice that three levels of sensitivity are recorded simultaneously ($\times 1$, $\times 10$, and $\times 100$) because of the wide variation in the height of the peaks. A convenient way of representing the raw data obtained from the three recordings is illustrated in Figure 16–17, a line diagram. Each peak has first been *normalized* by assigning the largest peak, the *base peak*, a value of 100 units or 100% relative abundance.

Molecular Ion or Parent Peak.   With an electron beam energy of 9 to 15 eV, the principal ion is the molecular ion produced by the loss of a single electron. This gives a very simple mass spectrum with essentially all of the ions appearing in the "parent peak." With organic compounds, because of the small, but observable, natural abundance of ^{13}C and ^2H, there is generally a small peak appearing one mass unit higher than the parent peak (the $M + 1$ peak); and if two heavy isotopes

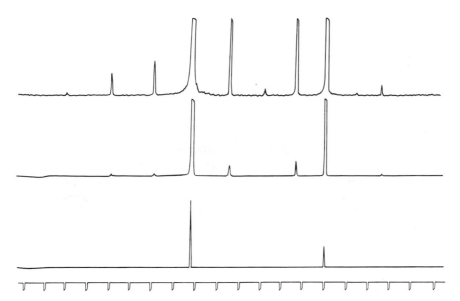

Figure 16–16 Typical mass spectrum recorded at three levels of sensitivity, ×1, ×10, and ×100.

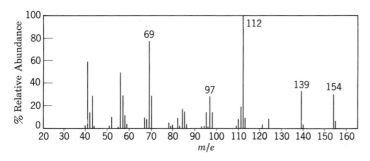

Figure 16–17 Line diagram of a typical mass spectrum.

happen to be in the same molecule, there is an even smaller peak at M + 2. Chlorine and bromine yield abnormally high M + 2 peaks (^{35}Cl, 75.8%; ^{37}Cl, 24.2%; ^{79}Br, 50.5%; ^{81}Br, 49.5%).

The stability of the molecular ion is increased if it contains a π-electron system which can more easily adjust to the loss of an electron than can a σ bond. Also, in a cyclic system the rupture of a bond does not split the molecule. The possibility of cleavage is related to both the bond strengths in the molecular ion and the stability of possible fragments. In general, the relative height of the parent peak decreases in the following order:

Aromatics > conjugated olefins > alicyclics > sulfides > unbranched hydrocarbons > ketones > amines > esters > ethers > carboxylic acids > branched hydrocarbons and alcohols

The parent peak gives the exact molecular weight of the sample. It is not always easy to identify the parent peak; CCl$_4$, for example, does not give any parent peak. The increased use of CI and FI techniques, however, allows easy recognition of a protonated molecular ion (M + 1)$^{\oplus}$ or the molecular ion M$^{\oplus}$.

Base Peak. With an electron beam energy of 70 eV, the original molecular ion splits into many fragments, and the parent peak is often quite small. The largest peak which is observed is called the "base peak," and all other peak heights are measured with respect to it. Thus ion abundances are given in terms of percent of the base peak, or as the percent of the total amount of ions produced.

Fragmentation Peaks. Under the usual conditions, the molecular ion originally produced is left with considerable excess energy. Both the energy and the charge are rapidly delocalized resulting in one or more *cleavages* with or without some rearrangements. One of the fragments retains the charge and the remaining fragments may be either stable molecules or radicals. The kinds of fragments obtained varies with the electron beam energy. The minimum voltage at which a given ion is observed in the spectrum is known as its "appearance potential." Fortunately, with electron beam energies of 50 to 80 eV, most organic molecules fragment in a reproducible manner. The relative abundance of the ions produced is related to the strength and chemical nature of the bonds which held the fragment to the rest of the molecule. Therefore, on the basis of the concepts of statistics, resonance, hyperconjugation, polarization, and inductive and steric effects, a set of general rules (i.e., favorable fragmentation processes) can be formulated to predict prominent peaks in the mass spectrum, as will be discussed in later sections.

Metastable Ions. For the most part, fragmentation occurs within the ion source before the ions are accelerated into the analyzer tube. Ions resulting from a fragmentation after acceleration do not have a kinetic energy equal to that of the molecular ion from which they were formed, nor do they have a kinetic energy equal to ions of the same m/e originating before acceleration. Such ions are called metastable ions, and usually appear in the spectrum at non-integral mass numbers. The relationship between the apparent m/e of the metastable ion and its parent is given from the following formulation:

$$M_1^{\oplus} \longrightarrow M_2^{\oplus} + M_0$$

$$M^* = \frac{(M_2)^2}{M_1} \tag{16–6}$$

where M_1 = mass of parent ion
  M_2 = mass of daughter (metastable) ion
  M_0 = mass of neutral fragment
  M^* = apparent mass of metastable ion

Metastable ions give peaks which are broad in comparison to regular peaks, and are also weaker in intensity. Such peaks are useful in studying the mechanism of fragmentation, but are not often used for the study of structure.

Multiply Charged Ions. The great majority of ions formed bear a single positive charge, but occasionally doubly charged ions are formed. These ions will appear at the same position as a singly charged ion of half the mass. Doubly charged ions of odd mass number must appear at half-integral masses and are easily recognized. Even numbered masses are more difficult to recognize when doubly charged.

The appearance potential for doubly charged ions is, of course, much higher than for singly charged ions—nearly always more than 30 eV. The removal of two electrons is much easier in the presence of a high π-electron density and in the absence of bonds which are easily broken. Therefore, it is more likely to be observed in heteroaromatic molecules.

Negative Ions. Although the most important ionization process produces positive ions, in about one out of a thousand collisions in the electron beam the electron is captured by the molecule and forms a negative ion. They will not be observed with the usual mass spectrometers without considerable modification and are ignored except for special studies.

16–8 INTERPRETATION OF MASS SPECTRA

The interpretation of the mass spectrum requires an understanding of the ionization processes that occur before the ions reach the collector. Although some of these phenomena have been treated theoretically, we must first observe the fragmentation patterns and then try to explain how and why they occurred. This empirical approach has been very fruitful in relating spectra to structure. A systematic description of this subject fills a complete book; all we can do here is to give some of the guiding principles and a few examples.

A complex organic compound will yield ions with almost every conceivable combination of its atoms, and its complete spectrum would be a nightmare to interpret. Fortunately, improbable fragments will have abundances too low to be observed. For most purposes, we need consider only major peaks (except in the region of the parent peak). In the case of unknowns, it is rare that we do not know, or cannot easily find out, something about the nature of the compound from other easily applied techniques.

In the following discussion we will first consider some general approaches used to arrive at the molecular formula, then explain the spectra of a few typical compounds, and finally reverse the procedure and try to identify compounds from their spectra. In order to simplify the discussion we will consider organic compounds containing only C, H, N, and O.

The Exact Molecular Weight. For a pure compound, it is usually not too difficult to determine the nominal (whole number) molecular weight from the identification of the parent peak. With a high-resolution instrument we can do much better and determine the molecular weight with considerable precision.

Example/Problem 16–1. We have found the molecular weight of an unknown compound to be 150.0681. Of the hundreds of molecules having the nominal value of 150, only $C_9H_{10}O_2$ has the exact value 150.0681. (This fact is quickly determined from Beynon's Tables; see References.) This is only the beginning, because the *Handbook of Chemistry and Physics* lists 39 compounds with this molecular formula, but it is a long way toward solving the problem. Unfortunately, the precise data required ($C_9H_{10}O_2$ would have to be distinguished from $C_4H_{10}N_2O_4$, Mol. Wt. = 150.0641) are not easy to obtain, but suitable high-resolution instruments are now commercially available.

The Isotope Effect. A second approach to the molecular formula of a compound is through the distribution of naturally occurring isotopes. Molecules containing heavy isotopes will show up in peaks at m/e one or more units higher than normal; thus there will be small peaks at $M + 1$ and $M + 2$ (one and two units higher than the parent peak.) The relative abundances of heavy isotopes of C, H, N, and O are well known. For example, for every 100 ^{12}C atoms, there are 1.12 ^{13}C isotopes in any naturally occurring carbon-containing sample. Likewise for every 100 ^1H atoms, there are 0.016 ^2H isotopes. Since the heavy isotopes occur in definite proportions, the probability of finding one or more in a given molecule can

Table 16–3 Intensities of Isotope
Peaks Relative to the Most Abundant
Isotope (Intensity = 100)

Isotope	M + 1	M + 2
^{13}C	1.12	—
^{2}H	0.016	—
^{15}N	0.38	—
^{17}O	0.04	—
^{18}O	—	0.20
^{33}S	0.78	—
^{34}S	—	4.40
^{37}Cl	—	32.0
^{81}Br	—	98.0

be calculated; thus we can predict the height of the M + 1 peak relative to the parent peak (Table 16–3).

Example/Problem 16–2. The parent peak for an unknown compound is at $m/e = 32$. The two most likely formulas are CH_4O and N_2H_4. What are the relative intensities of the M + 1 peaks for these two molecules?

For every 100 molecules of CH_4O that have Mol. Wt. of 32, there will be 1.12 molecules having one ^{13}C isotope. Since there are four H atoms in the molecule, the chances of finding a ^{2}H isotope are four times as great as if there were only one H atom in the molecule. For oxygen, only ^{17}O will contribute to the M + 1 peak; ^{18}O contributes to the M + 2 peak. In tabular form, for the two suggested formulas, we have

$$\begin{array}{ll} & CH_4O \\ ^{13}C & 1 \times 1.12\% = 1.12\% \\ ^{2}H & 4 \times 0.016\ \ = 0.06 \\ ^{17}O & 1 \times 0.04\ \ \ = 0.04 \\ & M+1\ \ \ \ \ = \overline{1.22\%} \end{array} \qquad \begin{array}{ll} & N_2H_4 \\ ^{15}N & 2 \times 0.38\% = 0.76\% \\ ^{2}H & 4 \times 0.016\ \ = 0.06 \\ & M+1\ \ \ \ \ \ \ \ = \overline{0.82\%} \end{array}$$

Thus we could distinguish between the two compounds by measuring the M + 1 peak.

Beynon has performed these calculations, and tabulated the results in an extensive table covering all reasonable combinations of C, H, N, and O, up to a molecular weight of 500. For each combination, an exact molecular weight is given, and the relative height to be expected for the M + 1 and M + 2 peaks. An M + 3 peak would require three heavy isotopes in the same molecule—a very unlikely occurrence unless the molecule has been deliberately labeled. A portion of Beynon's Table for mass 150 is reproduced in Table 16–4.

In the first example, rather than measure the exact molecular weight, we might have measured the relative heights of the M + 1 and M + 2 peaks. Suppose that we had found that the M + 1 peak was 10% and that the M + 2 peak was 0.9% of the M peak. From Beynon's Tables we could have quickly decided that $C_9H_{10}O_2$ was the most likely formula. Beynon's Tables include many combinations that are not stable molecules but which can occur as fragment ions in the spectrum. This technique of making use of the distribution of isotopes is a general approach. In addition to Beynon's compilation, there are other useful observations. We have

Table 16–4 Masses and Isotopic Abundance Ratios for Various Combinations of C, H, N, and O

Formula	M[a]	M + 1[b]	M + 2[c]
$C_4H_{10}N_2O_4$	0.0641	5.402	9.230
$C_4H_{14}N_4O_2$	0.1117	6.150	5.638
$C_6H_{12}NO_4$	0.0766	6.133	9.600
$C_6H_2N_2O_3$	0.0065	7.396	8.384
$C_6H_6N_4O$	0.0542	8.145	4.942
$C_7H_{10}N_4$	0.0905	9.250	3.840
$C_8H_8NO_2$	0.0555	9.233	7.797
$C_8H_{10}N_2O$	0.0793	9.607	6.138
$C_8H_{12}N_3$	0.1031	9.981	4.492
$C_9H_{10}O_2$	0.0681	9.964	8.447
$C_9H_{12}NO$	0.0919	10.338	6.816
$C_9H_{14}N_2$	0.1157	10.712	5.198
$C_{10}H_{14}O$	0.1045	11.069	7.547
$C_{11}H_{18}$	0.1408	12.175	6.769
$C_{12}H_6$	0.0469	13.062	7.832

[a] Mass in excess of 150.0000.
[b] Height of M + 1 peak relative to M = 100.
[c] Height of M + 2 peak relative to M = 1000.
[After Beynon and Williams, *Mass and Abundance Tables for Use in Mass Spectrometry*, Elsevier, 1963.]

already noted that Cl and Br will give abnormally large M + 2 peaks. Sulfur will do the same. On the other hand, F, I, and P have only a single isotope and will give abnormally low M + 1 and M + 2 peaks.

The Nitrogen Rule. All organic compounds having an even molecular weight must contain an even number (including zero) of nitrogen atoms, and all organic compounds with an odd molecular weight must have an odd number of nitrogen atoms. All *fragment* ions (formed by cleavage of one bond) have odd mass if they have an even number (0, 2, 4, . . .) of N atoms, but even mass if they have an odd number of N atoms. The rule applies to compounds in which all bonds are covalent and includes molecules containing any combination of C, H, N, O, S, Si, As, P, halogens, and alkaline earth metals. The molecular weight to be used is the sum of the most abundant isotopes. The rule is a consequence of the fact that the most abundant isotope of most elements of even valency has an even mass number, and vice versa. The only common exception is nitrogen.

The Ring Rule. Once the molecular formula is available, the number of "unsaturated sites" is apparent from the "ring rule." The number of unsaturated sites, R, is equal to the number of rings in the molecule plus the number of double bonds plus twice the number of triple bonds. For the molecule, $C_wH_xN_yO_z$:

$$R = w + 1 + \frac{y - x}{2} \qquad (16\text{–}7)$$

Example/Problem 16–3. In benzene, C_6H_6

$$R = 6 + 1 - 3 = 4 \qquad \text{(one ring, three double bonds)}$$

In diethyl ether, $C_4H_{10}O$

$$R = 4 + 1 - 5 = 0 \qquad \text{(no rings or double bonds)}$$

In diethyl ketone, $C_5H_{10}O$

$$R = 5 + 1 - 5 = 1 \qquad \text{(carbonyl group)}$$

In cyclohexylamine, $C_6H_{11}NH_2$

$$R = 6 + 1 - 6 = 1 \qquad \text{(one ring)}$$

The rule as given here is an alternate expression for the D.B.E. rule given in Section 11–6, and gives identical information.

16–9 BEHAVIOR OF CLASSES OF COMPOUNDS

Saturated Aliphatic Hydrocarbons. The fission of carbon-carbon bonds in saturated aliphatic systems will produce a series of peaks of odd mass $(C_nH_{2n+1})^{\oplus}$ corresponding to the process

$$[C_nH_{2n+1}\!-\!C_mH_{2m+1}]^{\oplus\cdot} \longrightarrow [C_nH_{2n+1}]^{\oplus} + {}^{\cdot}C_mH_{2m+1}$$

$$m/e \ 15, 29, 43, \ldots$$

Fragment ions of high mass have sufficient energy to decompose even further but as the value of n approaches 3, 4, and 5 there is less energy available and the above are the most abundant in the spectrum. The spectrum of n-dodecane (Figure 16–18) illustrates the typical spectrum of a long-chain hydrocarbon. In the case of branched hydrocarbons, however, fission is preferred at the tertiary site to produce, if possible, a tertiary carbonium ion. The position of such branching can often be inferred by the resulting carbonium ion and its enhanced abundance. Thus, in the case of 2-methylheptane (Figure 16–19) the anomalous abundance of $m/e = 99$ indicates fission occurring as follows:

$$[CH_3CH_2CH_2CH_2CH_2\overset{|}{\underset{\underset{CH_3}{|}}{CH}}\!-\!CH_3]^{\oplus\cdot} \longrightarrow [CH_3CH_2CH_2CH_2CH_2\overset{|}{\underset{\underset{CH_3}{|}}{CH}}]^{\oplus} + {}^{\cdot}CH_3$$

$$m/e \ 99$$

The resulting secondary carbonium ion then fragments as the straight-chain molecule would (i.e., m/e 29, 43, and 57 most abundant).

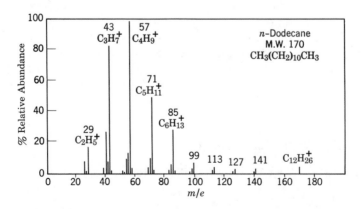

Figure 16–18 Mass spectrum of n-dodecane.

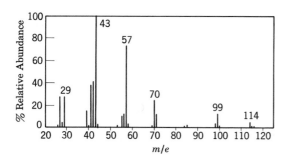

Figure 16–19 Mass spectrum of 2-methylheptane.

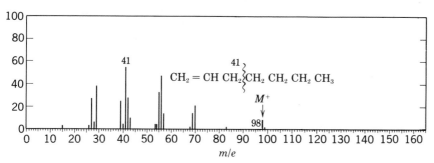

Figure 16–20 Mass spectrum of 1-heptene.

Unsaturated Aliphatic Hydrocarbons. Mono-olefins (Figure 16–20) tend to undergo fission at the carbon atom beta to the double bond system, the resulting carbonium ion being stabilized by resonance:

$$[CH_2{=}CHCH_2 \diagup CH_2CH_2CH_2CH_3]^{\oplus \cdot} \longrightarrow CH_2{=}CH\overset{\oplus}{C}H_2 \longleftrightarrow \overset{\oplus}{C}H_2CH{=}CH_2$$

$$m/e\ 98 \qquad\qquad\qquad\qquad\qquad\qquad m/e\,41$$

The ability of the double bond to migrate fairly easily in this type of system, however, makes the spectra of different olefinic isomers very similar.

Aromatic Hydrocarbons. Simple aromatic systems such as *n*-butyl-benzene (Figure 16–21) decompose via beta-fission initiated by the aromatic nucleus to produce the tropylium ion (*m/e* 91) together with further decomposition to *m/e* 65 with loss of acetylene.

m/e 135   *m/e* 91

m/e 91   *m/e* 65

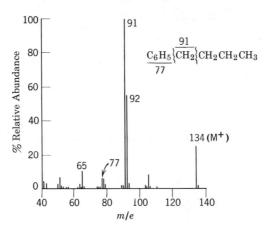

Figure 16–21 Mass spectrum of *n*-butylbenzene.

The beta-fission process may also occur via rearrangement of one hydrogen with production of a neutral olefin and *m/e* 92. Two distinct transition states exist for this decomposition, either four- or six-membered.

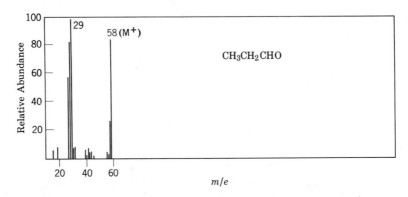

Aldehydes. The mass spectra of aldehydes, esters, and ketones have features in common, influenced by the presence of the carbonyl group.

One such feature includes the cleavage alpha to the carbonyl group. The peak *m/e* 29 is the base peak in the mass spectrum of acetaldehyde (CH₃CHO) since the formation of CHO⊕ plus a CH₃ radical is more favorable energetically than the formation of CH₃CO⊕ and a hydrogen radical; that is, in simple aliphatic

Figure 16–22 Mass spectrum of propionaldehyde.

aldehydes, alpha-cleavage with loss of the larger alkyl radical is preferred (Figure 16–22).

$$CH_3CH_2 \mathbin{\!/\mkern-5mu/\!} CHO^{\oplus \cdot} \longrightarrow CH_3CH_2^{\cdot} + CHO^{\oplus}$$

m/e 58   m/e 29

In the case of butyraldehyde (Figure 16–23) the peak at m/e 29 indicates the alpha-cleavage outlined above. However, the base peak at m/e 44 (even mass number, hence rearrangement peak) has arisen from the so-called *McLafferty rearrangement*—gamma hydrogen transfer to the carbonyl oxygen atom.

$$CH_3CH_2CH_2 \mathbin{\!/\mkern-5mu/\!} CHO^{\oplus \cdot} \longrightarrow CH_3CH_2CH_2^{\cdot} + CHO^{\oplus}$$

m/e 29

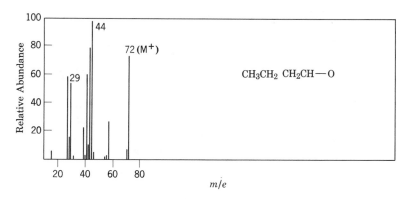

m/e 72   m/e 44

Figure 16–23   Mass spectrum of n-butyraldehyde.

In the case of n-hexanal (Figure 16–24) the base peak at m/e 44 indicates the preferred fragmentation via the McLafferty rearrangement while the m/e 29 is the alpha cleavage.

$$CH_3CH_2CH_2CH_2CH_2{-}CHO^{\oplus \cdot} \longrightarrow CH_3CH_2CH_2CH_2CH_2^{\cdot} + CHO^{\oplus}$$

m/e 100   m/e 29

m/e 100   m/e 44

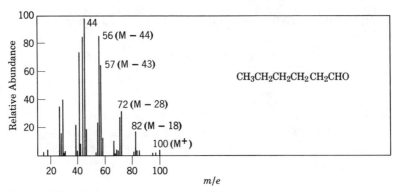

Figure 16–24 Mass spectrum of *n*-hexanal.

The m/e 56 and m/e 57 peaks can be explained in terms of fragments lost from the molecular ion (i.e., M−44 and M−43, respectively). The M−44 peak can arise from the McLafferty rearrangement with the charge remaining on the alkyl portion. The M−43 peak is a result of beta-cleavage with gamma-hydrogen transfer, the charge remaining on the alkyl group.

$$
\begin{array}{ccc}
\overset{\cdot\cdot}{\underset{|}{O}}: & & \overset{\cdot\cdot}{\underset{}{O}}: \\
\overset{|}{\underset{/}{CH^{\oplus}}} & \longrightarrow \quad C_3H_7^{\oplus} \quad + & \overset{}{\underset{}{CH_2}} \\
\overset{|}{\underset{|}{CH_2}} & \quad\quad m/e\ 57 & H_2C \\
\overset{|}{C_3H_7} & \quad\quad (M–43) &
\end{array}
$$

The remaining important peaks at m/e 72 (M − 28) and m/e 82 (M − 18) are elimination of small stable neutral fragments such as CO and H_2O from the molecular ion.

  Ketones. As with the aldehydes, alpha-cleavage is common in the case of ketones. This alpha-cleavage, however, may occur at either of the two bonds next to the carbonyl group. In general, loss of the larger alkyl fragment is preferred, that is, the base peak of methyl ketones should be m/e 43.

$$
CH_3-C{=}O^{\oplus\cdot} \longrightarrow CH_3-C{\equiv}O^{\oplus} + R^{\cdot}
$$
$$
\overset{}{\underset{R}{}} \quad\quad\quad m/e\ 43
$$

In the case of octan-4-one (Figure 16–25), therefore, the loss of $CH_3CH_2CH_2CH_2$ (M − 57 or m/e 71) should be preferred to the loss of $CH_3CH_2CH_2$ (M − 43 or m/e 85). As it happens, both processes take place, probably due to the small energy differences in either a C_3 or C_4 loss.

$$
\overset{O^{\oplus\cdot}}{\underset{\parallel}{}}
$$
$$CH_3CH_2CH_2-C{-}CH_2CH_2CH_2CH_3 \longrightarrow CH_3CH_2CH_2-C{\equiv}O^{\oplus}$$
$$+ {}^{\cdot}CH_2CH_2CH_2CH_3$$
$$m/e\ 128 \hspace{4cm} m/e\ 71$$

$$
\overset{O^{\oplus\cdot}}{\underset{\parallel}{}}
$$
$$CH_3CH_2CH_2-C{-}CH_2CH_2CH_2CH_3 \longrightarrow CH_3CH_2CH_2^{\cdot} + {}^{\oplus}O{=}CCH_2CH_2CH_2CH_3$$
$$m/e\ 128 \hspace{4cm} m/e\ 85$$

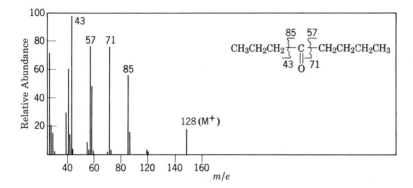

Figure 16–25   Mass spectrum of octan-4-one.

If there is a chain of three of more carbon atoms in each alkyl chain of the ketone (assuming gamma hydrogens are available), then a second McLafferty rearrangement becomes possible. In the case of octan-4-one the m/e 58 peak results from the rearrangement of the enolic product produced in the primary (first) arrangement.

m/e 86
(Primary enolic product)

m/e 58

Esters.   Esters follow the same fragmentation characteristics as other carbonyl functions. They exhibit peaks corresponding to alpha-cleavage on both sides of the carbonyl function yielding a variety of ions.

$$R_1 \!-\!\! \overset{\overset{\displaystyle O}{\|}}{C} \!-\! OR_2 \longrightarrow R_1^{\oplus},\ R_1\!-\!C\!\!\equiv\!\!O^{\oplus},\ [COOR_2]^{\oplus},\ ^{\oplus}OR_2$$

In the case of methyl butyrate (Figure 16–26), the products possible are m/e 71, 59, 31, and 43.

$$CH_3CH_2CH_2\!-\!\!\overset{\overset{\displaystyle O^{\oplus \cdot}}{/\!/}}{C}\!\!\overset{}{\diagdown}_{OCH_3} \longrightarrow CH_3CH_2CH_2^{\cdot} + ^{\oplus}O\!\!\equiv\!\!C\!-\!OCH_3$$
$$m/e\ 59$$

$$CH_3CH_2CH_2\!-\!\!\overset{\overset{\displaystyle O^{\oplus \cdot}}{/\!/}}{C}\!\!\overset{}{\diagdown}_{OCH_3} \longrightarrow CH_3CH_2CH_2C\!\!\equiv\!\!O^{\oplus} + ^{\cdot}OCH_3$$
$$m/e\ 71$$

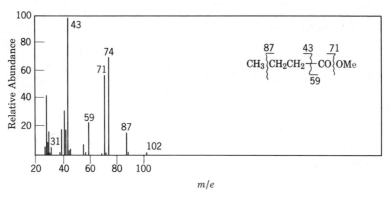

Figure 16–26 Mass spectrum of methyl butyrate.

Explanation of the m/e 74 can be found by employing a McLafferty rearrangement.

The remaining peak at m/e 87 (M − 15) must be loss of a methyl group.

   Aliphatic Alcohols.   The parent peak is small and sometimes absent entirely, especially for higher alcohols (Figures 16–27 and 16–28). Very often the M − 18 peak (loss of H_2O) is mistaken for the parent. The loss of H_2O occurs frequently since dehydration can occur by thermal decomposition prior to electron impact especially when a heated inlet system is used.

Elimination of water is frequently associated with the loss of C_2H_4 leading to the

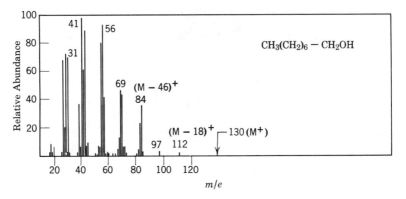

Figure 16–27 Mass spectrum of n-octanol, $C_8H_{17}OH$.

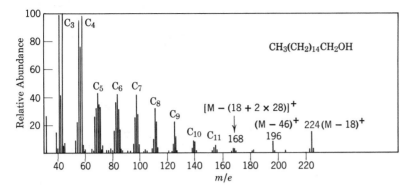

Figure 16–28 Mass spectrum of cetyl alcohol, $C_{16}H_{33}OH$.

$M - 46$ peak, which can be of substantial intensity.

Alcohols also exhibit $M - 1$, $M - 2$, and $M - 3$ peaks due to loss of hydrogen atoms, which are of the same order of intensity as the parent, $M^{\oplus}$.

$$RCH=\overset{\oplus}{O}H \qquad RCH=\overset{\oplus}{O} \qquad RC\equiv O^{\oplus}$$
$$M-1 \qquad\qquad M-2 \qquad\qquad M-3$$

In the case of a secondary alcohol the alpha-bond which has the highest possibility of breaking is the one that results in the loss of the largest side chain

$$\overset{\cdot\oplus}{OH}$$
$$CH_3CH_2{+}C{+}H$$
$$CH_3$$

$CH_3CH_2^{\cdot} + \overset{\oplus}{C}{-}H$	$CH_3{-}CH_2{-}\overset{\oplus}{C}{-}H$	$CH_3{-}CH_2{-}\overset{\oplus}{C}$ $+ H^{\cdot}$
CH_3	$+ {\cdot}CH_3$	CH_3
$m/e\ 45$	$m/e\ 59$	$m/e\ 73$

Relative abundance	100%	19%	1.2%

Ethers. As with the ketones, alpha-cleavage of the molecule may be expected to be the favored decomposition route. In an unsymmetrically substituted ether such as ethyl-sec-butyl ether (Figure 16–29), there are three possibilities depending

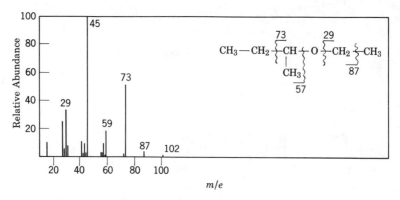

Figure 16–29   Mass spectrum of ethyl-*sec*-butyl ether.

on the site of the primary cleavage.

$$CH_3-CH_2\!\!\mid\!\!CH-\overset{\cdot\oplus}{O}-CH_2\!\!\mid\!\!CH_3 \longrightarrow CH_3CH_2^+ + \overset{\oplus}{CH}=\overset{}{O}-CH_2CH_3$$

$$\underset{m/e\ 73}{|}$$

with CH_3 branch below, leading down to:

$$CH_3CH_2-CH-\overset{\oplus}{O}=CH_2$$
$$\underset{CH_3}{|}$$
$$\underset{m/e\ 87}{}$$

$$+\ \dot{C}H_3$$

or

$$CH_3CH_2-\overset{\oplus}{C}H=O-CH_2-CH_3 \qquad +\dot{C}H_3$$
$$\underset{m/e\ 87}{}$$

These fragments (*m/e* 73 and *m/e* 87) may undergo further decomposition by rearrangement of one hydrogen atom and elimination of the neutral olefin, C_2H_4. The process requires the presence of at least an ethyl substituent.

$$CH=\overset{\oplus}{O}-CH_2CH_2 \longrightarrow CH=\overset{\oplus}{O}H + CH_2=CH_2$$
$$\underset{CH_3}{|} \qquad\qquad\qquad \underset{CH_3}{|}$$
$$\underset{m/e\ 73}{} \qquad\qquad\qquad \underset{m/e\ 45}{}$$

$$CH_3CH_2CH=\overset{\oplus}{O}-CH_2CH_2 \longrightarrow CH_3CH_2-CH=\overset{\oplus}{O}H + CH_2=CH_2$$
$$\underset{m/e\ 87}{} \qquad\qquad\qquad\qquad \underset{m/e\ 59}{}$$

Amines.   The parent peak of mono-amines occurs at an odd mass number and is weak for aliphatic amines and strong for aromatic amines. The most important process yielding the base peak is the rupture of the bond *beta* to the nitrogen atom, in which the loss of the largest hydrocarbon fragment is favored.

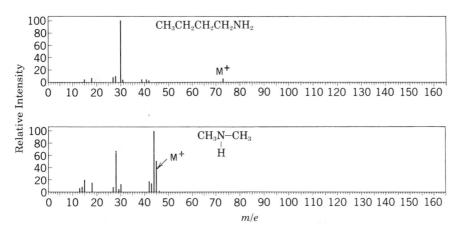

$$m/e \ 30, \ 44, \ 58, \ 72, \ 86$$

For primary amines, where R_2 and R_3 are H atoms, the base peak should be at 30. There is also a peak at 18 from the $NH_4^\oplus$ ion, which can be confused with a similar peak from $H_2O^\oplus$. Dimethyl amine and *n*-butyl amine (Figure 16–30) exhibit these features.

Figure 16–30   Mass spectra of *n*-butylamine and dimethylamine.

16–10   IDENTIFICATION OF UNKNOWNS

It is much easier to predict or explain the spectrum of a known compound than it is to identify a compound from its observed spectrum. However, rarely does one have to work from the mass spectrum alone. The sample usually comes with a history which may give some clues. Other spectral data (infrared, n.m.r.) should be taken, chromatographic retention behavior noted (for polarity), melting point and boiling points should be observed, and in general every available bit of information should be considered. The interpretation will be greatly simplified if the compound is as pure as possible.

We have seen that the mass spectrometer provides several kinds of information:

1. *The molecular weight.* A low resolution instrument gives only the nominal (nearest whole number) value, while a high resolution instrument can give the exact molecular weight to several decimal places. From the latter, we can ascertain the molecular formula.

2. *The isotope distribution.* The relative heights of the M + 1 and M + 2 peaks (if available) help to select the most probable formula if only the nominal molecular weight is known. An abnormally small M + 1 peak indicates fluorine, phosphorus or iodine. An abnormally large M + 2 peak indicates chlorine, bromine or sulfur.

3. *The fragmentation pattern.* This is unique for each compound, so that if one has a large file of labeled spectra and a workable routine for comparison of unknown to the knowns, the solution is straightforward, although perhaps tedious without a computer.

There is nothing like experience—the beginner must expect to make many wrong guesses. For the major peaks, and all peaks near the parent peak, list not only the m/e of the charged fragment, but also the mass of the neutral fragment lost to give that peak. The rest is a matter of comparing peaks to common fragments (Table 16–5), feeding in all other information as appropriate much the same as one solves a jigsaw puzzle. *Not all peaks need be identified.*

Table 16–5   Some Common Fragments

m/e	Fragment	m/e	Fragment
14	CH_2	44	CO_2, $C_2H_4NH_2$, CH_2=CHOH
15	CH_3	45	C_2H_4OH
16	O	46	NO_2
17	OH	55	C_4H_7
18	H_2O, NH_4	56	C_4H_8
19	F	57	C_4H_9, C_2H_5C=O
20	HF	58	$C_3H_6NH_2$, NCS
26	CN, C_2H_2	59	$(CH_3)_2COH$, $COOCH_3$
27	C_2H_3, HCN	59	$CH_2OC_2H_5$
28	C_2H_4, CO, N_2	69	C_5H_9, CF_3
29	C_2H_5, CHO	70	C_5H_{10}
30	CH_2NH_2, NO, CH_2O	71	C_5H_{11}
31	CH_2OH, OCH_3, CH_3NH_2	73	$OCOC_2H_5$
32	CH_3OH	77	C_6H_5
33	SH	79	^{79}Br
34	H_2S	81	^{81}Br
35	^{35}Cl	83	C_6H_{11}
36	HCl	84	C_6H_{12}
37	^{37}Cl	85	C_6H_{13}
41	C_3H_5	91	$C_6H_5CH_2$
42	C_3H_6	92	$C_6H_5CH_3$
43	C_3H_7, CH_3C=O		

Example/Problem 16–4.   Identify the compound whose spectrum is given in Figure 16–31a.

m/e	Relative Intensity	Preliminary Remarks
26	21	CN, C_2H_2
27	58	C_2H_3
28	69	C_2H_4, CO, N_2
29	100	C_2H_5, CHO
31	6	CH_2OH, OCH_3
39	4	
43	11	M − 15, loss of CH₃
57	10	M − 1, loss of H
58	38	Molecular ion

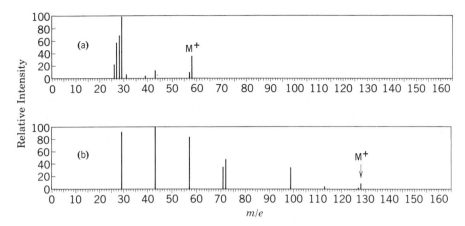

Figure 16–31 Mass spectra of example compounds discussed in text.

The molecular ion is intense; therefore, if it is an amine it must be aromatic. This is impossible because of the low molecular weight. Possible formulas involving C, H, and O are: $C_2H_2O_2$, C_3H_6O, and C_4H_{10}. Butane is unlikely because there is an unexplainable peak at $m/e = 31$. For $C_2H_2O_2$, the ring rule gives $2 + 1 - 1 = 2$ sites of unsaturation, which makes the compound glyoxal, OHC—CHO. For C_3H_6O, the ring rule gives $3 + 1 - 3 = 1$ site of unsaturation leading to the following compounds: methyl vinyl ether, $CH_3OCH{=}CH_2$; propionaldehyde, CH_3CH_2CHO; propylene oxide, CH_2CHCH_3; acetone, CH_3COCH_3

and allyl alcohol, $CH_2{=}CHCH_2OH$. Methyl vinyl ether is a reasonable guess except for the peak at $m/e = 28$. The intense peaks around $m/e = 29$ do not fit acetone. An alcohol should have a M − 18 peak larger than the parent peak. Since this is not the case, we eliminate allyl alcohol and look at propionaldehyde. All peaks except $m/e = 26$ fit predictions for an aldehyde; therefore the best identification seems to be propionaldehyde.

Example/Problem 16–5. Identify the compound whose spectrum is given in Figure 16–31b. Since we have neither the exact molecular weight nor the isotope distribution (M+1 and M+2 peaks) we must rely entirely on the nominal molecular weight and the fragmentation pattern.

m/e	Relative Intensity	Preliminary Remarks
128	9	Parent peak
127	1	M − 1, loss of H
113	4	M − 15, loss of CH_3
99	35	M − 29, loss of C_2H_5 or CHO
72	48	even m/e—rearrangement
71	36	C_5H_{11}
57	84	C_4H_9 or C_2H_5CO
43	100	C_3H_7 or CH_3CO
29	94	C_2H_5 or CHO

The compound appears not to contain N (no peaks typical of amines). Some likely formulas for molecular weight 128 include: $C_6H_8O_3$, $C_7H_{12}O_2$, $C_8H_{16}O$, C_9H_{20}, and $C_{10}H_8$. Next we propose reasonable structures for these formulas and see which of the predicted fragmentation patterns resemble the observed spectrum. For example, $C_{10}H_8$ (naphthalene) is aromatic and should give a very large parent peak; therefore $C_{10}H_8$ can be eliminated. Likewise, C_9H_{20} is a saturated hydrocarbon which would not give the large rearrangement peak observed at $m/e = 72$.

The ring rule aids in composing structures, for example $C_6H_8O_3$ has $6 + 1 - 4 = 3$ rings or double bonds and we can invent corresponding structures such as:

$$CH_3CCH_2CCH_2CH \qquad \text{or} \qquad HO-CH \begin{array}{c} CH_2-C \\ \\ CH_2-C \end{array} \begin{array}{c} O \\ \\ CH_2 \\ \\ O \end{array}$$

Neither of these structures will give the observed spectra. We continue proposing and predicting, gradually narrowing down the possibilities. Other spectral data would be a great help at this point. Reference spectra might be available for some of the proposed compounds, and these should be compared with the given spectrum.

Lack of an M − 18 and an M − 28 peak makes an aldehyde doubtful. $C_7H_{12}O_2$ can be eliminated because it would most likely be a diketone or an unsaturated ester (neither of which would give a rearrangement peak at $m/e = 72$). This brings us to $C_8H_{16}O$ for which we can propose several ketones. There must be at least a three-carbon chain next to the carbonyl in order to get a rearrangement peak. The $m/e = 71$ peak indicates a five-carbon chain, for which we can propose several isomers, each of which satisfies the given spectrum.

$$C_2H_5C-(CH_2)_4-CH_3, \qquad C_2H_5CCH_2CHC_2H_5, \qquad C_2H_5CCH_2CH_2CHCH_3$$
$$\underset{O}{} \qquad\qquad\qquad \underset{O\quad CH_3}{} \qquad\qquad \underset{O\quad CH_3}{}$$

This is as far as we can go with the data given; however, an n.m.r. spectrum would readily identify the compound.

16–11   ANALYSIS OF MIXTURES

We have seen that the interpretation of a spectrum of a pure compound is often a complex and sometimes doubtful process. The identification of compounds in a mixture from its mass spectrum is at least an order of magnitude more difficult. An experienced analyst may be able to recognize characteristic peaks, but it is a tedious process, unless there are enough peaks which are unique to each compound present.

If the identities of the compounds in the sample mixture are known, then quantitative analysis is fairly straightforward. The general principle is that contributions to each peak are additive for each compound present; that is, one records the sum of the spectra, in which each compound acts independently of all others present. Fortunately, at the very low pressures used in the spectrometer, this is a good approximation. Thus the total ion currents at each peak, I_1, I_2, I_3, etc., can be represented as the sum of the currents resulting from each compound:

$$I_1 = i_{1a}p_a + i_{1b}p_b + i_{1c}p_c + \cdots$$
$$I_2 = i_{2a}p_a + i_{2b}p_b + i_{2c}p_c + \cdots$$
$$I_3 = i_{3a}p_a + i_{3b}p_b + i_{3c}p_c + \cdots$$

where the numerical subscripts refer to various mass numbers and the letter subscripts refer to the various compounds present, that is, i_{2b} is the ion current at mass 2 due to component b at unit pressure (a calibration factor), and p_b is the actual partial pressure of component b in the sample. The values for i_{1a} etc. must be obtained from spectra of pure compounds at known pressures. Another equation is available by noting that the total pressure is equal to the sum of the partial pressures. A set of simultaneous equations (at least as many as there are unknowns) can be solved for the unknown partial pressures (percent composition)—an obvious job for a computer. Often the only reliable way to analyze a mixture is to separate the compounds by some other technique—gas chromatography is an obvious choice.

The presence of a mixture may be ascertained by noting the change in the spectrum with time. In the sample introduction device, the lighter molecules diffuse through the leak faster than the heavier; thus the ratio of light to heavy molecules will decrease with time.

A better approach to the identification of mixtures is by chemical or field ionization. With CI, usually the most abundant ions are the $(M + 1)^\oplus$ species and their singularly dominant appearance leaves little doubt in identifying the molecular weights of compounds that are present.

16–12   SOME ADDITIONAL APPLICATIONS

Isotope Labeling.   One of the most powerful means of determining the mechanism of a reaction or the structure of a complex compound is to use a reactant containing a labeled atom and then find out its exact location in the product. A radioactive isotope is easily spotted, but for many applications, particularly those in some biochemical systems, the radioactivity may be objectionable. In these cases, a stable heavy isotope can be used. The mass spectrometer can locate the position of the labeled isotope by the change it causes in the spectrum (some fragments will appear at a different mass number). In some cases, it may suffice to detect which fragment is the heavy one. In other cases, it may be necessary to determine the resulting ratio of heavy to normal isotope—for this the isotope-ratio spectrometer is ideal.

Nonvolatile Substances.   High polymers and many natural products have vapor pressures too low for ordinary mass spectrometry. Nevertheless, nearly all substances can be pyrolyzed and the decomposition products subjected to mass spectrometry. Such information may serve as a fingerprint and can be used for analytical purposes.

Physical Chemical Data.   With high resolution spectrometers, the exact determination of the masses of nuclides can be done with an accuracy of about one part in 10^8. Thus the masses of isotopes are known very precisely from mass spectrometry, and the limitation on exact atomic weights (in the chemical sense) is the uncertainty of the relative abundance of the isotopes. Even the relative abundances have been accurately measured by mass spectrometry, and the final remaining limitation is the variation of the abundance in nature.

Ionization potentials and bond strengths can be estimated from the appearance potentials, that is, the electron energy at which the ion is first produced.

Electromagnetic Separation. Although a mass spectrometer can hardly be classed as a preparative-scale machine, it is extremely useful in preparing small samples of pure isotopes of the elements. It is a very inefficient and expensive method, but it has been particularly valuable for this purpose in the atomic energy program.

QUESTIONS AND PROBLEMS

16–1. The mass spectrum of methane has a large peak (parent and base) at 16 for the $CH_4^{\oplus}$ ion, with smaller peaks at 15, 14, 13, and 12 for loss of H atoms and at 17 for the presence of ^{13}C or ^2H (= D). An equimolar mixture of CH_4 and CD_4 has the same number of D atoms as the amount of CH_2D_2 but the spectrum of the mixture $CH_4 + CD_4$ differs markedly from that of CH_2D_2. Describe and explain these two spectra.

16–2. The isotope contributions to the $CH_3^{\oplus}$ peak result from the normal abundance of ^{13}C and ^2D. On the C-12 scale $^1H = 1.007825$, $D = 2.014102$, and $^{13}C = 13.00354$. With ordinary resolution, only three peaks are seen at 15, 16, and 17. Calculate the relative heights of the peaks at 16 and 17 (height of the 15 peak = 100) from the following data:

Compound	Relative Abundance
$^{12}C^1H^1H^1H$	98.882
$^{13}C^1H^1H^1H$	1.074
$^{12}C^1H^1H^2D$	0.040
$^{13}C^1H^1H^2D$	0.004

16–3. From the data given in Problem 16–2, determine what resolution ($M/\Delta M$) would be required to separate the $^{13}C^1H_3$ peak from the $^{12}C^1H^1H^2D$ peak.

16–4. Explain the origin of as many peaks as you can in the spectra for the following compounds (R.I. is relative intensity):

(a) Iso-butane		(b) Iso-octane (2,2,4-trimethyl pentane)		(c) n-Octane	
m/e	R.I.	m/e	R.I.	m/e	R.I.
14	1.18	15	4.46	15	3.00
15	6.41	27	11.8	27	29.2
26	2.36	28	2.08	28	6.29
27	27.8	29	15.2	29	34.5
39	16.5	29.5	0.15	30	0.74
41	38.1	39	9.43	39	13.5
42	33.5	41	27.0	41	38.1
43	100	43	23.0	42	15.6
44	3.33	56	32.2	43	100

(a) Iso-butane		(b) Iso-octane (2,2,4-trimethyl pentane)		(c) n-Octane	
m/e	R.I.	m/e	R.I.	m/e	R.I.
57	3.0	57	100	44	3.33
58	2.73	58	4.34	55	10.1
59	0.11	99	4.56	56	18.0
		114	0.02	57	34.2
				70	12.3
				71	23.3
				84	5.96
				85	29.5
				114	6.74
				115	0.55
				116	0.03

(d) 2-Methyl-2-propanol		(e) Diisopropyl ether		(f) Methyl acetate	
m/e	R.I.	m/e	R.I.	m/e	R.I.
15	10.0	27	6.90	29	10.50
27	7.2	28	1.39	31	3.43
28	6.8	29	1.67	42	10.24
29	9.8	31	2.82	43	100
31	2.82	39	5.03	44	2.86
39	5.03	41	12.07	59	5.71
41	12.07	42	3.37	74	15.21
43	12.1	43	40.65	75	0.54
57	8.7	44	3.48		
59	100	45	100		
60	3.4	46	2.22		
74	0	59	10.61		
		69	3.27		
		87	22.26		
		102	0.54		

(g) Ethyl formate		(h) n-Propyl amine	
m/e	R.I.	m/e	R.I.
26	12.57	15	3.11
27	43.17	18	2.99
28	72.66	27	6.65
29	65.80	28	12.7
30	5.49	30	100
31	100	41	5.21
43	7.51	42	3.12
45	28.69	43	2.12
46	5.17	44	1.28
47	5.98	59	7.76
74	7.07		

16–5. Using Equation 16-5 and the data given in that section, compute the theoretical mass range of a time-of-flight mass spectrometer (assume that $k = 1$). Commercial instruments seldom are able to exceed a mass range of 200. Comment on the discrepancy. *Ans.* 10,000

16–6. A monoamine has the mass spectrum given below. What is the structure of this amine?

m/e	R.I.
115	17
100	50
58	100

16–7. Anthraquinone ($C_{14}H_8O_2$) gives major fragmentation peaks at m/e 180 and 152 as well as the molecular ion base peak at m/e 208. In addition, two metastable ions are present at m/e 155.8 and 128.3. What can be said regarding the origin of the fragment peaks?

16–8. Centrally labeled (^{13}C) neopentane upon electron impact gives $C_4H_9^{\oplus}$ and $C_3H_5^{\oplus}$ ions that are 100% and 90% labeled, but $C_2H_5^{\oplus}$ is only 47% labeled. Neopentane-1-^{13}C gives $C_4H_9^{\oplus}$ ions that are 76% labeled; $C_3H_5^{\oplus}$, 52% labeled; and $C_2H_5^{\oplus}$, 36% labeled. Rationalize this behavior, using structures where necessary.

$$CH_3-\overset{\displaystyle CH_3}{\underset{\displaystyle CH_3}{\overset{|}{\underset{|}{C}}}}-CH_3$$

neopentane

m/e	R.I.
72	0.01
57	100.0
41	41.5
39	12.7
29	38.5
27	14.9

16–9. The mass spectra for several unknown compounds are given below in terms of the relative intensity of the peak heights. Assign as many fragments as you can and try to identify the compounds.

			Compound		
m/e	A	B	C	D	E
14	5.9	a	a		
15	30.5	a	a	3.6	3.1
17				3.6	
18				12.8	3.0
26	4.9	9.1	21.5		
27	7.5	4.5	58.2	4.7	6.7
28		2.6	69.0		12.7
29		100	100	4.3	3.0
30			5.7	100	100
31			2.9	3.0	
41		3.9	2.0	3.0	5.2
42	6.8	9.2	3.8	6.2	3.1
43	100	26.7	11.2	5.3	
44		45.7	3.1	4.3	
45		1.24			

	Compound				
m/e	A	B	C	D	E
46		0.10			
57			10.8		
58	33.1		37.4		
59	1.3		1.5		7.8
60	0.10		0.10	2.4	0.32
61				0.10	

ᵃ Not observed.

16-10. Deduce the structure of each of the compounds whose mass spectra are given in Figures 16–32 to 16–37.

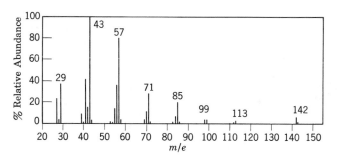

Figure 16–32 Mass spectrum of unknown compound A.

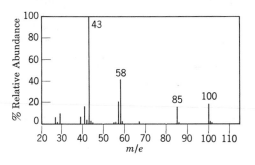

Figure 16–33 Mass spectrum of unknown compound B.

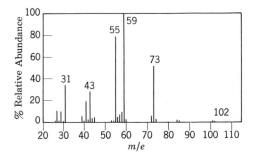

Figure 16–34 Mass spectrum of unknown compound C.

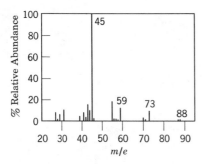

Figure 16–35 Mass spectrum of unknown compound D.

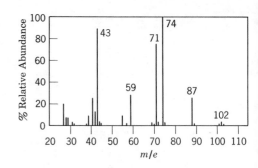

Figure 16–36 Mass spectrum of unknown compound E.

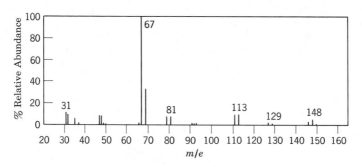

Figure 16–37 Mass spectrum of unknown compound F.

REFERENCES

R. M. Silverstein, G. C. Bassler, and T. C. Morrill, *Spectrometric Identification of Organic Compounds*, 3rd ed., Wiley, New York, 1974. Brief discussion in Chapter 2, and extensive reproduction of Beynon's Tables. Many spectra worked out in problems involving other methods as well.

K. Biemann, *Mass Spectrometry: Organic Chemical Applications*, McGraw-Hill, New York, 1962. Good discussion of instrumentation, fragmentation processes, and applications.

J. H. Beynon, *Mass Spectrometry and Its Applications to Organic Chemistry*, Elsevier, Amsterdam, 1960. Definitive but difficult, and not well organized for beginner.

H. Budzikiewicz, C. Djerassi, and D. H. Williams, *Mass Spectrometry of Organic Compounds*, Holden Day, San Francisco, 1967. Modern treatment of fragmentation of organic compounds.

R. W. Kiser, *Introduction to Mass Spectrometry and Its Applications*, Prentice-Hall, Englewood Cliffs, N.J., 1965. Heavy emphasis on instrumentation and handling of data. Many tables in Appendices.

J. H. Beynon and A. E. Williams, *Mass and Abundance Tables for Use in Mass Spectrometry*, Elsevier, Amsterdam, 1963. Tables for all reasonable combinations of C, H, N, and O with fragment masses up to 500. Data are based on the C-12 scale.

F. W. McLafferty, *Interpretation of Mass Spectra. An Introduction*, Benjamin, New York, 1966. A short, readable "self-study" text, including many example spectra to work on (with answers).

F. W. McLafferty and R. S. Gohlke, "Mass Spectrometry," *Chem. Eng. News*, May 18, 1964, p. 96. A review of applications with examples.

G. W. A. Milne, Ed., *Mass Spectrometry: Techniques and Applications*, Wiley, New York, 1971.

H. C. Hill, *Introduction to Mass Spectrometry*, Second Edition, revised by A. G. Loudon, Heyden & Son, London, 1975. Excellent introductory text for beginner.

17

SPARK SOURCE
MASS SPECTROMETRY

17-1 INTRODUCTION

From the beginning, chemists have been plagued by the limits of sensitivity of the "universal" methods of analysis that permit the determination of many elements in the same sample; for example, emission spectroscopy, X-ray fluorescence, and atomic absorption analysis. This has led to the adoption of more sensitive methods for a restricted number of elements, for example, neutron activation analysis. More recently, spark source mass spectrometry has provided the analytical chemist with the ability to cover the full range of elements in any sample in a single determination and the ability to detect these elements at the parts per billion (ppb) level. This technique is invaluable in semi-conductor technology where 10 ppb of copper in germanium or 2 ppb of gold in silicon changes the electrical properties of these semiconductors.

17-2 INSTRUMENTATION

A spark source mass spectrometer, like most other mass spectrometers, consists essentially of three parts that are housed in a highly evacuated system (Figure 17-1):

1. The source for ionizing the sample—a radio frequency spark.
2. The analyzer to separate these ions according to their mass to charge ratio (m/e)—a combination of both electrostatic and magnetic fields.
3. The detector to record the position and intensity of the ions—a photographic plate or an electronic collector.

A radio frequency (about one megahertz) voltage of several tens of kilovolts is applied in pulses across a small gap (a few thousandths of a centimeter) between the two sample electrodes in the source under high vacuum. The breakdown that occurs in this gap initiates an electrical vacuum discharge. The positive ions from this discharge, which are representative of the sample under study, are then accelerated through to the electrostatic analyzer and finally separated according to their m/e ratios. A monitor collector, placed between the electrostatic and magnetic analyzers (Figure 17-1) intercepts a small portion of the ion current and provides a measure of the intensity of the ion beam. The product of this ion current and the time of measurement gives the *exposure* for that particular run (in nanocoulombs, nC). Usually a series of about fourteen graded exposures (0.0001 to 100 nC) per sample are recorded on a photographic plate (Figure 17-2).

With the radio frequency spark source, crystal lattices are broken and the

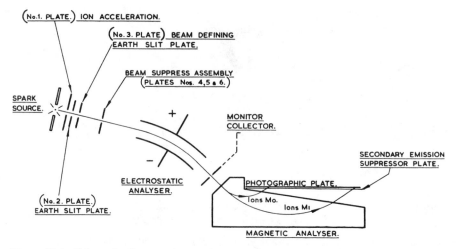

Figure 17–1 Schematic diagram of a spark source mass spectrometer. [Courtesy of A.E.I.].

elements are vaporized and ionized. Very few fragments containing more than one atom remain with the exception of some oxide and multi-carbon atom fragments. All elements are supposedly ionized with approximately equal sensitivity, within a factor of three. Two factors must be taken into consideration; first, that an ion beam with a large energy spread is produced, and second, that the output from such a spark is somewhat variable. The large energy spread is overcome by using a double focusing instrument which gives the required resolution (3000), and the output variation from the spark source is corrected by using an integrator to measure the ion beam intensities. Advantages of the photoplate method of detection are its simple integrating properties, the fact that it measures all ions at once, and the convenient record of the spectrum obtained.

17–3 SAMPLE HANDLING

In the case of drillings taken from a metallic sample, electrodes (20×2 mm) can easily be formed by compressing such drillings in a specially designed polytetrafluoroethylene (PTFE) slug. Lack of sufficient material to form a whole electrode can be overcome by "topping up" with the necessary amount of ultra pure silver powder. In many cases tipped electrodes such as these containing only a few milligrams can be prepared suitable for qualitative analysis.

More recently, in order to achieve very homogeneous electrodes the drillings are dissolved in acid and then added to ultra pure graphite powder. The resulting slurry is then evaporated to dryness and electrodes prepared in the normal manner. This new and novel technique can be used to prepare any concentration of the major components desired and so estimate them as if they were trace elements in the carbon matrix.

Electrodes made by compressing nonmetallic samples cannot be used because of their low electrical conductivity. The technique of incorporating a conducting powder must therefore be adopted. Ultra pure (99.999%) silver powder or graphite and the sample are mixed together in a PTFE capsule, containing a pestle of the same material, by means of a vibratory mixing machine. The resulting mixture is then compressed in the specially designed PTFE slug to form a solid electrode.

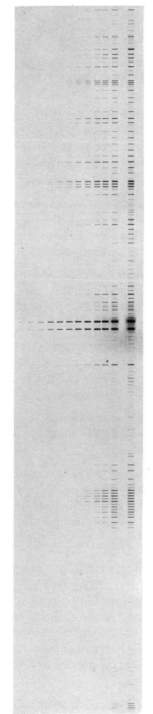

Figure 17–2 Exposed and developed plate for spark source mass spectrum.

Electrodes produced by this method have satisfactory mechanical strength and can be handled without difficulty. Furthermore, the use of enriched isotopes as internal standards (also added to the carbon as an acidic solution) has greatly improved the accuracy of the analysis. These new methods of sample preparation have greatly increased the accuracy for both the major and trace elements.

17-4   INTERPRETATION OF SPECTRA

Densitometer traces at the 3 nC level of exposure of a bronze sample are illustrated in Figures 17–3a, b, and c. This spectrum will be used to illustrate the salient features of a spark source mass spectrum.

The spectral lines in a spark source mass spectrum are caused by the following types of ions:

1. *Singly charged ions.* The most intense lines in the spectrum of a sample are caused by singly charged ions of the major component isotopes. Thus in Figures 17–3a and b, the most intense lines (i.e., peak heights or areas) are lines due to $^{56}Fe^+$, $^{63}Cu^+$, $^{65}Cu^+$, and $^{120}Sn^+$.

However, as the exposure is increased, lines due to the less abundant components appear. At the 3 nC level of exposure trace elements at the concentration level of 10 ppm are clearly visible; for example, bismuth at m/e 209 (Figure 17–3c).

2. *Multiply charged ions.* Lines caused by multiply charged ions are of much lesser intensity than those due to the singly charged species. In Figure 17–3a and 3b lines caused by multiply charged ions can be seen—Cu^{2+} at m/e 31.5 and 32.5; Pb^{2+} at m/e 103 and 104.

The intensity of multiply charged ions decreases with each degree of ionization (i.e., Pb^{3+} is of less intensity than Pb^{2+}) and the ratio of intensities depends on the prevailing spark conditions.

3. *Polyatomic ions.* Polyatomic ions are usually a minor feature in most mass spectra, but carbon and silicon are exceptions to the rule.
4. *Complex ions.* Complex ions, composed of two or more different elements, are occasionally observed, particularly the types XO^+ and XOH^+.

17-5   QUALITATIVE IDENTIFICATION OF ELEMENTS

Elements can usually be identified by their characteristic masses and isotopic patterns. In Figures 17–3b and 3c the elements lead, tin, molybdenum and copper may be identified by their characteristic groups of isotopes. The relative abundances of naturally occurring isotopes of these elements are given in Figure 17–4.

An element is reported as being present only when at least one of the following conditions is satisfied:

1. Lines corresponding to ions of at least two isotopes of the element in their correct relative intensity ratio are observed.
2. Lines due to multiply charged ions are observed at one-half or one-third of the masses of the major isotope. This condition is particularly stringent when lines due to such multiply charged ions occur at fractional masses.

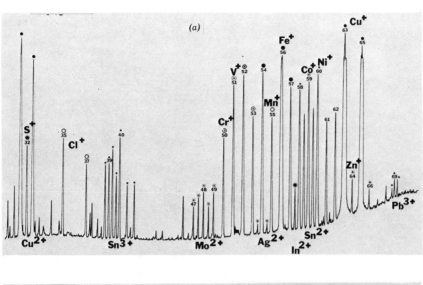

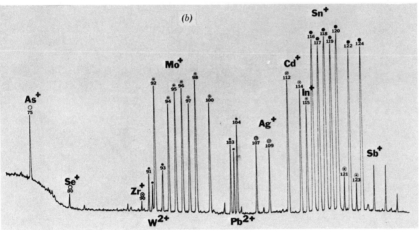

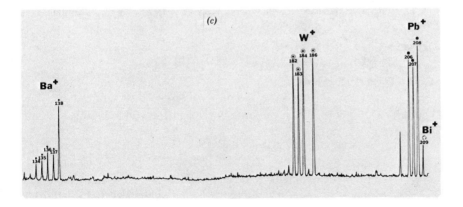

Figure 17–3 Spark source mass spectrum of a bronze sample at 3-nC level of exposure. Densitometer traces from photoplate recording.

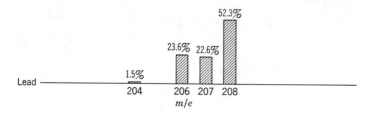

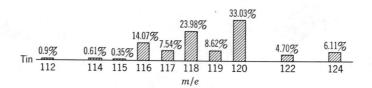

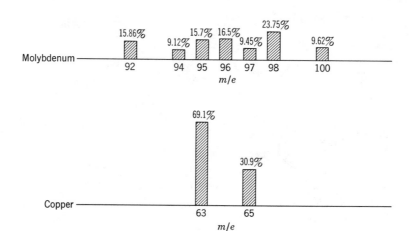

Figure 17–4   Relative abundances of naturally occurring isotopes of lead, tin, molybdenum, and copper.

17–6   QUANTITATIVE ANALYSIS

The approach adopted can be summarized in the following basic equation:

$$C_E = C_S \times \frac{\mathrm{Exp}_S}{\mathrm{Exp}_E} \times \frac{I_S}{I_E} \times \frac{1}{R} \tag{17-1}$$

where C_E = content of element E in electrode analyzed (ppm atomic, ppma)

C_S = content of a second element S, for example, internal standard (ppma)

Exp_S = exposure (in nC) required to give a line of chosen predetermined density (or peak height) for a chosen isotope of element S on a photoplate

Exp_E = exposure (in nC) required to give a line of same density for a chosen isotope of element E on the same photoplate

I_S = isotopic abundance of chosen isotope of S

I_E = isotopic abundance of chosen isotope of E

R = relative sensitivity factor—introduced as a measure of sensitivity of total recording procedure to the line of a given amount of element E used compared with sensitivity to line of the same amount of element S used, which is arbitrarily assigned a value of unity. Different sensitivities arise from different degrees of ionization of the elements depending on atomic weight, volatility, etc.

The exposures required to give the chosen density are obtained from a plot of exposure versus density (measured as peak height by a densitometer) as shown in Figure 17–5. Here the chosen density corresponds to a peak height of 90 mm, and the required exposures are read from the graph—all at the 90-mm height line (or any other chosen height line). This procedure ensures that any variables due to different line widths and shapes are cancelled out since the two elements or lines

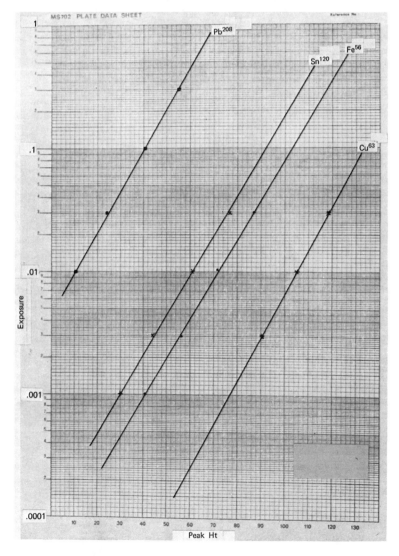

Figure 17–5 Plate data sheet (exposure versus density) for a bronze sample.

representing these elements are being compared at exactly the same integrated density on the plate. Isotopic abundances can be obtained for all elements from published tables. Relative sensitivity factors are predetermined using known standards.

Example/Problem 17–1. Calculate the weight % of the trace element Mo in a bronze sample known to contain 90 weight % Cu. An exposure of 100 nC for ^{63}Cu gives the same photoplate density as an exposure of 0.01 nC for ^{98}Mo. $R = 0.75$ for Mo with respect to Cu.

Solution. Equation 17–1 applies to concentrations given in ppma, and there is no way to convert ppma to weight % without knowing the complete composition of the sample. In this case Cu is the major constituent (90% by weight) and it will cause only a small error in the Mo result (a trace element) if we assume that $C_{Cu} = 9 \times 10^5$ ppma. It then follows from Equation 17–1:

$$C_{Mo} = C_{Cu} \times \frac{Exp_{Cu}}{Exp_{Mo}} \times \frac{I_{Cu}}{I_{Mo}} \times \frac{1}{R_{Mo}}$$

$$= 9 \times 10^5 \times \frac{0.01}{100} \times \frac{69.1}{23.75} \times \frac{1}{0.75} = 349 \text{ ppma}$$

The answer obtained gives the relative number of atoms of Mo and Cu in the sample. The weight % Mo can then be derived from the weight % Cu as follows:

$$\text{weight \% Mo} = \text{weight \% Cu} \times \frac{C_{Mo} \times \text{At. Wt. Mo}}{C_{Cu} \times \text{At. Wt. Cu}}$$

$$= 90 \times \frac{349}{9 \times 10^5} \times \frac{95.94}{63.55} = 0.053\%$$

The method used in Example 17–1 will lead to unacceptably large errors if applied to other major constituents. Equation 17–1 gives only the *relative* number of atoms of the two elements.

PROBLEMS

17–1. Calculate the ppma and the weight % of lead in the bronze sample represented by Figure 17–5, given that the weight % Cu is 85.7% and $R_{Pb} = 0.5$.   *Ans.* 0.41% Pb.

REFERENCES

A. J. Ahearn, Ed., *Mass Spectrometric Analysis of Solids*, Elsevier Publishing Co., New York, 1966. Basic introduction to spark source theory and techniques.

A. J. Ahearn, Ed., *Trace Analysis by Mass Spectrometry*, Academic Press, New York, 1972. Comprehensive review and appraisal of all aspects of trace analysis by mass spectrometry.

ELECTROANALYTICAL CHEMISTRY

18

ELECTROCHEMICAL CELLS AND POTENTIOMETRY

18–1   INTRODUCTION

Chemical reactions in which one or more electrons are transferred from one species to another provide the basis for electroanalytical chemistry. As the name implies, we will be dealing with electrical quantities, such as volts, ohms, amperes, and their relationship to chemical and physical properties of the species in a given system. Electrical quantities can be measured precisely and instantaneously with readily available instrumentation that may be automated or used to control the reaction being investigated. We are generally restricted to systems containing an electrolyte and a pair of electrodes (an electrochemical cell). Fortunately, a wide variety of reactions can be studied in electrochemical cells. In this chapter we examine a number of kinds of electrodes, the factors that determine their potentials, and some of the factors that determine the current flowing through the cell. Although most of the examples we consider might be called "inorganic," the same principles are applicable to "organic" examples. Water, because of relative inertness and high dielectric constant, is the most satisfactory solvent, but some organic solvents or mixed solvents are nearly as good. By far the most common use of electrochemical cells in chemical analysis is the determination of the concentration of one of the species in a solution, for example, the determination of hydrogen ion with the pH meter.

There are many ways in which an electrochemical cell can be constructed and operated giving rise to a variety of techniques for special purposes. We cannot cover all those in use, but some of the more important are:

Potentiometry: the potential of an electrode is used to measure the concentration of a given species in a nonreacting system.

Potentiometric titration: the concentration of a given species that changes during a titration is measured by potentiometry.

Electrolysis: the amount of a substance present is determined by reducing or oxidizing it at an electrode, measuring either changes in weight of the electrode (electrodeposition) or the number of coulombs required to complete the reaction (coulometry).

Coulometry: an electrolysis is performed at constant current or with one of the electrodes at a fixed potential and the number of coulombs measured.

Voltammetry: an electrolysis is performed in an unstirred solution and the current is measured as a function of the applied potential, or vice versa.

Polarography: a voltammetric method in which one of the electrodes is an easily polarized micro-electrode (usually a dropping mercury electrode) and the current is measured as a function of the applied potential and time.

Chronopotentiometry: an electrolysis is performed in an unstirred solution at constant current and the potential of an easily polarized electrode is measured as a function of time.

Conductimetry: the electrical resistance of a solution is used as a measure of the total ionic concentration in the solution.

Although some of these methods can no longer be considered "modern," applications in the areas of organic chemistry, biochemistry, and pollution monitoring continue to grow.

18–2   ELECTROCHEMICAL CELLS

The reaction between zinc metal and ferric ion is a typical redox reaction which can be carried out by dipping a piece of zinc metal in a solution of ferric ion, or by setting up an electrochemical cell as in Figure 18–1 in which this reaction takes place. One half-cell contains a solution of a zinc salt and a piece of metallic zinc; the other contains a solution of ferrous and ferric ion and a piece of platinum metal that serves as an inert electrode to supply or withdraw electrons from the solution.

In the zinc half-cell, zinc metal may dissolve as zinc ion, leaving the electrons behind to give the electrode a negative charge. Alternatively, zinc ion may deposit as zinc metal on the electrode, causing a deficiency of electrons on the electrode. For a given concentration of zinc ion, an equilibrium situation is established with a resultant charge on the electrode. Likewise in the ferrous-ferric half-cell, ferrous and ferric ions can exchange electrons with the platinum electrode establishing an equilibrium and a resultant charge on the platinum electrode. In this cell, the

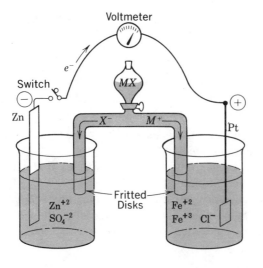

Figure 18–1   Schematic representation of Cell 1.

charge on the zinc electrode is more negative than the charge on the platinum electrode. If the two electrodes are connected with a wire, electrons must flow from zinc to platinum and the reaction just described presumably will occur. However, we cannot increase the concentration of zinc ion or change the initial ratio of ferrous to ferric ion without disturbing the electroneutrality of the solutions. Since we cannot accumulate electrical charge in either solution, the reaction cannot continue unless we provide a mechanism for transferring ions from one half-cell to the other. The salt bridge serves this function by allowing a small ion current to pass from one half-cell to the other in either direction with only negligible mixing of the two electrolytes. Some complications introduced by the salt bridge will be discussed later under liquid junction potentials.

When this cell operates spontaneously: (a) zinc metal is oxidized and dissolves, (b) electrons are transferred from the zinc electrode to the platinum electrode through the wire, (c) ferric ion is reduced by the electrons on the platinum electrode, (d) cations migrate from the zinc half-cell to the ferrous-ferric half-cell through the salt bridge, and (e) anions migrate in the reverse direction. By means of appropriate switches and resistances in the external circuit we can start or stop the cell reaction or control its rate. By means of appropriate meters we can measure its rate and the driving force (change in free energy) tending to make it go. With modern oscilloscopic and pulse techniques, we can also study the mechanisms and rates of fast electrode reactions.

The operation of a cell clearly requires a chemical reaction to take place. Thus the concentrations of the various species will necessarily change with time. For the most part we will be concerned only with cells that are not operating, that is, essentially no current is flowing. It is important to note that the charges on the electrodes are the result of chemical reactions; however, the amount of chemical reaction required to establish the charge is so slight that it cannot be measured and is safely neglected when compared to the total material in the cell. For example, a free charge of 10^{-17} mole of electrons will cause a potential of 1 volt at a distance of 1 cm. In these nonoperating cells, we shall assume that each half-cell reaction is *in equilibrium with its own electrode*. A difference in potential between two electrodes implies that the two half-cell reactions are not in equilibrium with each other. In fact, the potential difference measures the force tending to drive the system toward equilibrium.

Some Typical Electrodes. In shorthand style, a half-cell is designated by listing the species taking part in the half-reaction, separating the electrode from the electrolyte by a vertical bar. A few examples will illustrate some of the conventions.

1. Metal-metal ion, in which the metal of the electrode takes part in the half-reaction:

$$Zn|Zn^{+2} \text{ (as illustrated in Figure 18–1)}$$

2. Metal-complex ion, in which an excess of the complexing agent is added to regulate the particular form of the complex:

$$Cu|Cu(NH_3)_4^{+2}, NH_3$$

3. Metal-saturated solution of one of its salts, in which the presence of the solid salt is inconsequential except to insure that the solution is saturated:

$$Ag|AgCl, Cl^-$$

4. Gas-ion, in which the solution is saturated at a given pressure with the gas as

it is bubbled over the surface of an inert metal electrode:

$$\text{Pt}, \text{H}_2 | \text{H}^+$$

5. Ion-ion, in which both the reduced and oxidized forms are soluble in the electrolyte, and an inert metal electrode (usually platinum) is used to make electrical contact between the electrolyte and the external circuit:

$$\text{Pt} | \text{Fe}^{+3}, \text{Fe}^{+2} \text{ (as illustrated in Figure 18–1)}$$

6. Amalgam-ion, in which a reactive metal is amalgamated so that direct chemical reaction with the solvent is minimized:

$$\text{Na(Hg)} | \text{Na}^+$$

The order in which substances in the electrolyte are listed is irrelevant. When necessary to specify concentrations, this may be done within parentheses:

$$\text{Pt} | \text{Cr}^{+3}(0.5\ M), \text{Cr}_2\text{O}_7^{-2}(0.1\ M), \text{H}^+(1.5\ M)$$

Finally, if two half-cells are combined to make a whole cell, the electrodes are listed first and last; for example, the cell in Figure 18–1 would be given as:

Cell 1: $\text{Zn} | \text{Zn}^{+2} \| \text{Fe}^{+3}, \text{Fe}^{+2} | \text{Pt}$

The double bar signifies the salt bridge between the two half-cells. Unless otherwise indicated, the solvent is assumed to be water and it is not listed as one of the cell ingredients even though it may take part in the reaction. In so far as practical, we shall write the predominant form of a species in the solution; for example, Fe^{+3}, Fe(OH)^{+2}, FeCl_2^+, HOAc, OAc^-, etc.

The following examples illustrate some typical cells and cell reactions. The direction in which half-reactions and the corresponding complete reaction are to be written is arbitrary and we shall return to this question shortly in the discussion of sign conventions. It should be apparent that for every cell there is a specific cell reaction. As a corollary, for every chemical reaction there should be a corresponding electrochemical cell, if we are clever enough to devise suitable electrode systems. In all of the examples there is oxidation and reduction at the electrodes, although some of the cell reactions would not normally be recognized as redox reactions.

Cell 2: $\text{Pt}, \text{H}_2 | \text{H}^+ \| \text{Cu}^{+2} | \text{Cu}$
                   Right: $\text{Cu}^{+2} + 2e^- \rightleftharpoons \text{Cu}$
                   Left: $2\text{H}^+ + 2e^- \rightleftharpoons \text{H}_2$
                   Cell: $\text{H}_2 + \text{Cu}^{+2} \rightleftharpoons \text{Cu} + 2\text{H}^+$

Cell 3: $\text{Pt} | \text{Fe}^{+2}, \text{Fe}^{+3} \| \text{Cr}^{+3}, \text{Cr}_2\text{O}_7^{-2}, \text{H}^+ | \text{Pt}$
                   Right: $\text{Cr}_2\text{O}_7^{-2} + 14\text{H}^+ + 6e^- \rightleftharpoons 2\text{Cr}^{+3} + 7\text{H}_2\text{O}$
                   Left: $\text{Fe}^{+3} + e^- \rightleftharpoons \text{Fe}^{+2}$
                   Cell: $\text{Cr}_2\text{O}_7^{-2} + 6\text{Fe}^{+2} + 14\text{H}^+ \rightleftharpoons 2\text{Cr}^{+3} + 6\text{Fe}^{+3} + 7\text{H}_2\text{O}$

Cell 4: $\text{Pt}, \text{H}_2 | \text{OH}^- \| \text{H}^+ | \text{H}_2, \text{Pt}$
                   Right: $2\text{H}^+ + 2e^- \rightleftharpoons \text{H}_2$
                   Left: $2\text{H}_2\text{O} + 2e^- \rightleftharpoons 2\text{OH}^- + \text{H}_2$
                   Cell: $\text{H}^+ + \text{OH}^- \rightleftharpoons \text{H}_2\text{O}$

Cell 5: $\text{Ag} | \text{AgCl}, \text{Cl}^- \| \text{Cl}^-, \text{Hg}_2\text{Cl}_2 | \text{Hg}$
                   Right: $\text{Hg}_2\text{Cl}_2 + 2e^- \rightleftharpoons 2\text{Hg} + 2\text{Cl}^-$
                   Left: $\text{AgCl} + e^- \rightleftharpoons \text{Ag} + \text{Cl}^-$
                   Cell: $2\text{Ag} + \text{Hg}_2\text{Cl}_2 \rightleftharpoons 2\text{AgCl} + 2\text{Hg}$

Cell 6:   Pt, $H_2|H^+\|Cl^-|Cl_2$, Pt
            Right: $Cl_2 + 2e^- \rightleftharpoons 2Cl^-$
            Left: $2H^+ + 2e^- \rightleftharpoons H_2$
            Cell: $H_2 + Cl_2 \rightleftharpoons 2H^+ + 2Cl^-$

Cell 7: $Cu|CuY^{-2}, H_2Y^{-2}, H^+\|Cu^{+2}|Cu$
            Right: $Cu^{+2} + 2e^- \rightleftharpoons Cu$
            Left: $CuY^{-2} + 2H^+ + 2e^- \rightleftharpoons Cu + H_2Y^{-2}$
            Cell: $Cu^{+2} + H_2Y^{-2} \rightleftharpoons CuY^{-2} + 2H^+$

18–3   ELECTRODE POTENTIALS

The *charge* on an electrode results from an excess or deficiency of electrons on the metal. A large negative charge indicates the presence of a strong reducing agent (good electron donor).

The *potential* of an electrode is defined in an electrostatic sense. The absolute electric potential of a point is defined as the work needed to bring a unit of positive charge from an infinite distance in space to the point in question. Needless to say this is not a very practical measurement to make in the laboratory. We shall be concerned only with *potential differences* between two points; namely, the two electrodes of a cell. We should note that the point with the higher electric potential also has the higher positive charge. The potential difference thus compares the charge existing at two points, a relative measurement. We often use the term, "potential of an electrode," when we mean *potential difference relative to an arbitrary standard reference electrode*. Likewise, a "cell potential" is really the difference in potential between the two electrodes, and is more properly called the *cell voltage* or electromotive force (e.m.f. or EMF). We shall designate this term as E_{cell} rather than by ΔE.

Sign Conventions.   The charge on an electrode, the electrode potential, the difference between two electrode potentials (cell voltage) and the direction of electron flow in a wire connecting the electrodes are all physical facts, independent of any arbitrary sign conventions. Nevertheless, since it is impossible to measure the charge or potential of a single electrode, we must select a consistent set of rules for writing reactions and computing electrode potentials. With more experience, there is no problem in working with any convention, in fact several are in common use. Some of these are discussed by Laitinen and by Licht and deBethune in the references given at the end of this chapter. For this text, we shall use only a minimum of arbitrary rules:

1. The potential of and charge on an electrode and the driving force of the corresponding half-reaction have the same numerical value and sign. Thus, the electrode potential, designated by the symbol E, is also a measure of:
   (a) The charge on the electrode.
   (b) The tendency of the system to accept electrons, that is, its power to oxidize some other species.
2. E for an electrode or a half-reaction is compared to and measured relative to E for the standard hydrogen electrode (SHE) to be described shortly. Thus the charge on the electrode and the oxidizing power of a system may be either positive or negative relative to the SHE.
3. In every cell, one electrode is arbitrarily designated as the left electrode, E_L, and the other as the right electrode, E_R.

4. The cell voltage, E_{cell}, is the algebraic difference between the two electrodes, always taken in the direction: $E_R - E_L = E_{cell}$.

5. All half-reactions are written as reductions (electrons on the left).

6. The cell reaction is obtained by subtracting the left half-cell reaction from the right. It should be evident that the cell reaction cannot produce or consume electrons, it can only transfer them. If the number of electrons, n, is not the same in the two half-reactions, one or both half-reactions must be multiplied by a suitable factor to cancel electrons. Note that multiplying or dividing a half-reaction by a numerical factor does not change the electrode potential, a property which is independent of how the half-reaction is written.

7. A positive cell voltage indicates that:

   (a) The cell reaction proceeds *spontaneously* in the direction written.

   (b) The right electrode has a more positive (or less negative) potential than the left.

   (c) Reduction occurs in the right half-cell and oxidation in the left when the cell operates spontaneously.

   (d) Electrons flow from left to right in the external circuit.

8. A negative cell voltage corresponds to statements opposite to those just given.

The computation of cell voltages is aided by plotting and labeling the electrode potentials on a voltage scale as in Figure 18–2 for the example cells given on page 364. In this figure, it is assumed that all concentrations are 1 M and all gas pressures are 1 atm.

Standard Hydrogen Electrode. As we stated above, no one has yet measured a

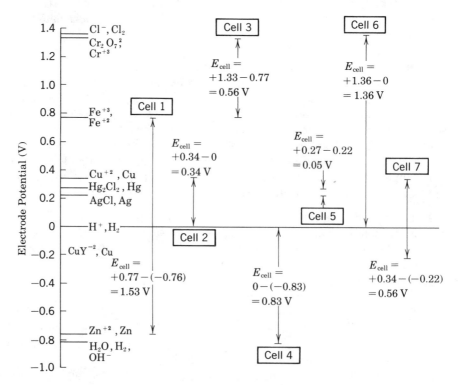

Figure 18–2 Electrode potentials and cell voltages for Example Cells 1 to 7.

single electrode potential. Nor is it likely that anyone will ever do so, because a second electrode is always necessary in order to complete the electrical circuit. The voltmeter (or other voltage measuring device) can measure only *differences* in electrode potentials. But we can assign an arbitrary value to a selected electrode. If this so-called "primary standard" electrode is connected with some other electrode, we can then measure the potential of the second electrode with respect to the standard. Chemists have generally agreed to the arbitrary assignment of zero volt to the standard hydrogen electrode (SHE). Actually, this is a hypothetical electrode which is illustrated in Figure 18–3. As we will see later, we cannot give directions for preparing a solution with a hydrogen ion of unit *activity*. In practice, the operation of obtaining a standard potential is performed by an extrapolation technique as described on page 374.

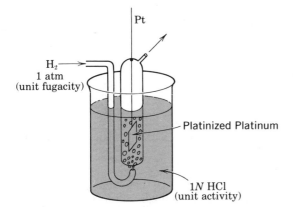

Figure 18–3 The (hypothetical) standard hydrogen electrode.

Any other electrode, once it has been calibrated against the SHE, will serve as a secondary standard. The silver–silver chloride electrode and the saturated calomel electrode are often used as reference electrodes.

Effect of Concentration. Let us consider the ferrous-ferric electrode in more detail. In Figure 18–4, the ferric ions and ferrous ions are in continuous motion and collide frequently with each other and with the electrode surface. For the ferric ions which reach the electrode surface, there is a finite probability that they will "pick up" an electron and become ferrous ions. The degree of the probability depends on the electron affinity of the ferric ion and the activity of electrons at the electrode surface. Similar considerations apply to the ferrous ions. Both processes occur simultaneously and for the most part cancel each other out. However, the equilibrium condition can be described by the ratio of the activities of the two ionic species and the activity of electrons at the surface of the electrodes. The latter is more easily described as the charge or potential of the electrode. The mathematical expression that represents the equilibrium situation is called the *Nernst equation*:

$$E = E^{\circ}_{Fe^{+3},\,Fe^{+2}} - \frac{RT}{nF} \ln \frac{(Fe^{+2})}{(Fe^{+3})} \tag{18-1}$$

in which $E^{\circ}_{Fe^{+3},\,Fe^{+2}}$ is a constant at a given temperature known as the *standard potential*, R the gas constant, T the absolute temperature, n the number of electrons transferred in the half-reaction and F the Faraday constant. The Nernst

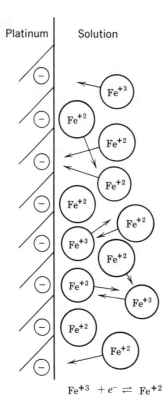

Platinum | Solution

$$Fe^{+3} + e^- \rightleftharpoons Fe^{+2}$$

Figure 18–4 The dynamic equilibrium at the ferrous-ferric electrode.

equation can be derived from thermodynamic considerations or it can be obtained empirically as the best fit to the experimental data.

If the activity of Fe^{+2} is increased, there is a greater probability for their oxidation at the electrode surface, a higher density of electrons accumulates, and the electrode becomes more negative. The opposite is true for an increase in the ferric ion activity. However, it is the *ratio* of activities that is important.

Standard Potentials. The constant term, $E°$, is the observed potential of the electrode when the activities of the two species, ferric and ferrous ion, are both unity; in which case the log term in Equation 18–1 is zero. $E°$ expresses the inherent tendency of the system to accept electrons from the electrode, and its value depends on the nature of the substances in the half-reaction. More accurately stated, the standard potential is the measured potential of an electrode (relative to the SHE) when all participating species are in their "standard states." Customarily, standard states for electrochemistry are defined as follows:

1. For all dissolved substances, activity equal to one molar.
2. For all gases, partial pressure equal to one atmosphere (more precisely, unit fugacity).
3. For all pure liquids and pure solids, the most stable form at 25°C, which by definition has unit activity.

In principle, standard potentials are measured by means of cells such as cell 2:

$$Pt, H_2|H^+||Cu^{+2}|Cu$$

with all species in their standard states; or by comparing the electrode to any secondary reference electrode which has been previously compared to the SHE.

In practice, this procedure is not as simple as it may seem. Some of the complicating factors to be considered in later sections include: (a) the difficulty of obtaining truly reversible electrode reactions, (b) the uncertainty of the junction potential between the two half-cells, and (c) nonideal behavior of solutes. Some $E°$ values cannot be measured at all and are computed indirectly from other kinds of related thermodynamic data. The potentials plotted in Figure 18–2 are, of course, $E°$ values. A more complete list is given in Table 18–1. Extensive lists are available in the References by Latimer and by Clark, and in handbooks.

Use of the Nernst Equation.   It is convenient to combine the constant factors in the log term of Equation 18–1 and convert natural logarithms (base e) to common logarithms (base 10). The simplified form of the Nernst equation is:

$$E = E° - \frac{0.059}{n} \log \frac{(\text{products})}{(\text{reactants})} \tag{18–2}$$

where 0.059 is valid for 25°C, 0.060 at 30°C, etc. The log term is set up much like the mass action expression for the reaction, but the numerical value is *not* equal to the equilibrium constant. Nernst equations for some of the electrodes used in cells 1 to 7 will serve as examples.

Table 18–1   Standard Electrode Potentials in Aqueous Solutions

$E°$, V	Couple
2.65	$F_2 + 2e^- \rightleftharpoons 2F^-$
2.07	$O_3 + 2H^+ + 2e^- \rightleftharpoons O_2 + H_2O$
1.77	$H_2O_2 + 2H^+ + 2e^- \rightleftharpoons 2H_2O$
1.695	$MnO_4^- + 4H^+ + 3e^- \rightleftharpoons MnO_2 + 2H_2O$
1.61	$Ce^{+4} + e^- \rightleftharpoons Ce^{+3}$
1.6	$H_5IO_6 + H^+ + 2e^- \rightleftharpoons IO_3^- + 3H_2O$
1.52	$BrO_3^- + 6H^+ + 5e^- \rightleftharpoons \frac{1}{2}Br_2 + 3H_2O$
1.51	$MnO_4^- + 8H^+ + 5e^- \rightleftharpoons Mn^{+2} + 4H_2O$
1.36	$Cl_2 + 2e^- \rightleftharpoons 2Cl^-$
1.33	$Cr_2O_7^{-2} + 14H^+ + 6e^- \rightleftharpoons 2Cr^{+3} + 7H_2O$
1.23	$MnO_2 + 4H^+ + 2e^- \rightleftharpoons Mn^{+2} + 2H_2O$
1.229	$O_2 + 4H^+ + 4e^- \rightleftharpoons 2H_2O$
1.195	$IO_3^- + 6H^+ + 5e^- \rightleftharpoons \frac{1}{2}I_2 + 3H_2O$
1.065	$Br_2 + 2e^- \rightleftharpoons 2Br^-$
1.06	$ICl_2^- + e^- \rightleftharpoons \frac{1}{2}I_2 + 2Cl^-$
1.00	$VO_2^+ + 2H^+ + e^- \rightleftharpoons VO^{+2} + H_2O$
0.87	$C_6H_5NO_2 + 7H^+ + 6e^- \rightleftharpoons C_6H_5NH_3^+ + 2H_2O$
0.799	$Ag^+ + e^- \rightleftharpoons Ag$
0.789	$Hg_2^{+2} + 2e^- \rightleftharpoons 2Hg$
0.771	$Fe^{+3} + e^- \rightleftharpoons Fe^{+2}$
0.73	$C_2H_2 + 2H^+ + 2e^- \rightleftharpoons C_2H_4$
0.699	$O{=}C_6H_4{=}O + 2H^+ + 2e^- \rightleftharpoons HOC_6H_4OH$
0.682	$O_2 + 2H^+ + 2e^- \rightleftharpoons H_2O_2$
0.586	$CH_3OH + 2H^+ + 2e^- \rightleftharpoons CH_4 + H_2O$
0.564	$MnO_4^- + e^- \rightleftharpoons MnO_4^{-2}$
0.536	$I_3^- + 2e^- \rightleftharpoons 3I^-$
0.5355	$I_2 + 2e^- \rightleftharpoons 2I^-$
0.521	$Cu^+ + e^- \rightleftharpoons Cu$

Table 18–1 (*continued*)

$E°$, V	Couple
0.52	$C_2H_4 + 2H^+ + 2e^- \rightleftharpoons C_2H_6$
0.44	$HOOCCH=CHCOOH + 2H^+ + 2e^- \rightleftharpoons HOOCCH_2CH_2COOH$
0.361	$VO^{+2} + 2H^+ + e^- \rightleftharpoons V^{+3} + H_2O$
0.36	$Fe(CN)_6^{-3} + e^- \rightleftharpoons Fe(CN)_6^{-4}$
0.337	$Cu^{+2} + 2e^- \rightleftharpoons Cu$
0.31	$H_2C_2O_4 + 6H^+ + 6e^- \rightleftharpoons CH_3COOH + 2H_2O$
0.2676	$Hg_2Cl_2 + 2e^- \rightleftharpoons 2Hg + 2Cl^-$
0.222	$AgCl + e^- \rightleftharpoons Ag + Cl^-$
0.20	$CH_3COCOOH + 2H^+ + 2e^- \rightleftharpoons CH_3CHOHCOOH$
0.192	$CH_3CHO + 2H^+ + 2e^- \rightleftharpoons C_2H_5OH$
0.19	$HCHO + 2H^+ + 2e^- \rightleftharpoons CH_3OH$
0.17	$S_4O_6^{-2} + 2e^- \rightleftharpoons 2S_2O_3^{-2}$
0.17	$SO_4^{-2} + 4H^+ + 2e^- \rightleftharpoons H_2SO_3 + H_2O$
0.15	$Sn^{+4} + 2e^- \rightleftharpoons Sn^{+2}$
0.152	$Sb_2O_3 + 6H^+ + 6e^- \rightleftharpoons 2Sb + 3H_2O$
0.153	$Cu^{+2} + e^- \rightleftharpoons Cu^+$
0.10	$TiO^{+2} + 2H^+ + e^- \rightleftharpoons Ti^{+3} + H_2O$
0.095	$AgBr + e^- \rightleftharpoons Ag + Br^-$
0.056	$HCOOH + 2H^+ + 2e^- \rightleftharpoons HCHO + H_2O$
0.000	$2H^+ + 2e^- \rightleftharpoons H_2$
−0.118	$CH_3COOH + 2H^+ + 2e^- \rightleftharpoons CH_3CHO + H_2O$
−0.126	$Pb^{+2} + 2e^- \rightleftharpoons Pb$
−0.13	$CrO_4^{-2} + 4H_2O + 3e^- \rightleftharpoons Cr(OH)_3 + 5OH^-$
−0.136	$Sn^{+2} + 2e^- \rightleftharpoons Sn$
−0.151	$AgI + e^- \rightleftharpoons Ag + I^-$
−0.196	$CO_2 + 2H^+ + 2e^- \rightleftharpoons HCOOH$
−0.255	$V^{+3} + e^- \rightleftharpoons V^{+2}$
−0.276	$H_3PO_4 + 2H^+ + 2e^- \rightleftharpoons H_3PO_3 + H_2O$
−0.403	$Cd^{+2} + 2e^- \rightleftharpoons Cd$
−0.41	$Cr^{+3} + e^- \rightleftharpoons Cr^{+2}$
−0.44	$Fe^{+2} + 2e^- \rightleftharpoons Fe$
−0.49	$2CO_2 + 2H^+ + 2e^- \rightleftharpoons H_2C_2O_4$
−0.50	$H_3PO_3 + 2H^+ + 2e^- \rightleftharpoons H_2PO_2 + H_2O$
−0.56	$Fe(OH)_3 + e^- \rightleftharpoons Fe(OH)_2 + OH^-$
−0.763	$Zn^{+2} + 2e^- \rightleftharpoons Zn$
−0.828	$2H_2O + 2e^- \rightleftharpoons H_2 + 2OH^-$
−1.18	$V^{+2} + 2e^- \rightleftharpoons V$
−1.66	$Al^{+3} + 3e^- \rightleftharpoons Al$
−2.25	$\frac{1}{2}H_2 + e^- \rightleftharpoons H^-$
−2.37	$Mg^{+2} + 2e^- \rightleftharpoons Mg$
−2.714	$Na^+ + e^- \rightleftharpoons Na$
−3.045	$Li^+ + e^- \rightleftharpoons Li$

Examples.

Pt, $H_2|H^+$

$$2H^+ + 2e^- \rightleftharpoons H_2$$

$$E = 0 - \frac{0.059}{2} \log \frac{P_{H_2}}{(H^+)^2}$$

$Cu|Cu^{+2}$

$$\dot{C}u^{+2} + 2e^- \rightleftharpoons Cu$$

$$E = E^°_{Cu^{+2}, Cu} - \frac{0.059}{2} \log \frac{1}{(Cu^{+2})}$$

$Pt|Cr^{+3}, Cr_2O_7^{-2}, H^+$ $Cr_2O_7^{-2} + 14H^+ + 6e^- \rightleftharpoons 2Cr^{+3} + 7H_2O$

$$E = E^\circ_{Cr_2O_7^{-2}, Cr^{+3}} - \frac{0.059}{6} \log \frac{(Cr^{+3})^2}{(Cr_2O_7^{-2})(H^+)^{14}}$$

$Pt, H_2|OH^-$ $2H_2O + 2e^- \rightleftharpoons 2OH^- + H_2$

$$E = E^\circ_{OH^-, H_2} - \frac{0.059}{2} \log P_{H_2}(OH^-)^2$$

$Ag|AgCl, Cl^-$ $AgCl + e^- \rightleftharpoons Ag + Cl^-$

$$E = E^\circ_{AgCl, Ag} - 0.059 \log (Cl^-)$$

Example/Problem 18–1. What is the cell voltage for cell 3 if all activities are 0.1 M?

For the ferric-ferrous electrode,

$$E = E^\circ_{Fe^{+3}, Fe^{+2}} - 0.059 \log \frac{(Fe^{+2})}{(Fe^{+3})}$$

$$= +0.77 - 0.059 \log \frac{0.1}{0.1} = +0.77 \text{ V}$$

For the dichromate-chromic electrode,

$$E = E^\circ_{Cr_2O_7^{-2}, Cr^{+3}} - \frac{0.059}{6} \log \frac{(Cr^{+3})^2}{(Cr_2O_7^{-2})(H^+)^{14}}$$

$$= +1.33 - \frac{0.059}{6} \log \frac{(0.1)^2}{(0.1)(0.1)^{14}} = +1.20 \text{ V}$$

$$E_{cell} = 1.20 - 0.77 = 0.43 \text{ V}$$

Note that E_{cell} is very sensitive to (H^+). For example if (H^+) were 0.001 M, other activities remaining 0.1 M, E_{cell} would be 0.14 V.

Activity Coefficients. Strictly speaking, the Nernst equation requires the use of activities rather than concentrations. The activity or effective concentration differs from the actual concentration because of interionic attractions and repulsions, as well as the finite space occupied by the various species and their solvation shells. To some extent, thermal motions modify the electrical effects. The activity coefficient expresses the ratio of activity to concentration

$$f_i = a_i/C_i \quad \text{or} \quad a_i = f_i C_i \tag{18-3}$$

where the subscript "i" stands for any given species. Although some may think f stands for "fudge factor," actually we know a great deal about this factor that measures the departure of real systems from ideal behavior. There is, however, no way to be sure that we have included every possibility since there is no way to measure the activity coefficient of a single ion. Consider the problem of the activity coefficients in the following cell:

Cell 8: $Pt, H_2|H^+, Cl^-, AgCl|Ag$

This cell requires only a single electrolyte with both electrodes immersed in the same solution of HCl—there is no salt bridge. The voltage for this cell is obtained from the two half-cell potentials:

Left: $$E_L = E^\circ_H - \frac{0.059}{2} \log \frac{P_{H_2}}{(H^+)^2}$$

Right:
$$E_R = E^\circ_{AgCl, Ag} - 0.059 \log (Cl^-)$$

$$E_{cell} = E^\circ_{AgCl, Ag} + \frac{0.059}{2} \log P_{H_2} - 0.059 \log (H^+)(Cl^-) \qquad (18\text{–}4)$$

Here we have the *product* of two activities. It is impossible to construct a cell in which there is only one ion in the cell reaction. While we can measure the *concentration* of each ion (for example, by titration or precipitation) we can measure only the *product of their activities* and thus only the product of their activity coefficients. If both ions have the same numerical charge, it is customary to assign the same value to each activity coefficient. This clearly must be the square root of the product, or the geometric mean. The *mean ionic activity coefficient* is thus defined:

$$f_\pm = \sqrt{f_+ \times f_-} \qquad (18\text{–}5)$$

Although the *mean* ionic activity coefficient is the only one we can measure, the Debye-Hückel limiting law provides a means to estimate the value of an *individual* ionic activity coefficient.

$$-\log f_i = 0.512\, Z_i^2 \sqrt{\mu} \qquad (18\text{–}6)$$

where Z_i is the charge on the ion and μ is the ionic strength of the solution (see below). Equation 18–6 as given applies reasonably well to dilute aqueous solutions with concentrations no greater than about 0.05 M. Equation 18–6 has been modified by many workers in an effort to approach the behavior of real systems at higher concentrations. One such equation takes into account the effect of the finite size of the ions.

$$-\log f_\pm = 0.512\, Z_i^2 \left[\frac{\sqrt{\mu}}{1 + Ba\sqrt{\mu}} \right] \qquad (18\text{–}7)$$

where B is a function of the dielectric constant and the temperature ($B = 0.328$ for water at 25°C) and a is an ion size parameter (Table 18–2). Equation 18–7 extends the applicability of the Debye-Hückel limiting law up to concentrations of 0.1 M.

Selected values of individual ionic activity coefficients are given in Table 18–2.

Table 18–2 Single Ion Activity Coefficients

Ion Size, a	Ion	Ionic Strength			
		0.005	0.01	0.05	0.1
9	H^+	0.933	0.914	0.86	0.83
6	Li^+, $C_6H_5COO^-$	0.929	0.907	0.84	0.80
4	Na^+, IO_3^-, HSO_3^-, $H_2PO_4^-$	0.927	0.901	0.82	0.77
3	K^+, Cl^-, Br^-, CN^-, NO_3^-	0.925	0.899	0.81	0.76
2.5	Cs^+, NH_4^+, Ag^+	0.924	0.898	0.80	0.75
8	Mg^{+2}, Be^{+2}	0.755	0.69	0.52	0.45
6	Ca^{+2}, Cu^{+2}, Zn^{+2}, Mn^{+2}, $C_6H_4(COO)_2^{-2}$	0.749	0.675	0.49	0.41
4	Hg_2^{+2}, SO_4^{-2}, CrO_4^{-2}, HPO_4^{-2}	0.740	0.660	0.445	0.355
9	Al^{+3}, Fe^{+3}, Cr^{+3}	0.54	0.445	0.245	0.18
4	PO_4^{-3}, $Fe(CN)_6^{-3}$	0.505	0.395	0.16	0.10
5	$Fe(CN)_6^{-4}$	0.31	0.20	0.05	0.02

[From Kielland, *J. Am. Chem. Soc.*, **59**, 1675 (1937).]

A detailed discussion of ionic activity coefficients is given in physical chemistry texts. For our purposes here, we should note that the values depend on the charge and size of the ion, and on the ionic strength, μ, of the solution which is equal to one-half of the sum of the concentration of each ion multiplied by the square of its charge

$$\mu = \frac{1}{2} \sum C_i Z_i^2 \tag{18–8}$$

The effects of these variables are more apparent in the plot of the data from Table 18–2 in Figure 18–5.

The definition of activity coefficients, Equations 18–3 and 18–5, is used in rewriting Equation 18–4:

$$E_{\text{cell}} = E^\circ_{\text{AgCl, Ag}} + \frac{0.059}{2} \log P_{\text{H}_2} - 0.059 \log C^2_{\text{HCl}} - 0.059 \log f^2_{\pm} \tag{18–9}$$

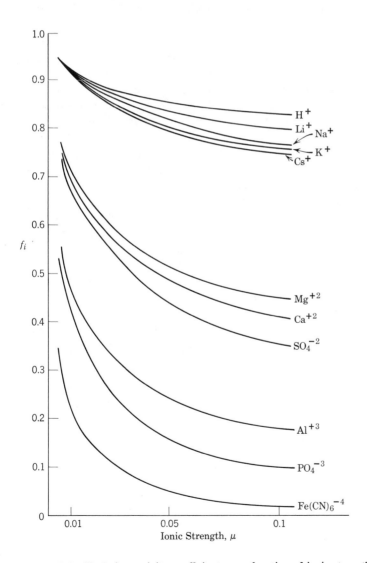

Figure 18–5 Single ion activity coefficients as a function of ionic strength of the solution.

All measurable terms are gathered on the left:

$$E_{cell} - \frac{0.059}{2} \log P_{H_2} + 0.118 \log C_{HCl} = E^{\circ}_{AgCl, Ag} - 0.118 \log f_{\pm} \qquad (18\text{--}10)$$

It is found from experiment that a plot of the left side of Equation 18–10 versus $\sqrt{\mu}$ becomes nearly a straight line at low ionic strength, Figure 18–6. At zero concentration, or zero ionic strength, $\log f_{\pm}$ also becomes zero, and the intercept gives the value of $E^{\circ}_{AgCl, Ag}$. *This is the fundamental method of measuring E° values.*

Now that E° has been determined we can use Equation 18–10 or Figure 18–6 to determine $f_{\pm}$ for HCl at any ionic strength.

For many of the discussions in this text, the distinction between activity and concentration is not important, and we will often use concentration where activity is more properly correct. Where the distinction is important, we will point this out and make such correction as we can. In general, the use of brackets will mean that activities should be used, but that concentrations will suffice. Where a distinction is required, we use the symbols a_i and C_i.

Liquid Junction Potential. Cell 8 is a special type of cell with only a single electrolyte and therefore no liquid junction (salt bridge). Most cells require two electrolytes with a junction between them. This introduces another uncertainty into our measurements. Whenever two solutions of different compositions are in contact, there is a tendency for diffusion of species from one solution to the other in order to equalize the concentrations. In general, the diffusion rates of the various species will not be the same and this will cause a slight separation of charges near the boundary between the two solutions. The electrical potential generated by this separation is called the *junction potential*, E_j. It will, of course, be included in the measured cell potential. In a few simple cases, the magnitude of this junction potential can be calculated from known diffusion coefficients. In general, the problem is too complex to handle. The junction potential can be reduced (but not altogether eliminated) by the use of a saturated solution of KCl in the salt bridge. This is so because the diffusion rates of K^+ and Cl^- are nearly equal. The effects of the diffusion of relatively large and nearly equal amounts of K^+ and Cl^- will overshadow the diffusion of the less concentrated constituents of the half-cells. Even so the uncertainty may be of the order of several millivolts. The use of $\|$ indicates that we have reduced E_j to a minimum and will ignore it.

Reversibility. In the discussion of the Nernst equation, we have tacitly assumed that the electrode system behaves reversibly. By this we mean that the

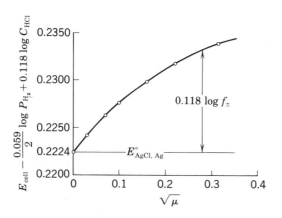

Figure 18–6 Plot of Equation 18–10 to yield the standard potential of the silver–silver chloride electrode and the mean ionic activity coefficient of HCl solutions.

components within the half-cell system are always at equilibrium with the electrode. In other words, an infinitesimal change in the charge on the electrode results in an immediate change in concentrations to maintain the equilibrium; or, conversely, that an infinitesimal change in the concentrations results in an immediate change in charge on the electrode. This represents an ideal system; real systems approach this behavior to varying degrees. A few of the soft metals (Ag, Hg, Pb, Zn, Cu, etc.) with their respective ions meet this criterion very well. Some others behave reversibly if the metal is amalgamated with mercury (Na, Cd, Bi, Tl, etc.). On the other hand, the mechanisms of most electrode reactions are very complicated indeed. The transfer of electrons between the electrode and electro-active species in solution can take place rapidly and reversibly only when the reduced and oxidized forms have essentially identical structures. Any rearrangement in the solvation shell or a transfer of atoms from one species to another will require promotion of the species to a higher energy state (activated complex). Under these circumstances the electrode reaction may take place very slowly and irreversibly or not at all, and the Nernst equation will be a poor approximation of the observed behavior. It is not surprising that many inorganic and most organic electrode reactions are irreversible. Many of these do, however, give reproducible behavior and are useful on an empirical basis.

Formal Potentials. Many redox reactions involve species other than those between which electrons are transferred. For example, many oxidants contain oxygen, requiring that hydrogen or hydroxyl ions and water appear in the reaction. In other cases, complexing agents may be consumed or released. Consider, for example, the reaction between copper(I) and permanganate in hydrochloric acid:

$$10Cl^- + 5CuCl_2^- + MnO_4^- + 8H^+ \rightleftharpoons 5CuCl_4^{-2} + Mn^{+2} + 4H_2O$$

We can imagine this as a cell reaction and write each half-cell potential separately:

$$E_{Mn} = E^\circ_{MnO_4^-, Mn^{+2}} - \frac{0.059}{5} \log \frac{[Mn^{+2}]}{[MnO_4^-][H^+]^8} \qquad (18-11)$$

$$E_{Cu} = E^\circ_{CuCl_4^{-2}, CuCl_2^-} - 0.059 \log \frac{[CuCl_2^-][Cl^-]^2}{[CuCl_4^{-2}]} \qquad (18-12)$$

When we use this reaction, we employ an excess of hydrochloric acid and its concentration stays relatively constant as the reaction proceeds. Under these conditions, $[H^+]$ in Equation 18–11 is a constant and the log term can be separated:

$$E_{Mn} = \underbrace{E^\circ_{MnO_4^-, Mn^{+2}} + \frac{0.059}{5} \log [H^+]^8}_{} - \frac{0.059}{5} \frac{[Mn^{+2}]}{[MnO_4^-]} \qquad (18-13)$$

$$E_{Mn} = \qquad E'_{MnO_4^-, Mn^{+2}} \qquad - \frac{0.059}{5} \log \frac{[Mn^{+2}]}{[MnO_4^-]} \qquad (18-14)$$

Equations 18–13 and 18–14 define the *formal potential, E'*, for this couple. E' is a constant which has a different value for each concentration of hydrogen ion. E' can be described as an effective or practical value of E°, and in this case it is seen that E' becomes more positive by $(8/5) \times 0.059$ V as $[H^+]$ is increased tenfold. Thus changing the pH is a simple and powerful means for altering the oxidizing or reducing power of a reagent. This is a common situation for many inorganic and most organic reactions.

The same argument can be applied to define a formal potential for the copper half-cell which includes the effect of the concentration of Cl^-:

$$E_{Cu} = \underbrace{E^{\circ}_{CuCl_4^{-2},\,CuCl_2^-} - 0.059 \log [Cl^-]^2 - 0.059 \log \frac{[CuCl_2^-]}{[CuCl_4^{-2}]}} \qquad (18\text{--}15)$$

$$E_{Cu} = \qquad\qquad E'_{CuCl_4^{-2},\,CuCl_2^-} \qquad\qquad -0.059 \log \frac{[CuCl_2^-]}{[CuCl_4^{-2}]} \qquad (18\text{--}16)$$

Formal potentials include the effects of ionic strength, complexation, hydrolysis, the pH effect and any other effect which can be attributed to the composition of the electrolyte other than the ratio of reduced to oxidized forms. A few formal potentials are listed in Table 18–3. Unfortunately, there are only a limited number of formal potentials available. The whole idea departs from the beauty of a simple tabulation of fundamental data, but in a practical sense, the answer obtained using a formal potential is more likely to be correct.

Table 18–3   Formal Potentials

Couple	E°, V	E', V		
		1 F HClO$_4$	1 F HCl	1 F H$_2$SO$_4$
V^{+3}/V^{+2}	-0.255	-0.21	—	—
$H^+/\frac{1}{2}H_2$	0.000	0.005	0.005	—
$Fe(CN)_6^{-3}/Fe(CN)_6^{-4}$	0.36	0.72	0.71	0.72
VO^{+2}/V^{+3}	0.361	—	—	0.360
Fe^{+3}/Fe^{+2}	0.771	0.732	0.700	0.68
Ag^+/Ag	0.7991	0.792	0.228	0.77
VO_2^+/VO^{+2}	1.00	1.02	1.02	1.0
Ce^{+4}/Ce^{+3}	1.61	1.70	1.28	1.44

[After Swift, *A System of Chemical Analysis*, Prentice-Hall, 1940; p. 540.]

Biochemical Formal Potentials.   The redox couples of interest to organic chemists, and especially to biochemists, often include hydrogen ion as a reactant or product. The potentials of such systems are thus a function of the pH of the medium, which in the case of biochemical systems is normally close to pH 7. In order to make comparisons of oxidizing or reducing power for biochemical purposes, it is convenient to set up a new table of "standard" potentials adjusted to pH 7. The conversion term is $0.059\,(n_H/n_e)\log[H^+]$, where n_H and n_e are the number of protons and electrons respectively appearing in the half-reaction. In general, $n_H = n_e$, so that the potential becomes approximately 0.41 V more negative (better reducing agents) at pH 7 than at pH 0. Typical "adjusted" potentials, symbolized by E'_7, are found in Table 18–4.

A simple example is the quinone-hydroquinone system:

or

$$Q \qquad\qquad + 2H^+ + 2e^- \rightleftharpoons \qquad\qquad H_2Q$$

Table 18-4 Formal Potentials at pH 7

Couple	$E°$, V	E'_7, V
$H^+/\frac{1}{2}H_2$	0	-0.413
Riboflavin/leucoriboflavin	0.21	-0.20
Pyruvate/lactate	0.22	-0.19
Oxalacetate/malate	0.247	-0.166
Fumarate/succinate	0.444	0.031
Q/H_2Q	0.6994	0.286
O_2/H_2O	1.229	0.816

for which the Nernst equation is:

$$E = E° - 0.059/2 \log [H_2Q]/[Q][H^+]^2 \tag{18-17}$$

or

$$E = \underbrace{E° + 0.059 \log [H^+]}_{E'_7 \text{ if } [H^+] = 10^{-7} M} - 0.059/2 \log [H_2Q]/[Q] \tag{18-18}$$

The quinone-hydroquinone system forms the quinhydrone electrode (see below) which measures pH. This follows from a different combination of terms in Equation 18-17:

$$E = \underbrace{E° - 0.059/2 \log [H_2Q]/[Q]}_{\substack{E' \text{ for quinhydrone electrode} \\ \text{used to measure pH.}}} + 0.059 \log [H^+] \tag{18-19}$$

The pyruvate-lactate couple is another typical example:

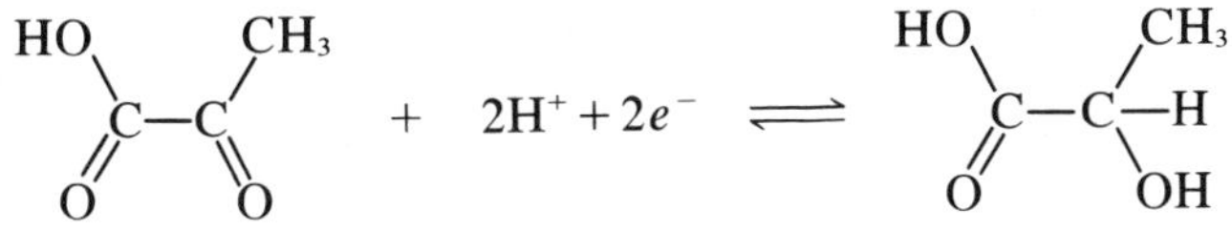

or

$$py \quad + \quad 2H^+ + 2e^- \quad \rightleftharpoons \quad lac$$

for which:

$$E = \underbrace{E° + 0.059/2 \log [H^+]^2}_{E'_7 \text{ if } [H^+] = 10^{-7} M} - 0.059/2 \log [lac]/[py] \tag{18-20}$$

The pyruvate-lactate couple is one of many couples in a long series of redox reactions associated with the metabolism of carbohydrates. Tables of E'_7 data are far more convenient and pertinent to the study of these complex systems.

18-4 TYPES OF ELECTRODES

Electrodes are classified in two broad groups: indicator and reference electrodes. A reference electrode must be easy to construct, and must maintain a constant, reproducible potential even if small currents are passed. An indicator electrode must respond to changes in activity of the species to be measured. Some electrodes are especially useful for one or both of these purposes.

The Hydrogen Electrode.   Although this electrode is the standard of reference (see Figure 18–3 and related discussion), it is inconvenient to construct and difficult to maintain. A source of highly purified hydrogen is required, and its partial pressure within the cell must be regulated. The platinum surface is coated with a thin layer of platinum black to increase its surface area and catalyze the electrode reaction. Traces of impurities in either the gas or solution may poison the electrode. The electrode responds to most other redox couples and therefore strong oxidants or reductants must not be present. Its use is limited mostly to calibrating other electrodes and to measuring hydrogen ion activities for solutions in which no other electrode will operate satisfactorily.

The Quinhydrone Electrode.   The composition and function of this electrode have just been discussed, and it is only necessary to add that a solution containing equal concentrations of quinone and hydroquinone is easily prepared from solid quinhydrone, a molecular addition compound of quinone and hydroquinone:

$$2QH \rightleftharpoons Q + H_2Q$$

Equation 18–19 then becomes

$$E = 0.6994 - 0.059\,pH \tag{18–21}$$

Hydroquinone may be oxidized in the presence of strong oxidants and it will dissociate as a weak acid ($pK_a \sim 10$) in appreciable amounts above pH 8.5. Either of these processes will disturb the ratio of $[H_2Q]/[Q]$ and give erroneous results for pH.

Smooth Platinum Electrode.   A piece of smooth, bright platinum wire or plate responds to most redox couples when all species are present in the solution (including dissolved gases). In moderately strong acid solutions, we would expect the platinum to respond to pH (as in the SHE). However, with *smooth* platinum, the reduction of hydrogen ion is somewhat irreversible and the electrode is normally insensitive to pH. Very strong reductants, for example, Cr^{+2}, may be oxidized by hydrogen ion in the presence of platinum, giving erroneous redox potentials.

Saturated Calomel Electrode (SCE).   This popular reference electrode is easily constructed from readily available materials. A solution is saturated with both calomel (mercurous chloride) and potassium chloride and placed over a layer of mercury which serves as the electrode. A commercial version of this electrode is shown in Figure 18–7.

$$Hg|Hg_2Cl_2, KCl(sat)$$

The potential can be derived from the primary reaction:

$$Hg_2^{+2} + 2e^- \rightleftharpoons 2Hg$$

and the corresponding Nernst equation:

$$E = E_{Hg}^{\circ} - \frac{0.059}{2}\log\frac{1}{[Hg_2^{+2}]} \tag{18–22}$$

But the concentration of Hg_2^{+2} is in turn controlled by the concentration of Cl^-, which is essentially the same as the solubility of KCl. Chloride ion from Hg_2Cl_2 is

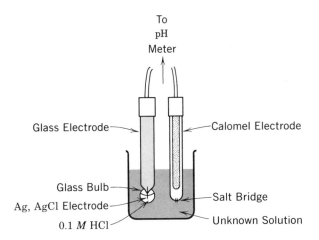

Figure 18–7 Schematic diagram of the glass and calomel electrodes.

negligible compared to that from KCl.

$$E = E_{Hg}^{\circ} - \frac{0.059}{2} \log \frac{[Cl^-]^2}{K_{Hg_2Cl_2}} \qquad (18-23)$$

where $K_{Hg_2Cl_2}$ is the solubility product of Hg_2Cl_2.

Rearranging Equation 18–23, we define a new E° term for this electrode reaction:

$$E = \underbrace{E_{Hg}^{\circ} + \frac{0.059}{2} \log K_{Hg_2Cl_2}}_{E^{\circ}_{Hg_2Cl_2,Hg}} - 0.059 \log [Cl^-] = 0.242 \text{ v.} \qquad (18-24)$$

The temperature coefficient of the SCE is practically the same as that of the solubility of KCl; namely,

$$E_{SCE} = +0.242 - 7.6 \times 10^{-4} (t - 25), \qquad (18-25)$$

where t is the temperature in °C. Because the potential is controlled by a species that does not take part in the electrode reaction, this is known as an *electrode of the second order*.

Silver Chloride Electrode. A silver wire in contact with chloride ion is readily coated with a layer of AgCl. It is an excellent indicator for chloride ion concentration and a convenient reference electrode.

$$Ag|AgCl, Cl^-$$

$$E = E_{Ag^+, Ag}^{\circ} - 0.059 \log \frac{1}{[Ag^+]}$$

$$E = \underbrace{E_{Ag^+, Ag}^{\circ} + 0.059 \log K_{AgCl}}_{E^{\circ}_{AgCl, Ag}} - 0.059 \log [Cl^-] \qquad (18-26)$$

Glass Electrode. No doubt the most common potentiometric measurement is the determination of hydrogen ion with the glass electrode and a pH meter. Figure 18–7 is a schematic representation of typical glass and calomel electrodes available commercially. Inside the glass electrode is a silver wire dipping into a solution of *ca.* 0.1 M HCl. The bulb at the tip of the electrode is made of a special glass composition. The inner surface of this glass membrane contacts the 0.1 M HCl solution and the outer surface contacts the solution in which the pH is to be measured. At each surface the glass membrane absorbs water forming a gel layer. The hydrogen ions from the solution can diffuse through the gel layer and replace (ion exchange) sodium or other metal ions in the glass structure. The net result of the diffusion and exchange processes is that a phase boundary potential is set up on each side of the glass membrane with the magnitude determined by the activity of hydrogen ion in the contacting solution.

The overall potential of the glass electrode consists of several parts: (a) the potential of the internal silver-silver chloride electrode, (b) the potential developed at the inner glass surface, (c) the asymmetry potential (caused by strains and imperfections in the glass membrane) and (d) the potential developed at the outer surface. However, potentials (a), (b), and (c) are all constant for a given electrode, and it is only (d) which changes with pH of the test solution. In practice, it is convenient to combine the constant potential terms and write the potential of the electrode in the usual Nernst form:

$$E = E'_G + 0.059 \log [\text{H}^+] \tag{18-27}$$

Equation 18–27 can be derived from Donnan membrane theory or simply stated as the observed response of the glass electrode.

The glass electrode is a most convenient electrode to use but it has a few drawbacks. The commercial version is moderately expensive and somewhat fragile—the sensitive tip is easily scratched and ruined. The electrical resistance of the glass membrane is exceedingly high ($ca.$ 30 megohms) requiring the use of high impedance amplifiers (see below) and special care to prevent electrical leakages in other parts of the circuit. In highly alkaline solution, the concentration of H^+ is so low that other ions in the solution may interfere. For example, in strong NaOH solutions, the electrode may function as a sodium ion indicator electrode to some extent and a correction is necessary for accurate pH measurements. Special glass compositions with low sodium content are available at extra cost. Other special glasses with high sodium content have been developed specifically for measuring Na^+ concentrations, and glasses for measuring some other metal ions are on the market.

Ion-Selective Electrodes. The glass electrode just described is one example of a membrane electrode that is more or less selective in response to a single type of ion. We noted that at high pH, or high concentrations of sodium hydroxide, the glass electrode functions partly as a sodium ion electrode. If the composition of the glass is altered, for example, by increasing the proportion of Al_2O_3, the electrode responds more selectively to sodium ion activity. Several types of membranes are used in a variety of ion-selective electrodes, a few of which are illustrated in Figure 18–8.

These electrodes function to a large extent because of the ion exchange properties of the membrane. By their nature, ion exchange sites are not specific for a given ion. Nevertheless, the sites are somewhat selective and will exhibit a preference for some ions over others, depending primarily on the size and charge of the ions. Selectivities are usually expressed as a *selectivity ratio*, K, defined as the ratio of concentration of the two ions which will give identical potentiometric response under otherwise identical conditions. As an example, a certain membrane electrode is 100 times more sensitive to Br^- than to Cl^-, that is $K_{\text{Br,Cl}} = 10^{-2}$ for chloride ion. Then

$$E = \text{constant} + \frac{RT}{F} \ln (a_{\text{Br}} + K_{\text{Br,Cl}} a_{\text{Cl}}) \tag{18-28}$$

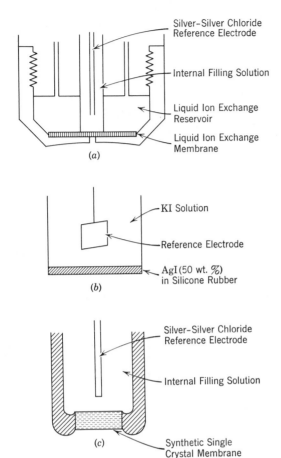

- Silver–Silver Chloride Reference Electrode
- Internal Filling Solution
- Liquid Ion Exchange Reservoir
- Liquid Ion Exchange Membrane

(a)

- KI Solution
- Reference Electrode
- AgI (50 wt. %) in Silicone Rubber

(b)

- Silver–Silver Chloride Reference Electrode
- Internal Filling Solution

(c)

- Synthetic Single Crystal Membrane

Figure 18–8 Types of ion-selective electrodes: (a) liquid–liquid membrane; (b) precipitate impregnated membrane; (c) solid state membrane.

$K_{Br,Cl}$ can also be defined in terms of the potentials measured in two separate solutions, one containing 0.1 M KBr and the other 0.1 M KCl. Then

$$E_{KCl} - E_{KBr} = \frac{RT}{F} \ln K_{Br,Cl} \times \frac{a_{Cl}}{a_{Br}}$$

Unfortunately, selectivity ratios often vary with the composition of the solution. Therefore they are useful only as a first approximation and are not reliable for making extensive corrections for interfering ions. They are used primarily as an indication of the suitability of an electrode for a particular determination. Since membrane electrodes are continually being modified, the manufacturer's current literature should be consulted.

The selective response of several types of glass membranes is given in Table 18–5.

Silicone Rubber Matrix Electrodes. Only cations appear to be exchanged at glass membranes. Sparingly soluble salts of the anion to be measured are suitable candidates for anion-selective electrodes. It is not easy to prepare coherent electrode membranes with favorable mechanical properties. Pungor and his coworkers in Hungary have polymerized a mixture of a finely divided silver halide precipitate (grain size 5 to 10 μm) and silicone rubber monomer. When incorporated in these heterogeneous membranes, each silver halide shows a distinct preference for exchanging its own anion with the test solution. Cations are not involved in the potential determining step, therefore the membrane electrode is

Table 18–5   Ion Selectivity of Various Glasses

Ion To Be Measured	Glass Composition	Selectivity
Na^+	11% Na_2O, 18% Al_2O_3, 71% SiO_2	$\begin{cases} Na^+/K^+ \sim 2800 \ (pH \ 11) \\ Na^+/K^+ \sim 300 \ (pH \ 7) \end{cases}$
Na^+	10.4% Li_2O, 22.6% Al_2O_3, 67% SiO_2	$Na^+/K^+ \sim 10^5$
K^+	27% Na_2O, 5% Al_2O_3, 68% SiO_2	$K^+/Na^+ \sim 20$
Li^+	15% Li_2O, 25% Al_2O_3, 60% SiO_2	$\begin{cases} Li^+/Na^+ \sim 3 \\ Li^+/K^+ \sim 1000 \end{cases}$
Ag^+	11% Na_2O, 18% Al_2O_3, 71% SiO_2	$Ag^+/Na^+ \sim 1000$

[From G. A. Rechnitz, *Chem. Eng. News* **45**, 146, June 12, 1967.]

free from redox interference (strong oxidants or reductants) as is the case with silver metal-silver halide electrodes discussed previously.

Solid-State Electrodes.   Suitably doped single crystals of lanthanum fluoride are used as fluoride electrodes. The detailed composition of the crystal is proprietary information (traces of europium salts are often added). A number of similar electrodes are available for sulfide and the other halides.

Selectivity in solid membranes results from the way that ions "move" through the crystal. Ion conduction proceeds through a defect mechanism in which a mobile ion moves into a vacancy in the crystal structure. Thus a new vacancy is created and the charge is propagated through the crystal even though a given ion moves only a small distance. Such vacancies will admit only a certain type of mobile ion.

Liquid–Liquid Electrodes.   A large number of highly selective ion exchange materials are available only in the liquid state. However, liquid–liquid electrodes present mechanical problems; for example, the liquid exchanger must be in electrolytic contact with the sample solution with minimal mixing of the two liquid phases. Its viscosity should be high enough to inhibit flow across the interface. The interface between the two liquid phases can be maintained by using a cellulose dialysis membrane which is permeable to ions but not to solvent molecules. Alternatively, the liquid exchanger can be trapped with a Millipore filter disk.

One of the most successful liquid–liquid electrodes is the calcium electrode. The substance used as a liquid ion exchanger is an organic derivative of phosphoric acid, $(RO)_2POOH$, where R is a long chain organic group. A calcium reference solution contacts the inner surface of the membrane, and the test solution containing calcium contacts the outer surface, thus the measured potential is really the difference between two liquid junction potentials. At the junctions, the organic phosphate group forms a salt selectively with calcium, with selectivity coefficients of: Ca^{+2}/Sr^{+2} 70, Ca^{+2}/Mg^{+2} 200, Ca^{+2}/Ba^{+2} 630, Ca^{+2}/Na^+ 3200, Ca^{+2}/K^+ 3200. Similar electrodes are available for potassium and ammonium ions.

Some liquid exchangers are selective because of chelate formation. In $R{-}S{-}CH_2{-}COO^-$, both the sulfur and carboxylate groups participate in chelate formation with metal ions. Copper and lead electrodes are based on this principle.

Because of their selectivity, sensitivity, and rapid, convenient operation, ion-selective electrodes have found many applications. Ions, such as fluoride, are not easily determined by other methods. Previously the determination of fluoride in toothpaste involved steam distillation followed by titration with thorium nitrate. With the fluoride electrode, no separation is necessary and the fluoride can be determined by direct potentiometry. Similarly, the calcium electrode is used extensively in biomedical research. Numerous applications are discussed in *Ion-Selective Electrodes*, edited by Durst.

A selection of commercially available electrodes is given in Table 18–6.

Table 18–6 Characteristics of Some Membrane and Solid-State Ion Selective Electrodes

A. Liquid Ion Exchange Membrane Electrodes

Ion	pM Range
Ca^{+2}	0 to 5
Cu^{+2}	1 to 5
Pb^{+2}	2 to 5
BF_4^-	1 to 5
Cl^-	0 to 5
ClO_4^-	1 to 5
NO_3^-	1 to 5

B. Heterogeneous Membrane Electrodes

Ion	pM Range
I^-	1 to 7
Br^-	1 to 6
Cl^-	1 to 5
S^{-2}	1 to 17

C. Solid State Electrodes

Ion	pM Range
F^-	0 to 6
Cd^{+2}	1 to 7
Cu^{+2}	0 to 8
Pb^{+2}	1 to 7
CN^-	2 to 6
SCN^-	0 to 5
S^{-2}	0 to 17
Ag^+	0 to 17

18–5 CELL VOLTAGE MEASUREMENTS

Ordinary voltmeters and ammeters require small but appreciable currents to deflect the needle. Since any current flowing through the cell will necessarily change the concentrations of the reacting species, the simple voltmeter is not a suitable device for accurately measuring a cell voltage. Two alternatives are available.

Electronic Voltmeter. A high input impedance vacuum tube voltmeter requires as little as 10^{-14} A to give an accurately measurable reading. Modern meters are transistorized and require a negligible amount of power from the source to be measured.

Potentiometer Circuits. In a potentiometer circuit, the voltage to be measured is exactly balanced by a known variable voltage as shown in Figure 18–9. The accuracy of determining the null point is limited by the current necessary to detect

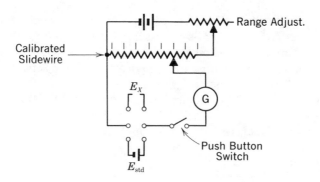

Figure 18–9 Schematic potentiometer circuit.

the off-balance. The use of a high-impedance voltage amplifier in a potentiometric circuit provides the ultimate accuracy in voltage measurement. The convenient automatic recording potentiometer is discussed in Section 25–2.

18–6   DEFINITION OF pH AND THE pH METER

We have deferred discussing this seemingly very elementary concept until we have developed the necessary background to present the practical scale of pH numbers. In all of chemistry "pH" is one of the most used and misused terms. First, let us note the difference between concentration of total titratable hydrogen ion (including that of undissociated acids present), and the concentration of *free* hydrogen ion. The former can be determined by titration, but the latter cannot because any addition of base will cause further dissociation of any acids present. To determine free hydrogen ion concentration we must make the measurement without disturbing the existing equilibria. The term "pH" refers to free hydrogen ion concentration.

Operational Definition of pH.   The definition employed should be capable of theoretical interpretation and practical measurement. Of all the techniques available to measure free hydrogen ion in a solution of unknown composition, none will measure either concentration or activity. The simple definitions, $pH = -\log C_H$ and $pH = -\log a_H$, are therefore not sufficient. Since pH is most often measured with a glass electrode and a pH meter, let us restrict our considerations to this technique. (The problems are similar with all other methods.) The cell to be used is represented:

Cell 9:   Glass|H$^+$‖SCE

$$E_L = E'_G + 0.059 \log (\text{H}^+) \text{ and } E_R = E_{SCE}$$

$$E_{cell} = \underbrace{E_{SCE} + E_j - E'_G} - 0.059 \log (\text{H}^+)$$

or, combining constants over the bracket

$$E_{cell} = \qquad E^* \qquad - 0.059 \log a_H$$

$$= \qquad E^* \qquad - 0.059 \log C_H f_H \qquad (18\text{–}28)$$

In our discussion of activity coefficients, it was shown that it is not possible to measure or compute f_H by itself. However, by convention f_H is equal to $f_\pm$ which can be measured in cells such as Cell 8. (See page 371.)

Cell 8: $$Pt, H_2|H^+, Cl^-, AgCl|Ag$$

Therefore, for practical purposes, pH is defined as:

$$pH = -\log C_H f_{\pm} \tag{18-29}$$

or in terms of the voltage of either cell 8 or 9 just described

$$E_{cell} = E^* + 0.059\,pH \tag{18-30}$$

or

$$pH = \frac{E_{cell} - E^*}{0.059} \quad \text{at } 25°C \tag{18-31}$$

where E^* is a different constant for each cell.

Equation 18-31 is the *operational definition of pH*. Thus two approximations are tacitly included in the definition of pH: (*a*) the junction potential, E_j, remains constant, and (*b*) $f_H = f_{\pm}$. The second approximation is reasonably good in dilute aqueous solution, but the first is good only if the value of E_j is small (a few millivolts at most), as it should be with an effective salt bridge.

The pH Scale.   The E^* term contains so many complications that it can only be determined empirically. A solution of known pH is required. Such a solution has been specified by the National Bureau of Standards using Equation 18-28 after a careful study of the most nearly ideal systems available. Actually several standard buffers have been specified, because no glass electrode follows the Nernst equation precisely over a wide range of pH values. Thus the working pH scale is an arbitrary set of numbers defined as described above. The pH value is close, but not exactly equal, to $-\log a_H$. When one considers the approximations and experimental limitations, it is evident that it is very difficult indeed to reduce the systematic error below ± 0.01 pH unit, although small changes in pH can be measured at the ± 0.001 level if the random error (electronic noise) can be made quite small.

The pH Meter.   The pH meter is basically an electronic voltmeter designed for use with a glass electrode system. From Equation 18-30 we see that 1 pH unit is equivalent to 59 mV at 25°C. Therefore, the same meter can be used with two scales to read either mV or pH units.

The basic circuit diagram of a typical pH meter is given in Figure 18-10. The sensitivity of the meter (scale divisions/mV) is adjustable by a temperature control knob to allow for a change in the 0.059 factor with temperature (or to allow for nonideal response of the electrode). To read absolute mV rather than pH, the shunt used for temperature control is disconnected entirely. The zero calibration knob changes the zero setting of the meter by introducing a bias voltage into the meter circuit so that the meter reading will correspond to the pH of the standard buffer.

pH meters are often constructed so that the true zero of the voltmeter is at or near pH 7 on the scale. In use, the meter should first be standardized with a pH 7 buffer, adjusting the zero calibration to give the correct reading. Then a second buffer, pH 4 or 10, is substituted and the meter is adjusted to give the correct reading with the temperature control. This procedure establishes the correct linear relationship between mV and pH. At least two points are required to establish the line as shown in Figure 18-11, although for rough measurements, one point and the theoretical slope of 0.059 may be adequate for calibration.

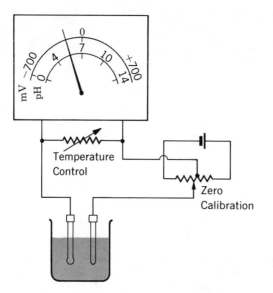

Figure 18–10   Schematic diagram of a pH meter.

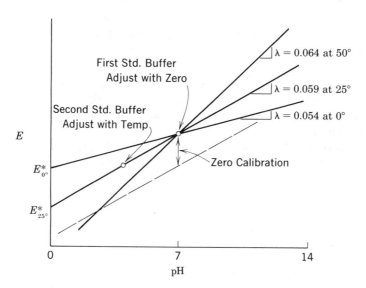

Figure 18–11   Plot of Equation 18–30, the basic equation of a pH meter. (Not to scale.) Zero calibration raises or lowers response line. Temperature control changes slope of response line.

18–7   EQUILIBRIUM CONSTANTS

A useful relation is obtained by considering a cell at equilibrium. As we have already observed, the half-cell reactions are always at equilibrium with the corresponding electrode, but the overall cell reaction, in general, is not at equilibrium, unless the cell voltage happens to be zero. Under the latter condition (zero cell voltage) the ratio of activities of products to activities of reactants must satisfy the equilibrium constant, K, for the cell reaction and the following equations pertain:

$$E_{cell} = E_{cell}^{\circ} - \frac{0.059}{n} \log \frac{\text{(products)}}{\text{(reactants)}}$$

At equilibrium,

$$E_{cell} = 0 \quad \text{and} \quad \frac{\text{(products)}}{\text{(reactants)}} = K$$

$$0 = E_{cell}^{\circ} - \frac{0.059}{n} \log K$$

or

$$\log K = \frac{nE_{cell}^{\circ}}{0.059} \quad \text{at 25°C.} \tag{18-32}$$

where E_{cell}° and K must correspond to the same reaction. If E_{cell}° is positive, then K is greater than unity, and vice versa. To avoid confusion over signs of E° values, remember that half-reactions (always written as reductions) should be subtracted in a direction to give the required whole reaction, and that E° values corresponding to these half-reactions must be subtracted in the same direction. Thus the equilibrium constant for a reaction can be calculated from a table of E° values (Table 18-1).

Example/Problem 18-2. Calculate the equilibrium constant for the reaction:

$$Fe^{+2} + Ag^{+} \rightleftharpoons Ag + Fe^{+3}$$

The two half-reactions are:

$$Fe^{+3} + e^{-} \rightleftharpoons Fe^{+2}; \quad E_{Fe}^{\circ} = 0.771 \text{ V}$$
$$Ag^{+} + e^{-} \rightleftharpoons Ag; \quad E_{Ag}^{\circ} = 0.799 \text{ V}$$

The reaction given is obtained by subtracting the first half-reaction from the second; E_{react}° must be obtained in the same way:

$$E_{react}^{\circ} = E_{Ag}^{\circ} - E_{Fe}^{\circ} = 0.799 - 0.771 = 0.028 \text{ V}$$
$$\log K = \frac{nE^{\circ}}{0.059} = \frac{0.028}{0.059} = 0.475$$
$$K = 2.98$$

Example/Problem 18-3. Calculate the equilibrium constant for the reaction:

$$Fe^{+2} + AgCl \rightleftharpoons Fe^{+3} + Ag + Cl^{-}$$

The two half-reactions are:

$$Fe^{+3} + e^{-} \rightleftharpoons Fe^{+2} \quad\quad E_{Fe}^{\circ} = 0.771 \text{ V}$$
$$AgCl + e^{-} \rightleftharpoons Ag + Cl^{-} \quad E_{AgCl}^{\circ} = 0.222 \text{ V}$$

Again, we must subtract the first from the second to obtain the given reaction

$$E_{react}^{\circ} = E_{AgCl}^{\circ} - E_{Fe}^{\circ} = 0.222 - 0.771 = -0.549 \text{ V}$$
$$\log K = \frac{-0.549}{0.059} = -9.30$$
$$K = 5.0 \times 10^{-10}$$

The negative value of E_{react}° and the very small K both show that the given reaction will be spontaneous in the reverse direction.

Example/Problem 18-4. The voltage of the following cell is -0.753 V:

$$Ag|Ag^{+}(0.1 \text{ } M)\|I^{-}(0.01 \text{ } M), AgI|Ag$$

From this information, calculate the solubility product of AgI.

Left half-reaction: $\qquad\qquad\qquad$ $Ag^+ + e^- \rightleftharpoons Ag$

$$E_L = E^\circ_{Ag^+,Ag} - 0.059 \log 1/[Ag^+]_L$$

Right half-reaction: $\qquad\qquad\qquad$ $Ag^+ + e^- \rightleftharpoons Ag$

where $[Ag^+] = K_{AgI}/[I^-]$,

$$E_R = E^\circ_{Ag^+,Ag} - 0.059 \log [I^-]/K_{AgI}$$

$$E_{cell} = E_R - E_L$$

$$-0.753 = 0.059 \log K_{AgI} - 0.059[I]_R - 0.059 \log [Ag^+]_L$$

$$\log K_{AgI} = -\frac{0.753}{0.059} + \log 0.1 + \log 0.01$$

$$= -12.8 - 1.0 - 2.0 = -15.8$$

$$K_{AgI} = 1.6 \times 10^{-16}$$

Alternatively, the solubility product can be obtained from the E° value for the half-reaction $AgI + e^- \rightleftharpoons Ag + I^-$ (Table 18–1) and the definition given in Equation 18–24. The details are left as an exercise for the student.

Example/Problem 18–5. Cell 4 is assembled with 0.0500 M HCl in the right half-cell and 0.0200 M NaOH in the left. The pressure of hydrogen is 0.923 atm at both electrodes.

$$Pt, H_2|OH^-\|H^+|H_2, Pt$$

The measured cell voltage is +0.6528 V. From this information, calculate K_w.
The left half-cell reaction is:

$$2H_2O + 2e^- \rightleftharpoons 2OH^- + H_2$$

but it is equally correct to separate the half-reaction into two steps with the dissociation of water followed by the reduction of hydrogen ion:

$$2H_2O \rightleftharpoons 2H^+ + 2OH^-$$

$$2H^+ + 2e^- \rightleftharpoons H_2$$

With this approach we have the same electrode reaction occurring in both half-cells; therefore, the same Nernst equation but with two different values for the hydrogen ion concentration.

Left side: $\quad E_L = E^\circ_H - \dfrac{0.059}{2} \log \dfrac{P_{H_2}}{[H^+]^2_L}$

$$= E^\circ_H - \frac{0.059}{2} \log \frac{P_{H_2}[OH^-]^2_L}{K^2_w}$$

$$= E^\circ_H - \frac{0.059}{2} \log P_{H_2} - 0.059 \log [OH^-]_L + 0.059 \log K_w$$

Right side: $\qquad\qquad E_R = E^\circ_H - \dfrac{0.059}{2} \log \dfrac{P_{H_2}}{[H^+]^2_R}$

$$= E^\circ_H - \frac{0.059}{2} \log P_{H_2} + 0.059 \log [H^+]_R$$

$$E_{cell} = E_R - E_L$$

The E° terms and the hydrogen pressure terms cancel.

$$E_{cell} = 0.059 \log [H^+]_R + 0.059 \log [OH^-]_L - 0.059 \log K_w$$

The data given justify using 0.05915 rather than 0.059.

$$\frac{0.6528}{0.05915} = \log 0.0500 + \log 0.0200 - \log K_w$$

$$\log K_w = -11.03 - 3.00 = -14.03$$

$$K_w = 9.33 \times 10^{-15}$$

For many electrode systems, E° cannot be measured directly because many electrodes do not behave reversibly; however, various combinations of the relationships expressed above allow us to calculate E, E°, or K from other thermodynamic data.

18–8   POTENTIOMETRIC TITRATIONS

In a potentiometric titration the course of the titration reaction is followed by measuring the concentration of one or more of the species potentiometrically. The titration beaker becomes one of the half-cells of an electrochemical cell, along with a convenient reference electrode for the other half-cell. It is not necessary that the titration reaction be a redox reaction as long as one of the species is, or can be made to be, a part of a redox couple.

It is important to distinguish between the titration reaction and the cell reaction. The reactants and products of the titration are all in the same half-cell, and the titration reaction is normally always at equilibrium. On the other hand the cell reaction consists of either of two possible half-reactions in the titration half-cell and the half-reaction in the reference half-cell. The cell reaction is normally not at equilibrium because it is not allowed to take place except for brief periods while measuring the cell voltage.

The Iron(II)-Cerium(IV) Titration.   Let us study the titration of Fe^{+2} with Ce^{+4}. The experimental features are shown in Figure 18–12 in which the cell is represented:

$$Hg|Hg_2Cl_2, KCl_{(sat)}||Fe^{+2}, Fe^{+3}, Ce^{+4}, Ce^{+3}|Pt$$

Titration Half-cell: $Fe^{+3} + e^- \rightleftharpoons Fe^{+2}$;

$$E = E^\circ_{Fe} - 0.059 \log \frac{[Fe^{+2}]}{[Fe^{+3}]} \tag{18–33}$$

or

$$Ce^{+4} + e^- \rightleftharpoons Ce^{+3};$$

$$E = E^\circ_{Ce} - 0.059 \log \frac{[Ce^{+3}]}{[Ce^{+4}]} \tag{18–34}$$

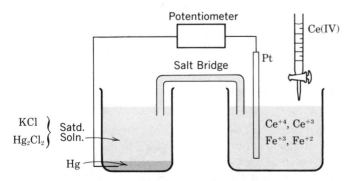

Figure 18–12   Apparatus for the potentiometric titration of iron(II) with cerium(IV).

Reference Half-cell: $Hg_2Cl_2 + 2e^- \rightleftharpoons 2Hg + 2Cl^-$

Cell reaction: $2Fe^{+3} + 2Hg + 2Cl^- \rightleftharpoons 2Fe^{+2} + Hg_2Cl_2$

or

$$2Ce^{+4} + 2Hg + 2Cl^- \rightleftharpoons 2Ce^{+3} + Hg_2Cl_2$$

Titration reaction: $Fe^{+2} + Ce^{+4} \rightleftharpoons Fe^{+3} + Ce^{+3}$

If we perform the titration, measuring the cell voltage at many points along the way, the data are conveniently plotted as shown in Figure 18–13. Now we shall examine the theory that explains this curve and would allow us to predict this curve without ever having to do the experiment. It is convenient to divide the titration into several regions.

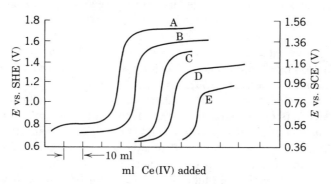

Figure 18–13 Potentiometric titration curves for iron(II) with cerium(IV) in various acids: A, Theoretical; B, HClO₄; C, HNO₃; D, H₂SO₄: E, HCl. [From Smith, *Cerate Oxidimetry*, Smith Chemical Co., 1942, p. 23.]

1. *At the beginning* the solution contains only Fe^{+2} plus traces of Fe^{+3} due to oxidation by air or hydrogen ion. If the concentration of Fe^{+3} were really zero, the Nernst equation would give us an infinitely negative value for the potential. This would indicate an infinitely powerful reducing agent that would surely reduce water. The initial potential can be measured but not calculated without knowing the extent of this extraneous oxidation. Since the initial potential is of no consequence to our titration, we will ignore the problem.

2. *Once the titration is begun,* all four species are present. However, up to the equivalence point, there remains an excess of Fe^{+2} in the half-cell, and the concentration of Ce^{+4} must remain extremely small as a consequence of the very large equilibrium constant for the titration reaction. In this region, the amount of Fe^{+2} converted to Fe^{+3} is directly proportional to the amount of Ce^{+4} added. It is easy to calculate the ratio $[Fe^{+2}]/[Fe^{+3}]$ from the volume of Ce^{+4} added, V, compared to the additional volume increment needed to reach the equivalence point, $V_{ep} - V$.

$$\frac{[Fe^{+2}]}{[Fe^{+3}]} = \frac{V_{ep} - V}{V}$$

3. *At the equivalence point,* there is no excess of either reactant. By definition, the same number of equivalents of both Fe^{+2} and Ce^{+4} have been added. The large equilibrium constant indicates that each is converted essentially completely to Fe^{+3} and Ce^{+3}, respectively. But the equilibrium constant, though large, is not infinite, and therefore the concentrations of Fe^{+2} and Ce^{+4} cannot be zero. The concentrations of Fe^{+2} and Ce^{+4} could be computed from the value of the equilibrium constant, but there is a more elegant way of calculating the potential without needing to know these concentrations. Remembering that both Nernst equations (Equations 18–33 and 18–34) are valid, we simply add the two together to obtain another equation which is true throughout the titration, but is useful only at the equivalence point:

$$2E = E^{\circ}_{Fe} + E^{\circ}_{Ce} - 0.059 \log \frac{[Fe^{+2}][Ce^{+3}]}{[Fe^{+3}][Ce^{+4}]} \tag{18–35}$$

Equation 18–35 may resemble a Nernst equation, but it is only a bit of algebraic maneuvering performed with Equations 18–33 and 18–34. We combined the two Nernst equations in this manner only because it leads to a very simple expression for the equivalence point potential. At this point,

$$[Fe^{+2}] = [Ce^{+4}] \quad \text{and} \quad [Fe^{+3}] = [Ce^{+3}]$$

Thus we can cancel out all concentrations in Equation 18–35, leaving a value of unity for the concentration ratio and zero for the log term. Therefore,

$$E_{ep} = \tfrac{1}{2}(E^{\circ}_{Fe} + E^{\circ}_{Ce}) \tag{18–36}$$

Observe that E_{ep} is independent of concentrations and equal to the arithmetic mean of the two standard potentials. A bit of reflection will show that if n_{ox} for the oxidant and n_{red} for the reductant in the titration reaction are not the same, we must modify our treatment. The two log terms of the individual Nernst equations cannot be combined unless the factor preceding the log is the same for both. The two factors can be made the same if we multiply the Nernst equation for the oxidant by n_{ox} and multiply that for the reductant by n_{red}. Thus the E_{ep} will become a weighted mean:

$$E_{ep} = \frac{n_{ox}E^{\circ}_{ox} + n_{red}E^{\circ}_{red}}{n_{red} + n_{ox}} \tag{18–37}$$

For some unsymmetric titration reactions, for example, those involving H^{+} or complexing agents, not all concentrations will cancel. In these cases, E_{ep} will depend on the concentration of such species.

4. *Beyond the equivalence point,* an excess of Ce^{+4} has been added. The amount of the excess is clearly that which has been added after the equivalence point has been reached, and its concentration can be easily calculated from the total volume of the solution. Likewise, the amount of Ce^{+3} in the solution can be calculated from the amount of Ce^{+4} required to reach the equivalence point—essentially all of it was reduced to Ce^{+3} by the Fe^{+2} in the original solution.

$$\frac{[Ce^{+3}]}{[Ce^{+4}]} = \frac{V_{ep}}{V - V_{ep}}$$

The calculations for the titration curve, E versus ml of titrant, are summarized in Table 18–7. The concentrations of the various species present have also been tabulated, although it is not necessary to know the concentrations of each in order to compute E. All species are present in the same volume; therefore, the ratio of concentrations is the same as the more readily calculated ratio of amounts. However, it is always very helpful to know what's in the beaker. As a visual aid in keeping track of "what's going on" during the titration, the concentration curves of Figures 18–14 and 18–15 should be carefully compared with the titration curves in Figure 18–13.

Table 18–7 Titration of 50.00 ml of 0.1 M Fe^{+2} with 0.1 M Ce^{+4}

V, ml	f	[Fe^{+2}] M	[Fe^{+3}] M	[Ce^{+3}] M	[Ce^{+4}] M	E versus SHE, V
0	0	0.100	~0	0	0	—
1	0.02	0.096	0.002	0.002	2.3×10^{-19}	0.67
5	0.10	0.082	0.0091	0.0091	5.6×10^{-18}	0.71
10	0.20	0.067	0.0167	0.0167	2.3×10^{-17}	0.73
25	0.50	0.033	0.033	0.033	1.8×10^{-16}	0.77
40	0.80	0.011	0.045	0.045	1.0×10^{-15}	0.79
45	0.90	5.3×10^{-3}	0.047	0.047	2.3×10^{-15}	0.83
49	0.98	5.0×10^{-4}	0.049	0.049	2.5×10^{-14}	0.87
49.9	0.998	1.0×10^{-4}	0.050	0.050	1.4×10^{-13}	0.93
50	1.000	3.8×10^{-9}	0.050	0.050	3.8×10^{-9}	1.19
50.1	1.002	1.4×10^{-13}	0.050	0.050	1.0×10^{-4}	1.45
51	1.02	1.4×10^{-14}	0.050	0.050	9.9×10^{-4}	1.51
55	1.10	2.7×10^{-15}	0.048	0.048	4.8×10^{-3}	1.55
75	1.50	4.4×10^{-16}	0.040	0.040	2.0×10^{-2}	1.59
100	2.00	1.8×10^{-16}	0.033	0.033	3.3×10^{-2}	1.61

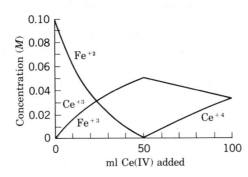

Figure 18–14 Actual concentration of species during the iron(II)–cerium(IV) titration.

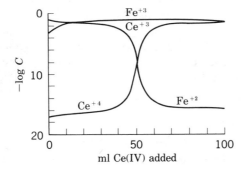

Figure 18–15 Logarithmic concentration of species during the iron(II)–cerium(IV) titration.

The theoretical titration curve is plotted in Figure 18–13 along with several actual titration curves obtained experimentally in different solutions. There are two important reasons why the actual curves differ from the theoretical curve. All four of the species involved in this titration readily form complexes and are subject to hydrolysis and dimerization. Furthermore, highly charged ions have activity coefficients which deviate greatly from unity. A different theoretical curve would have been obtained if we had used the appropriate formal potentials. A second cause of the difference between theoretical and actual curves is the irreversible behavior of many electrodes.

On the other hand, theoretical curves do approximate the actual curves in many important respects, and are useful in obtaining a graphical picture of what happens during a titration. The shape of a titration curve in the vicinity of the equivalence point, and in particular the slope of the curve at that point, tells us how sharp the end point will be, and within what range of potential we must stop in order to obtain a given accuracy.

The calculations performed in Table 18–7 should be noted.

1. Observe that 50 ml of Ce^{+4} are required to reach the equivalence point.
2. The concentrations at any point in the titration can be readily computed by introducing a new variable, f, the fraction titrated. $f = V/50$, or volume of titrant added/volume required at the equivalence point

$$[Fe^{+2}] = (1-f) \times 0.1 \times \frac{50}{50+V} \qquad (0 < f < 1)$$

$$\begin{array}{ccc} \text{fraction} & \text{original} & \text{dilution} \\ \text{remaining} & \text{conc.} & \text{factor} \end{array}$$

$$\begin{aligned}
&[Fe^{+3}] = f \times 0.1 \times \frac{50}{50+V} & 0 < f < 1 \\
&[Ce^{+3}] = [Fe^{+3}] & \text{all } f \\
&[Ce^{+4}] = [Ce^{+3}][Fe^{+3}]/K[Fe^{+2}] & 0 < f < 1 \\
&[Fe^{+2}] = [Ce^{+3}][Fe^{+3}]/K[Ce^{+4}] & f > 1 \\
&[Fe^{+3}] = 0.1 \times 50/(50+V) & f > 1 \\
&[Ce^{+4}] = 0.1(V-50)/(50+V) & f > 1
\end{aligned}$$

3. E is computed in one of three ways (no concentrations are required).

$$\begin{aligned}
&E = E^{\circ}_{Fe} - 0.059 \log (1-f)/f & 0 < f < 1 \\
&E = \tfrac{1}{2}(E^{\circ}_{Fe} + E^{\circ}_{Ce}) & f = 1 \\
&E = E^{\circ}_{Ce} - 0.059 \log 1/(f-1) & f > 1
\end{aligned}$$

The potentials listed in Table 18–7 refer to the platinum indicator electrode versus the standard hydrogen electrode. In the actual cell, a saturated calomel electrode was used as the reference electrode. The conversion of one scale to the other is a simple matter of subtracting 0.24 V from the value versus SHE to obtain the value versus SCE; that is, potentials measured versus SCE are 0.24 V more negative than versus SHE.

Redox Indicator. It is not always necessary to set up the potentiometric equipment and actually measure an electrode potential in order to determine the end point of a titration. A redox indicator may be used to give a visual indication that this point has been reached. The color change takes place upon oxidation or reduction of the indicator itself. If the standard potential of the indicator system is close to the equivalence point potential of the titration reaction, the indicator will exist half in each form at the end point and display a color intermediate to the two forms.

Ferroin is a very stable, intensely red complex of 1,10-phenanthroline with Fe(II):

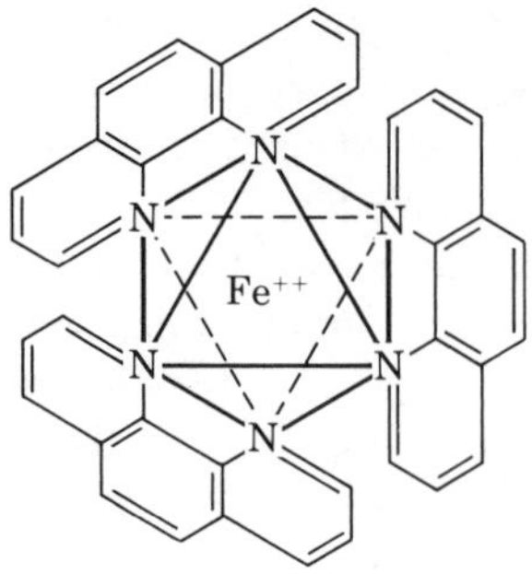

or $(Phen)_3Fe^{+2}$

When oxidized, the ferric complex is faintly blue:

$$Phen)_3Fe^{+3} + e^- \rightleftharpoons (Phen)_3Fe^{+2}; \qquad E' = 1.06 \text{ V in } 0.5 \ M \ H_2SO_4$$
$$\text{blue} \qquad\qquad\qquad \text{red}$$

$$E = 1.06 - 0.059 \log \frac{\text{(red form)}}{\text{(blue form)}}$$

With the Nernst equation, it is easy to show that this indicator exists 90% in the red form at $E = 1.00$ V, 50% in each form at $E = 1.06$ V, and 90% in the blue form at $E = 1.12$ V. The transition range is sufficiently close to the equivalence point potential (1.19 V) of the ferrous-ceric titration so that the end-point error is less than 0.1%. Most redox indicators exhibit complicated reaction mechanisms and do not exactly follow the Nernst equation. Nevertheless, a comparison of the formal potential of the indicator with the titration curve is very useful in choosing the best indicator. A few such indicators are listed in Table 18–8.

Table 18–8   Transition Potentials of Redox Indicators

Indicator	Transition Potential, V	Conditions
Phenosafranine	−0.25	pH 7
	0.28	pH 0
Indigo tetrasulfonate	−0.05	pH 7
	0.36	pH 0
Methylene blue	0.01	pH 7
	0.53	pH 0
Diphenylamine	0.76	pH 0
Diphenylamine sulfonic acid	0.85	pH 0
Ferroin	1.06	pH 0
Nitroferroin	1.25	pH 0

18–9   COUPLED REACTION SYSTEMS

Stepwise Oxidation and Reduction.   In a number of systems, it is possible to oxidize or reduce a substance in several stages in a fashion analogous to titrating protons one at a time from a polyprotic acid. Vanadium, for example, has stable

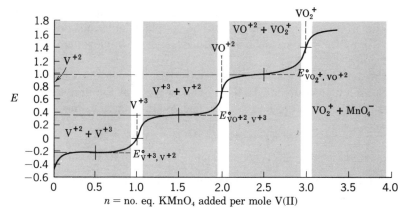

Figure 18-16 Potentiometric titration curve of vanadium(II) with permanganate.

oxidation states of (II), (III), (IV) and (V). Permanganate added in small increments to a solution of V(II) will oxidize the vanadium stepwise through each of the intermediate states. The composition and potential of the solution are represented in Figure 18–16. The curve is, of course, the titration curve one would obtain with a platinum indicator electrode. The details of the calculations will not be given, but it should be noted that up to 3 equivalents of permanganate added, the curve is independent of the nature of the oxidizing agent (titrant) used. In each of the first three regions, the potential is determined by the appropriate vanadium couple:

Eq. of MnO_4^- per mole of V	Appropriate Nernst Equation	
0–1	$E = E^\circ_{V^{+3}, V^{+2}} - 0.059 \log [V^{+2}]/[V^{+3}]$	(18–38)
1–2	$E = E^\circ_{VO^{+2}, V^{+3}} - 0.059 \log [V^{+3}]/[VO^{+2}][H^+]^2$	(18–39)
2–3	$E = E^\circ_{VO_2^+, VO^{+2}} - 0.059 \log [VO^{+2}]/[VO_2^+][H^+]^2$	(18–40)
>3	$E = E^\circ_{MnO_4^-, Mn^{+2}} - 0.059/5 \log [Mn^{+2}]/[MnO_4^-][H^+]^8$	(18–41)

The calculation of E at exactly 1 and 2 equivalents added is perhaps not obvious. Consider the composition of the solution when exactly 1 equivalent of oxidant is added. The equilibrium constant for the reaction may be large but not infinite and therefore a small amount of V^{+2} must remain. But since a total of 1 equivalent of oxidant was added, the V^{+2} must be balanced by a like amount of VO^{+2}. To put it another way, 1 equivalent of V^{+2} was oxidized to V^{+3} which then partially disproportionated to V^{+2} and VO^{+2}:

$$2V^{+3} + H_2O \rightleftharpoons VO^{+2} + V^{+2} + 2H^+$$

Among other things, this indicates that at the first equivalence point:

$$[VO^{+2}] = [V^{+2}] \qquad (18\text{–}42)$$

Now, Equations 18–38 to 18–41 are each valid throughout all regions and in particular at 1 equivalent. The trick is to add Equations 18–38 and 18–39, then substitute from Equation 18–42, yielding:

$$E_{n=1} = \tfrac{1}{2}(E^\circ_{V^{+3}, V^{+2}} + E^\circ_{VO^{+2}, V^{+3}}) + 0.059 \log [H^+]^2 \qquad (18\text{–}43)$$

A similar argument applies to $E_{n=2}$.

In the vanadium system, the $E°$ values of successive couples are separated by several tenths of a volt (see Table 18–1). The simple calculations described above must be modified for systems where the successive $E°$ values are closer together than ~ 0.2 volt.

Coupled Reactions. The series of stepwise reactions in the vanadium system could be classified as coupled reactions. The product of the first step becomes a reactant in the second step, and so on. The overall reaction has a tremendous driving force:

$$5V^{+2} + 3MnO_4^- + 4H^+ \rightleftharpoons 5VO_2^+ + 3Mn^{+2} + 2H_2O$$

$$E° = 1.15 \text{ V}; \qquad K \sim 10^{290}; \qquad \Delta G° = -4000 \text{ kcal}$$

In a series of stepwise processes, the free energy is utilized in smaller packages in a series of reactions:

$$V^{+2} + VO^{+2} + 2H^+ \rightleftharpoons 2V^{+3} + H_2O; \qquad \Delta G° = -147 \text{ kcal}$$

$$V^{+3} + VO_2^+ \rightleftharpoons 2VO^{+2}; \qquad \Delta G° = -153 \text{ kcal}$$

$$5VO^{+2} + MnO_4^- + H_2O \rightleftharpoons 5VO_2^+ + Mn^{+2} + 2H^+; \qquad \Delta G° = -600 \text{ kcal}$$

When the three preceding reactions are multiplied by the proper coefficients (5, 10, and 3, respectively) and added, the same overall reaction is obtained with the same total $\Delta G°$ as above. Now, suppose we add other redox couples to the vanadium-permanganate system. Suppose Fe^{+3} is added to V^{+2}. The Fe^{+3} is reduced to Fe^{+2} by the V^{+2}, forming Fe^{+2} and V^{+3}. Now if permanganate is added, it makes no difference to the final result whether the MnO_4^- reacts with Fe^{+2} or V^{+3} since any VO^{+2} formed will oxidize the Fe^{+2} back to Fe^{+3}. The exact course of the reactions will depend on relative rates of the several possible reactions and the possible presence of specific catalysts.

Far more interesting and complicated examples of coupled reactions are found in the metabolic processes of living organisms. The tremendous oxidizing power of oxygen is transferred in a series of stepwise oxidative processes producing many intermediates between the original reducing agent (foodstuff) and the final products ($CO_2 + H_2O$). In the complete oxidation of a mole of glucose, 688 kcal of energy are released. No organism could withstand the release of so much energy in one step. (It is something like touching a match to the gasoline tank of your car!) In metabolic processes, this energy is released in small packages. through a long series of reactions each of which involves small energy differences. The details of these are beyond the scope of this text, but the principles are essentially the same as in the vanadium system.

For example, in the respiratory chain of the mitochondrion, a series of compounds are intermediates in the oxidation of nicotinamide adenine dinucleotide (NAD). The direct oxidation of NAD with molecular oxygen would release 52.4 kcal per mole, all of which would be lost as destructive heat were it not for the sequential reactions of the respiratory chain. In the steps where ΔG is greater than 7 kcal, there is sufficient energy released to convert one molecule of adenosine diphosphate (ADP) to adenosine triphosphate (ATP) in a side reaction. Three of the sequential steps listed in Table 18–9 release enough energy to produce one ATP molecule each. Thus 3×7 kcal, or 21 kcal, of the 52 kcal released when one NAD molecule is oxidized are conserved by the simultaneous conversion of ADP to ATP. The ATP molecules so produced are the prime carriers of energy throughout living organisms. Each compound in the series listed in Table 18–9 participates in an electron transport sequence in which the end result is the transfer of electrons from NAD to oxygen.

Table 18–9 Respiratory Chain of the Mitochondrion

E'_7, V	Couple	Stepwise Reaction	$\Delta E'_7$, V	$-\Delta G$ kcal/mole
	O_2, H_2O			
0.8				
0.7				
0.6				
		$O_2 + Cyt\ a_{red} \rightarrow H_2O + Cyt\ a_{ox}$	0.53	24.4[a]
0.5				
0.4				
	Cyt a_{ox}, Cyt a_{red}			
0.3		Cyt a_{ox} + Cyt $c_{red} \rightarrow$ Cyt a_{red} + Cyt c_{ox}	0.03	1.4
	Cyt c_{ox}, Cyt c_{red}			
0.2				
		Cyt c_{ox} + Cyt $b_{red} \rightarrow$ Cyt c_{red} + Cyt b_{ox}	0.22	10.1[a]
0.1				
	Cyt b_{ox}, Cyt b_{red}			
0		Cyt b_{ox} + Flpr$_{red} \rightarrow$ Cyt b_{red} + Flpr$_{ox}$	0.09	4.1
	Flpr$_{ox}$, Flpr$_{red}$			
−0.1				
		Flpr$_{ox}$ + NAD$_{red} \rightarrow$ Flpr$_{red}$ + NAD$_{ox}$	0.29	12.4[a]
−0.2				
−0.3				
	NAD$_{ox}$, NAD$_{red}$			
		Overall Reaction O_2 + NAD$_{red} \rightarrow H_2O$ + NAD$_{ox}$	1.16	52.4

[a] Steps with sufficient energy to convert ADP → ATP.

QUESTIONS AND PROBLEMS

18–1. Describe the cell for which the following are cell reactions:
(a) $Ag^+ + Cl^- \rightleftharpoons AgCl$
(b) $Cd + 2Ag^+ \rightleftharpoons Cd^{+2} + 2Ag$
(c) $6Ti^{+3} + Cr_2O_7^{2-} + 2H^+ \rightleftharpoons 6TiO^{+2} + 2Cr^{+3} + H_2O$
(d) $CH_3COO^- + H^+ \rightleftharpoons CH_3COOH$
(e) $2Ag + Hg_2Cl_2 \rightleftharpoons 2AgCl + 2Hg$
(f) $Cu^{+2} + H_2 \rightleftharpoons Cu + 2H^+$
(g) $Cu^{+2} + 4NH_3 \rightleftharpoons Cu(NH_3)_4^{+2}$ (log $K_f = 13.3$)

Ans. Ag|AgCl, Cl⁻‖Ag⁺|Ag

18–2. If the concentrations of all soluble species in the cells of Problem 18–1 are 0.1 M and gases are at unit pressure, what is the voltage of each cell? Which electrode bears the positive charge? *Ans.* (a) 0.459 V, (c) 1.22 V, (d) 0.222 V

18–3. Write the cell reactions and compute the cell voltage for each of the following cells:
(a) $Cd|Cd^{+2}(0.1\ M)\|Cd^{+2}(0.5\ M)|Cd$
(b) $Pt|V^{+3}(0.1\ M), VO^{+2}(0.01\ M), H^+(0.1\ M)\|SCE$
(c) $Ag|Ag^+(0.01\ M)\|Fe^{+2}(0.1\ M), Fe^{+3}(0.01\ M)|Pt$
(d) $SCE\|H^+(10^{-7}\ M), Q(10^{-5}\ M), H_2Q(10^{-5}\ M)|Pt$
   Q = quinone
(e) $Sb|Sb_2O_3, H^+(10^{-7}\ M)\|H^+(10^{-7}\ M)|H_2(1\ atm), Pt$

Ans. (a) 0.021 V, (c) 0.031 V, (e) 0.152 V

18–4. Compute K for each of the reactions in Problem 18–1. *Ans.* (a) 10^{125}

18–5. The lead storage battery is represented as

$$Pb|PbSO_4, H_2SO_4|PbO_2$$

Write the electrode reactions and the cell reaction. Show that the density of the electrolyte is a measure of the "charge condition" of the battery.

18–6. If the saturated calomel electrode had been selected as the standard reference electrode, $E_{SCE} = 0$, what would be the standard potentials of each of the following couples?

(a) $Zn^{+2} + 2e^- \rightleftharpoons Zn$
(b) $V^{+3} + e^- \rightleftharpoons V^{+2}$
(c) $Sn^{+4} + 2e^- \rightleftharpoons Sn^{+2}$
(d) $2H^+ + 2e^- \rightleftharpoons H_2$
(e) $Hg_2Cl_2 + 2e^- \rightleftharpoons 2Hg + 2Cl^-$

Ans. (a) -1.005 V, (c) -0.09 V

18–7. What is the voltage of a cell consisting of a quinhydrone electrode in a pH 4.00 buffer and a saturated calomel reference electrode? *Ans.* 0.22 V

18–8. The voltage of the cell given below is -0.500 V. What is the pH of the unknown solution?

$$Ag|AgCl, Cl^-(1.00\ M)||H^+(unknown)|H_2, (0.50\ atm), Pt$$

Ans. pH $= 8.63$

18–9. Calculate the equilibrium constant for the reaction:

$$2H_2 + O_2 \rightleftharpoons 2H_2O\ (liq)$$

Describe the cell for which this would be the cell reaction. How would the cell voltage depend on the pH of the electrolyte? *Ans.* 2.5×10^{83}

18–10. An excess of silver metal is added to 250 ml of 0.01 M Fe^{+3}. What is the final concentration of Fe^{+3} in the solution if it contained in addition to the Fe^{+3}: (a) 1 M HNO_3 (b) 1 M HCl? *Ans.* (a) 2.7×10^{-4} M, (b) 5×10^{-12} M

18–11. On a single graph, plot the electrode potential versus log C of the ionic species for the following:

(a) $Ag|Ag^+$
(b) $Cu|Cu^{+2}$
(c) $Pt, H_2|H^+$
(d) $Al|Al^{+3}$
(e) $Fe|Fe^{+2}$
(f) $Ag|AgCl, Cl^-$
(g) $Zn|Zn^{+2}$
(h) $Pt|Q, H_2Q, H^+$

18–12. Zinc metal will reduce vanadic solutions according to the reaction:

$$Zn + 2V^{+3} \rightleftharpoons Zn^{+2} + 2V^{+2}$$

If an excess of zinc metal is added to 100 ml of 0.150 M VCl_3, what will be the concentrations of V^{+3}, V^{+2}, and Zn^{+2} ions at equilibrium?

Ans. 1.01×10^{-10} M, 0.15 M, 0.075 M

18–13. A metal becomes a better reducing agent if the solution in contact with it contains a complexing agent capable of complexing the metal ion. Explain why this is so.

18–14. Cupric ion is an oxidizing agent and ferrous ion is a reducing agent. What happens when the two are mixed? Explain.

18–15. Compute E_7' for the MnO_4^-, Mn^{+2}, and the $Cr_2O_7^{-2}$, Cr^{+3} couples. *Ans.* 0.85 V, 0.37 V

18–16. Compute the solubility product for AgBr from the $E°$ values in Table 18–1.

18–17. Describe the experimental set-up required for the potentiometric titration of V^{+2} with MnO_4^-.

18–18. A 50.0-ml aliquot of 0.0500 M Sn^{+2} is titrated with 0.100 M Ce^{+4}. Calculate the potential of a platinum indicator electrode immersed in the titration mixture after the addition of 10, 25, 45, 50, and 60 ml of titrant.

Ans. 0.13 V, 0.15 V, 0.18 V, 0.64 V, 1.57 V

18–19. Calculate the equivalence point potential in each of the following titration reactions all in 1 M H_2SO_4:

(a) $Ce^{+4} + Cr^{+2} \rightleftharpoons Ce^{+3} + Cr^{+3}$

(b) $VO_2^+ + Fe^{+2} + 2H^+ \rightleftharpoons VO^{+2} + Fe^{+3} + H_2O$

(c) $VO^{+2} + Cr^{+2} + 2H^+ \rightleftharpoons V^{+3} + Cr^{+3} + H_2O$

(d) $5Sn^{+2} + 2MnO_4^- + 16H^+ \rightleftharpoons 5Sn^{+4} + 2Mn^{+2} + 8H_2O$

Ans. (a) 0.60 V, (b) 0.90 V, (c) 0.01 V, (d) 1.16 V

18–20. What is the molarity of a permanganate solution, if 35.00 ml is needed to titrate 50.00 ml of an oxalic acid solution? Another 50.00 ml aliquot of the oxalic acid solution required 45.00 ml 0.200 M NaOH in an acid–base titration. *Ans.* 0.0514 M

18–21. A 0.500-g sample of iron ore containing both FeO and Fe_2O_3 is dissolved in acid, reduced with silver metal to Fe^{+2} and titrated with 0.0100 M $KMnO_4$, requiring 25.00 ml. What is the % Fe in the ore sample? *Ans.* 13.96% Fe

18–22. What weight of an iron ore should be used for analysis in order that the volume of 0.1000 N $K_2Cr_2O_7$ used in its titration will be numerically the same as the % Fe_2O_3 in the ore? *Ans.* 0.798 g

REFERENCES

J. J. Lingane, *Electroanalytical Chemistry*, 2nd ed., Wiley-Interscience, New York, 1958. A very readable and definitive discussion of the theory and applications of potentiometry, various kinds of electrolysis, and coulometry.

W. M. Latimer, *Oxidation Potentials*, 2nd ed., Prentice-Hall, Englewood, N.J., 1952. The most extensive and widely quoted source of standard potentials and related thermodynamic data. It includes much descriptive chemistry of the elements.

W. M. Clark, *Oxidation-Reduction Potentials of Organic Systems*. Williams and Wilkins, Baltimore, Md., 1960. An extensive discussion of theory as applied to organic chemistry with many tables of data.

A. L. Lehninger, *Bioenergetics*, Benjamin, New York, 1965. A lucid discussion of energy transport and storage in living systems.

T. S. Licht and A. J. deBethune, "Recent Developments Concerning the Signs of Electrode Potentials," *J. Chem. Educ.*, **34**, 433 (1957).

R. A. Durst, "Mechanism of the Glass Electrode Response," *J. Chem. Educ.*, **44**, 175 (1967).

H. A. Laitinen and W. E. Harris, *Chemical Analysis*, 2nd ed., McGraw-Hill, New York, 1975.

R. G. Bates, *Determination of pH, Theory and Practice*, Wiley, New York, 1973.

G. A. Rechnitz, "Ion Selective Electrode," *Chem. Eng. News*, June 12, 1967, p. 146; Jan. 27, 1975, p. 29.

19

SOME OTHER
ELECTROCHEMICAL
TECHNIQUES

The introduction to electrochemical cells, electrode potentials, and potentiometry given in Chapter 18 covered only a small portion of electroanalytical chemistry. There are many other methods based on electrical measurements. Electrochemical methods are very useful because the measurements can be made quickly, easily, and precisely, and are readily automated.

19-1 ELECTRODEPOSITION

Electrolysis. For potentiometry, the cell is not allowed to operate. With zero current, it is assumed that the electrode is in equilibrium with the solution and that concentrations do not change with time. Such cells, however, possess an e.m.f. whether or not they are operating, and are sources of electrical energy if the circuit is completed and current allowed to flow. With a finite current, the concentrations must change as the cell reaction proceeds. The e.m.f. gradually approaches zero as the concentrations approach the equilibrium state. Now let us add another component to the circuit. The cell in Figure 19–1 is connected to an external battery or power supply, E_{appl}, so that the two e.m.f.'s oppose each other. If E_{appl} is greater than E_{cell}, then the current is reversed. Likewise, the cell reaction is reversed from its normal spontaneous direction. Instead of copper metal dissolving into the solution, copper is deposited on the electrode surface. The normal spontaneous cell reaction for this cell is:

$$\text{Pt Electrode:} \quad O_2 + 4H^+ + 4e^- \rightleftharpoons 2H_2O$$
$$\text{Cu Electrode:} \quad Cu^{+2} + 2e^- \rightleftharpoons Cu$$
$$\text{Cell:} \quad 2Cu + O_2 + 4H^+ \rightleftharpoons 2Cu^{+2} + 2H_2O$$

with corresponding potentials:

$$E_{Pt} = E^{\circ}_{O_2, H_2O} - \frac{0.059}{4} \log \frac{1}{P_{O_2}[H^+]^4}$$

$$= 1.23 - \frac{0.059}{4} \log \frac{1}{(0.2)(0.2)^4} = 1.18 \text{ V}$$

$$E_{Cu} = E^{\circ}_{Cu^{+2}, Cu} - \frac{0.059}{2} \log \frac{1}{[Cu^{+2}]}$$

$$= 0.34 - \frac{0.059}{2} \log \frac{1}{0.1} = 0.31 \text{ V}$$

$$E_{cell} = E_{Pt} - E_{Cu} = 1.18 - 0.31 = 0.87 \text{ V}$$

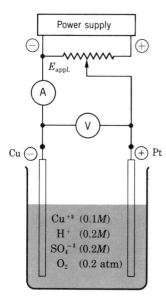

Figure 19–1 Apparatus for the electrodeposition of copper. A is an ammeter, V is a voltmeter.

The cell voltage is often called the "back e.m.f." If E_{appl} just equals 0.87 V, the two e.m.f.'s cancel each other and no current flows. But if E_{appl} exceeds 0.87 V, the cell reaction is reversed, and *electrolysis* of the solution takes place. In actual cases, an E_{appl} slightly greater than E_{cell} is required to drive the cell reaction backwards. The extra voltage, called the *overvoltage*, is usually caused by the irreversibility of the electrode reactions. The basic equation for electrolysis is:

$$E_{appl} = E_{cell} + E_{ov} + iR \qquad (19-1)$$

in which all quantities are considered to be positive. E_{ov} is the sum of the overvoltages (if any) at the electrodes, i is the current, and R is the total resistance of the circuit. With current flowing, the concentrations change with time which results in changes in all three terms on the right side of Equation 19–1. There are many kinds of electrolysis which make use of Equation 19–1. One or more of the terms in this equation may be held constant giving rise to several special techniques which will be described briefly.

Electrodeposition. If the electrolysis of copper just described is continued for any length of time, the concentration of cupric ion in the electrolyte must decrease. In fact, copper ion consumed at the cathode is being replaced by hydrogen ion produced in the anode reaction. The net result is that the cell voltage (back e.m.f.) must increase; therefore, the applied voltage must be increased in order to maintain the current flow. If the current is continued long enough to be sure that the cell reaction is essentially completed, the original concentration of copper ion can be determined from the increase in weight of the electrode. This has been the classical way to determine copper and a few other metals, but it finds little use in organic analysis.

19–2 CONTROLLED CURRENT ELECTROLYSIS

If E_{appl} is continuously adjusted to maintain a constant current, and if conditions are such that only a single electrode reaction takes place at the electrode (100%

current efficiency), then it is not necessary to weigh the electrode. The amount of the reaction can be readily determined from Faraday's law:

$$F \times \text{equiv reacted} = i \times t \tag{19-2}$$

where t is the time in seconds and F is the number of coulombs per equivalent (96493) or the Faraday constant. However, we have noted in the electrolysis of copper, that the cathode potential must be made more negative with time in order to continue the reduction of copper ion. Concurrently the hydrogen ion concentration is increasing, and eventually the cathode potential will be sufficiently negative to reduce hydrogen ion as well as copper ion. This is a typical situation—that is to say, it is rarely possible to maintain a single electrode reaction. The method, as such, finds little use, but is the basis for coulometric titrations to be described shortly.

Example/Problem 19–1.   Copper is to be electroplated from a solution containing 0.1 M CuSO$_4$ and 0.1 M H$_2$SO$_4$ in a cell similar to Figure 19–1. The initial E_{appl} required must be at least 0.87 V. The overvoltage for deposition of soft metals at the cathode is usually very small, but the discharge of oxygen on the platinum anode may require an extra 0.4 V. If we wish to maintain a current of 0.25 A and the total resistance is 0.2 ohms, the iR drop will require an additional 0.05 V.

$$E_{appl} = \quad 0.87 \quad + \quad 0.4 \quad + \quad 0.05 \quad = \quad 1.32 \text{ V}$$
$$\qquad\quad \text{back e.m.f} \qquad \text{overvoltage} \qquad iR \text{ drop}$$

What concentration of copper ion remains in solution at the point when hydrogen ion begins to discharge? First we must note that the overvoltage required for the discharge of hydrogen on copper is ~ 0.4 V. Therefore, as a first guess, nearly all of the copper ion can be reduced before any hydrogen ion is reduced. But an equivalent amount of hydrogen ion is produced at the anode so that the concentration of H$^+$ increases to 0.2 (original) + 0.2 (added) = 0.4 M. The potential required to discharge H$^+$ is

$$E_H = E_{ov} + E^{\circ}_{H^+, H_2} - \frac{0.059}{2} \log \frac{P_{H_2}}{[H^+]^2}$$

$$= -0.4 + 0 - \frac{0.059}{2} \log \frac{1}{(0.4)^2} = -0.4 \text{ V}$$

Note that the overvoltage always acts in a direction to make the process more difficult to perform. In this case H$^+$ is to be reduced which requires electrons from the electrode. Because of overvoltage, the potential of the electrode must be more *negative* by 0.4 V in order to reduce H$^+$ on copper; therefore, E_{ov} is given a negative sign.

The copper ion is in equilibrium with the same electrode at the same potential. Its concentration is obtained from:

$$E_H = E_{Cu} = E^{\circ}_{Cu^{+2}, Cu} - \frac{0.059}{2} \log \frac{1}{[Cu^{+2}]}$$

$$-0.4 = 0.34 - \frac{0.059}{2} \log \frac{1}{[Cu^{+2}]}$$

$$[Cu^{+2}] = 10^{-25} \text{ } M$$

The fraction of the original copper remaining in solution is only $10^{-25}/0.1$ or $10^{-22}\%$.

19–3   CONTROLLED CATHODE (OR ANODE) POTENTIAL ELECTROLYSIS

If a (third) auxiliary reference electrode is placed within the electrolysis cell and connected through a voltage measuring device, then the potential of (either) one of the working electrodes can be monitored, as shown in Figure 19–2. The E_{appl} is then adjusted throughout the electrolysis to maintain a constant cathode or anode potential. In this way, we can be fairly sure that only a single reaction is taking place and that the number of coulombs passed is directly related to the substance to be determined. However, the current decreases with time and approaches zero asymptotically. Therefore, Equation 19–2 must be modified:

$$F \times \text{equiv reacted} = \int_0^t i \, dt \qquad (19\text{–}3)$$

In addition to analytical applications, controlled potential electrolysis is useful in identifying reaction products and the number of electrons taking part in a reaction. (These are not always obvious in complex organic reactions!) The method also serves as a preparative tool. The electrode acts as an oxidant or a reductant with a constant known potential of any value one desires. Some preparations for which no suitable reagents are available can be easily carried out.

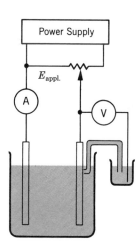

Figure 19–2   Apparatus for electrolysis with controlled cathode (or anode) potential. Signal from voltmeter may be used to adjust E_{appl} automatically as required.

19–4   SECONDARY COULOMETRIC TITRATIONS

We have seen that with controlled cathode (or anode) potential electrolysis it is relatively easy to maintain a single electrode reaction, or 100% current efficiency. However, the current necessarily decreases with time in a logarithmic fashion so that the area under the i versus t curve must be found by integration. Constant current electrolysis allows an easy determination of coulombs, $i \times t$, but since the electrode potential necessarily changes with time, other electrode reactions must occur. Therefore we cannot maintain 100% current efficiency with respect to the desired reaction.

Consider the electrolytic oxidation of ferrous ion at a platinum anode. If the current is kept constant, the extent of reaction is directly proportional to the time of electrolysis, but the electrode potential must increase and another reaction will eventually take over. On the other hand, if the potential is kept constant, only the desired reaction will occur, but the rate of reaction will decrease exponentially.

The advantages of both these techniques are neatly combined in a *secondary coulometric titration.* In this method the substance to be determined is indirectly "titrated" with a reagent that is electrolytically generated *in situ.* Instead of oxidizing ferrous ion at the electrode, as in the above examples, let us add a large excess of cerous ion prior to the electrolysis. When the current is started, the ferrous ion in the vicinity of the anode is exhausted. The potential tends to increase rapidly, but is prevented from doing so because of the large amount of cerous ion available for oxidation. The ceric ion so produced is immediately mixed with the solution and reacts with the remaining ferrous ion. The concentration of ceric ion stays at the trace level until all of the ferrous ion is oxidized. The end point of the ferrous reaction can be monitored by a visual (redox) indicator, by potentiometry using an extra set of electrodes and a separate circuit, or by amperometry (to be described later). Thus it is possible to maintain a reasonably constant electrode potential for the duration of the electrolysis that is carried out at constant current.

A typical coulometric titration apparatus includes a titration beaker and two independent circuits as shown in Figure 19–3. The generator circuit requires a source of constant current (which should be adjustable over a range of values) and a pair of electrodes, one of which generates the titrant and the other of which is separated from the solution by a fritted glass connection. For precision and convenience, the switch controlling the current source also controls a timer. The detector circuit requires a suitable indicator electrode, reference electrode, and potentiometer. The signal from the potentiometer can be used to stop the titration and the timer. For repetitive determinations, the timer can be calibrated in grams, equivalents, or even % composition of a particular weight of sample.

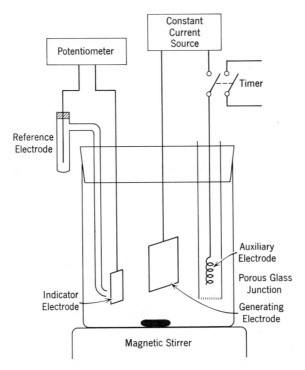

Figure 19–3　Coulometric titration apparatus with potentiometric end point.

Table 19–1 Typical Titrants
Generated Coulometrically

Solution	Titrant
Na_2SO_4 (H_2O)	H^+ and OH^-
$NaCl$ (H_2SO_4)	Cl_2
$NaBr$ (H_2SO_4)	Br_2
KI (H_2SO_4)	I_2
$Ce_2(SO_4)_3$ (H_2SO_4)	Ce^{+4}
$CuSO_4$ (HCl)	$CuCl_2^-$
$CrCl_3$ $(NaOH)$	CrO_4^{-2}
$FeNH_4$ $(SO_4)_2$ (H_2SO_4)	Fe^{+2}
$AgNO_3$ (HNO_3)	Ag^{+2}
$TiCl_4$ (HCl)	Ti^{+3}
$NaSCN$ $(HClO_4)$	SO_4^{-2}
HgY^{-2} (NH_3, NH_4^+)	Y^{-4}
$KAg(CN)_2$ $(NaOH)$	CN^-

Many types of titrants normally used in volumetric analysis can be generated with 100% current efficiency at an electrode. These include oxidants, reductants, acids, bases, precipitants, complexing agents, and several metal ions. There is no need to prepare or store standard solutions. Thus, chlorine and chromous ion are as easy to use as sodium chloride. Some of the titrants that have been successfully generated are given in Table 19–1.

Chemical primary standards are replaced by electrical measurements, or, stated another way, the electron becomes the universal primary standard. Small currents and short times can be measured accurately. In principle, the accuracy is limited by the uncertainty in the value of the Faraday. In practice, the accuracy is limited by the reproducibility of the end point.

Acid–base titrations are readily performed by coulometry. Both acid and base can be generated either internally within the solution, or externally as shown in Figure 19–4. The external apparatus is particularly convenient for generating small amounts of carbonate-free base as needed.

A large number of organic compounds react with the halogens that are easily generated from the corresponding halide at a platinum electrode. The determinations of 8-hydroxyquinoline with electrolytically generated bromine and the bromine number of unsaturated compounds are two examples.

The determination of traces of pesticide residues in foods is a difficult but important analytical problem. The sensitivity of most analytical methods is not adequate, and even with the best gas chromatographs an ultra-sensitive, specific detector is required. Many common pesticides contain chlorine or sulfur which are readily converted to HCl and H_2S in reactors built into the end of the chromatographic column. After selective absorption of these gases in an electrolytic cell, the coulometric titrator with electrolytic generation of silver ion is an ideal detector for this problem.

The advantages of coulometric titration can be summarized as follows:

1. Small amounts of substances can be determined accurately.
2. No standard solutions are required.
3. Unstable intermediates can be used as titrants.
4. The apparatus required is inexpensive and easily automated.

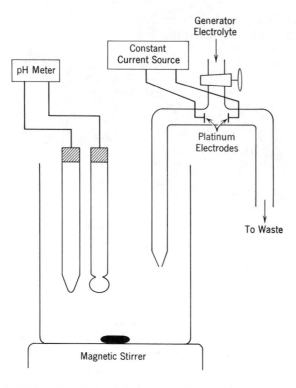

Figure 19–4 Coulometric apparatus for external generation of acid or base.

19–5   POLAROGRAPHY

If one of the electrodes in the electrolysis cell of Figure 19–1 is replaced with a micro-electrode, the current may become limited by the rate at which the reacting substance can reach the surface of the electrode. Such an electrode is said to be "polarized" because the current no longer depends on the potential, but rather on the diffusion rate of the electroactive species. Jaroslav Heyrovsky, a Czech chemist, first described this technique in 1922. Since then it has become a routine method for determining metal ions and electroactive organic compounds in the concentration range of 10^{-6} to 10^{-2} M.

The most commonly used micro-electrode is shown in Figure 19–5. It consists of small droplets of mercury issuing from a piece of small bore (inside diameter of about 0.06–0.08 mm) capillary tubing. Mercury is easily purified and in this arrangement its surface is continually renewed as the drop grows. An external reference electrode completes the electrolysis cell. The electrical circuit provides a source of continuously variable applied voltage and a device for measuring the very small electrolysis currents (10^{-8} to 10^{-4} A). The contact is driven across the slide-wire by a synchronous motor that increases the applied voltage at a constant rate; and the current is measured by means of an automatic recorder. On modern instruments the applied voltage is supplied by a combination of operational amplifier modules (see Section 25–9).

Polarography is the study of the current-voltage relationships obtained with this apparatus. With the proper conditions, the *polarogram* (current versus voltage) so obtained can be interpreted to yield both the nature and concentration of the

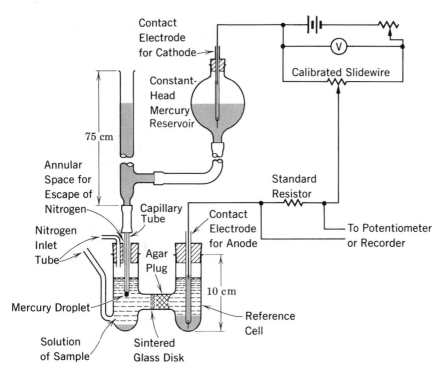

Figure 19–5 Polarographic cell (after Lingane and Laitinen) and basic electrical circuit required to obtain current-voltage curves.

reacting species. A typical polarogram for a solution containing 10^{-3} M Pb^{+2} and 10^{-3} M Zn^{+2} in 1 M KCl is shown in Figure 19–6.

The polarogram is obtained by slowly increasing the applied voltage so that the dme (dropping mercury electrode) becomes more negative. The polarographic current plotted in Figure 19–6 is apparently controlled by different factors in the regions of fairly constant current and the regions of sharp increase, the so-called "polarographic waves." At potentials less negative than -0.3 V (in the example

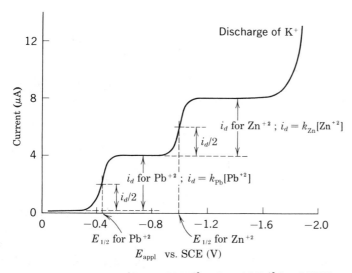

Figure 19-6 Polarogram for 1 mM Pb^{+2} and 1 mM Zn^{+2} in 1 M KCl.

given) no electrode process can take place, and therefore essentially no current flows. A small background current is usually observed because reducible impurities (especially oxygen) are often present. Also, a small current is required to charge each drop of mercury (which carries away its charge when it falls). All current measurements should be corrected for the background, or *residual current*, which varies slightly with applied potential.

At potentials between -0.3 and -0.5 V the current increases rapidly, because the lead ion in the solution can now be reduced to lead metal that forms an amalgam with the mercury electrode. In the region of the polarographic wave, the electrode becomes depolarized; that is, it obeys the Nernst equation. The current which flows is the result of the reduction of lead ion to give the appropriate ratio of $[Pb]_{amalg}/[Pb^{+2}]$ *at the electrode surface.* Because of the small surface area available, only a small amount of electrolysis occurs, and the concentration in the bulk solution does not change appreciably. (The solution must *not* be stirred.)

As the potential is made more negative, the ratio $[Pb]_{Hg}/[Pb^{+2}]$ should increase. However, the rate at which Pb^{+2} ion can reach the surface of the electrode is limited by the diffusion rate for Pb^{+2} in this particular solution. Thus, at potentials between -0.5 and -0.9 V, the current is limited by the rate of diffusion and is called the *diffusion current*, i_d. During the lifetime of a drop, 2 to 8 sec, the diffusion layer does not exceed a thickness of 0.05 mm and the concentration gradient through this layer is uniform. Under these conditions, the rate of diffusion, and the resulting current is directly proportional to the bulk concentration:

$$i_d = k[Pb^{+2}] \tag{19-4}$$

The proportionality constant, k, contains many factors and is more completely written as the Ilkovic equation:

$$i_d = 605 \, nCD^{1/2}m^{2/3}t_d^{1/6} \tag{19-5}$$

where i_d is the average current during the life of a drop (μA), n is the number of electrons in the half-reaction, C is the bulk concentration of the reacting species (mmole/liter), D is the diffusion coefficient (cm^2/sec), m is the rate at which mercury flows through the capillary electrode (mg/sec) and t_d is the average drop time (sec). Each reducible species in the solution creates its own wave at the proper potential, so that the polarogram may consist of a series of independent but additive steps. The *increase* in diffusion current at each step is directly proportional to the concentration of the corresponding substance.

The final current rise near -2.0 V is caused by the reduction of potassium ion in the solution. A large excess (at least 100-fold) of a supporting electrolyte greatly simplifies the interpretation of polarographic data. The supporting electrolyte, which can be an ionic substance less easily reduced (or oxidized) than the substance to be determined, serves a number of purposes. (a) The current is carried through the solution mostly by the ions of the supporting electrolyte. Thus, the substances reacting at the electrode move only through a thin diffusion layer because of the concentration gradient and not because of electrical attraction (migration). This is essential for the limiting current to be proportional to concentration. (b) The electrical resistance of the solution can usually be neglected. Under these circumstances, the potential of the dme is a linear function of the applied potential. It is customary to use a saturated calomel reference electrode and by polarographic convention, $E_{SCE} = 0$. Therefore, if both the iR drop and the junction potential are small, $E_{appl} = E_{dme}$. (c) The supporting electrolyte may consist of, or contain, a buffer or a complexing agent to change and control the nature of the electrode reaction.

Quantitative polarographic analysis requires attention to a number of details. All current measurements must be corrected for residual current. This can often be done by extrapolating the straight portion of the polarogram. The supporting electrolyte may contain reducible impurities that can cause errors. Likewise, dissolved oxygen is reducible and should be removed by purging the solution with high purity nitrogen for 15 min before taking a polarogram. The drop time and the rate of flow of mercury depend on the dimensions of the capillary and height (pressure) of the mercury column. Both of these factors must be controlled for reproducible results. The temperature coefficient of the diffusion current is about 1 to 2% per degree and consequently the cell should be thermostated to ±0.1°. Variations in all of these factors, plus uncertainties in the diffusion coefficient, require that standard solutions be run to calibrate a given determination. Not all substances yield well-formed polarograms with sharp waves and level plateaus. Thus there may be uncertainty in measuring the diffusion current. The instantaneous current changes continuously with the area of the growing drop and reaches a maximum at the time the drop falls, as shown in Figure 19–7. Either the average of the current oscillations or an envelop of the maximum values can be used as long as one is consistent. The several uncertainties just described mean that the usual error in polarographic analysis is 1 to 5%, although well-behaved systems allow an error of only 0.1%.

Qualitative analysis by polarography is based on the "half-wave potential", $E_{1/2}$, which is the potential of the electrode at a point where the current is one-half of the diffusion current for a given substance. For reversible systems, it can be shown that the half-wave potential is independent of concentration and is nearly

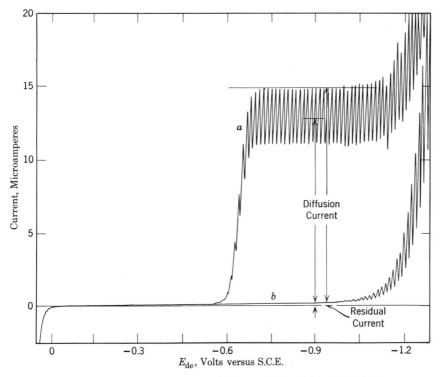

Figure 19–7 Polarograms of (a) 0.5 mM Cd^{+2} in 1 M HCl, and (b) 1 M HCl alone.

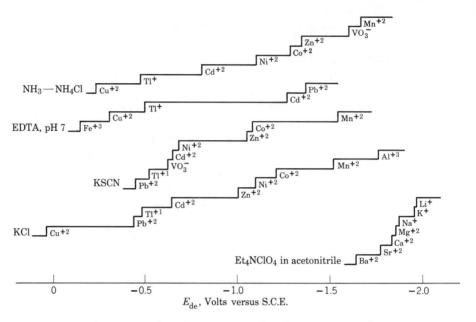

Figure 19–8 Half-wave potentials of selected metal ions in several supporting electrolytes.

equal to the $E°$ for the half-reaction. Irreversible systems tend to give drawn-out waves and the half-wave potential is less useful in identification. Like $E°$ values, $E_{1/2}$ values are shifted by pH, complexing agents, or any species that takes part in the half-reaction. Tables of half-wave potentials in various media are compiled in handbooks. Figure 19–8 gives selected values in graphical form.

Applications of polarography are limited to those species that can be oxidized or reduced at electrodes. The potential region is limited in the positive direction by the oxidation of the electrode material, about $+0.2$ V for mercury, and in the negative direction by the reduction of the supporting electrolyte or the solvent, about -2.0 V for KCl in water. A platinum micro-electrode will extend the anodic region to about $+1.0$ V and the use of nonaqueous solvents (acetonitrile) and organic salt-type (e.g., tetraethyl ammonium iodide) supporting electrolytes will extend the cathodic region to -2.5 V. The optimum concentration range is 10^{-2} to 10^{-4} M. The upper limit is set by the need for a 100-fold excess of supporting electrolyte, and the lower limit is set by the magnitude of the residual current which should not exceed the diffusion current. Special instrumentation is available to compensate for the residual current and allow determinations as low as 10^{-7} M.

Most of the metal ions and many inorganic anions are readily determined alone or in mixtures. Several classes of organic compounds are amenable to polarography. These include conjugated double bonds, aromatic ring systems, halogen compounds, carbonyls, quinones, dicarboxylic acids, keto acids, as well as many compounds which contain carbon–nitrogen, nitrogen–nitrogen, nitrogen–oxygen, oxygen–oxygen, sulfur–sulfur, and carbon–sulfur bonds.

Because of the reproducibility of the dropping mercury electrode and the ease of acquiring data, polarographic techniques are valuable for studying fundamental electrode processes. There are many variations in the electrodes and circuits used that are described in more advanced and specialized texts.

Related Techniques. A complete polarogram can be obtained during the lifetime of a single drop by making a *rapid-scan* of the applied voltage using an electronic *ramp generator*. The peak shapes are distorted from the usual waves, because the diffusion process is unable to supply reducible material to the electrode surface to satisfy the rapidly increasing potential.

In *cyclic voltammetry*, a triangular shaped voltage is applied during the last second or so before the drop falls. This gives a complete scan, forward and backward, so that the products of reduction can be reoxidized before they can diffuse away from the electrode surface. This is a convenient way of studying the degree of reversibility of electrode reactions.

The slowly increasing ramp voltage can be modulated by superimposing a small alternating potential of a few millivolts. This leads to an ac component of the current that reaches a maximum at the half-wave potential. *Ac polarography* makes it easier to resolve overlapping waves.

In order to eliminate the effect of the residual current, the ramp voltage may be modulated with a small *square wave*. During each pulse, the residual current (required to charge the drop) dies out quickly and the current is measured only at the end of the pulse. Other types of pulses have been used to increase resolvability and for studies of electrode kinetics.

A two-step method, called *anodic stripping analysis*, is useful for trace quantities of metals. In the first step, the metals are concentrated by cathodic deposition at a solid micro-electrode or on a single, hanging mercury droplet. In the second step, the applied voltage is reversed, and the metals are determined by a reverse (anodic) polarographic or coulometric analysis. A typical procedure includes a pre-electrolysis of about 15 min which is sufficient to reduce nearly all of the metal ions in the entire sample (a few ml). This greatly increases the sensitivity over conventional polarography in which only a negligible amount of the sample is actually used.

QUESTIONS AND PROBLEMS

19–1. Copper in an alloy is to be determined by electrodeposition from a solution of 0.5 M HCl (neglect the presence of chloro-complexes). If the alloy contains approximately 75% Cu, 10% Sn, and 15% Pb, what fraction of the Cu can be deposited before either of the other metals begins to deposit?   *Ans.* 6.6×10^{-16}%

19–2. Silver is to be deposited from a 0.010 M AgNO$_3$ solution which also contains 2.0 M HNO$_3$. The electrolysis cell includes two smooth platinum electrodes and its resistance is 0.25 ohm. The overvoltages at the anode and cathode are 0.85 and 0.05 V, respectively. What voltage must be applied in order to obtain an initial current of 0.75 A?   *Ans.* 1.65 V

19–3. Phenol can be titrated coulometrically by anodic generation of bromine as a titrant. The titration reaction is:

How many mg of phenol is present in a sample which requires 300 sec to titrate with a current of 25 mA?   *Ans.* 1.22 mg

19–4. The complete reduction of picric acid to triaminophenol would require 18 electrons per molecule. In an electrolytic reduction at a mercury cathode held at a constant potential of -0.65 V versus SCE, Lingane found that 65.7 coulombs were needed to reduce 0.0399 millimole of picric acid. What was the product of the electrolysis under these conditions?

Ans.

19–5. Copper in a brass tack was determined by coulometric titration with a constant current of 20.0 mA. If a 10.0 mg sample requires 400 sec, what is the % Cu in the tack?   *Ans.* 26.3% Cu

19–6. A zinc solution of unknown concentration gives a polarographic diffusion current of 5.00 μA. A 5.00-ml portion of 0.0010 M Zn^{+2} is added to 10.00 ml of the unknown solution, and the polarogram is taken again giving a diffusion current of 12.00 μA. What is the concentration of Zn^{+2} in the unknown solution?   *Ans.* 1.9×10^{-4} M

19–7. A solution of 2×10^{-3} M $CdCl_2$ in 0.1 M KNO_3 gives a diffusion current of 25.0 μA. The drop-time of the capillary is 15 drops per min and the weight of 25 drops is 100 mg. What is the diffusion coefficient of Cd^{+2} in this solution?

Ans. 1.1×10^{-6} cm^2/sec

19–8. The polarographic diffusion current of a 3.0×10^{-3} M $NiCl_2$ solution is 15 μA. If the solution volume is 15 ml, how long will it take to reduce 1% of the Ni^{+2} ion in the cell?

Ans. 5790 sec

REFERENCES

W. C. Purdy, *Electroanalytical Methods in Biochemistry*, McGraw-Hill, New York, 1965. Elementary discussion of techniques.

J. J. Lingane, *Electroanalytical Chemistry*, 2nd ed., Wiley-Interscience, New York, 1958.

R. N. Adams, *Electrochemistry at Solid Electrodes*, Decker, New York, 1969.

G. A. Rechnitz, *Controlled Potential Analysis*, Macmillan, New York, 1963.

I. M. Kolthoff and J. J. Lingane, *Polarography*, 2nd ed., Wiley-Interscience, New York, 1952.

L. Meites, *Polarographic Techniques*, 2nd ed., Wiley, New York, 1965.

P. Zuman, *Organic Polarographic Analysis*, Macmillan, New York, 1964.

P. Zuman and C. L. Perrin, *Organic Polarography*, Wiley-Interscience, New York, 1969.

ACIDS, BASES AND THEIR SALTS AND COMPLEXES

20

MONOPROTIC SYSTEMS

20–1 INTRODUCTION

Probably the most commonly used reagents in chemical laboratories are aqueous solutions of acids and bases. The properties and reactions of these substances are fundamental to all branches of chemistry and related sciences. The free hydrogen ion concentration dominates many chemical reactions by determining the extent to which the reaction proceeds, the rate at which it goes, or the detailed mechanism by which it takes place.

20–2 ACID–BASE CONCEPTS

The titration of a base with an acid is surely one of the simplest of all analytical determinations, and is one of the first experiments to be encountered by the student. Since detailed discussions of the stoichiometry and equilibria involved are found in all general chemistry texts, we provide here only a summary of the definitions and elementary principles, and then consider the complex equilibria that prevail in solutions of polyprotic systems, and the use of solvents other than water.

Table 20–1 summarizes several common definitions for acid–base systems. For our purposes, the Brønsted concept is the most useful. In this model, when an *acid* donates its proton (to some other species) the remaining fragment becomes the

Table 20–1 Some Definitions of Acids and Bases

Originator	Acid	Base	Neutralization
Arrhenius	Provides H^+ in water	Provides OH^- in water	$H^+ + OH^- \rightleftharpoons H_2O$
Brønsted-Lowry	H^+ donor	H^+ acceptor	Proton transfer
Lewis	Electron pair acceptor	Electron pair donor	Formation of a coordinate–covalent bond

conjugate base of the original acid. If the conjugate base accepts a proton, it reverts to the parent acid. Likewise, every base has a *conjugate acid.*

$$\text{Acid} \rightleftharpoons \text{H}^+ + \text{Base}$$
$$\text{HCl} \rightleftharpoons \text{H}^+ + \text{Cl}^-$$
$$\text{NH}_4^+ \rightleftharpoons \text{H}^+ + \text{NH}_3$$

20-3   ROLE OF THE SOLVENT

Free protons do not exist under normal conditions; therefore, an acid cannot donate its proton unless a base is present to accept it. Many solvents, including water, possess basic properties, that is, they can accept protons. Thus, when pure HCl is dissolved in water, a proton transfer reaction takes place:

conjugate pairs

$$\text{HCl} + \text{H}_2\text{O} \rightarrow \text{H}_3\text{O}^+ + \text{Cl}^-$$
acid 1 base 2   acid 2   base 1

H_2O—H_3O^+ and HCl—Cl^- are examples of conjugate pairs. Acid–base reactions involve the transfer of a proton; however, usually this is only a partial transfer and there is a competition between two bases for the proton:

$$\text{HOAc} + \text{OH}^- \rightleftharpoons \text{OAc}^- + \text{HOH}$$
acid 1   base 2   base 1   acid 2

Acetic acid, CH_3COOH, is commonly abbreviated as HOAc. In the above reaction, the OH^- and the OAc^- ions compete for the proton. The OH^- ion is a much stronger base than OAc^- ion, therefore the equilibrium lies far to the right. A comparison of the two reactions just given shows that water can act as either an acid or a base depending on the acidic or basic nature of the other partner in the reaction—water is an *amphoteric* substance, and it is also an *amphiprotic* solvent.

Amphiprotic solvents undergo *autoprotolysis*, or self-ionization, a proton transfer reaction where one solvent molecule (acting as an acid) donates a proton to another solvent molecule (acting as a base). Water is the most familiar example, but there are many others:

$$\text{H}_2\text{O} + \text{H}_2\text{O} \rightleftharpoons \text{H}_3\text{O}^+ + \text{OH}^-$$
acid$_1$ base$_2$ acid$_2$ base$_1$
$$\text{C}_2\text{H}_5\text{OH} + \text{C}_2\text{H}_5\text{OH} \rightleftharpoons \text{C}_2\text{H}_5\text{OH}_2^+ + \text{C}_2\text{H}_5\text{O}^-$$
$$\text{NH}_3 + \text{NH}_3 \rightleftharpoons \text{NH}_4^+ + \text{NH}_2^-$$
$$\text{HOAc} + \text{HOAc} \rightleftharpoons \text{H}_2\text{OAc}^+ + \text{OAc}^-$$
$$\text{SH} + \text{SH} \rightleftharpoons \text{SH}_2^+ + \text{S}^-$$

where SH is any amphiprotic solvent. The extent of the autoprotolysis of a solvent is very small and is measured by the ion product, or autoprotolysis constant, K_w or K_s:

$$[\text{H}_3\text{O}^+][\text{OH}^-] = K_w = 1.00 \times 10^{-14} \text{ at } 25°\text{C}$$
$$[\text{CH}_3\text{OH}_2^+][\text{CH}_3\text{O}^-] = K_s = 2 \times 10^{-17}$$
$$[\text{SH}_2^+][\text{S}^-] = K_s$$

Classification of Solvents. Solvents are classified with respect to acid–base properties as either *aprotic* (inert) or *amphiprotic*. Aprotic solvents are neither acidic nor basic (e.g., benzene and carbon tetrachloride). Amphiprotic solvents act as both proton acceptors and proton donors. There are gradations among the amphiprotic solvents, from predominantly acidic to predominantly basic. Glacial acetic acid is very acidic, whereas liquid ammonia is very basic. Water and ethanol are neither strongly acidic nor strongly basic. *Protogenic* amphiprotic solvents (e.g., sulfuric and formic acids) exhibit very strong acidic properties and very weak basic properties. Autoprotolysis constants for protogenic solvents are usually larger than that of water (e.g., $K_s = 10^{-6}$ for formic acid, whereas $K_s = 10^{-14}$ for water). Intermediate amphiprotic solvents possess weakly acidic protons and can also act as very weak bases. Autoprotolysis constants for nonaqueous intermediate amphiprotic solvents tend to be smaller than that of water (e.g., $K_s = 10^{-19}$ for ethanol). *Protophilic* amphiprotic solvents (e.g., ethylenediamine) exhibit very weak acidic properties and relatively strong basic properties; their autoprotolysis constants are usually less than that of water. A list of solvents which are commonly used for acid–base titrations is presented in Table 20–2. The dielectric constants and autoprotolysis constants are indicated where they are known.

Table 20–2 Solvents Commonly Used for Acid–Base Titrations

Solvent Classification	Subclassifi- cation	Name	Dielectric Constant (Debye units)	Autoprotolysis Constant pK_s
Aprotic		Chlorobenzene	5.8	—
		Acetonitrile	37.5	—
		Acetone	20.7	—
		Chloroform	4.8	—
		Methyl ethyl ketone	18.5	—
		1,4-Dioxane	2.2	—
Amphiprotic	Protogenic	Acetic acid	6.2	14.45
		Formic acid	58.0	6.2
	Intermediate	Water	78.5	14.0
		Methanol	32.6	16.7
		Ethanol	24.3	19.1
		Isopropanol	18.3	—
		Ethylene glycol	37.7	—
	Protophilic	Ethylenediamine	14.2	—
		Pyridine	12.3	—
		Ammonia	17.0	—
		Aniline	6.9	—
		N,N-Dimethyl formamide	34.8	—

Dielectric Constant Effects. The strength of an acid, HB, in a given solvent, SH, is defined in terms of the extent to which the reaction $HB + SH \rightleftharpoons SH_2^+ + B^-$ proceeds. This reaction is a combination of two steps, ionization and dissociation:

$$HB + SH \;\rightleftharpoons\; SH_2^+B^- \;\rightleftharpoons\; SH_2^+ + B^-$$
$$\quad\;\; \text{ionization} \quad \text{ion-pair dissociation}$$

In aqueous solutions both of these steps occur rapidly and aqueous acid–base reactions are thus very fast reactions (e.g., much faster than aqueous precipitation, complexometric, and oxidation–reduction reactions). In any solvent, the extent of the ionization step depends on the relative strength of the conjugate acid–conjugate base pairs. The extent of the dissociation step depends on the charge type of the members of the ion-pair and the polarity of the solvent. The dielectric constant is a measure of this polarity—the higher the dielectric constant, the more polar the solvent. The extent of dissociation of ion-pair aggregates increases with the dielectric constant of the solvent. Clearly then, solvents with high dielectric constants are necessary for complete dissociation. The dielectric constant of a vacuum is arbitrarily defined as zero. The very polar water molecule has a dielectric constant of 78.5 whereas the value for the slightly polar acetic acid molecule is 6.2. Consequently, in water all products of acid–base reactions at moderate to low concentrations are essentially completely dissociated into solvated ions, whereas in glacial acetic acid these products exist principally as ion-pairs or ion-pair aggregates. Most of the more elementary techniques used to study acid–base reactions give a measure of the concentrations of one or more of the solvated ions. Thus, the overall dissociation constant, K_a, is the value normally determined, but the dielectric constant of the solvent can have a pronounced effect on the extent of dissociation.

20–4 RELATIVE STRENGTHS OF ACIDS AND BASES

The strengths of different acids can be measured by comparing their ability to donate a proton to some common base. Likewise the strengths of different bases can be measured by comparing their ability to accept a proton from some common acid. Water is selected as both the common acid and the common base for these comparisons. The quantitative measure of these strengths is the equilibrium constant for the appropriate proton transfer reaction. For acids we use K_a, the *acid dissociation or "acidity" constant*. For bases we use K_b, the *base dissociation or "basicity" constant*:

$$HCl + H_2O \rightarrow H_3O^+ + Cl^-$$

or

$$HCl \qquad \rightarrow H^+ \quad + Cl^-; \quad K_a = \frac{[H^+][Cl^-]}{[HCl]} \gg 1$$

$$HOAc + H_2O \rightleftharpoons H_3O^+ + OAc^-$$

or

$$HOAc \qquad \rightleftharpoons H^+ \quad + OAc^-; \quad K_a = \frac{[H^+][OAc^-]}{[HOAc]} = 1.75 \times 10^{-5}$$

$$C_6H_5OH + H_2O \rightleftharpoons H_3O^+ + C_6H_5O^-$$

or

$$C_6H_5OH \qquad \rightleftharpoons H^+ \quad + C_6H_5O^-; \quad K_a = \frac{[H^+][C_6H_5O^-]}{[C_6H_5OH]} = 1 \times 10^{-10}$$

$$NH_3 + H_2O \rightleftharpoons NH_4^+ + OH^-; \quad K_b = \frac{[NH_4^+][OH^-]}{[NH_3]} = 1.78 \times 10^{-5}$$

$$C_6H_5NH_2 + H_2O \rightleftharpoons C_6H_5NH_3^+ + OH^-; \quad K_b = \frac{[C_6H_5NH_3^+][OH^-]}{[C_6H_5NH_2]} = 1.5 \times 10^{-9}$$

Note that water is also the solvent and is present in large excess. Therefore its concentration (and activity) remains essentially constant. By convention, the activity of a pure solvent is taken as unity, and its concentration does not appear in the mass action expression.

The relative values of the K_a's and K_b's indicate that hydrochloric acid is a stronger acid than acetic acid, acetic acid is stronger than phenol (carbolic acid), and ammonia is a stronger base than aniline.

The dissociation of an acid produces its conjugate base, which is, of course, capable of accepting a proton from the solvent (or any other acid present). For example:

$$HBO_2 \rightleftharpoons H^+ + BO_2^-; \quad K_a = \frac{[H^+][BO_2^-]}{[HBO_2]} = 5.8 \times 10^{-10}$$

$$BO_2^- + H_2O \rightleftharpoons HBO_2 + OH^-; \quad K_b = \frac{[HBO_2][OH^-]}{[BO_2^-]} = 1.7 \times 10^{-5}$$

This reaction of borate ion with water is often called the *hydrolysis* of the borate ion, but it is in fact entirely analogous to the proton transfer reaction of any base with water, producing its conjugate acid and a hydroxide ion. There is a simple relationship between the K_a for any acid and the K_b for its conjugate base. For example, if we multiply the K_a for HBO_2 by the K_b for BO_2^-, the result is K_w.

$$K_a \times K_b = \frac{[H^+][BO_2^-]}{[HBO_2]} \times \frac{[HBO_2][OH^-]}{[BO_2^-]} = [H^+][OH^-] = K_w$$

Therefore the numerical value of K_b for BO_2^- is

$$K_b = \frac{K_w}{K_a} = \frac{1.00 \times 10^{-14}}{5.8 \times 10^{-10}} = 1.7 \times 10^{-5}$$

In other words, the basic strength of the borate ion is comparable to that of ammonia. Similarly, the numerical value of K_a for NH_4^+ is

$$K_a = \frac{K_w}{K_b} = \frac{1.00 \times 10^{-14}}{1.78 \times 10^{-5}} = 5.62 \times 10^{-10}$$

Thus, the ammonium ion has about the same acid strength as boric acid, a very weak acid. Since the product of K_a and K_b for a conjugate pair always equals K_w, a constant, it follows that the stronger the acid, the weaker its conjugate base, and vice versa. From another point of view, a weak acid has little tendency to give up its protons, but its conjugate base has a great attraction for protons.

Tables of the numerical values of K_a and K_b for acids and bases pertain to water as the solvent by convention. These values are not directly transferable to other solvents although the relative order of their strengths is comparable. Some qualitative observations on the effect of the solvent are given in the next section.

Differentiating and Leveling Effects of Solvents. The most important characteristic of a solvent affecting the strength of an acid (or a base) is the acidity (or basicity) of the solvent itself. In aqueous systems, perchloric, nitric, and hydrochloric acids are all considered to be "strong acids," because the reactions

$$HCl + \underset{\text{solvent}}{H_2O} \rightarrow H_3O^+ + Cl^- \quad \text{(strong)}$$

$$HNO_3 + \underset{\text{solvent}}{H_2O} \rightarrow H_3O^+ + NO_3^-, \text{etc.} \quad \text{(strong)}$$

all proceed essentially 100% from left to right. Water as a solvent does not differentiate between the inherent strengths of these acids because water, even though it is a poor base, is nevertheless a stronger base than the conjugate base anions: ClO_4^-, NO_3^-, Cl^-, and so on. In water, therefore, these acids are *leveled* to the strength of the solvated proton, H_3O^+; the H_3O^+ ion is the strongest acid that can exist in aqueous solution. For any solvent, S, the protonated solvent molecule, SH^+, called its *lyonium* ion, is the strongest acid that can exist in that solvent.

If a more acidic solvent than water is used, the strong acids mentioned above are less able to transfer their protons to solvent molecules.

$$HClO_4 + \underset{\text{solvent}}{HCOOH} \rightleftharpoons HCOOH_2^+ + ClO_4^- \qquad \text{(strong)}$$

$$HNO_3 + \underset{\text{solvent}}{HCOOH} \rightleftharpoons HCOOH_2^+ + NO_3^- \qquad \text{(weak)}$$

The first reaction is far more complete than the second, that is, perchloric acid is a stronger acid than nitric acid in a formic acid solvent. Formic acid *differentiates* strong acids, whereas water *levels* the strengths of the acids. Solvents which are less acidic (i.e., more basic) than water are even better leveling solvents for acids. For example, benzoic acid which is a weak acid in water, appears to have the same strength as "strong acids" in liquid NH_3. That is, the reactions

$$C_6H_5COOH + \underset{\text{solvent}}{NH_3} \rightarrow NH_4^+ + C_6H_5COO^- \qquad \text{(strong)}$$

$$HCl + \underset{\text{solvent}}{NH_3} \rightarrow NH_4^+ + Cl^- \qquad \text{(strong)}$$

both go completely to the right; liquid NH_3 levels the strength of any acid stronger than NH_4^+. A basic solvent will enhance the apparent acidity of any acid; for example, acetic acid, which is a weak acid in water, behaves like a strong acid in liquid NH_3. Phenol ($pK_a = 10.0$ in water) is too weak to titrate in aqueous solution, but becomes only moderately weak and easily titratable in liquid NH_3 or ethylenediamine.

There are analogous leveling and differentiating effects for bases. The strongest base that can exist in any solvent is the *lyate* ion (anion of the solvent); strong bases are leveled to the same strength as that of the lyate ion (OH^- ion in water, NH_2^- ion in liquid NH_3, OAc^- ion in glacial acetic acid).

In water:

$$NaOH + H_2O \rightarrow Na^+ + OH^- \qquad \text{(strong)}$$
$$Na_2O + H_2O \rightarrow 2Na^+ + 2OH^- \qquad \text{(strong)}$$
$$NaOC_2H_5 + H_2O \rightarrow Na^+ + C_2H_5OH + OH^- \qquad \text{(strong)}$$
$$NaNH_2 + H_2O \rightarrow Na^+ + NH_3 + OH^- \qquad \text{(strong)}$$
$$NH_3 + H_2O \rightleftharpoons NH_4^+ + OH^- \qquad \text{(weak)}$$

In liquid NH_3:

$$NaOH + NH_3 \rightleftharpoons Na^+ + H_2O + NH_2^- \qquad \text{(weak)}$$
$$NaNH_2 \rightarrow Na^+ + NH_2^- \qquad \text{(strong)}$$
$$Na_2O + 2NH_3 \rightleftharpoons 2Na^+ + H_2O + 2NH_2^- \qquad \text{(weak)}$$

In glacial acetic acid:

$$NaOH + HOAc \rightarrow Na^+ + H_2O + OAc^- \qquad \text{(strong)}$$
$$NH_3 + HOAc \rightarrow NH_4^+ + OAc^- \qquad \text{(strong)}$$

Briefly summarized, the leveling and differentiating effects tell us that:

1. An amphiprotic solvent *levels* the strength of acids that are stronger than its lyonium ion, and *levels* the strength of bases that are stronger than its lyate ion.
2. An amphiprotic solvent *differentiates* the strengths of acids that are weaker than its lyonium ion; and *differentiates* the strengths of bases that are weaker than its lyate ion.
3. To make a weak acid appear stronger, use a basic solvent.
4. To make a weak base appear stronger, use an acidic solvent.

At the end of this chapter we will show that some nonaqueous solvents offer distinct advantages in the titration of very weak acids and bases and for mixtures of several acids or bases.

20–5 EQUILIBRIA IN MONOPROTIC SYSTEMS

We turn now to the problem of calculating the hydrogen ion concentration (or pH) in a number of acid–base systems in water. These calculations give us a means of estimating the pH in various solutions including buffers and titration mixtures.

The following conventions are used in this chapter.

1. The hydronium (lyonium) ion, $H\cdot(H_2O)_n^+$ or H_3O^+, is used interchangeably with the hydrogen ion, H^+.
2. Brackets around a chemical formula indicate the molar concentration of that species. The activity coefficient is ignored (i.e., assumed to be unity), because in practice numerical values of activity coefficients are rarely available.
3. Equilibrium constants are assumed to be constant even though the distinction between activity and concentration is not made. These approximations are adequate for present purposes. The temperature is assumed to be 25°C.
4. $pH = -\log[H^+]$; $pOH = -\log[OH^-]$; $pK_w = -\log K_w$

$$\text{therefore } pH + pOH = pK_w = 14.00 \text{ at } 25°C.$$

5. Values for K_a and K_b are given in Tables 20–3 and 20–4.

pH Calculations in Simple Systems. In calculating the $[H^+]$ or pH of solutions of monoprotic acids, bases, and their salts, or mixtures thereof, the main difficulty arises in deciding which of the various possible species are the principal constituents of the solution. Once this has been determined, the appropriate equation is used:

1. Solution of a strong acid, HX

$$[H^+] = C_{HX} \tag{20–1}$$

2. Solution of a weak acid, HB

$$[H^+] = \sqrt{K_a(C_{HB} - [H^+])} \approx \sqrt{K_a C_{HB}} \tag{20–2}$$

3. Solution of the salt of a weak acid, NaB

$$[OH^-] = \sqrt{K_w/K_a(C_{B^-} - [OH^-])} \approx \sqrt{\frac{K_w}{K_a} C_{B^-}} \tag{20–3}$$

Table 20-3 Monoprotic Weak Acid Dissociation Constants at 25°C

Acid Name	Formula	K_a	pK_a
Acetic	CH_3COOH	1.75×10^{-5}	4.76
Benzoic	C_6H_5COOH	6.3×10^{-5}	4.20
Boric	HBO_2	5.8×10^{-10}	9.24
Chloroacetic	$ClCH_2COOH$	1.38×10^{-3}	2.86
Cyanic	$HOCN$	2.0×10^{-4}	3.70
Formic	$HCOOH$	1.77×10^{-4}	3.75
Glycolic	$HOCH_2COOH$	1.32×10^{-4}	3.88
Hydrocyanic	HCN	4.8×10^{-10}	9.32
Hydrofluoric	HF	6.75×10^{-4}	3.17
Hypochlorous	$HOCl$	2.95×10^{-8}	7.53
Lactic	$CH_3CHOHCOOH$	1.38×10^{-4}	3.86
Nitrous	HNO_2	5.1×10^{-4}	3.29
Phenol	C_6H_5OH	1.05×10^{-10}	9.98
Propionic	CH_3CH_2COOH	1.34×10^{-5}	4.87
Ammonium ion	NH_4^+	5.62×10^{-10}	9.25
Pyridium ion	$C_5H_5NH^+$	6.2×10^{-6}	5.21

Table 20-4 Dissociation Constants of Weak Bases at 25°C

Name	Formula	K_b	pK_b
Ammonia	NH_3	1.78×10^{-5}	4.75
Aniline	⬡—NH_2	4.2×10^{-10}	9.38
Dimethylamine	$(CH_3)_2NH$	1.18×10^{-3}	2.93
Ethylamine	$CH_3CH_2NH_2$	4.7×10^{-4}	3.33
Methylamine	CH_3NH_2	5.25×10^{-4}	3.28
Pyridine	⬡N	1.5×10^{-9}	8.82
Trimethylamine	$(CH_3)_3N$	8.1×10^{-5}	4.09
Acetate ion	OAc^-	5.75×10^{-10}	9.24
Benzoate ion	$C_6H_5COO^-$	1.6×10^{-10}	9.80
Borate ion	BO_2^-	1.7×10^{-5}	4.76
Bicarbonate ion	HCO_3^-	2.2×10^{-8}	7.65
Phenolate ion	$C_6H_5O^-$	1.0×10^{-4}	4.01

4. Solution containing both a weak acid and its salt, $HB + NaB$, a buffer

$$[H^+] = K_a \frac{C_{HB}}{C_{B^-}} \tag{20-4}$$

or

$$pH = pK_a + \log \frac{C_{B^-}}{C_{HB}} \tag{20-5}$$

5. Solution of a weak base, BOH

$$[OH^-] = \sqrt{K_b(C_{BOH} - [OH^-])} \approx K_b C_{BOH} \qquad (20\text{–}6)$$

6. Solution of the salt of a weak base, BX

$$[H^+] = \sqrt{\frac{K_w}{K_b}(C_{BX} - [H^+])} \approx \sqrt{\frac{K_w}{K_b} C_{BX}} \qquad (20\text{–}7)$$

7. Solution containing both a weak base and its salt, BOH + BX, a buffer

$$[OH^-] = K_b \frac{C_{BOH}}{C_{BX}} \qquad (20\text{–}8)$$

or

$$pOH = pK_b + \log \frac{C_{BX}}{C_{BOH}} \qquad (20\text{–}9)$$

Equations 20–5 and 20–9 are known as the *Henderson-Hasselbalch equations.* All of the above equations neglect the contribution to $[H^+]$ and $[OH^-]$ provided by the autoprotolysis of water. These contributions, which can never exceed $10^{-7}\ M$, are too small to take into account except in highly dilute solutions. These are encountered in practice, but it is not usually necessary to calculate their pH.

Several examples will illustrate the application of the equations given above. When in doubt, use approximate equations first, then check the error caused by the approximation.

Example/Problem 20–1. Calculate the pH of a solution of 0.0100 F acetic acid, HOAc.

Solution. The principal species is HOAc, a weak acid which dissociates according to

$$HOAc \rightleftharpoons H^+ + OAc^-; \quad K_a = \frac{[H^+][OAc^-]}{[HOAc]} = 1.75 \times 10^{-5}$$

$$[H^+] = \sqrt{K_a(C_{HOAc} - [H^+])} \approx \sqrt{K_a C_{HOAc}}$$

$$\approx \sqrt{1.75 \times 10^{-5} \times 0.0100}$$

$$\approx 4.2 \times 10^{-4}\ M$$

$$pH = -\log(4.2 \times 10^{-4}) = 3.38$$

Note that a more accurate value for [HOAc] in the equilibrium mixture is $0.0100 - 0.00042\ M$. Neglecting the second term leads to an error of only 2% in the value of $[H^+]$, which is acceptable for most applications.

Example/Problem 20–2. Calculate the pH of a solution of $1.34 \times 10^{-4}\ F$ propionic acid, HOPr.

Solution. This problem is similar to Example/Problem 20–1, except that the solution is more dilute.

$$[H^+] = \sqrt{K_a(C_{HOPr} - [H^+])} \approx \sqrt{K_a C_{HOPr}}$$

$$\approx \sqrt{1.34 \times 10^{-5} \times 1.34 \times 10^{-4}}$$

$$\approx 4.24 \times 10^{-5}\ M \qquad \text{(first approx.)}$$

In this case, a considerable fraction $(4.24 \times 10^{-5}/1.34 \times 10^{-4}$ or $\sim 32\%)$ of the

HOPr is dissociated, and we must use the more exact expression for $[H^+]$. The first approximate answer, $[H^+] = 4.24 \times 10^{-5}\ M$, may be reasonably close to the correct value, and we can use it to get a more accurate second approximation

$$[H^+] \approx \sqrt{1.34 \times 10^{-5}\,(1.34 \times 10^{-4} - 4.24 \times 10^{-5})}$$
$$\approx 3.50 \times 10^{-5}\ M \qquad \text{(second approx.)}$$

The process is continued, using the second approximation

$$[H^+] \approx \sqrt{1.34 \times 10^{-5}\,(1.34 \times 10^{-4} - 3.50 \times 10^{-5})}$$
$$\approx 3.64 \times 10^{-5}\ M \qquad \text{(third approx.)}$$

Continuing the process will lead to little further change in $[H^+]$.

$$\text{Therefore pH} = -\log{(3.64 \times 10^{-5})} = 4.44$$

Note that the exact expression of Equation 20–2 is in the form of a quadratic equation that could be solved equally well by the standard formula:

$$ax^2 + bx + c = 0; \quad x = \frac{-b \pm \sqrt{b^2 - 4ac}}{2a} \tag{20–10}$$

Rearrangement of the exact expression for $[H^+]$ gives

$$[H^+]^2 + 1.34 \times 10^{-5}[H^+] - 1.80 \times 10^{-9} = 0$$

$$[H^+] = \frac{-1.34 \times 10^{-5} \pm \sqrt{1.80 \times 10^{-10} + 7.20 \times 10^{-9}}}{2}$$

$$= 3.62 \times 10^{-5}\ M$$

The minus sign before the radical gives a negative answer that is ignored.

Example/Problem 20–3. If 50 ml of 0.0300 M HOAc is added to 75 ml of 0.0200 M NaOH, what is the pH of the mixture?
  Solution. 50 ml $\times$ 0.0300 M = 1.50 mmole HOAc
           75 ml $\times$ 0.0200 M = 1.50 mmole NaOH

Neither reagent is present in excess, so the mixture contains 1.50 mmole of NaOAc in 125 ml. The pH is calculated as it would be for any weak base, in this case the hydrolysis of a salt of a weak acid. $K_b = K_w/K_a$

$$[OH^-] = \sqrt{K_b(C_{OAc^-} - [OH^-])} \approx \sqrt{\frac{K_w}{K_a}\,C_{OAc^-}}$$

$$\approx \sqrt{\frac{1.00 \times 10^{-14}}{1.75 \times 10^{-5}} \times \frac{1.50}{125}} = 2.62 \times 10^{-5}\ M$$

$$\text{pOH} = 4.58 \text{ and pH} = 9.42$$

Example/Problem 20–4. If 50 ml of 0.100 M NH$_4$OH is added to 30 ml of 0.200 M HNO$_3$, what is the pH of the mixture?

  Solution. 50 ml $\times$ 0.100 M = 5.00 mmole NH$_4$OH
         30 ml $\times$ 0.200 M = 6.00 mmole HNO$_3$
         _____
                 = 1.00 mmole HNO$_3$ + 5.00 mmole NH$_4$NO$_3$

The mixture contains both a strong acid, HNO_3, and a weak acid, NH_4^+. The strong acid is the principal source of H^+ and suppresses the dissociation of the weak acid.

$$[H^+] \approx C_{HNO_3}$$

$$\approx \frac{1.00}{80} = 1.25 \times 10^{-2} \ M$$

$$pH = 1.90$$

Example/Problem 20–5. What is the pH of a solution of 1.0 M $(NH_4)_2SO_4$? Ammonium ion is the conjugate acid of ammonia, for which $K_b = 1.78 \times 10^{-5}$. Thus, K_a for $NH_4^+ = K_w/K_b = 5.62 \times 10^{-10}$. Because $C_{NH_4^+} \gg K_a \gg K_w$, all approximations will be valid.

$$K_a = \frac{[NH_3][H^+]}{[NH_4^+]} \approx \frac{[H^+]^2}{C_{NH_4^+}}$$

$$5.62 \times 10^{-10} = \frac{[H^+]^2}{2.0}$$

$$[H^+] = 3.36 \times 10^{-5} \ M \text{ or } pH = 4.47$$

Example/Problem 20–6. One-tenth mole of the salt of a weak base is dissolved in 500 ml of water. If the pH of this solution is 5.50, what is K_b of the base?

Solution. The salt of a weak base is its conjugate acid which hydrolyzes in water to give an acidic solution.

$$B^+ + H_2O \rightleftharpoons BOH + H^+$$

$$[H^+] \approx \sqrt{K_a C_{B^+}} \quad \text{(approx.)}$$

$$3.16 \times 10^{-6} = \sqrt{K_a \times \frac{0.100}{0.500}}$$

$$K_a = 5.0 \times 10^{-11}$$

$$K_b = K_w/K_a = 2.0 \times 10^{-4}$$

Example/Problem 20–7. Calculate the pH of a buffer solution containing both 0.300 M HOAc and 0.200 M NaOAc

$$pH = pK_a + \log \frac{C_{OAc^-}}{C_{HOAc}}$$

$$= 4.76 + \log \frac{0.200}{0.300} = 4.58$$

Buffer Solutions. Many chemical reactions generate free hydrogen or hydroxyl ions. If these ions remained in the system, there would be significant changes in the pH. Buffer solutions contain constituents that react with both strong acids and strong bases in such a way that the free hydrogen ion concentration of the system remains relatively constant. Buffers typically consist of mixtures of weak acids and their conjugate bases, or of weak bases and their conjugate acids. To illustrate how a buffer solution functions, consider the following problem.

Example/Problem 20–8. A buffer solution contains 0.100 F sodium acetate and 0.100 F acetic acid. A chemical system generates 10.0 mmole of protons in one liter of the buffer solution. (a) What was the original pH of the buffer? (b) What is the pH after the reaction? (c) What would be the pH if the same reaction had occurred in a 1-liter solution of HCl which had an original pH equal to that of the 0.100 F acetic acid—0.100 F sodium acetate solution?

Solution.   Part (a): Find the pH of the original acetic acid-sodium acetate solution.

$$[H^+] = K_a \frac{C_{HOAc}}{C_{OAc^-}} = 1.75 \times 10^{-5} \left(\frac{0.100}{0.100}\right) = 1.75 \times 10^{-5} \ M$$

and

$$pH = 4.76$$

Part (b): What happens in the system when 10.0 mmole of hydrogen ions are added? As the protons are generated, they will react with the strongest base present, in this case acetate ions:

$$OAc^- + H^+ \rightleftharpoons HOAc$$

The mass action expression is the inverse of the dissociation constant

$$\frac{[HOAc]}{[OAc^-][H^+]} = \frac{1}{K_a} = 5.72 \times 10^4$$

The large magnitude of this equilibrium constant indicates that the reaction proceeds essentially to completion. Next we determine how much acetate and acetic acid were present in the original system and then how much of each of these species is present in the final system.

$$\text{original mmole HOAc} = (1000 \text{ ml})(0.100 \text{ mmole/ml}) = 100.0 \text{ mmole}$$

$$\text{original mmole OAc}^- = (1000 \text{ ml})(0.100 \text{ mmole/ml}) = 100.0 \text{ mmole}$$

$$\text{final mmole OAc}^- = 100.0 \text{ mmole} - 10.0 \text{ mmole} = 90.0 \text{ mmole}$$
$$\text{(original)} \qquad \text{(used in reaction)}$$

$$\text{final mmole HOAc}^- = 100.0 \text{ mmole} + 10.0 \text{ mmole} = 110.0 \text{ mmole}$$
$$\text{(original)} \qquad \text{(generated in reaction)}$$

Thus, the hydrogen ion which was generated has been "absorbed" by the acetic acid-sodium acetate solution. The free hydrogen ion concentration is again determined by the equilibrium concentrations of acetic acid and acetate ion in the *final* solution.

$$C_{HOAc} = \frac{110.0 \text{ mmole}}{1000 \text{ ml}} = 0.1100 \ F$$

and

$$C_{NaOAc} = \frac{90.0 \text{ mmole}}{1000 \text{ ml}} = 0.0900 \ F$$

therefore

$$[H^+] = 1.75 \times 10^{-5} \frac{0.110}{0.0900} = 2.14 \times 10^{-5} \ M$$

and

$$pH = 4.67$$

So, even though a considerable amount of strong acid has been added to this system, there has been very little change in pH, only 0.09 unit.

Part (c): In the 1-liter HCl solution of pH = 4.76, $[H^+] = 1.75 \times 10^{-5}$ M. If 10.0 mmole of H^+ were added to this system, the total number of mmole of H^+ would be,

$$(1000 \text{ ml})(1.75 \times 10^{-5} \text{ mmole/ml}) + 10.0 \text{ mmole} \approx 10.0 \text{ mmole}$$
$$\text{original} \qquad\qquad\qquad \text{generated}$$

and thus the final H^+ concentration would be

$$[H^+] = \frac{10.0 \text{ mmole}}{1000 \text{ ml}} = 1.00 \times 10^{-2} \ M$$

and the final

$$pH = 2.00.$$

Therefore in this system, the addition of 10.0 mmole of hydrogen ion would cause a significant change in pH, 2.76 units. The acetic acid-sodium acetate system is a well-buffered solution, whereas the HCl solution represents an unbuffered solution.

If free hydroxide ions are added to the buffer solution discussed in Example/Problem 20–8, the hydroxide ion is absorbed by the system due to the reaction

$$HOAc + OH^- \rightleftharpoons OAc^- + H_2O$$

Thus, when strong base is added to the buffer solution, the formal concentration of the weak acid decreases and the formal concentration of its conjugate base increases. However, the free hydrogen ion concentration is still determined by the slight dissociation of the weak acid.

For a weak base buffer system, the hydrogen ion concentration is given by

$$[H^+] = K_a \frac{C_{BH^+}}{C_B} \quad \text{or} \quad [OH^-] = K_b \frac{C_B}{C_{BH^+}}$$

We can generalize the action of a buffer by noting that the final $[H^+]$ can be expressed as

$$[H^+]_{\text{final}} = K_a \frac{VC_{HB}^\circ + n_{\text{acid}} - n_{\text{base}}}{VC_B^\circ + n_{\text{base}} - n_{\text{acid}}} \tag{20–11}$$

where V is the volume of the buffer, C_{HB}° and C_B° are the initial formal concentrations of HB and B, and n_{acid} and n_{base} are the number of equivalents of strong acid and/or strong base added.

The most effective buffer for any system consists of a $1:1$ ratio of weak acid or weak base to its conjugate. Therefore, if we wish to prepare a buffer solution of a certain pH, we try to select a weak acid with a pK_a value which is close to the desired pH value. If the pK_a does not exactly equal the desired pH, the C_B/C_{HB} ratio may be adjusted so that the value of $(pK_a + \log C_B/C_{HB})$ does equal the desired value. Let us consider some problems which illustrate how to prepare buffer solutions.

Example/Problem 20–9. Describe a method of preparation of a 1-liter buffer with pH = 5.00 in which the *total acid + conjugate base* concentration is 0.200 F. Pick a system such that the acid and conjugate base concentrations are as nearly equal as possible.

Solution. From Equation 20–5,

$$pH = pK_a + \log \frac{C_{B^-}}{C_{HB}}$$

If $C_{B^-} \approx C_{HB}$,

$$pH \approx pK_a$$

From Table 20–3, the weak acid with a pK_a closest to 5.00 is propionic acid

$(pK_a = 4.87)$. Therefore,

$$5.00 = 4.87 + \log \frac{C_{B^-}}{C_{HB}}$$

But, from the stated problem

$$C_{HB} + C_{B^-} = 0.200 \quad \text{or} \quad C_{HB} = 0.200 - C_{B^-}$$

Therefore

$$5.00 = 4.87 + \log \frac{C_{B^-}}{0.200 - C_{B^-}}$$

$$\log \frac{C_{B^-}}{0.200 - C_{B^-}} = 0.13 \quad \text{and} \quad \frac{C_{B^-}}{0.200 - C_{B^-}} = 10^{0.13} = 1.35$$

Thus $C_{B^-} = 0.115\ F$ and $C_{HB} = 0.085\ F$.

So, to prepare the desired buffer take 0.115 mole of sodium propionate and 0.085 mole of propionic acid and dilute to one liter.

Example/Problem 20–10. How many grams of solid ammonium chloride must be added to 500 ml of 0.0200 F NH_3 to prepare a 1.000-liter buffer solution of pH = 9.30 after dilution with distilled water? (Mol. Wt. NH_4Cl = 53.49.)

Solution. First, we must determine the concentration of ammonium ion required to attain pH 9.30 (pOH = 4.70) in the final solution. From Equation 20–9,

$$pOH = 14.00 - 9.30 = 4.70 = pK_b + \log \frac{C_{NH_4^+}}{C_{NH_3}} = 4.75 + \log \frac{C_{NH_4^+}}{C_{NH_3}}$$

We know the concentration of NH_3, that is, we started with 500 ml of 0.0200 F NH_3 and have diluted this to 1.000 liter. Therefore, the final NH_3 concentration is

$$C_{NH_3} = \frac{(500\ \text{ml})(0.0200\ \text{mmole/ml})}{(1000\ \text{ml})} = 0.0100\ F$$

$$\log \frac{C_{NH_4^+}}{C_{NH_3}} = -0.05 \quad \text{and} \quad \frac{C_{NH_4^+}}{C_{NH_3}} = 0.89$$

Therefore,

$$C_{NH_4^+} = (0.89)(0.0100) = 0.0089\ F$$

Since the added NH_4Cl is the principal source of NH_4^+ in this solution, we must add (1 liter $\times$ 0.0089 mole/liter) = 0.0089 mole or 0.476 g solid NH_4Cl.

Buffer Capacity. Buffer solutions resist changes in pH upon the addition of strong acids or strong bases. In Example/Problem 20–8 the addition of 10 mmole of strong acid caused a change in pH of only 0.09 unit. The buffer capacity of a system often is defined as the moles of strong acid or strong base required to change the pH of 1 liter of the buffer solution by 1 unit. The larger the buffer capacity, the better the buffer, that is, the more acid or base it can consume without significant changes in pH. Clearly, more concentrated buffer solutions have higher capacity than more dilute solutions. For example, even though a solution that is 0.0100 F in both acetic acid and acetate ion has the same pH as a solution which is 0.100 F in both of these constituents, the more concentrated solution can consume ten times as much strong acid or strong base for the same change in pH. Also, it can be shown that the minimum change of pH with an added increment of strong acid or strong base is achieved in a solution which has a 1 : 1 ratio of weak acid to conjugate base.

Example/Problem 20–11.   What is the buffer capacity of 500 ml of a buffer which contains 0.0800 M NH$_4$Cl and 0.0800 M NH$_4$OH?

Solution.   Buffer capacity is based on one liter of solution

$$\text{Initial pH} = pK_a + \log \frac{C_{NH_3}}{C_{NH_4^+}}$$

$$= 9.25 + \log \frac{0.08}{0.08} = 9.25$$

After the addition of n moles of a strong acid, the pH should be (from the statement in the problem) $9.25 - 1.00 = 8.25$.

$$\text{Final pH} = 8.25 = 9.25 + \log \frac{0.08 - n}{0.08 + n}$$

$$n = 0.065 \text{ mole/liter (buffer capacity)}$$

20–6   ACID–BASE TITRATIONS IN AQUEOUS SOLUTION

One of the principal reasons for mastering pH calculations is to be able to use them in predicting and analyzing titration curves for acid–base reactions. A titration curve shows how the pH of a solution changes upon addition of acid or base; it is a plot of pH versus volume of titrant added. The region near the equivalence point is of special interest. The inflection point, which occurs at the steepest portion of the curve, coincides with the equivalence point. The steeper the curve, the sharper the end point and the more accurate the titration will be. A study of the titration curve will aid in the selection of the most suitable indicator. Titration curves can, of course, be obtained experimentally by measuring the pH with a pH meter after adding successive increments of titrant. Nevertheless, it is instructive, and sometimes easier and quicker, to calculate a theoretical curve.

The routine steps to follow in plotting any titration curve are:

1. Write the chemical reaction for the titration.
2. Select the initial concentration, $C°$, and volume, $V°$, of the sample to be titrated.
3. Select several volumes of titrant added, V, and calculate the fraction, f, of the sample which is titrated at each point, $f = V/V_{ep}$, where V_{ep} is the volume of titrant required to reach the equivalence point. It is convenient to select values of V such that $f = 0$, 0.1, 0.5, 0.9, 1.0, 1.1, and 1.5.
4. Determine the concentration of the principal species present at each f value selected and compute the pH.
5. Tabulate and plot the points selected.

Although our discussion of acid–base systems has been detailed and fairly rigorous, these refinements are seldom necessary for plotting theoretical curves. At most we need a half-dozen points, and these can be selected for compositions that are simple to handle. Whenever possible, we select reagents and conditions yielding large equilibrium constants. Thus, most of the simplified formulas for pH calculation are adequate.

Titration curves for strong and weak acids and bases are plotted in Figures 20–1 to 20–4. In calculating points for these curves, it is helpful to use generalized formulas such as the following, in which $V_A°$, $V_B°$, $C_A°$, $C_B°$ are the initial volumes

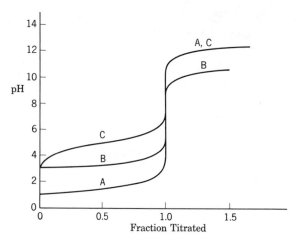

Figure 20–1   Titration curves for acids with strong base:

A: $C_A^\circ = 0.1\ M$ $C_B^\circ = 0.1\ M$

B: $C_A^\circ = 0.001\ M$ $C_B^\circ = 0.001\ M$

C: $C_A^\circ = 0.1\ M$ $(K_a = 1 \times 10^{-5})$ $C_B^\circ = 0.1\ M$

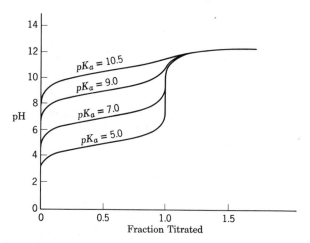

Figure 20–2   Titration curves for weak acids with strong base:

$C_A^\circ = 0.1\ M.\ C_B^\circ = 0.1\ M$ pK_a noted on curves

and concentrations of acid or base, and V_A and V_B are the volumes of titrant (acid or base) added at any point. (See table on page 430.)

Values of f between 0.95 and 1.05 should be avoided because the simplified formulas are usually not valid (except for the equivalence point where $f = 1.000$). Fortunately, titration curves are always smooth and can be drawn through this region without numerous points. In an experimental curve, however, this is the very region where a large number of points should be taken.

A number of useful observations can be gleaned from these titration curves. For example, note that the pH change near the equivalence point is considerably smaller in the weak acid titration than in the strong acid case. The magnitude of the pH break depends on the concentration of reagents in the strong acid case;

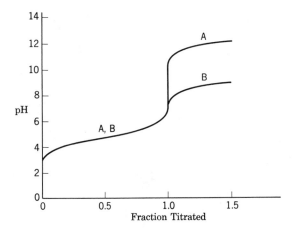

Figure 20–3 Titration curves for acetic acid, $pK_a = 4.76$, with strong and weak base:

A: $C_A^0 = 0.1\ M$ $C_B^0 = 0.1\ M$
B: $C_A^0 = 0.1\ M$ $C_B^0 = 0.1\ M\ (pK_b = 4.74)$

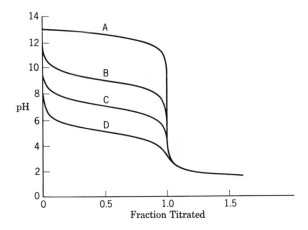

Figure 20–4 Titration curves for $0.1\ M$ bases with $0.1\ M$ strong acid:

A: strong base
B: $pK_b = 5.0$
C: $pK_b = 7.0$
D: $pK_b = 9.0$

whereas, for weak acids, the magnitude of the pH break depends on the K_a of the acid, unless the solution is very dilute. Titration of a weak acid with a weak base (Figure 20–3) is unsatisfactory and unnecessary, since a strong titrant can always be used. These curves clearly indicate that if K_a or K_b is less than 10^{-8}, there will be difficulty in locating the equivalence point even with a pH meter. Such weak acids and bases are best titrated in nonaqueous solvents, to be discussed later.

Acid–Base Indicators. Solutions of the common acids and bases are colorless, thus colored indicators are useful to detect the end points of their titration. Acid–base indicators are intensely colored weak acids (or weak bases) whose conjugate form has a different color. A small amount of the indicator is added to the sample and is titrated along with it. An indicator is selected that is just half-titrated at the pH of the equivalence point of the main titration. The indicator

Formulas for calculations described on pages 427 and 428.

Titration Region	Strong Acid versus Strong Base	Weak Acid versus Strong Base	Weak Base versus Strong Acid
$f = 0$	$[H^+] = C_A^o$	$[H^+] = \sqrt{K_a C_A^o}$	$[OH^-] = \sqrt{K_b C_B^o}$
$0 < f < 1$	$[H^+] = C_A^o(1-f)\dfrac{V_A^o}{V_A^o + V_B}$	$[H^+] = K_a\dfrac{1-f}{f}$	$[OH^-] = K_b\dfrac{1-f}{f}$
$f = 1$	$[H^+] = [OH^-] = \sqrt{K_w}$	$[OH^-] = \sqrt{\dfrac{K_w}{K_a} \times C_A^o \times \dfrac{V_A^o}{V_A^o + V_{ep}}}$	$[H^+] = \sqrt{\dfrac{K_w}{K_b} \times C_B^o \times \dfrac{V_B^o}{V_B^o + V_{ep}}}$
$f > 1$	$[OH^-] = C_B^o\dfrac{V_B - V_{ep}}{V_A^o + V_B}$	$[OH^-] = C_B^o\dfrac{V_B - V_{ep}}{V_A^o + V_B}$	$[H^+] = C_A^o\dfrac{V_A - V_{ep}}{V_B^o + V_A}$

will exist half in its acid form and half in its basic form and exhibit an intermediate color. This results if the pK_a of the indicator is close to the pH at the equivalence point, pH_{ep}

$$pH_{ep} = pK_a + \log \frac{C_B}{C_A}$$

The log term equals zero if $C_A = C_B$. For a two-color indicator, the observed color will gradually change as the ratio C_B/C_A changes. If this ratio is <0.1 or >10, only one color is observed. In the intermediate range where $0.1 < C_B/C_A < 10$ and $pH = pK_a \pm 1$, the sharpness of the transition from one color to the other depends on the slope of the titration curve. Typical indicators are listed in Table 20–5.

Table 20–5 Typical Acid–Base Indicators

Indicator	Acid Form	pH Transition		Base Form
Paramethyl red	red	1.0 to	3.0	yellow
2,6-Dinitrophenol	colorless	2.0	4.0	yellow
Bromophenol blue	yellow	3.0	4.6	blue
Methyl orange	red	3.1	4.4	yellow
Bromocresol green	yellow	3.8	5.4	blue
Methyl red	red	4.2	6.2	yellow
Bromocresol purple	yellow	5.2	6.8	purple
Bromothymol blue	yellow	6.0	7.6	blue
Phenolphthalein	colorless	8.0	9.6	red
Thymolphthalein	colorless	9.3	10.6	blue
Alizarin yellow R	yellow	10.1	12.0	violet
2,4,6-Trinitrotoluene	colorless	12.0	14.0	orange

20–7 ACID–BASE TITRATIONS IN NONAQUEOUS SOLVENTS

Fundamental studies concerning the nature of titration reactions in nonaqueous solvents represent a new frontier in chemistry where most of the important progress has been made since about 1957. The majority of nonaqueous titrations involve neutralizations of organic bases or acids and the direct determinations of acidic and basic functional groups. In addition to their importance in organic analysis, nonaqueous acid–base titrations are used extensively in pharmaceutical analysis, for example, to determine constituents present in antihistamines, antibiotics, and sulfonamides. Direct titrations are possible in nonaqueous acid–base reactions since they proceed to completion within a few microseconds.

There are numerous advantages offered by acid–base titrations in nonaqueous media, the most important of which is that a much larger number of acids and bases can be titrated in nonaqueous solvents than in aqueous solution. For example, a weak acid with a pK_a of 9 or greater cannot be determined accurately in water because of the competition of the solvent for the strong base titrant (leveling effect). But the weak acid can be titrated in ethylenediamine, which is considerably more basic than water. The availability of a large variety of nonaqueous solvents allows the choice of a solvent that will assist but not interfere with a specific acid–base titration. A properly formulated nonaqueous acid–base titration procedure provides results that are very accurate and often more precise than those obtained from a corresponding titration in aqueous media. We will consider briefly some of the practical considerations that apply to nonaqueous acid–base titrations. The student is referred to the excellent monograph by Kucharský and Šafařik for a more comprehensive review of the multitude of applications of nonaqueous titrations. Kolthoff and Bruckenstein present a rigorous theoretical treatment of acid–base chemistry in nonaqueous media in the *Treatise on Analytical Chemistry.*

Titrants. Since perchloric acid is the strongest mineral acid, acidic titrants normally consist of solutions of perchloric acid in either anhydrous acetic acid or dioxane. Acetic acid solutions that are 0.1 to 1.0 F in perchloric acid commonly are used. Perchloric acid is essentially 100% ionized in acetic acid; it exists as the ion pair, $H_2OAc^+ClO_4^-$, as well as the solvated ion, H_2OAc^+.

Basic titrants include quaternary tetraalkyl ammonium hydroxides, which are stronger bases than the alkali metal hydroxides. Although the product salts of alkali metal ions are quite insoluble in nonaqueous media, quaternary ammonium salts are very soluble. Tetrabutylammonium hydroxide in isopropanol is the most commonly used base titrant. To insure that the titrant solvent does not interfere with the principal reaction, the concentration of the titrant solution is usually high relative to the concentration of the substance titrated. Only a small volume of titrant is added compared to the volume of the titration mixture.

Choice of Solvent for Acid–Base Titration. The following considerations are pertinent to the choice of a solvent for a specific nonaqueous acid–base titration.

1. The solvent should permit a *large change* in the solvated proton concentration near the equivalence point. Other things being equal, the smaller the autoprotolysis constant, the better the end point.
2. The substance to be titrated must be soluble, either in the solvent or in an excess of the titrant which then may be back-titrated.
3. The product of the titration must be soluble, or if it is a precipitate, it must be compact and crystalline and not gelatinous. Gelatinous precipitates tend to interfere with accurate end-point determinations.
4. The solvent should not introduce interfering side reactions with either the substance to be titrated or the titrant.
5. Preferably, the solvent should be inexpensive and easily purified. Many of these solvents are toxic and require special precautions in handling.

When several acids or bases in a mixture are "leveled" to the same strength they may not be differentiated by titration. For example, the concentration of each constituent in a mixture of perchloric acid and sulfuric acid cannot be determined in a titration with a strong base in aqueous solution. Both of these acids are leveled to hydronium ion which is then titrated with the base. As a result, only the *total amount* of strong acid may be determined. In glacial acetic acid, perchloric acid is approximately 235 times as strong as sulfuric acid. Even this large difference is not enough to give two distinct equivalence points in a titration of this mixture in glacial acetic acid. Furthermore, the bisulfate ion is too weak to be titrated in this solvent. However, in a titration of a perchloric acid and sulfuric acid mixture in a methyl isobutyl ketone solvent, three distinct breaks in "pH" occur when t-butyl ammonium hydroxide in isopropanol is used as the titrant. The breaks correspond to the successive neutralizations of perchloric acid, the first proton of sulfuric acid and finally the bisulfate ion. Curve A of Figure 20–5 shows the titration curve obtained with a glass electrode-calomel electrode cell. The e.m.f. of this electrode system is approximately proportional to the "pH" of the

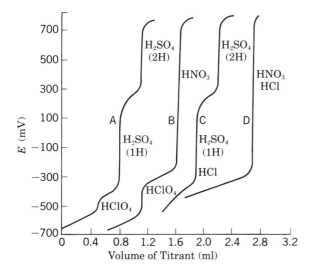

Figure 20–5 Titration curves of acid mixtures in the solvent methylisobutyl ketone using 0.2 F t-butylammonium hydroxide as the titrant and a glass electrode-saturated calomel electrode cell. [After Bruss and Wyld, *Anal. Chem.*, **29**, 232 (1957).]

system. Additional acid-differentiating titrations also are illustrated in Figure 20–5.

Inert solvents may be added to either protogenic or protophilic solvents to modify differentiating titrations. The presence of the inert solvent decreases the amphiprotic character of both acidic and basic solvents. For example, butylamine and pyridine are both titrated as strong bases in glacial acetic acid, but in a solvent which consists of 10% acetic acid and 90% chloroform, two distinct equivalence points are noted corresponding to the neutralizations of butylamine and then pyridine.

Normally, the titration of a weak acid in an inert solvent involves the addition of a small volume of concentrated base dissolved in an amphiprotic solvent such as an alcohol. Thus, after the first addition of titrant, the system contains a small amount of amphiprotic solvent (the alcohol) in an inert solvent. The "effective" autoprotolysis constant for the system is that of the amphiprotic solvent mixed with the inert solvent. When the amount of amphiprotic solvent is small, the "effective" autoprotolysis constant is small and the "pH" break is large near the equivalence point. This is the reason why highly concentrated solutions of bases are used as titrants. Isopropyl alcohol is commonly used as the solvent for tetra-alkylammonium hydroxides because it is the least acidic of the amphiprotic solvents.

In summary, for titrations of weak acids and weak bases, inert solvents provide the largest break in "pH" near the equivalence point. Inert solvents are also preferable in successive titrations of acids and bases in mixtures. When solubility problems are encountered with inert solvents, protogenic solvents are used for weak base titrations and protophilic solvents are used in weak acid titrations.

End Point Determination in Nonaqueous Titrations. Unfortunately, in nonaqueous media we do not have accurate values for the dissociation constants of chemical indicators, nor do we have quantitative relationships which relate the e.m.f. of electrochemical cells to constituent concentrations. So how can we find accurate end points in nonaqueous titrations? Let us briefly examine some of the indicator systems for nonaqueous titrations.

Potentiometric End Point. In a typical nonaqueous potentiometric acid–base titration, we might use this cell:

$$\text{Ag} \mid \text{AgCl, HCl (0.10 } N) \mid \begin{array}{c} \text{Glass} \\ \text{Membrane} \\ \hline \text{Nonaqueous} \\ \text{Solution} \end{array} \mid\mid \begin{array}{c} \text{Saturated Calomel} \\ \text{Electrode} \end{array}$$

As in aqueous solution, in most nonaqueous solvents the glass electrode appears to respond to the difference between the activity of the solvated proton in the solution inside the glass membrane and the activity of solvated proton in the solution on the outside. However, a quantitative interpretation of this potential difference is not obvious. In the first place, ion activities in one solvent cannot be simply related to ion activities in another solvent. In addition, when we measure the e.m.f. of the cell just shown, we include a sizable, unknown, liquid junction potential between the aqueous calomel electrode and the nonaqueous solution. This liquid junction potential may vary considerably during the course of a nonaqueous titration, depending on the nature of the solvent and the titrants. Liquid junction potentials must remain reasonably constant during the course of the titration for high precision. In many cases the e.m.f. of the above cell does vary approximately linearly with the pSH_2 of the nonaqueous solution. It should be pointed out, however, that the e.m.f. variation in no way corresponds to the theoretical variation derived for aqueous solutions from the Nernst equation. In other words, a *change* in 1.00 unit of pSH_2 *does not* correspond to a change in e.m.f. of 59.16 mV at 25°C in nonaqueous solvents. As a result, the pH scale of a pH meter is meaningless in nonaqueous titrations and the millivolt scale(s) should be used. So, for titrations in ethanol or glacial acetic acid, the glass-calomel electrode system appears to respond to changes in $-\log[\text{EtOH}_2^+]$ and $-\log[\text{H}_2\text{OAc}^+]$, respectively (or the "pH" of these solutions). When an acid is titrated in an inert solvent (e.g., a ketone) with *t*-butyl ammonium hydroxide in isopropanol, the glass electrode responds to changes in the isopropanol solvated proton concentration. In highly basic solvents, such as butylamine and ethylenediamine, the glass electrode does not function as an indicator electrode and an antimony electrode is used instead. Some examples of potentiometric nonaqueous titration curves are illustrated in Figures 20–6 through 20–8.

Chemical Indicators. Indicators for nonaqueous acid–base titrations behave similarly to those that are employed in aqueous titrations in either giving up or accepting protons and changing color in the process. An indicator, therefore, may dissociate in solvent SH

$$\underset{\text{color A}}{\text{HIn}} + \text{SH} \rightleftharpoons \text{SH}_2^+ + \underset{\text{color B}}{\text{In}^-}$$

Clearly the predominant form of the indicator depends on the SH_2^+ concentration. So far this story is no different from that for an indicator in aqueous solution. However, in nonaqueous media many solvents are available and the behavior of a particular indicator depends on the acidity, basicity, or inertness of the solvent as well as its dielectric constant.

Some indicators have simple color changes; others successively pass through a wide range of color shades. For example, as crystal violet is increasingly protonated in glacial acetic acid, it changes from violet to blue–green to green to yellow. Methyl violet passes through similar changes. To determine the color change that best indicates the end point, a potentiometric titration should be run with the indicator present. The color change that occurs nearest the potentiomet-

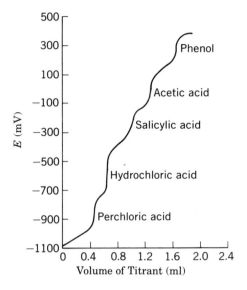

Figure 20–6 Titration curve for strong, weak, and very weak acid mixture in methylisobutyl ketone using 0.2 F t-butylammonium hydroxide in isopropanol as the titrant. Indicating electrode system: glass-platinum (in titrant). [After Bruss and Wyld, *Anal. Chem.*, **29**, 232 (1957).]

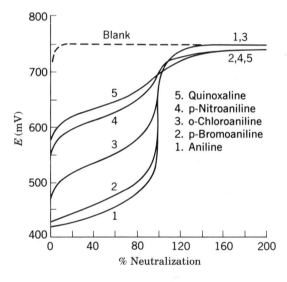

Figure 20–7 Titration curves for various aromatic amines in anhydrous acetic acid using 0.10 F perchloric acid in glacial acetic acid as the titrant. Indicating electrode system: glass-calomel. [From Fritz and Hammond, *Quantitative Organic Analysis*, Wiley, 1957, p. 32.]

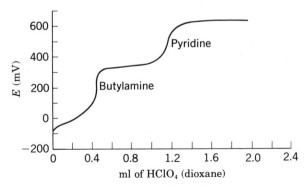

Figure 20–8 Titration curve for butylamine-pyridine mixture in acetonitrile using 0.10 F perchloric acid in dioxane as the titrant. Indicating electrode system: glass-calomel placed very close together. [From Fritz, *Anal. Chem.*, **25**, 407 (1953).]

Table 20–6 Indicators Commonly Used in Nonaqueous Titrations

Indicator	Constituent Titrated	Solvent
Crystal violet or Methyl violet	Weak bases	Acetic acid; acetonitrile
Methyl red	Weak bases	Dioxane; glycol-isopropanol
Dibenzalacetone	Weak bases	Nitromethane
Azo-violet or Thymol blue or o-Nitroaniline	Weak acids	Dimethylformamide

ric equivalence point should then be used as the end point in future titrations of the same kind. A list of indicators for titrations of weak acids and weak bases is given in Table 20–6.

QUESTIONS AND PROBLEMS

Note: In all the following problems, assume that all solutions are at 25°C and that all activity coefficients are unity.

20–1. What is the pH of the following solutions?
   (a) 500 ml of 2.45×10^{-3} F HCl   *Ans.* 2.61
   (b) 300 ml of 1.76×10^{-2} F NaOH   *Ans.* 12.25
   (c) A solution prepared by mixing together 250 ml of 0.106 M $HClO_4$ and 350 ml of 5.62×10^{-2} M HNO_3.   *Ans.* 1.11
   (d) A solution prepared by mixing together 325 ml of HCl with a pH of 2.700 and 475 ml of HCl with a pH of 3.20.   *Ans.* 2.92
   (e) A solution prepared by mixing together 225 ml of 0.125 F NaOH, 325 ml of 0.110 F HCl and 250 ml of distilled (degassed) water.   *Ans.* 2.02

20–2. What is the pH of the final solution when 15.00 ml of 0.250 F NaOH is added to 30.00 ml of 0.125 F HCl?   *Ans.* 7.00

20–3. What is the pH of the final solution when 14.90 ml of 0.250 F NaOH is added to 30.00 ml of 0.125 F HCl?   *Ans.* 3.25

20–4. 125 ml of 0.255 F propionic acid is diluted to 500 ml in a volumetric flask. What is the pH of the final solution?   *Ans.* 3.04

20–5. 100 ml of 0.325 F trimethylamine is diluted to 250 ml in a volumetric flask with distilled water. What is the pH of the final solution?   *Ans.* 11.50

20–6. 15.0 g of sodium propionate, C_2H_5COONa, is added to 500 ml of 0.250 F propionic acid. The mixture is diluted to one liter with distilled water. What is the pH of the final solution?   *Ans.* 4.97

20–7. 10.0 g of trimethylammonium chloride, $(CH_3)_3NHCl$, is added to 250 ml of 0.250 F trimethylamine. This mixture then is diluted to 500 ml with distilled water. What is the pH of the final solution?   *Ans.* 9.68

20–8. (a) What is the pH of 250 ml of 0.110 F ammonia?   *Ans.* 11.15
   (b) What is the pH of a solution composed of 250 ml of the solution in part (a), 7.54 g of NH_4Br and water of dilution to a final volume of 500 ml?   *Ans.* 8.80

20–9. What is the pH of the following solutions?
   (a) 200 ml of 0.124 F acetic acid.   *Ans.* 2.83

(b) 200 ml of 0.124 F acetic acid plus 50.0 ml of 0.110 N NaOH.   *Ans.* 4.21
(c) Solution in part (b) plus 100 ml of 0.110 N NaOH.   *Ans.* 5.06
(d) 200 ml of 0.124 F acetic acid plus 100 ml of 0.248 N NaOH.   *Ans.* 8.84
(e) 200 ml of 0.124 F acetic acid plus 105 ml of 0.248 N NaOH.   *Ans.* 11.60
(f) 9.15 g of sodium acetate (CH_3COONa) dissolved in distilled water and diluted to
     1 liter.   *Ans.* 8.90

20–10. How much solid sodium acetate would have to be added to 2 liters of 0.125 F acetic acid to prepare final buffer solution of pH 4.00? (Assume no volume increase on addition of salt.)   *Ans.* 3.59 g

20–11. How much 0.100 N NaOH must be added to a solution initially 0.200 F in acetic acid and 0.0200 F in sodium acetate to make the pH of the resulting solution be 5.00? Assume the volume of the initial acid–salt solution was 500 ml.   *Ans.* 600 ml

20–12. What is the pH of a solution formed by mixing 300 ml of a solution 0.240 F in NH_3 and 0.120 F in NH_4Cl with 200 ml of 0.150 N HCl?   *Ans.* 9.05

20–13. A 0.200 F solution of a weak base has a pH of 11.00. What is its pK_b?
                                                                              Ans. 5.30

20–14. A 0.200 F solution of a weak base has a pH of 12.30. What is its pK_b?   *Ans.* 2.65

20–15. A 0.100 F solution of a weak acid is studied spectroscopically and found to be 1.5% dissociated.
(a) What is the K_a of the acid?   *Ans.* 2.28×10^{-5}
(b) At what formal concentration would the acid be 2% dissociated?   *Ans.* 0.056 F

20–16. Select the best indicator from Table 20–5 for the following titrations:
(a) 10^{-4} M HCl with 0.01 M NaOH.
(b) 0.01 M C_6H_5ONa with 0.1 M HCl.
(c) 0.1 M $NaBO_2$ with 0.1 M HCl.
(d) 0.01 M $C_6H_5N \cdot HCl$ with 0.1 M NaOH.

20–17. A student titrates a vinegar sample (dilute HOAc) with NaOH and erroneously uses bromothymol blue as an indicator, stopping the titration at pH 7.00. If the vinegar actually contained 5.00% HOAc, what answer will he get?   *Ans.* 4.97%

20–18. A 1-liter bottle of a commercial ammonia solution has a pH of 11.86. How many grams of NH_3 does the bottle contain? Assume density of solution is 0.99 g/ml.
                                                                              Ans. 50.2 g

REFERENCES

J. N. Butler, *Ionic Equilibrium*, Addison-Wesley, Reading, Mass., 1964. A very comprehensive treatment of the field of solution equilibrium calculations.

E. E. Conn and P. K. Stumpf, *Outlines of Biochemistry*, 2nd ed., Wiley, New York, 1966.

H. Freiser and Q. Fernando, *Ionic Equilibria in Analytical Chemistry*, Wiley, New York, 1963.

A. J. Bard, *Chemical Equilibrium*, Harper and Row, New York, 1966.

E. J. King, *Acid–Base Equilibria*, Macmillan, New York, 1965.

J. Kurcharský and L. Šafařik, *Titrations in Non-Aqueous Solvents*, Elsevier, Amsterdam, 1965.

H. H. Sisler, *Chemistry in Non-Aqueous Solvents*, Van Nostrand Reinhold, New York, 1961.

21

POLYPROTIC SYSTEMS

21-1 INTRODUCTION

Compounds with more than one acidic proton are referred to as polyprotic acids. The dissociation of each proton from a polyprotic acid consists of a separate step; consequently, the number of equilibrium constants corresponds to the number of acidic protons. The stepwise dissociation constants for a number of polyprotic acids are listed in Table 21–1.

21-2 PHOSPHORIC ACID SYSTEM

The dissociation of phosphoric acid is represented by the three dissociation equilibria:

$$H_3PO_4 \rightleftharpoons H^+ + H_2PO_4^-; \quad K_1 = \frac{[H^+][H_2PO_4^-]}{[H_3PO_4]} = 5.89 \times 10^{-3} \qquad (21\text{-}1)$$

$$H_2PO_4^- \rightleftharpoons H^+ + HPO_4^{-2}; \quad K_2 = \frac{[H^+][HPO_4^{-2}]}{[H_2PO_4^-]} = 6.16 \times 10^{-8} \qquad (21\text{-}2)$$

and

$$HPO_4^{-2} \rightleftharpoons H^+ + PO_4^{-3}; \quad K_3 = \frac{[H^+][PO_4^{-3}]}{[HPO_4^{-2}]} = 4.79 \times 10^{-13} \qquad (21\text{-}3)$$

The structure of phosphoric acid is represented by

$$\begin{array}{c} OH \\ | \\ HO\!\!-\!\!P\!\!=\!\!O \\ | \\ OH \end{array}$$

with the phosphorus atom located in the center of a tetrahedral arrangement of oxygen atoms. When one of the protons is removed, the remaining anion has one negative charge that is delocalized over all four of the oxygen atoms. As a result, the removal of a second proton from an oxygen that assumes a partially negative charge is more difficult, and the removal of the third proton from the dinegatively charged species is even more difficult. These trends are reflected in the values of the successive dissociation constants ($K_1 = 5.89 \times 10^{-3}$, $K_2 = 6.16 \times 10^{-8}$, $K_3 = 4.79 \times 10^{-13}$) which differ by large amounts. On the other hand, for a molecule like succinic acid, $HOOC\!\!-\!\!CH_2\!\!-\!\!CH_2\!\!-\!\!COOH$, the two acidic protons are far removed from each other and the negative charge formed when the first proton leaves is not delocalized into the vicinity of the other proton. As a result, the dissociation constants are nearly the same ($K_1 = 6.45 \times 10^{-5}$ and $K_2 = 3.31 \times 10^{-6}$).

Table 21–1 Polyprotic Acid Dissociation Constants at 25°C

Acid Name	Formula	K_1 (pK_1)	K_2 (pK_2)	K_3 (pK_3)	K_4 (pK_4)
Carbonic acid	$CO_2 + H_2O$	4.46×10^{-7} (6.35)	5.62×10^{-11} (10.25)		
Hydrogen sulfide	H_2S	1.0×10^{-7} (7.0)	1×10^{-14} (14.0)		
Oxalic acid	HOOCCOOH	5.62×10^{-2} (1.25)	5.25×10^{-5} (4.28)		
Tartaric acid	HOOCCHOHCHOHCOOH	9.1×10^{-4} (3.04)	4.26×10^{-5} (4.37)		
Succinic acid	$HOOC(CH_2)_2COOH$	6.45×10^{-5} (4.19)	3.31×10^{-6} (5.48)		
Adipic acid	$HOOC(CH_2)_4COOH$	3.80×10^{-5} (4.42)	3.89×10^{-6} (5.41)		
Phthalic acid	$C_6H_4(COOH)_2$	1.12×10^{-3} (2.95)	3.89×10^{-6} (5.41)		
Phosphoric acid (ortho)	H_3PO_4	5.89×10^{-3} (2.23)	6.16×10^{-8} (7.21)	4.79×10^{-13} (12.32)	
Citric acid	HOOCCH$_2$ĊCH$_2$COOH (with OH above and COOH below central C)	1.15×10^{-3} (2.94)	7.25×10^{-5} (4.14)	1.51×10^{-6} (5.82)	
Ethylenediamine-tetraacetic acid (EDTA)	$[(HOOCCH_2)_2NCH_2{-}]_2$	1.01×10^{-2} (1.99)	2.13×10^{-3} (2.67)	6.90×10^{-7} (6.16)	5.47×10^{-11} (10.26)

21–3 DISTRIBUTION OF SPECIES DIAGRAMS

The molecular or anionic forms that a polyprotic acid assumes in aqueous solution depend upon the pH of the system or, conversely, the pH of a polyprotic acid system is determined by the forms of the polyprotic acid that are present. A plot of the fraction of the total polyprotic acid present as each species versus pH is referred to as a distribution diagram. This graphic representation of a polyprotic system is extremely helpful in understanding the nature of the equilibria involved. A distribution diagram for the phosphoric acid system is derived in Example/Problem 21–1.

Example/Problem 21–1.   Determine the fraction of the total phosphoric acid which is present as H_3PO_4, $H_2PO_4^-$, HPO_4^{-2}, and PO_4^{-3} as a function of pH.

Solution.   The mass action expressions for the three dissociations are given in Equations 21–1 through 21–3. The material balance for this system is

$$C_{PO_4} = [H_3PO_4] + [H_2PO_4^-] + [HPO_4^{-2}] + [PO_4^{-3}] \tag{21–4}$$

where C_{PO_4} is the *total analytical concentration* of all phosphate species. The fraction of each species is equal to the concentration of that species divided by the total analytical concentration. The fraction is denoted by the symbol α (alpha) with a numerical subscript corresponding to the number of acidic protons on the species, for example,

$$\alpha_3 = \frac{[H_3PO_4]}{C_{PO_4}}, \quad \alpha_2 = \frac{[H_2PO_4^-]}{C_{PO_4}}, \quad \alpha_1 = \frac{[HPO_4^{-2}]}{C_{PO_4}}, \quad \text{and} \quad \alpha_0 = \frac{[PO_4^{-3}]}{C_{PO_4}}$$

where $\alpha_3 + \alpha_2 + \alpha_1 + \alpha_0 = 1.00$.

Determination of α_3:

From the material balance expression and the definition of α_3,

$$\alpha_3 = \frac{[H_3PO_4]}{[H_3PO_4] + [H_2PO_4^-] + [HPO_4^{-2}] + [PO_4^{-3}]} \tag{21-5}$$

The right side of this equation has four unknowns. Substitution of the appropriate mass action expressions in the denominator gives

$$\alpha_3 = \frac{[H_3PO_4]}{[H_3PO_4] + \dfrac{K_1[H_3PO_4]}{[H^+]} + \dfrac{K_1K_2[H_3PO_4]}{[H^+]^2} + \dfrac{K_1K_2K_3[H_3PO_4]}{[H^+]^3}}$$

Further rearrangement gives

$$\alpha_3 = \frac{[H^+]^3}{[H^+]^3 + K_1[H^+]^2 + K_1K_2[H^+] + K_1K_2K_3} \tag{21-6}$$

Determination of α_2:

$$\alpha_2 = \frac{[H_2PO_4^-]}{C_{PO_4}} = \frac{K_1\dfrac{[H_3PO_4]}{[H^+]}}{C_{PO_4}}$$

and since $[H_3PO_4] = \alpha_3 C_{PO_4}$,

$$\alpha_2 = \frac{K_1\alpha_3}{[H^+]} \tag{21-7}$$

Determination of α_1:

$$\alpha_1 = \frac{[HPO_4^{-2}]}{C_{PO_4}} = \frac{K_2\dfrac{[H_2PO_4^-]}{[H^+]}}{C_{PO_4}}$$

Since $[H_2PO_4^-] = \alpha_2 C_{PO_4}$,

$$\alpha_1 = \frac{K_2\alpha_2}{[H^+]} = \frac{K_1K_2\alpha_3}{[H^+]^2} \tag{21-8}$$

In a similar manner it may be shown that

$$\alpha_0 = \frac{[PO_4^{-3}]}{C_{PO_4}} = \frac{K_3\alpha_1}{[H^+]} = \frac{K_1K_2K_3\alpha_3}{[H^+]^3} \tag{21-9}$$

Therefore, Equations 21-6 through 21-7 may be used to derive the distribution curves for the phosphoric acid system. These are shown in Figure 21-1.

For any polyprotic acid system where H_nB represents the *most* protonated form of the acid,

α_n = the fraction of the *total* "B" existing as the *most* protonated form, H_nB

$$\alpha_n = \frac{[H^+]^n}{[H^+]^n + K_1[H^+]^{n-1} + K_1K_2[H^+]^{n-2} + K_1K_2K_3[H^+]^{n-3}} \tag{21-10}$$

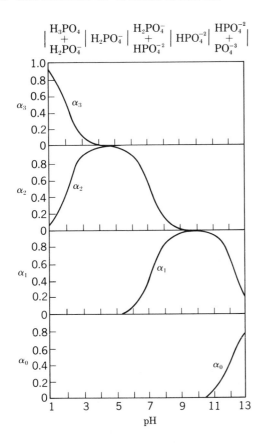

Figure 21-1 Distribution curves for the phosphoric acid system.

$\alpha_{n-1} =$ the fraction of the total "B" existing as $H_{n-1}B$

$$\alpha_{n-1} = \frac{K_1[H^+]^{n-1}}{[H^+]^n + K_1[H^+]^{n-1} + K_1K_2[H^+]^{n-2} + K_1K_2K_3[H^+]^{n-3}} \qquad (21\text{-}11)$$

$\alpha_{n-2} =$ the fraction of the total "B" existing as $H_{n-2}B$

$$\alpha_{n-2} = \frac{K_1K_2[H^+]^{n-2}}{[H^+]^n + K_1[H^+]^{n-1} + K_1K_2[H^+]^{n-2} + K_1K_2K_3[H^+]^{n-3}} \qquad (21\text{-}12)$$

$\alpha_{n-3} =$ the fraction of the total "B" existing as $H_{n-3}B$

$$\alpha_{n-3} = \frac{K_1K_2K_3[H^+]^{n-3}}{[H^+]^n + K_1[H^+]^{n-1} + K_1K_2[H^+]^{n-2} + K_1K_2K_3[H^+]^{n-3}} \qquad (21\text{-}13)$$

Although the equations just given may appear to be complex, they give a complete description of a polyprotic acid system in terms of the dissociation constants and a single variable, the pH of the solution. Analogous equations are used to describe stepwise complex ion dissociation equilibria (see Section 22–4).

The distribution curves shown in Figure 21–1 can be used to determine either the relative amounts of species needed to prepare a solution of a given pH, or to predict the pH of the system, given the amounts of the various species. The curves in Figure 21–1 indicate that at a given pH there are a maximum of two *principal* phosphate species present. This is true in any polyprotic acid system in which the successive dissociation constants differ by a factor of $\sim 10^3$ or more. The fraction of each species at a particular pH is determined as follows. At pH = 3.00, α_1 and α_0 are negligible; therefore $\alpha_3 + \alpha_2 = 1.00$. From the graph, $\alpha_3 = 0.127$ and $\alpha_2 = 0.873$.

Therefore, if it is desired to prepare a solution of pH = 3.00 for which the total phosphate concentration is 0.1000 F, the solution is made up so that $[H_3PO_4] =$ 0.0127 F and $[H_2PO_4^-] = 0.0873$ F. At pH = 7.21, α_3 and α_0 are insignificantly small and $\alpha_2 + \alpha_1 = 1.00$. At pH = 7.21, both α_2 and α_1 equal 0.500; thus, an equimolar mixture of $H_2PO_4^-$ and HPO_4^{-2} has a pH of 7.21 (neglecting activity effects, which, unfortunately, are important in phosphate systems).

Equations 21–10 through 21–13, representing the fractional distribution of species of a polyprotic acid system, apply to all polyprotic systems, whether their dissociation constants are close together or far apart.

When the polyprotic acid dissociation constants are significantly different (i.e., K_1/K_2 and $K_2/K_3 > 10^3$), the system may be treated as independent monoprotic acids of much different strengths. Example problems which illustrate this are now presented.

21–4   CALCULATIONS FOR POLYPROTIC SYSTEMS

Example/Problem 21–2.   Calculate the pH of a solution of 0.100 F H_3PO_4.
Solution.   The possible equilibria are:

$$H_3PO_4 \rightleftharpoons H^+ + H_2PO_4^-; \quad K_1 = 5.89 \times 10^{-3}$$
$$H_2PO_4^- \rightleftharpoons H^+ + HPO_4^{-2}; \quad K_2 = 6.16 \times 10^{-8}$$

From the magnitude of the equilibrium constants, the principal equilibrium is the dissociation of H_3PO_4. The $H_2PO_4^-$ ions in this system result from the partial dissociation of H_3PO_4 and the extent of dissociation of $H_2PO_4^-$ is relatively small. This is graphically illustrated in the distribution curves in Figure 21–1 demonstrating that the amount of HPO_4^{-2} in a system containing H_3PO_4 and $H_2PO_4^-$ is insignificant. Therefore, the mass action expression,

$$\frac{[H^+][H_2PO_4^-]}{[H_3PO_4]} = 5.89 \times 10^{-3}$$

is used to determine the pH. Since the dissociation of H_3PO_4 is the only source of $H_2PO_4^-$,

$$[H^+] = [H_2PO_4^-] \quad \text{and} \quad [H_3PO_4] = C_{H_3PO_4} - [H^+]$$

These relationships are analogous to those for a solution of a monoprotic acid, HB; therefore

$$[H^+] = \sqrt{K_1(C_{H_3PO_4} - [H^+])} \qquad (21\text{–}14)$$

The relatively large value of K_1 for H_3PO_4, 5.89×10^{-3}, indicates that the degree of dissociation is large enough so that it is not valid to neglect $[H^+]$ under the square root sign. A quadratic solution of the rearranged equation

$$[H^+]^2 = 5.89 \times 10^{-3}(0.100 - [H^+])$$

yields $$[H^+] = 2.13 \times 10^{-2} \, M \quad \text{or} \quad pH = 1.67$$

Example/Problem 21–3.   Calculate the pH of the solution which results from the addition of 30.0 ml of 0.100 N NaOH to 50.0 ml of 0.100 F H_3PO_4.
Solution.   There are originally (50.0 ml)(0.100 mmole/ml) = 5.00 mmole of H_3PO_4 before the base is added. Since OH^- is a *much* stronger base than $H_2PO_4^-$, the reaction $H_3PO_4 + OH^- \rightarrow H_2O + H_2PO_4^-$ is essentially quantitative. Therefore, in the final solution there are 2.00 mmole of H_3PO_4 left and 3.00 mmole of $H_2PO_4^-$ formed

since $(30.0 \text{ ml})(0.100 \text{ mmole/ml}) = 3.00 \text{ mmole}$ of base were added. The possible equilibria are:

$$H_3PO_4 \rightleftharpoons H^+ + H_2PO_4^-; \quad K_1 = 5.89 \times 10^{-3}$$
$$H_2PO_4^- \rightleftharpoons H^+ + HPO_4^{-2}; \quad K_2 = 6.16 \times 10^{-8}$$

In this solution the principal source of hydrogen ions is the slight dissociation of phosphoric acid. However, there are now two sources of $H_2PO_4^-$, the amount formed from the neutralization reaction and the amount contributed from the dissociation of H_3PO_4. The latter amount is the same as the $[H^+]$ in the solution. Therefore,

$$[H_3PO_4] = \frac{2.00 \text{ mmole}}{80.0 \text{ ml}} - [H^+]$$

and

$$[H_2PO_4^-] = \frac{3.00 \text{ mmole}}{80.0 \text{ ml}} + [H^+]$$

Substituting into the mass action expression for K_1,

$$\frac{[H^+](0.0375 + [H^+])}{0.0250 - [H^+]} = 5.89 \times 10^{-3} \tag{21–15}$$

Considering the value of the dissociation constant, $[H^+]$ will *not* be $\ll 0.025$. Therefore, Equation 21–15 must be solved by successive approximations or quadratically. Using successive approximations, assuming $[H^+] \ll 0.025$

$$[H^+] = 5.89 \times 10^{-3} \times \frac{2}{3} = 3.93 \times 10^{-3} \, M \qquad \text{(First approx)}$$

This value is substituted into the additive and subtractive terms of Equation 21–15. A second approximation is

$$[H^+] = 5.89 \times 10^{-3} \times \frac{0.0211}{0.0414} = 3.00 \times 10^{-3} \, M \qquad \text{(Second approx)}$$

The third approximation is

$$[H^+] = 5.89 \times 10^{-3} \times \frac{0.0220}{0.0405} = 3.20 \times 10^{-3} \, M \qquad \text{(Third approx)}$$

The third approximation is reasonably close to the second, so we stop here. Therefore

$$pH = 2.50$$

Example/Problem 21–4. Calculate the pH of a solution containing $0.0500 \, F$ NaH_2PO_4 and $0.0600 \, F$ Na_2HPO_4.

Solution. Clearly, this is a buffer in which $H_2PO_4^-$ is the acid and HPO_4^{-2} is its conjugate base. The ratio C_B/C_A is $0.0600/0.0500 = 1.20$. From the Henderson-Hasselbalch equation (Equation 20–5, where $pK_a = pK_2$).

$$pH = pK_2 + \log \frac{C_B}{C_A} = 7.21 + \log 1.20 = 7.29$$

Check this answer by comparing with the distribution diagram in Figure 21–1, reading the pH for $\alpha_2 = 5/11$ or $\alpha_1 = 6/11$.

Example/Problem 21–5. Calculate the pH of a solution that is prepared by dissolving 9.84 g of Na_3PO_4 (Mol. Wt. = 164.0.) in 250 ml of water and diluting to 500.0 ml with distilled water.

Solution.   There is only one base (other than water) in the solution, and no acids. Therefore the principal equilibrium is

$$PO_4^{-3} + H_2O \rightleftharpoons HPO_4^{-2} + OH^-; \quad K_b = K_w/K_3 = 2.09 \times 10^{-2}$$

The phosphate ion is obviously a fairly strong base. By analogy to fairly strong acids (e.g., H_3PO_4), we can calculate $[OH^-]$ by solving a quadratic equation

$$\frac{[HPO_4^{-2}][OH^-]}{[PO_4^{-3}]} = \frac{[OH^-]^2}{C_{PO_4} - [OH^-]} = K_b$$

$$\frac{[OH^-]^2}{0.120 - [OH^-]} = 2.09 \times 10^{-2}$$

$$[OH^-] = 4.08 \times 10^{-2} \, M; \quad pOH = 1.39$$

$$pH = 12.61$$

Example/Problem 21-6.   Calculate the pH of a solution of 0.100 F NaH_2PO_4. The equilibria showing the amphoteric behavior are

$$H_2PO_4^- \rightleftharpoons H^+ + HPO_4^{-2}; \quad K_2 \qquad\qquad (21\text{--}16)$$

$$H_2PO_4^- + H_2O \rightleftharpoons H_3PO_4 + OH^-; \quad K_b = K_w/K_1 \qquad (21\text{--}17)$$

or

$$H_2PO_4^- + H^+ \rightleftharpoons H_3PO_4; \quad 1/K_1 \qquad\qquad (21\text{--}18)$$

In this solution, $[H^+]$ produced in the ionization of $H_2PO_4^-$ (Equation 21–16) is not equal to $[HPO_4^{-2}]$, because some of the H^+ combines with $H_2PO_4^-$ to produce H_3PO_4 (Equation 21–18). Stating the above as an equation:

$$[H^+] = [HPO_4^{-2}] - [H_3PO_4] \qquad\qquad (21\text{--}19)$$

The first term on the right side measures the amount of H^+ produced in Equation 21–16 and the second term measures the amount of H^+ consumed in Equation 21–18. All of the mass action expressions for the stepwise dissociation of H_3PO_4 must be valid in any solution containing phosphate species at equilibrium. Therefore

$$[HPO_4^{-2}] = K_2 \frac{[H_2PO_4^-]}{[H^+]}$$

and

$$[H_3PO_4] = \frac{[H^+][H_2PO_4^-]}{K_1}$$

Substitution of these values into Equation 21–19 gives

$$[H^+] = K_2 \frac{[H_2PO_4^-]}{[H^+]} - \frac{[H^+][H_2PO_4^-]}{K_1}$$

Rearrangement of this equation gives

$$[H^+]^2 = \frac{K_1 K_2 [H_2PO_4^-]}{K_1 + [H_2PO_4^-]} \qquad\qquad (21\text{--}20)$$

To solve this equation for $[H^+]$ would require a second independent equation involving the two unknown quantities, $[H^+]$ and $[H_2PO_4^-]$. However, in most cases of practical interest, $[H_2PO_4^-] \gg K_1$. If K_1 in the denominator can be neglected, then the unknown $[H_2PO_4^-]$ cancels, yielding the simple result

$$[H^+] = \sqrt{K_1 K_2} \qquad\qquad (21\text{--}21)$$
$$= \sqrt{5.89 \times 10^{-3} \times 6.16 \times 10^{-8}}$$
$$= 1.91 \times 10^{-5} \, M$$
$$pH = 4.72$$

The analogous expression for a solution of Na_2HPO_4 is

$$[H^+] = \sqrt{K_2 K_3}$$

Equation 21–21 is useful in finding the pH of many amphoteric systems, such as: $NaHCO_3$, $NaHPO_4$, glycine, NH_4OAc, and so on, using the appropriate acidity constants. The accuracy decreases, however, if the concentration is not much greater than the larger acidity constant, or if the smaller acidity constant is not much greater than K_w.

To summarize, in Example/Problems 21–2 through 21–6 above, we have considered a typical polyprotic acid system in which the stepwise dissociation constants differ by factors of 100 or more. Example calculations were worked out in detail for several combinations of constituents.

1. A solution with *only* H_3PO_4, the completely protonated form, present. The equilibrium concentrations are calculated on the basis of a simple first step dissociation using the K_1 mass action expression.
2. A solution containing *both* H_3PO_4 and $H_2PO_4^-$. The equilibrium concentrations are calculated using the mass action expression containing both of these species, K_1. Because K_2 is so much smaller than K_1, the extent of dissociation of $H_2PO_4^-$ is negligible compared to the extent of the dissociation of H_3PO_4.
3. A system containing *both* $H_2PO_4^-$ and HPO_4^{-2}. The equilibrium concentrations are calculated using the mass action expression that includes both of these species, K_2. The extent of dissociation of HPO_4^{-2} is considered to be small compared to that of $H_2PO_4^-$.
4. A solution containing *only* PO_4^{-3}, the unprotonated form of the acid, which is in fact a moderately strong base.
5. A solution containing *only* $H_2PO_4^-$ or *only* HPO_4^{-2}. These species are amphoteric, that is, they act as both acids and bases. When present with another acid or base, the amphoteric character can be neglected, but when present alone, we must consider both their proton donating and proton accepting character.

21–5   TITRATIONS WITH POLYPROTIC SYSTEMS

If the successive pK's of a diprotic acid differ by 3 or more, the titration curve is a composite of two monoprotic acid curves as shown in Curves A and B of Figure 21–2. As the pK values approach each other, the two curves merge into a single curve and the equivalence point of the first proton disappears as in Curves C and D. Carbonic acid, $pK_1 = 6.35$, $pK_2 = 10.25$, (Curve A, Figure 21–3) exhibits a fairly sharp end point at pH 8.3 when the first proton has been titrated, giving a solution of HCO_3^-. The second equivalence point is poor because the second proton is too weakly acidic ($pK_2 = 10.35$) to titrate successfully in water. Note that the reverse titration of Na_2CO_3 (Curve B) exhibits two well-defined end points.

In order to obtain a satisfactory end point with a polyprotic acid system, it is necessary that at least two successive pK values differ by at least 3 and that the end

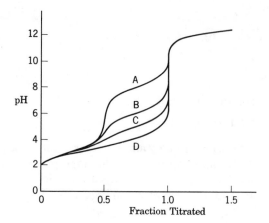

Figure 21-2   Titration curves for 0.1 M diprotic acids with 0.1 M strong base:

A: $pK_1 = 3.0$ $pK_2 = 8.0$
B: $pK_1 = 3.0$ $pK_2 = 6.0$
C: $pK_1 = 3.0$ $pK_2 = 5.0$
D: $pK_1 = 3.0$ $pK_2 = 4.0$

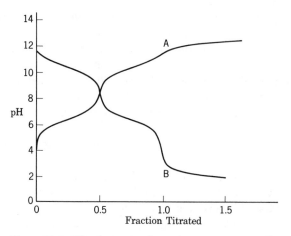

Figure 21-3   Titration curve for carbonate systems; $pK_1 = 6.35$, $pK_2 = 10.25$:

A: 0.05 M H_2CO_3 with 0.10 M NaOH
B: 0.05 M Na_2CO_3 with 0.05 M HCl

point occur within the pH range 4 to 10. These constraints effectively limit the number of end points obtainable in any aqueous system to two, as shown by the titration curves for a typical triprotic acid, H_3PO_4, and tetraprotic acid, ethylenediaminetetraacetic acid, H_4Y, in Figure 21-4.

21-6   POLYAMINE AND AMINO ACID SYSTEMS

Polyamine Systems.   Polyamines, which function as weak bases, are commonly used as buffering agents and as complexing agents for metal ions. In complexing media, there is a direct competition between available protons and the metal ions to

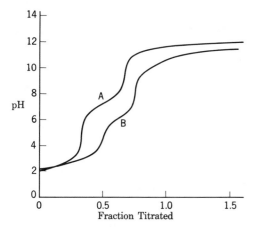

Figure 21–4 Titration curves of polyprotic acids:

A: 0.02 M H_3PO_4 with 0.02 M NaOH
$pK_1 = 2.15$, $pK_2 = 7.20$, $pK_3 = 12.36$

B: 0.01 M H_4Y with 0.01 M NaOH
$pK_1 = 2.00$, $pK_2 = 2.68$, $pK_3 = 6.16$, $pK_4 = 10.1$

bond to basic sites of the polyamine. Thus a knowledge of the nature of the equilibrium acid–base chemistry of these types of molecules is essential for an understanding of their use and limitations as metal complexing agents.

Polyamines are treated somewhat the same as polyprotic acids. Consider the system ethylenediamine-water. The possible reactions include

$$H_2NCH_2CH_2NH_2 + H_2O \rightleftharpoons H_3^+NCH_2CH_2NH_2 + OH^-$$

for which the mass action expression is

$$\frac{[enH^+][OH^-]}{[en]} = K_{b1}$$

where en = ethylenediamine, and the reaction

$$H_3^+NCH_2CH_2NH_2 + H_2O \rightleftharpoons H_3^+NCH_2CH_2NH_3^+ + OH^-$$

with the mass action expression

$$\frac{[enH_2^{+2}][OH^-]}{[enH^+]} = K_{b2}$$

K_{b1} and K_{b2} represent base dissociation constants. The conjugate acid dissociation constants are more useful for calculation purposes, that is,

$$enH_2^{+2} \rightleftharpoons enH^+ + H^+$$

with

$$K_{a1} = \frac{[enH^+][H^+]}{[enH_2^{+2}]} = \frac{K_w}{K_{b2}} = \frac{1.00 \times 10^{-14}}{7.09 \times 10^{-8}} = 1.41 \times 10^{-7}$$

and

$$enH^+ \rightleftharpoons en + H^+$$

with

$$K_{a2} = \frac{[en][H^+]}{[enH^+]} = \frac{K_w}{K_{b1}} = \frac{1.00 \times 10^{-14}}{8.55 \times 10^{-5}} = 1.17 \times 10^{-10}$$

Acid–base calculations for these systems are carried out in the same manner as those that have been described for other polyprotic acid systems (e.g., phosphoric acid).

Example/Problem 21-7. Calculate the concentration of ethylenediamine dihydrogen chloride required to produce a buffer solution of pH = 7.00 which is 0.0100 F in ethylenediamine monohydrogen chloride.

Solution. The pH = 7.00; therefore, $[H^+] = 1.00 \times 10^{-7}$ M. From the equilibrium constants at this hydrogen ion concentration

$$\frac{[enH^+]}{[enH^{+2}]} = \frac{K_{a1}}{[H^+]} = \frac{1.41 \times 10^{-7}}{1.00 \times 10^{-7}} = 1.41$$

and

$$\frac{[en]}{[enH^+]} = \frac{K_{a2}}{[H^+]} = \frac{1.17 \times 10^{-10}}{1.00 \times 10^{-7}} = 1.17 \times 10^{-3}$$

Thus $[en] \ll [enH^+]$. Therefore, the principal species in the solution are enH^+ and enH_2^{+2}. Since $[enH^+] = 0.0100$ F, $[enH_2^{+2}] = 0.0100/1.41 = 0.0071$ F or the solution must contain 0.0071 F en(HCl)$_2$ to give a buffer of pH = 7.00.

Amino Acids.   An amino acid contains at least one acidic and one basic group. For example, the structure of glycine is represented by

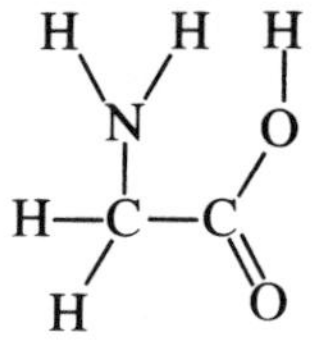

glycine

The α-amine group acts as a base and the carboxyl group behaves as an acid. Because the amine group is a stronger base ($K_b \sim 5 \times 10^{-4}$) than the acetate group ($K_b \sim 5 \times 10^{-10}$), the proton is more strongly attracted to the amine end of the molecule. As a matter of fact, when glycine is dissolved in water the conductivity indicates that glycine molecules have a very large dipole moment caused by significant charge separation in the molecules. This property suggests that the amino acid is an internally ionized molecule, commonly referred to as a *zwitterion*. Thus, in aqueous solution, glycine actually assumes the structure

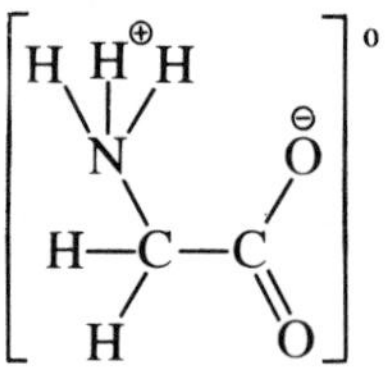

GH = glycine

The negatively charged carboxylate group in the glycine "zwitterion" clearly has basic character. As a result, the molecule might accept a proton to form,

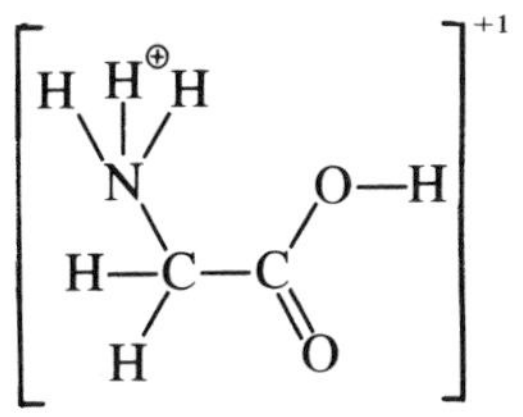

$$\left[\begin{array}{c} H \ \overset{\oplus}{H} H \\ \diagdown \ | \ \diagup \\ N \\ | \qquad \qquad O{-}H \\ H{-}C{-}C \diagup \\ | \qquad\quad \diagdown \\ H \qquad\quad O \end{array} \right]^{+1}$$

GH$_2^+$ = glycinium ion

On the other hand, a strong base might strip the acidic proton from the amine segment of the "zwitterion" to yield

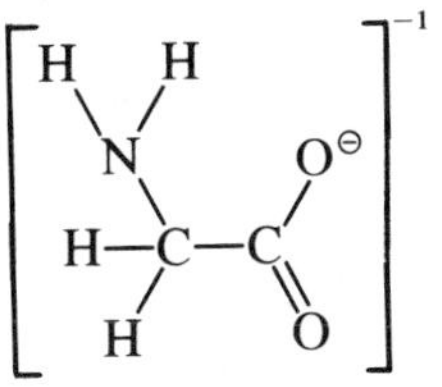

$$\left[\begin{array}{c} H \qquad H \\ \diagdown \quad \diagup \\ N \\ | \qquad\quad O^{\ominus} \\ H{-}C{-}C \diagup \\ | \qquad\quad \diagdown \\ H \qquad\quad O \end{array} \right]^{-1}$$

G$^-$ = glycinate ion

The overall acid–base equilibria between species GH, GH$_2^+$, and G$^-$ are given by

$$GH_2^+ \rightleftharpoons GH + H^+; \quad K_{a1} = \frac{[H^+][GH]}{[GH_2^+]}$$

and

$$GH \rightleftharpoons G^- + H^+; \quad K_{a2} = \frac{[H^+][G^-]}{[GH]}$$

A list of the acid dissociation constants of various amino acids is presented in Table 21–2.

At different pH values, an amino acid may exist principally as a positively charged ion, a neutral molecule (zwitterion) or a negatively charged ion. This is why electrophoresis separates amino acids that have different acid dissociation

Table 21–2 Acid Dissociation Constants for Representative Amino Acids

Name	Neutral Structure	K_{a1} (pK_{a1})	K_{a2} (pK_{a2})	K_{a3} (pK_{a3})
Glycine	$\overset{\displaystyle NH_2}{\underset{\displaystyle |}{H_2CCOOH}}$	4.46×10^{-3} (2.35)	1.86×10^{-10} (9.77)	
Serine	$\overset{\displaystyle NH_2}{\underset{\displaystyle |}{HOCH_2CHCOOH}}$	6.30×10^{-3} (2.20)	5.61×10^{-10} (9.25)	
Cysteine	$\overset{\displaystyle NH_2}{\underset{\displaystyle |}{HSCH_2CHCOOH}}$	1.10×10^{-2} (1.96)	4.36×10^{-9} (8.36)	5.25×10^{-11} (10.28)
Aspartic acid	$\overset{\displaystyle NH_2}{\underset{\displaystyle |}{HOOCCH_2CHCOOH}}$	8.30×10^{-3} (2.08)	1.15×10^{-4} (3.94)	1.05×10^{-10} (9.98)

constants. At a given pH, different amino acids exist as either positive ions, neutral molecules, or negative ions. The positive ion species migrate toward the negatively charged electrode, the neutral species are not deflected and the negative ion species migrate toward the positively charged electrode. This same variation in charge type with pH also is used as the basis for amino acid separations in ion exchange columns. Thus amino acid equilibria are extremely important both from theoretical and practical points of view. Acid–base calculations involving amino acids may be performed by using the same methods which have been developed for other polyprotic acid systems.

The pH at which an amino acid exists principally as the "zwitterion" neutral species and where the number of cations equals the number of anions is called the *isoelectric point*. For glycine, this is the point at which essentially all of the amino acid is in the form of GH, and where $[GH_2^+] = [G^-]$.

Example/Problem 21–8.   Calculate the pH of 0.0150 F glycine.
Solution.   The possible equilibria are:

$$GH + H_2O \rightleftharpoons GH_2^+ + OH^-; \quad K_b = K_w/K_{a1} = 2.24 \times 10^{-12}$$

base dissociation

$$GH \rightleftharpoons H^+ + G^-; \quad K_{a2} = 1.86 \times 10^{-10}$$

acid dissociation

$$GH + GH \rightleftharpoons GH_2^+ + G^-; \quad K_{eq} = K_{a2}/K_{a1} = 4.17 \times 10^{-8}$$

disproportionation

Since GH represents the intermediate species of the polyprotic acid system, we can use Equation 21–20:

$$[H^+]^2 = \frac{K_{a1}K_{a2}[GH]}{K_{a1} + [GH]}$$

Solving,

$$[H^+]^2 = 6.41 \times 10^{-13}$$

Therefore,

$$[H^+] = 8.0 \times 10^{-7} M$$

and

$$pH = 6.10$$

Example/Problem 21–9.   Determine the isoelectric point pH for the system glycine–water.
Solution.   At the isoelectric point by definition essentially all of the glycine is in the form of the neutral species and in addition

$$[GH_2^+] = [G^-]$$

Multiplying the mass action expressions K_{a1} and K_{a2}, we obtain

$$\frac{[H^+]^2[G^-]}{[GH_2^+]} = K_{a1}K_{a2}$$

Since $[GH_2^+] = [G^-]$,

$$[H^+]^2 = K_{a1}K_{a2} = 8.30 \times 10^{-13}$$

Thus

$$[H^+] = 9.1 \times 10^{-7} M$$

and the isoelectric point pH = 6.04.

From Example/Problems 21–8 and 9, the glycine isoelectric point pH is slightly lower than the pH of a 0.0150 F solution of glycine. Thus a *small* amount of strong acid would have to be added to a glycine solution to produce an "isoelectric solution."

21-7 AMINO ACID TITRATIONS

Amino acids may be titrated with either strong acids or strong bases. The titration curves for L-alanine, $CH_3-\overset{\overset{\displaystyle NH_2}{|}}{\underset{\underset{\displaystyle H}{|}}{C}}-\overset{\overset{\displaystyle OH}{\diagup}}{\underset{\underset{\displaystyle O}{\diagdown\!\!\diagdown}}{C}}$, are shown in Figure 21–5.

Aspartic acid, $HOOC-CH_2-\overset{\overset{\displaystyle NH_2}{|}}{C}HCOOH$, is representative of amino acids that have more than one carboxyl or amine group. The various ionic and molecular states of the compound are illustrated in this manner:

$$\left[\; HO\overset{O}{\diagup}C-CH_2-C\overset{\overset{H\;\overset{\oplus}{H}\;H}{\diagdown N\diagup}}{\underset{H}{|}}-C\overset{OH}{\diagdown\!\!\diagdown O}\;\right]^{+1} \quad \text{or} \quad H_3Ap^+$$

$$\left[\; HO\overset{O}{\diagup}C-CH_2-C\overset{\overset{H\;H\;\overset{\oplus}{H}}{\diagdown N\diagup}}{\underset{H}{|}}-C\overset{O^{\ominus}}{\diagdown\!\!\diagdown O}\;\right]^{0} \quad \text{or} \quad H_2Ap$$

$$\left[\; {}^{\ominus}O\overset{O}{\diagup}C-CH_2-C\overset{\overset{H\;\overset{\oplus}{H}}{\diagdown N\diagup}}{\underset{H}{|}}-C\overset{O^{\ominus}}{\diagdown\!\!\diagdown O}\;\right]^{-1} \quad \text{or} \quad HAp^-$$

$$\left[\; {}^{\ominus}O\overset{O}{\diagup}C-CH_2-C\overset{\overset{H\;H}{\diagdown N\diagup}}{\underset{H}{|}}-C\overset{O^{\ominus}}{\diagdown\!\!\diagdown O}\;\right]^{-2} \quad \text{or} \quad Ap^{-2}$$

The equilibria for the aspartic acid-water system are represented by:

$$H_3Ap^+ \rightleftharpoons H^+ + H_2Ap \qquad K_{a1} = 8.3 \times 10^{-3}$$
$$H_2Ap \rightleftharpoons H^+ + HAp^- \qquad K_{a2} = 1.15 \times 10^{-4}$$

and

$$HAp^- \rightleftharpoons H^+ + Ap^{-2} \qquad K_{a3} = 1.05 \times 10^{-10}$$

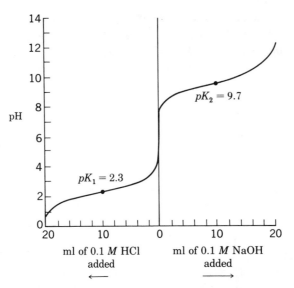

Figure 21–5 Titration curves for the reaction of 20.0 ml of 0.100 F L-alanine with 0.100 N NaOH and 0.100 N HCl. [From Conn and Stumpf, *Outlines of Biochemistry*, Wiley, 1963, p. 67.]

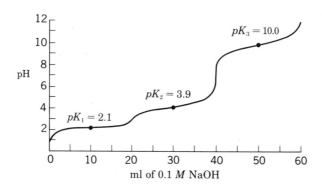

Figure 21–6 Titration curve for the reaction of 20.0 ml of 0.100 F aspartic hydrochloride with 0.100 N NaOH. [From Conn and Stumpf, *Outlines of Biochemistry*, Wiley, 1963, p. 69.]

A titration curve obtained in the reaction of 20.0 ml of 0.100 M aspartic hydrochloride, $H_3Ap^+Cl^-$, with 0.100 N NaOH is given in Figure 21–6.

Example/Problem 21–10. Determine the isoelectric point pH for the system aspartic acid-water.

Solution. At the isoelectric point,

$$[H_2Ap] \gg [H_3Ap^+] = [HAp^-] \gg [Ap^{-2}]$$

Multiplying the mass action expressions for K_{a1} and K_{a2},

$$\frac{[H^+]^2[HAp^-]}{[H_3Ap^+]} = K_{a1}K_{a2}$$

Thus

$$[H^+]^2 = K_{a1}K_{a2}$$

And

$$[H^+] = 9.8 \times 10^{-4}\ M$$

Therefore, the isoelectric point pH = 3.01.

QUESTIONS AND PROBLEMS

21–1. Calculate the pH of the solutions prepared by mixing the following components:
(a) 50 ml of 0.200 F NaHCO$_3$ with 50 ml H$_2$O.   *Ans.* 8.30
(b) 50 ml of 0.300 F Na$_3$PO$_4$ with 50 ml 6.0 M HCl.   *Ans.* − 0.40
(c) 50 ml of 0.300 F Na$_3$PO$_4$ with 50 ml 6.0 M NaOH   *Ans.* 14.48
(d) 50 ml of 0.300 F Na$_3$PO$_4$ with 50 ml 0.300 F H$_3$PO$_4$.   *Ans.* 7.21
(e) 50 ml of 0.500 F H$_3$PO$_4$ with 50 ml 1.00 F Na$_3$PO$_4$.   *Ans.* 9.77
(f) 50 ml of 0.02 M HOAc with 50 ml 0.02 M NH$_4$OH.   *Ans.* 7.01

21–2. A buffer is prepared by taking 20.0 ml of 10.0 F H$_3$PO$_4$, adding 175 ml of 2 M NaOH, diluting to 1 liter. The pH is adjusted by adding 10 ml of 5.0 M HCl.
(a) Calculate the pH of the buffer as adjusted.   *Ans.* 7.21
(b) What is the buffer capacity of this buffer?   *Ans.* 0.082 mole/liter

21–3. 25.00 ml of a certain solution which contains *either* a mixture of HCl and H$_3$PO$_4$ *or* H$_3$PO$_4$ and NaH$_2$PO$_4$ was titrated with 0.100 N standard NaOH. The pH was monitored with a pH meter. 28.0 ml of base were required to reach the H$_2$PO$_4^-$ equivalence point and a *total* of 36.0 ml were required to reach the HPO$_4^{-2}$ equivalence point. What were the components and their respective concentrations in the unknown?   *Ans.* [HCl] = 0.0800 M; [H$_3$PO$_4$] = 0.0320 M.

21–4. A 50.0 ml solution that is 0.100 F in NaH$_2$PO$_4$ *and* 0.0500 F in Na$_2$HPO$_4$ is titrated with 0.250 N NaOH. What is the pH:
(a) Before any NaOH has been added?   *Ans.* 6.91
(b) After 20.0 ml of NaOH have been added?   *Ans.* 9.73
(c) After 30.0 ml of NaOH have been added?   *Ans.* 11.86

21–5. It is desired to prepare a buffer solution of pH = 7.00 starting with 500 ml of a solution which is 0.400 F in Na$_2$HPO$_4$. Standardized 0.200 N solutions of both HCl and NaOH are available. Indicate the quantity (in ml) of which of the two reagent solutions you would add to the 500 ml Na$_2$HPO$_4$ solution to prepare the desired buffer.   *Ans.* 619 ml of 0.200 N HCl

21–6. A 50.0 ml solution that is 0.100 F in H$_3$PO$_4$ and 0.0500 F in NaH$_2$PO$_4$ is titrated with 0.250 N NaOH. What is the pH:
(a) Before any NaOH has been added?   *Ans.* 2.04
(b) After 20.0 ml of NaOH have been added?   *Ans.* 4.72
(c) After 30.0 ml of NaOH have been added?   *Ans.* 6.91

21–7. Determine the general shape of the titration curve for the titration of 50.0 ml of 0.100 F serine hydrochloride with 0.200 N NaOH (K_a's given in Table 21–2).

21–8. Select the best indicator from Table 20–5 for the following titrations:
(a) 0.1 F Na$_2$HPO$_4$ with 0.1 N HCl (to NaH$_2$PO$_4$).
(b) 0.05 F Na$_2$H$_2$Y with 0.1 N NaOH (to Na$_3$HY).

21–9. Determine the isoelectric point pH for the amino acid cysteine.   *Ans.* 5.16

21–10. Calculate [S^{-2}] in a 0.050 F H$_2$S solution in which the pH has been adjusted to 3.0.   *Ans.* 5×10^{-17} M

21–11. Calculate [S^{-2}] in a solution made by saturating water with H$_2$S gas.   *Ans.* 1×10^{-14} M

21–12. A 7.38 g sample of an unknown acid of unknown purity is dissolved in one liter of water and a 100-ml aliquot is titrated with 0.100 M NaOH. From the following titration data, identify the acid and calculate the purity of the original sample, assuming that the impurities are neither acidic nor basic.   *Ans.* purity = 90%

ml NaOH	pH	ml NaOH	pH	ml NaOH	pH
10	2.46	45	4.56	80	8.88
20	2.95	50	4.93	85	11.43
30	3.42	60	5.41	90	11.72
35	3.80	70	5.90	100	12.00
40	4.18	75	6.26		

22

METAL ION COMPLEXES

22-1 INTRODUCTION

Among the earliest methods of determining the presence and concentration of metal ions is the titration technique in which a metal complex is formed; for example:

$$Hg^{+2} + 4I^- \rightleftharpoons HgI_4^{-2}$$

$$Ag^+ + 2CN^- \rightleftharpoons Ag(CN)_2^-$$

However, there are a limited number of complexing reactions that proceed according to simple exact stoichiometry, with an equilibrium constant large enough to insure completion at the equivalence point. Visual detection of the end point also presents many problems specific to each titration. With the introduction of ethylenediaminetetraacetic acid (EDTA) and related complexing (chelating) agents, compleximetric titrations have developed rapidly since 1945. We will examine some of the properties and reactions of these new reagents, and their important analytical applications.

A complexed metal ion is an example of a *coordination compound* because it includes a *coordinate–covalent bond*. The donor, or *ligand*, provides the electron pair that is shared between the ligand and the *central atom* (in this case the metal ion). The ligand must have at least one pair of free (unbonded) electrons, although the metal ion may be able to form two or more coordinate bonds. Each metal ion has a characteristic maximum number of coordinate bonds it can form under normal conditions. This maximum number is called the *coordination number*, usually two, four, or six; for example,

$$Ag(NH_3)_2^+, ZnCl_4^{-2}, Fe(CN)_6^{-3}$$

Other coordination numbers are known but are relatively uncommon.

22-2 MONODENTATE LIGANDS

Copper-Ammonia System. If a ligand has only a single pair of free electrons to share, it will require 2, 4, or 6 ligands to satisfy completely the coordination sphere of the central atom. Such ligands are called *monodentate* (single tooth). Some ligands, such as ethylenediamine, $H_2\ddot{N}CH_2CH_2\ddot{N}H_2$, have two or more pairs of electrons to share. They are called *bidentate, tridentate, quadridentate, sexadentate,* or *multidentate* ligands.

We will first consider the reactions between a central metal ion, Cu^{+2}, and a monodentate ligand, NH_3. The formation of a complex actually involves the replacement of water molecules from the solvated metal ion, but for this discussion we will ignore the hydrated nature of ions in solution. Since the Cu^{+2}

ion has a normal coordination number of four, it can react with four ammonia molecules, and does so, one at a time.

$$Cu^{+2} + NH_3 \rightleftharpoons Cu(NH_3)^{+2} \qquad\qquad K_1 = 2.0 \times 10^4$$
$$Cu(NH_3)^{+2} + NH_3 \rightleftharpoons Cu(NH_3)_2^{+2} \qquad K_2 = 4.7 \times 10^3$$
$$Cu(NH_3)_2^{+2} + NH_3 \rightleftharpoons Cu(NH_3)_3^{+2} \qquad K_3 = 1.1 \times 10^3$$
$$Cu(NH)_3^{+2} + NH_3 \rightleftharpoons Cu(NH_3)_4^{+2} \qquad K_4 = 2.0 \times 10^2$$

Each of the K values represents the *stepwise formation constant* of the given complex from the next lower complex; that is,

$$K_3 = \frac{[Cu(NH_3)_3^{+2}]}{[Cu(NH_3)_2^{+2}][NH_3]}$$

As for any stepwise process, the overall reaction is the sum of the partial reactions, and the overall formation constant is the product of the stepwise constants. (The terms "formation constant" and "stability constant" are synonymous.)

$$Cu^{+2} + 4NH_3 \rightleftharpoons Cu(NH_3)_4^{+2}; \quad K_f = K_1 K_2 K_3 K_4 = 2.1 \times 10^{13}$$

The large value of K_f for $Cu(NH_3)_4^{+2}$ suggests that the formation reaction is quantitative and should provide a useful method of titrating copper ion with ammonia. The complex equilibria encountered here are similar to those in the acid–base titration of a polyprotic acid. Rather than go through the tedious calculations of deriving a theoretical titration curve, we present the results in Figure 22–1, Curve A, here plotted as $-\log[Cu^{+2}]$, or pCu, as a function of ml of

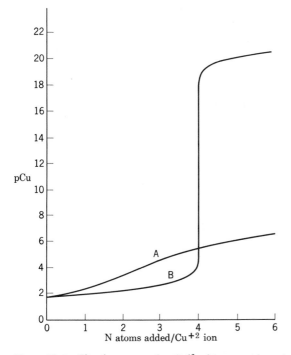

Figure 22–1 Titration curves for Cu^{+2} with ammonia and trien:
A: 0.02 M Cu^{+2} with 0.02 M NH_3
B: 0.02 M Cu^{+2} with 0.02 M trien

added titrant. No break is observed at the equivalence point, and the titration is useless. What happened? The explanation is that the ammonia does not add to the copper ions in an orderly fashion. The formation of $Cu(NH_3)^{+2}$ is never completed because the addition of a second NH_3 to a $Cu(NH_3)^{+2}$ is almost as probable as the addition of the first NH_3 to the remaining free Cu^{+2} ions; K_2 is almost as large as K_1. This behavior is more apparent in the distribution diagram shown in Figure 22–2. As we approach the end point of the overall titration reaction (pNH$_3$ ca. 3), there are still significant amounts of $Cu(NH_3)_2^{+2}$ and $Cu(NH_3)^{+2}$ remaining in the solution, and the stoichiometry is fuzzy indeed. All of the stepwise reactions occur simultaneously.

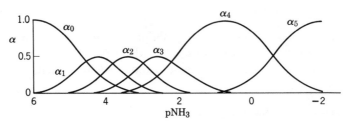

Figure 22–2 Distribution diagram for copper–ammonia complex ions:

$$\alpha_0 = Cu^{+2}, \alpha_1 = Cu(NH_3)^{+2}, \alpha_2 = Cu(NH_3)_2^{+2}, \text{etc.}$$

22–3 METAL CHELATES

Copper-Trien System.   The above titration would be improved if there were some way of simultaneously adding four ammonia molecules to each copper ion; that is, we need a method of adding ammonia molecules in bundles of four. There is, in fact, just such a reagent, triethylenetetramine (abbreviation: trien), in which four ammonia molecules are "tied" together with ethylene bridges:

$$H_2\ddot{N} \diagdown_{CH_2-CH_2} \overset{\ddot{N}}{\underset{H}{\diagup|\diagdown}}_{CH_2-CH_2} \overset{\ddot{N}}{\underset{H}{\diagup|\diagdown}}_{CH_2-CH_2} \diagup \ddot{N}H_2$$

The trien molecule is a tetradentate ligand possessing four pairs of free electrons. The coordinate bonds it can form are similar to the Cu—NH$_3$ bonds. Furthermore, its flexible structure permits it to "wrap around" a copper ion so that a single trien molecule satisfies the coordination sphere of the metal ion.

$$\begin{bmatrix} & CH_2-CH_2 & \\ & & \overset{H}{\underset{|}{N}}-CH_2 \\ H_2N & & \\ & Cu & \\ H_2N & & N-CH_2 \\ & & \overset{|}{\underset{H}{}} \\ & CH_2-CH_2 & \end{bmatrix}^{+2}$$

When a trien molecule reacts with a copper ion, all four coordinate bonds are formed almost simultaneously—an all-or-nothing process. The formation of the first bond brings the next N-atom in close proximity to the Cu ion; thus, the formation of the second bond is more probable than is the case with ammonia molecules, where the second molecule must diffuse from a greater distance. Likewise the third and fourth bonds are easily and rapidly formed.

A multidentate ligand is called a *chelating agent* from the Greek word "chele" meaning claw. The complex formed is called a *chelate*; it contains *chelate rings* involving the central atom as one of the members of the ring structure. Let us compare the stability of the copper-trien complex with that of the ammonia complex.

$$Cu^{+2} + trien \rightleftharpoons Cu(trien)^{+2}; \quad K_f = 2.5 \times 10^{20}$$
$$Cu^{+2} + 4NH_3 \rightleftharpoons Cu(NH_3)_4^{+2}; \quad K_f = 2.1 \times 10^{13}$$

Both complexes involve the formation of four Cu—N bonds, but K_f for Cu(trien)$^{+2}$ is 10^7 greater than K_f for Cu(NH$_3$)$_4^{+2}$. The added stability of the former results from the greater probability of forming a ring as compared to individual NH$_3$ molecules reacting (an entropy effect). This phenomenon is common to all chelates and is called the *chelate effect*. The chelate effect is at a maximum when five- or six-membered rings can be formed. Smaller rings are highly strained and large rings are less likely to form.

EDTA Complexes. The titration reaction of copper with trien proceeds in a well-defined manner; the large equilibrium constant provides a sharp end point as shown in Figure 22–1, Curve B. Trien reacts similarly with a few other metals, such as cobalt, nickel, cadmium, and zinc. It is not, however, a widely used titrant, because it is difficult to purify and unpleasant to work with. Chelating agents containing both nitrogen and oxygen atoms are far more versatile. One of the best and most popular of these is ethylenediaminetetraacetic acid (abbreviation: EDTA, symbol H$_4$Y). Under appropriate conditions EDTA forms complexes with nearly all metal ions, including weak complexes with some alkali metals. The six coordination sites are located in positions that favor the formation of five-membered chelate rings; thus, the EDTA can "wrap around" a metal ion. Co^{+2}, for example, has a coordination number of six with an octahedral configuration. With metal ions having a coordination number of four, like Cu^{+2} and Zn^{+2}, EDTA is a quadridentate chelating agent.

There is always a competition in the solution between the metal ion and hydrogen ions seeking the negative sites on the EDTA molecule. The equilibrium situation is determined by relative bond strengths, and the relative concentrations of the various species. Formation constants for several EDTA complexes are given in Table 22–1.

Table 22–1 Formation Constants of Metal-EDTA Complexes

Metal Ion	log K_f	Metal Ion	log K_f
Fe^{+3}	25.1	Al^{+3}	16.1
Cr^{+3}	23	Fe^{+2}	14.3
Hg^{+2}	21.8	Mn^{+2}	13.8
Cu^{+2}	18.8	Ca^{+2}	10.7
Ni^{+2}	18.6	H$^+$	10.34
Pb^{+2}	18.0	Mg^{+2}	8.7
Cd^{+2}	16.5	Ba^{+2}	7.8
Zn^{+2}	16.5	Na$^+$	1.7
Co^{+2}	16.3	—	—

22–4   CONDITIONAL FORMATION CONSTANTS

The formation constants in Table 22–1 pertain to reactions of a metal ion with the completely dissociated EDTA anion, Y^{-4}. However, EDTA is most commonly used as the disodium salt, $Na_2H_2Y \cdot 2H_2O$, and its solutions contain the dinegatively-charged anion, H_2Y^{-2}. In order to compute the extent of complex formation in such a solution, we can use one of two alternative approaches.

First, let us consider the formation of the CuY^{-2} complex, which requires three steps:

$$
\begin{array}{ll}
H_2Y^{-2} \rightleftharpoons H^+ + HY^{-3} & K_3 = 6.9 \times 10^{-7} \\
HY^{-3} \rightleftharpoons H^+ + Y^{-4} & K_4 = 5.5 \times 10^{-11} \\
\dfrac{Cu^{+2} + Y^{-4} \rightleftharpoons CuY^{-2}}{Cu^{+2} + H_2Y^{-2} \rightleftharpoons CuY^{-2} + 2H^+} & \dfrac{K_f = 6.3 \times 10^{18}}{K_f^* = 2.4 \times 10^2}
\end{array}
$$

where

$$K_f^* = \frac{[CuY^{-2}][H^+]^2}{[Cu^{+2}][H_2Y^{-2}]}$$

The "*effective*" *formation constant*, K_f', for the formation of CuY^{-2} from H_2Y^{-2} and Cu^{+2} is

$$K_f' = \frac{[CuY^{-2}]}{[Cu^{+2}][H_2Y^{-2}]} = \frac{K_f^*}{[H^+]^2}$$

From our earlier discussion of the pH of solutions of intermediate salts of polyprotic acids, the pH of a solution of Na_2H_2Y is equal to

$$1/2(pK_2 + pK_3) = 1/2(2.7 + 6.2) = 4.45. \text{ Thus, } K'_f = K^*_f/[H^+]^2$$
$$= 2.4 \times 10^2/(3.5 \times 10^{-5})^2 = 2.0 \times 10^{11} \quad \text{(at pH 4.45)}$$

The fact that K'_f is much less than K_f shows that H^+ competes with Cu^{+2} for coordination sites on the EDTA molecule, and even at pH 4.45 the effect is considerable.

A second approach to calculating K'_f is to determine the actual $[Y^{-4}]$ at the pH of interest. Here is a use for the distribution of species diagrams discussed in Section 21–3. In this case, we need to know the fraction of EDTA species that exist as Y^{-4} at pH 4.45. Distribution diagrams for H_4Y are derived in the same manner as for H_3PO_4, and are plotted in Figure 22–3. The value of α_0 at pH 4.45 is very close to zero and cannot be read from this plot. Since we are often interested in values close to zero, a logarithmic plot is far more useful. The same data are replotted in Figure 22–4 as $\log \alpha$ vs. pH. At pH 4.45, $\alpha_0 = 4 \times 10^{-8}$. Therefore,

$$K'_f = \frac{[CuY^{-2}]}{[Cu^{+2}][H_2Y^{-2}]} = \frac{[CuY^{-2}]}{[Cu^{+2}][Y^{-4}]} \times \frac{[Y^{-4}]}{[H_2Y^{-2}]} = K_f\alpha_0$$
$$= 6.3 \times 10^{18} \times 4 \times 10^{-8} = 2.5 \times 10^{11}$$

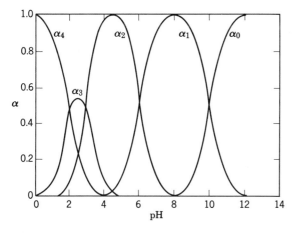

Figure 22–3 Distribution diagram for EDTA species as a function of pH.

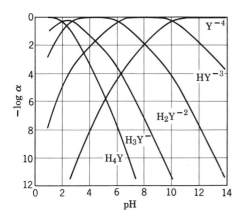

Figure 22–4 Distribution diagram for EDTA species, plotted as $\log \alpha$ versus pH.

The "effective" formation constant, K'_f, is also called the *conditional formation constant*, because its numerical value depends on the conditions, especially the pH of the solution. The pH determines the distribution of EDTA species and may also affect the state of the metal ion (e.g., hydrolysis).

Example/Problem 22–1. What is the conditional formation constant of the ZnY^{-2} complex in a solution of 0.02 F EDTA at pH 5.00?

From Table 22–1, $\hspace{4cm} K_f = 3.2 \times 10^{+16}$

From Figure 22–4, $\hspace{4cm} \alpha_0 = 5.0 \times 10^{-7}$

$$K'_f = K_f \alpha_0 = 3.2 \times 10^{16} \times 5.0 \times 10^{-7} = 1.6 \times 10^{10}$$

Titrations carried out at pH > 5 with Na_2H_2Y solutions normally release H^+ during the course of the reaction

$$M^{+2} + H_2Y^{-2} \rightleftharpoons MY^{-2} + 2H^+$$

If the pH of the titration mixture changes, the conditional formation constant will change and the shape of the titration curve, as well as the equivalence point, will be altered. Figure 22–5 shows the effect of pH on the titration of Ca^{+2} with EDTA; a value of $K'_f > 10^8$ is required to obtain a satisfactory break at the equivalence point. Obviously, it is necessary to use buffer mixtures in order to get accurate and reproducible results. Unfortunately (or fortunately, in some cases) many buffers are also complexing agents themselves. The resulting equilibria are extremely complicated and beyond the scope of this discussion. In general, it is possible to handle these situations by further adjusting the conditional formation constant, but it should be apparent that EDTA titrations require close control of all conditions.

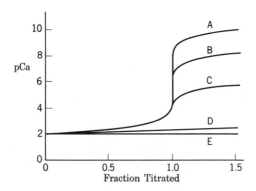

Figure 22–5 Titration curves for calcium with EDTA at various pH values.

A: pH 10	$K'_f = 1.75 \times 10^{10}$
B: pH 8	$K'_f = 2.55 \times 10^8$
C: pH 6	$K'_f = 1.10 \times 10^6$
D: pH 4	$K'_f = 1.65 \times 10^2$
E: pH 2	$K'_f = 1.85 \times 10^{-3}$

22–5 TITRATION EQUILIBRIA

For the titration of Zn^{+2} with EDTA, it is desirable to maintain the pH at approximately 9. In order to prevent $Zn(OH)_2$ from precipitating, it is convenient to use an ammonia–ammonium chloride buffer in which the Zn^{+2} is complexed as $Zn(NH_3)_4^{+2}$. At any point in this titration, the total concentration of zinc is:

$$C_{Zn} = [Zn^{+2}] + [ZnY^{-2}] + [Zn(NH_3)^{+2}] + [Zn(NH_3)_2^{+2}] + [Zn(NH_3)_3^{+2}] + [Zn(NH_3)_4^{+2}]$$

It will simplify matters somewhat if we consider the total concentration of zinc *not* including that complexed with EDTA:

$$C'_{Zn} = [Zn^{+2}] + [Zn(NH_3)^{+2}] + [Zn(NH_3)_2^{+2}] + [Zn(NH_3)_3^{+2}] + [Zn(NH_3)_4^{+2}]$$

and then define the fraction of C'_{Zn} which exists as free Zn^{+2}

$$\alpha_{Zn} = \frac{[Zn^{+2}]}{C'_{Zn}}$$

Now we can further adjust the conditional formation constant to take into account the Zn—NH_3 complexes:

$$K_f = \frac{[ZnY^{-2}]}{[Zn^{+2}][Y^{-4}]} = \frac{[ZnY^{-2}]}{\alpha_{Zn}C'_{Zn}\alpha_0 C_{EDTA}}$$

$$K'_f = K_f\alpha_{Zn}\alpha_0 = \frac{[ZnY^{-2}]}{C'_{Zn}C_{EDTA}}$$

Thus the conditional formation constant, K'_f, which governs the success of the titration, depends on α_{Zn} and α_0. The value of α_{Zn} is determined by the concentration of free ammonia in the solution, and the stepwise formation constants of the Zn—NH_3 complexes. The value of α_0 is determined by the pH of the solution, and the stepwise dissociation constants of EDTA. Graphical and computer techniques are available to simplify the calculations in deriving titration curves. We can also predict that α_{Zn} will decrease as $[NH_3]$ is increased. Thus K'_f will decrease with increasing $[NH_3]$ and the titration will be less satisfactory at high concentrations of ammonia. This behavior is shown in Figure 22–6.

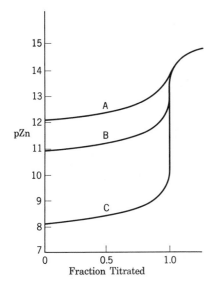

Figure 22–6 Titration curves for zinc with EDTA at various concentrations of ammonia. The concentration of NH_4Cl is also adjusted to maintain constant pH = 9.26.

$$A: [NH_3] = 1.0\ F$$
$$B: [NH_3] = 0.5\ F$$
$$C: [NH_3] = 0.1\ F$$

22–6   COMPLEXIMETRIC INDICATORS

If compleximetric methods are to be widely used, good visual indicators should be available; the success of EDTA methods has in fact been due largely to the development of metallochromic indicators. In general these indicators are highly colored organic dyes that form chelates of a different color with the metal ion in question. Thus, they perform the same function in much the same manner as an acid–base indicator does in an acid–base titration. However, the equilibria are much more complicated because the dyes combine not only with metal ions but also with hydrogen ions. They could be used as acid–base indicators but metal ion titrations are usually performed at constant pH.

Eriochrome Black T is typical of these indicators. It has one strong sulfonic acid proton and two weak phenolic protons which can be replaced by a metal ion:

The pertinent equilibria are:

$$H_2In^- \rightleftharpoons H^+ + HIn^{-2} \qquad K_2 = 5.0 \times 10^{-7}$$
$$\text{red} \qquad\qquad \text{blue}$$

$$HIn^{-2} \rightleftharpoons H^+ + In^{-3} \qquad K_3 = 2.5 \times 10^{-12}$$
$$\text{blue} \qquad\quad \text{orange}$$

$$Zn^{+2} + HIn^{-2} \rightleftharpoons ZnIn^- + H^+ \qquad K'_f$$
$$\text{blue} \qquad\quad \text{red}$$

$$ZnIn^- + HY^{-3} \rightleftharpoons ZnY^{-2} + HIn^{-2} \qquad K \gg 1$$
$$\text{red} \qquad\qquad\qquad \text{blue}$$

The stability of the $ZnIn^-$ must be greater than the stability of the $Zn(NH_3)_4^{+2}$ complex, but less than the stability of the ZnY^{-2} complex. Let us consider the titration of a zinc solution with EDTA in an ammonia buffer at pH ~ 9. When the buffer is added to the zinc solution, the Zn^{+2} forms $Zn(NH_3)_4^{+2}$ which is colorless. Addition of the indicator forms a small amount of highly colored red $ZnIn^-$. As the solution is titrated, the principal reaction is:

$$Zn(NH_3)_4^{+2} + HY^{-3} \rightleftharpoons ZnY^{-2} + 3NH_3 + NH_4^+$$

The small amount of $ZnIn^-$ remains as such until the equivalence point is reached, when the first excess EDTA displaces the indicator from its zinc complex. In this case the color changes from red to blue. Under the proper conditions the end point is sharp and easily recognized.

Other indicators of similar structure are available for a variety of metal-chelate determinations. Each of these requires a particular set of conditions for a specific metal ion:

(a) Eriochrome Black T for Ba, Ca, Cd, Mg, Zn.
(b) Calmagite, for same metals as Eriochrome Black T.
(c) PAN for Cu, Cd and Zn.
(d) Murexide for Co, Cu and Ni.
(e) Pyrocatechol Violet for Al, Bi, Ga, Tb, Fe, Mn.

22-7 SCOPE OF EDTA APPLICATIONS

Procedures are available for the determination of nearly every metal ion with EDTA. In addition to the direct titration with a visual indicator that we have just discussed, there are numerous alternative methods.

1. *Indirect or back titrations.* Some metal ions react very slowly with EDTA and their titration would require an inconveniently long time or elevated temperature. Metals such as Cr(III), Fe(III), Al(III), and Ti(IV) may be determined by adding an excess of EDTA, followed by back-titration of the excess with a standard Zn^{+2} or Mg^{+2} solution.

2. *Photometric titration.* Many chelates are highly colored and are likely to have distinctive absorption spectra that are useful in photometric titrations. The accuracy of the end point is often enhanced by following the color change in a spectrophotometric titration cell.

3. *Potentiometric titration.* A mercury electrode becomes a versatile indicator in chelate titrations. A small amount of mercury in a J-tube serves as an indicator electrode whose potential is a function of the concentration of free mercuric ions.

$$E = \text{constant} - \frac{0.059}{2} \log \frac{1}{[Hg^{+2}]} \qquad (22\text{--}1)$$

Hg(II) forms an extremely stable EDTA complex ($\log K_f = 21.8$). In a solution containing Hg^{+2}, Y^{-4}, and another metal ion (e.g., Ca^{+2}) the equilibrium will be:

$$HgY^{-2} + Ca^{+2} \rightleftharpoons Hg^{+2} + CaY^{-2}$$

$$\frac{[Hg^{+2}][CaY^{-2}]}{[HgY^{-2}][Ca^{+2}]} = \frac{K_{f,\,CaY}}{K_{f,\,HgY}} \qquad (22\text{--}2)$$

During the course of a titration of Ca^{+2} with EDTA, a small amount of added Hg(II) will be essentially completely complexed and the $[HgY^{-2}]$ will be nearly constant. Therefore Equation 22-2 can be rearranged, combining all constants, to give:

$$[Hg^{+2}] = K * \frac{[Ca^{+2}]}{[CaY^{-2}]} \qquad (22\text{--}3)$$

Substituting Equation 22-3 into 22-1, we obtain:

$$E = \text{constant}' - \frac{0.059}{2} \log \frac{[CaY^{-2}]}{[Ca^{+2}]} \qquad (22\text{--}4)$$

Although it would be possible to compute the $[Ca^{+2}]$ from the measured potential during the titration, this is not necessary. The ratio $[CaY^{-2}]/[Ca^{+2}]$ will change drastically in the region of the equivalence point, thus abruptly changing the potential of the mercury electrode. This procedure is applicable to any metal ion, provided the formation constant of its EDTA complex is considerably less than that of the HgY^{-2} complex.

4. *Masking agents.* Because EDTA forms complexes with nearly all metal ions, samples containing several metal ions cause complications. Some selectivity can be achieved by proper selection of pH; for example, conditional formation constants show that most trivalent metal ions can be titrated below pH 4 while many divalent metals require a pH greater than 8. Further selectivity can be achieved by adding auxiliary complexing agents. The object is to "mask" interfering metal ions without affecting the ion of interest. Cyanide ion is widely used for this purpose because it forms highly stable complexes with Cd, Co, Ni, Zn, and Fe; fluoride ion will "mask" the behavior of Fe, Al, and Mg, and so on. When all else fails, chemical separations are possible (extraction, ion exchange) but these should be used only as a last resort.

5. *Metal ion buffers.* A solution containing roughly equal amounts of a complexed metal ion and excess ligand acts to "buffer" the concentration of free metal ion in much the same way as a pH buffer maintains a constant $[H^+]$.

$$M + L \rightleftharpoons ML \qquad \text{(charges omitted)}$$

$$K_f' = \frac{[ML]}{[M][L]}$$

$$[M] = \frac{[ML]}{K_f'[L]}$$

$$\log [M] = -\log K_f' + \log \frac{[ML]}{[L]}$$

$$pM = \log K_f' + \log \frac{[L]}{[ML]} \qquad (22\text{--}5)$$

By appropriate selection of experimental conditions, the pM of a solution can be maintained close to the log K_f' value for the conditions extant, even though the metal ion may be generated or consumed by other reactions occurring in the system. Note that Equation 22–5 is similar in form to the Henderson–Hasselbalch equations (20–5 and 20–9) for pH buffers.

QUESTIONS AND PROBLEMS

22–1. An EDTA solution is to be standardized with pure $CaCO_3$. A 0.500-g sample is dissolved in hydrochloric acid and diluted to 1 liter. A 50.00-ml aliquot of the Ca solution requires 35.00 ml of the EDTA solution. Calculate the formality of the EDTA solution.   *Ans. 7.14×10^{-3} F*

22–2. The hardness of water can be determined by titrating with EDTA solution. Calculate the hardness in ppm of $CaCO_3$ if 30.0 ml of 0.0050 M Na_2H_2Y are required for 100 ml water.   *Ans. 150 ppm*

22–3. A Zn solution, made by dissolving 0.817 g of Zn dust in HCl and diluting to 250 ml, is used to standardize an EDTA solution. A 50-ml aliquot of the Zn solution requires 40.00 ml of the EDTA solution for titration. If the EDTA solution is subsequently used to titrate iron samples, how many mg of Fe is each ml of EDTA solution equivalent to?   *Ans. 2.79 mg Fe/ml EDTA*

22–4. Calculate pM for each of the following solutions:
(a) 0.0123 F $CaCl_2$ *Ans. 1.91*
(b) 1.5×10^{-7} M Na_2SO_4 *Ans. 6.52*
(c) 0.050 F $FeSO_4 \cdot (NH_4)_2SO_4 \cdot 5H_2O$ *Ans. 1.30*

(d) 0.10% NaCl   *Ans.* 1.77

(e) 2.0 ppm KI   *Ans.* 4.92

22–5. A certain solution contains $1.0 \times 10^{-2}\,F$ Cu(II) and $1.0 \times 10^{-3}\,M$ free NH_3 (excess). Estimate the concentrations of the various copper-ammine complexes that are present using Figure 22–2.

22–6. List in order of decreasing concentration the formulas of the various copper ions present in a solution containing $10^{-3}\,F$ Cu(II) and $10^{-4}\,M$ free NH_3.

22–7. Nitrilotriacetic acid (NTA or H_3Y) is a triprotic acid which forms complexes similar to EDTA. Its formula is $N(CH_2COOH)_3$ and its three pK_a values are 1.9, 2.5, and 9.7.

(a) Draw a structural formula of the Cu–NTA complex.

(b) What is the pH of a solution of Na_2HY?

22–8. If a solution contains $0.10\,F$ EDTA, at what pH will the $[HY^{-3}]$ be $0.06\,M$?

Ans. 6.5 and 9.7

22–9. A 5.0-ml portion of $0.10\,M$ $CoCl_2$ is added to one liter $0.20\,F$ EDTA buffered at pH 10.0. Calculate pCo in this mixture.   *Ans.* 18.6

22–10. The Ni^{+2} and Zn^{+2} in a 25.0-ml sample were complexed by adding 50.0 (an excess) of $0.0125\,F$ EDTA. The excess EDTA was back-titrated with 7.50 ml of $0.0100\,M$ $MgCl_2$ solution. At this point an excess of 2,3-dimercapto-1-propanol (BAL) was added, which displaced only Zn^{+2} from its EDTA complex:

$$ZnY^{-2} + BAL \rightarrow ZnBAL + H_2Y^{2-}$$

Titration of the liberated H_2Y^{-2} required an additional 21.5 ml of the $MgCl_2$ solution. Calculate the concentrations of Zn^{+2} and Ni^{+2} in the original solution.

Ans. $[Zn^{+2}] = 0.0086\,M$

$[Ni^{+2}] = 0.0134\,M$

22–11. Calculate the value of pM at the equivalence point when $0.02\,M$ metal ion is titrated with $0.02\,F$ EDTA at pH 4.5.

(a) $NiCl_2$ *Ans.* 6.60

(b) $CdCl_2$ *Ans.* 5.55

(c) $FeCl_2$ *Ans.* 4.20

(d) $CaCl_2$ *Ans.* 2.65

22–12. A metal ion solution contains 50 mmole Cu(II) and 200 mmole EDTA per liter. The solution is also buffered at pH 7.00 by the excess EDTA present. Calculate pCu in this solution.   *Ans.* 16.0

22–13. At what pH would the buffer in Problem 22–12 have a pCu of 10.0?   *Ans.* pH = 3.5

22–14. What is the minimum pH at which magnesium ion can be titrated with EDTA, assuming that a conditional formation constant of at least 10^8 is necessary?

Ans. 10.2

22–15. Calculate values of K_f' for the PbY^{-2} complex at pH 2, 4, 6, 8, 10, and 12. Plot log K_f' versus pH and determine the value of log K_f' at pH 7.4.

REFERENCES

H. Flaschka, *EDTA Titrations*, 2nd ed., Pergamon Press, New York, 1964.

G. Schwarzenbach and H. Flaschka, *Complexometric Titrations*, 2nd ed., Methuen, London, 1969.

F. J. Welcher, *The Analytical Uses of Ethylenediaminetetraacetic Acid*, Van Nostrand Reinhold, New York, 1958.

A. Ringbom, *Complexation in Analytical Chemistry*, Wiley-Interscience, New York, 1963.

J. N. Butler, *Ionic Equilibrium, A Mathematical Approach*, Addison-Wesley, Reading, Mass., 1964.

RADIOCHEMISTRY

23

RADIOCHEMICAL METHODS

23-1 INTRODUCTION

Most of the techniques for studying radioactive isotopes were developed in the investigation of the naturally occurring radioactive elements. However, with the advent of the cyclotron, the atomic pile, and other "atom smashing" devices, unstable radioactive isotopes of all the elements can be produced, and thus radio-counting techniques are of general analytical applicability. Over 1000 radionuclides are known.

Although the intelligent use of radiochemical methods requires some knowledge of nuclear chemistry (what holds the nucleus together, why it may be unstable, how the disintegration takes place), we will concern ourselves primarily with the resulting radiation. We will study how the radiation is detected and measured, and how these measurements are used.

23-2 RADIOACTIVE DISINTEGRATION

Radioactive Isotopes. The stability of an atomic nucleus is a complex function of the number of protons in it, and the ratio of neutrons to protons. For many of the light elements (e.g., $^{12}_{6}C$, $^{16}_{8}O$, $^{32}_{16}S$), an equal number of neutrons and protons is a stable configuration, whereas those isotopes having a ratio other than unity (e.g., $^{3}_{1}H$, $^{14}_{6}C$, $^{15}_{8}O$, $^{19}_{8}O$, $^{35}_{16}S$) are likely to be unstable. For the heavier elements, relatively more neutrons are required for stability (e.g., $^{208}_{82}Pb$). For the still heavier elements all isotopes are radioactive, regardless of the neutron–proton ratio, although $^{238}_{92}U$ and $^{232}_{90}Th$ with half-lives of 10^{10} years are considered stable for most purposes.

As we have already indicated, most of the radioactive isotopes we deal with are produced artificially. But the decay process of an artificially produced isotope is no different from that of a natural isotope, except that it will probably occur sooner.

The Disintegration Process. All atoms can be broken apart by sufficiently energetic projectiles. On the other hand, "radioactive" atoms are so unstable that they break up spontaneously. All kinds of nuclear fragments are produced, but except for fission processes, the detectable radiation consists of only a few common fragments. For our purposes we will consider only these common fragments and assume that the rest of the nucleus "gets lost" somewhere in the surroundings. These common fragments are: alpha particles (helium nuclei), beta particles (positrons or electrons), gamma rays (high-energy photons), and neutrons.

The disintegration of a nucleus is regulated by the laws of probability. We cannot predict when an individual nucleus will distintegrate, but we can predict the average behavior of a large number of identical nuclei. In order to obtain meaningful rates, we must observe large numbers of disintegrations. For example, there is no way of predicting when a certain ^{32}P nucleus will disintegrate, but we can say with nearly absolute certainty that if we have 10 billion ^{32}P atoms, approximately 5 billion of them will disintegrate in 14.3 days and another 2.5 billion will disintegrate in the next 14.3 days.

In a large group of radioactive atoms the average lifetime before disintegration is apparently a fixed number for each type of atom. The rate at which this process occurs, or the rate of decay, is proportional to the number, N, of potentially radioactive atoms present at any given instant.

$$\text{Rate} = \lambda N \tag{23–1}$$

where λ is a proportionality constant or rate constant which is called the *decay constant*. Radioactive decay is a *first-order process* because the rate depends on the concentration (in this case, amount) of only one species. There is only a single reactant yielding two or more products.

$$A \rightarrow B + C + \cdots$$

Both the concentration of A in a given sample and the rate of decay must decrease with time. Therefore, it is convenient to express the rate as a derivative—the rate of change with time of the concentration of A.

$$\text{Rate} = -\frac{d[A]}{dt} = \lambda [A] \tag{23–2}$$

Separating the variables in Equation 23–2 yields

$$-\frac{d[A]}{[A]} = \lambda \, dt$$

which on integration yields

$$-\int_{[A]_0}^{[A]} \frac{d[A]}{[A]} = \lambda \int_0^t dt$$

or

$$\ln \frac{[A]_0}{[A]} = \lambda t \tag{23–3}$$

Equation 23–3 is written in a more useful form (changing from the concentration of A to the more conventional symbol, N, the number of potentially radioactive atoms in a sample) as

$$N = N_0 e^{-\lambda t} = N_0 10^{-\lambda t / 2.303} \tag{23–4}$$

where N_0 is the number present at time $t = 0$. A radioactive isotope is frequently characterized by its *half-life*, $t_{1/2}$, the time required for one-half of the atoms present at a given instant to disintegrate. At $t = t_{1/2}$, $N/N_0 = 1/2$. From Equation 23–3:

$$\ln 2 = \lambda t_{1/2}$$

or

$$t_{1/2} = \frac{0.693}{\lambda} \tag{23–5}$$

Table 23–1   Typical Radioactive Nuclides

Nuclide	Abundance, %	Half-life	Radiations, MeV
^{3}H	—	12.26 y	β^- 0.0186
^{11}C	—	20.4 m	β^+ 0.99
^{14}C	—	5720 y	β^- 0.155
^{13}N	—	10.0 m	β^+ 1.19
^{22}Na	—	2.58 y	β^+ 0.544; γ 1.274
^{32}P	—	14.3 d	β^- 1.71
^{35}S	—	87 d	β^- 0.167
^{36}Cl	—	3×10^5 y	β^- 0.714
^{40}K	0.012	10^9 y	β^- 1.32; γ 1.46
^{45}Ca	—	165 d	β^- 0.255
^{59}Fe	—	45 d	β^- 0.46, 0.27; γ 1.10, 1.29
^{60}Co	—	5.26 y	β^- 0.32; γ 1.173, 1.333
^{64}Cu	—	12.8 h	β^- 0.573; β^+ 0.656
^{90}Sr	—	29 y	β^- 0.54
^{131}I	—	8.06 d	β^- 0.60; γ 0.364
^{137}Cs	—	39 y	β^- 0.51
^{226}Ra	100	1622 y	α 4.78
^{227}Ac	—	22 y	β^- 0.046
^{232}Th	100	10^{10} y	α 4.01, 3.95
^{233}U	—	10^5 y	α 4.82, 4.77
^{234}U	0.0056	2×10^5 y	α 4.77, 4.72
^{235}U	0.7205	7×10^8 y	α 4.39; γ 0.18, 0.14
^{238}U	99.27	5×10^9 y	α 4.19

Observed half-lives vary from 10^{-22} seconds to over 10^{10} years. Selected radioisotopes are listed in Table 23–1.

The *curie* is a basic unit of radioactivity. In former times, a curie denoted an activity (rate of disintegration) equal to that of 1 gram of radium. Its value varied with refinements in measurement. It is now defined as the quantity of nuclide in which the number of disintegrations per second is 3.700×10^{10}. For example, 0.224 mg of ^{14}C is 1 mc (millicurie) because it produces 3.700×10^7 disintegrations per second. The weight of a millicurie of ^{14}C could also be obtained from its half-life of 5720 years:

$$\lambda = \frac{0.693}{5720 \times 3.156 \times 10^7} = 3.83 \times 10^{-12} \text{ sec}^{-1}$$

$$-\frac{dN}{dt} = \lambda N = \lambda \frac{W}{14} \times 6.02 \times 10^{23} = 1.65 \, W \times 10^{11} \text{ sec}^{-1}$$

For 1 mc,

$$-\frac{dN}{dt} = 3.700 \times 10^7$$

$$W = \frac{3.700 \times 10^7}{1.65 \times 10^{11}} = 0.224 \times 10^{-3} \text{ g}$$

The weight of a curie thus depends on the half-life and the atomic or molecular weight.

Activity is thus measured in disintegrations per unit time, or curies. For many purposes, the term *specific activity* is useful; the activity per gram of sample, or per gram of a particular element in the sample.

23–3   INTERACTION OF RADIATION WITH MATTER

Alpha Particles.   All alpha particles from a given type of nucleus have the same energy or, in some cases, are distributed within a few monoenergetic groups. Alpha particles from different nuclides have various energies, but all are very high, in the range of 1 to 10 MeV. The unit of energy here is $1\,MeV = 10^6$ electron-volts $= 1.6 \times 10^{-6}$ erg. Remember, this is the energy per particle.

When alpha particles pass through matter, they lose energy chiefly by excitation or dissociation of the electrons of the molecules in their pathway. Because of their relatively high mass, alpha particles tend to travel in air in straight lines, leaving a trail of ions produced by collisions. To produce an ion-pair in air requires an average energy of 35 eV. Thus the total number of ions produced, or the length of the path, is a measure of the initial energy of the alpha particles. Since alpha particles from a given source are monoenergetic, they all have roughly the same penetrating power as shown in Figure 23–1. As a rough guide, 5 MeV alpha particles produce 25,000 ion-pairs/cm in air and spend their entire energy in about 3.5 cm of air. These particles are stopped by a single sheet of paper.

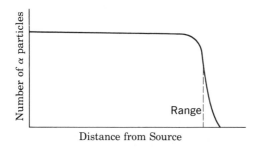

Range

Distance from Source

Figure 23–1   Number of alpha particles from a source as a function of distance from the source.

Beta Particles.   The energies of beta particles emitted from a given kind of isotope have a continuous energy distribution from zero up to a maximum value. The average energy is about one-third of the maximum energy. For a given isotope the maximum energy has a sharp cut-off, but the various beta-emitters give energies varying from 15 keV to 15 MeV.

Beta particles react with matter in much the same way as alpha particles—they produce ions at a cost of about 35 eV per ion-pair. However, there are significant differences. The primary ionization by beta particles accounts for only about one-quarter of the total ionization; the remainder results from secondary ionization. For a given energy, the velocity of a beta particle is much larger than for an alpha particle. In air, ionization stops when the energy has been reduced to 12.5 eV, the ionization potential of an oxygen molecule. On the other hand, the beta particle may lose a large fraction of its energy in a single collision. Beta particles are widely scattered by collisions with atomic nuclei, but energy loss is caused almost entirely by interactions with electrons. The range in air of beta particles is about 100 cm, during which a 0.5-MeV particle will produce 60 ion-pairs/cm. An idealized curve for the absorption of beta particles is given in Figure 23–2. Actual curves do not exhibit such a sharp cut-off because of secondary ionization effects. The penetrating power of beta particles is commonly expressed in terms of the mg/cm^2 of aluminum required for 100% absorption. With absorbers of low atomic number, the penetrating power is independent of the particular substance used.

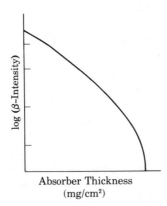

Figure 23–2 Idealized absorption curve for beta particles.

Gamma Rays. When a nucleus in an excited state returns to its ground state, the most common way involves the emission of electromagnetic radiation. The energies are large (10 keV to 7 MeV) and the radiation is called gamma rays.

The specific ionization caused by gamma rays is much less than for alpha or beta particles; therefore, gamma rays have much greater ranges. Again, the production of ions costs about 35 eV per ion-pair, although for gamma rays the ionization is due almost entirely to secondary processes:

1. *Photoelectric effect.* This is the most important process at low energies. The gamma ray ejects a bound electron from an atom or molecule. The entire energy of the gamma rays is used up—partly in releasing the electron and partly in the kinetic energy it takes with it.

2. *Compton effect.* This is the predominant process in the 0.6- to 4-MeV range. Instead of giving up its entire energy to the departing electron, only a part of its energy is transferred. Thus the gamma ray now has much less energy and is also traveling in a different direction.

3. *Pair-production process.* This process takes place only if the energy exceeds 1.02 MeV and increases with increasing energy. Above this energy threshold, an electron-positron pair can be created by conversion of energy into "mass." The positron produced has an extremely short lifetime—its fate is immediate annihilation with an electron with the simultaneous emission of two 0.51-MeV photons. Thus pair-production always results in low-energy secondary radiation.

There is no well-defined range for gamma rays. All three processes lead to exponential absorption of the form given by Beer's law. In this case, we write:

$$I_d = I_o e^{-\mu d} \tag{23–6}$$

where I_o is the original intensity, I_d is the intensity transmitted through a thickness d, and μ is an absorption coefficient. The half-thickness, $d_{1/2}$, is defined as the thickness required to reduce I_d to $I_o/2$, or $d_{1/2} = 0.693/\mu$. Half-thickness values may be translated into photon energies, the details of which are beyond the scope of this discussion.

Neutrons. The interaction of neutrons (which carry no charge) with electrons is practically negligible. Primary ionization by neutrons is not observed. It is only because of secondary effects resulting from collisions with atomic nuclei that we are able to detect neutrons.

"Fast" neutrons are of little interest to us, but after they have been slowed down, the resulting "thermal" neutrons are very efficient at producing nuclear reactions. This not only gives a means of detecting neutrons, but also provides a source of artificially created isotopes.

23-4 DETECTION AND MEASUREMENT OF RADIOACTIVITY

The several kinds of particles and rays produced during the disintegration of a radioactive nuclide leave a large number of ions along their paths. It is these ions that are normally detected and measured. Several techniques are used, any one of which may be the best for a particular application.

Photographic Emulsion. An ionizing particle or ray will cause an activation and subsequent darkening of a photographic plate. Thus, the path becomes visible and the degree of darkening is a measure of the total activity. This method is used to locate the exact distribution of radioactive material in a thin slice of the sample (e.g., a tracer in a slice of tissue). The most common application is for film badges used to monitor total exposure of laboratory personnel exposed to radiation. The blackening of photographic plates that had been placed close to samples of uranyl sulfate led to the historic discovery of radioactivity by Becquerel in 1896.

The Ionization Chamber. Many detectors make use of the electrical conductivity of a gas that has been partially ionized by radiation passing through it. In the ionization chamber, an electric field is applied between two electrodes separated by a gas. The kind of gas used depends on the particular application and the geometry of the electrodes is selected to give the optimum performance. These details will be discussed below for the Geiger counter—a common example of an ionization chamber.

Charged particles or rays passing into the chamber will produce ion-pairs that, in the absence of an electric field, will recombine and will not be detected. As the voltage between the electrodes is increased, the electrons will be collected at the anode and positive ions at the cathode. At low voltages, the magnitude of this "pulse" of ion current is approximately proportional to the applied voltage as shown in Figure 23-3 in the region up to V_1. At voltages greater than V_1 there is a region of nearly constant pulse height, indicating that all of the ions produced by the radiation are being collected before they have a chance to recombine. This is called the *saturation region.* There is a range of voltage over which the exact voltage setting is not critical and systems designed to operate in this region are called *ionization chambers.* Some of these instruments are designed to give a separate signal or "pulse" for each disintegration. The height of the pulse depends in part on the number of ion-pairs produced. Other instruments incorporate a large time constant so that the current is averaged out to a more or less steady state. In this way the average rate of disintegration is determined, rather than the individual events. The currents produced in ionization chambers are of the order of 10^{-15} amp. Vibrating-reed electrometers are commonly used as amplifiers.

Proportional Counters. As the voltage in the ionization chamber is increased beyond V_2 in Figure 23-3, the current again increases because the electrons moving through the higher electric field acquire sufficient energy to cause secondary ionization. In this way, the original ionization is multiplied and a larger pulse is obtained.

In the region between V_2 and V_3 (Figure 23-3), the *pulse height* increases with applied voltage, but more important, it *is proportional to the initial ionization.* Note, however, that the *counting rate* does not depend on voltage in the

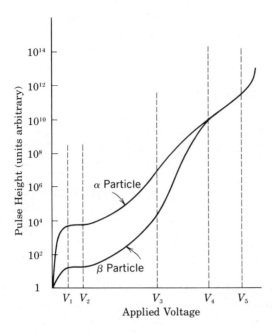

Figure 23–3 Variation of pulse height with applied voltage in an ionization chamber or counter. [From Friedlander, Kennedy, and Miller, *Nuclear and Radiochemistry*, Wiley, 1964, p. 142.]

proportional region. Thus we can have a voltage plateau for alpha particles that produce a large number of primary ions. Under the same conditions we might not see beta particles, which are 100 times less ionizing. At a higher voltage, where the multiplication factor is greater, both the alpha and the beta particles will be counted. This kind of performance is illustrated in Figure 23–4. Compared to the Geiger counter (see below), the discharge is limited to the immediate region of the entering particle and the paths traversed by the secondary ions. Thus the dead-time (recovery time between pulses) is very short, about 1 μsec, and the counter is useful for rates up to 200,000 counts/sec. Multiplication factors vary up to 10^4, but this still requires considerable external amplification.

The Geiger Counter. The proportional region of a counter, as just described, is limited at the upper voltage end by the onset of photoionization. At high voltages, the electrons formed in the initial and secondary ionization are accelerated so much that photons are emitted when the electrons strike the central anode. The photons so released spread the discharge along the wire, resulting in an "avalanche" of electrons reaching the anode. Typical geometry of a Geiger counter is illustrated in Figure 23–5. The interior of the tube may be filled with a mixture of argon (which provides the ionizable substance) and some heavier gas such as alcohol, methane, or chlorine (which quenches the avalanche).

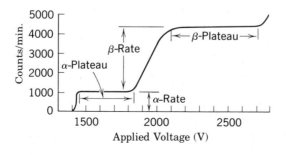

Figure 23–4 Effect of applied voltage on the counting rate of a proportional counter receiving both alpha and beta particles.

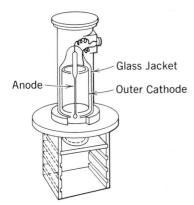

Figure 23–5   Schematic diagram of a Geiger counter and sample holder.

Each ionizing particle entering the tube thus triggers an enormous discharge, or pulse, which no longer depends on the number of primary ion-pairs produced. The electrons reach the central anode very quickly because of the extremely high field around the wire. Typically, this takes about 5×10^{-7} sec. But then the wire is surrounded by a sheath of positively charged fragments, effectively reducing the voltage gradient below the level needed for ion multiplication. Thus the counter is inoperative and must recover before another pulse can take place. It takes about 100 to 500 μsec for the positive ions to reach the outer cathode, and this is the inherent dead-time of the counter after each pulse. When the positive ions strike the cathode they might well release secondary electrons and set off a new (unwanted) pulse if it were not for the *quenching effect* of the ether, alcohol, or methane contained in the filling gas. These polyatomic molecules manage to absorb much of the energy of the positive ions by collisions.

The plateau region of a Geiger counter is shorter and less well-defined than for a proportional counter—the region between V_4 and V_5 in Figure 23–3. The optimum voltage should be determined for each instrument. As indicated by the steep rise in the curve beyond V_5, if the upper voltage limit is exceeded, the quenching effect breaks down and the tube goes into a continuous discharge condition. Counters that use an organic vapor for quenching deteriorate after about 10^8 counts. Tubes using halogen vapors as quenching agents last indefinitely.

Geiger counters produce a very large pulse (1 to 10 V) that requires little if any amplification for detection. There is, of course, no possibility of determining the energy of the original ionizing particle, because the pulse is an "all or nothing" phenomenon. Because of the long dead-time, counting rates are limited to about 15,000 counts/min.

Scintillation Counters.   Many of the early studies of natural radioactivity were done by observing the fluorescence produced when an alpha particle struck a screen impregnated with zinc sulfide. This crude method produced considerable eye strain, but it has been highly refined by using better scintillators and photomultiplier detectors.

For counting beta particles, crystals of anthracene or stilbene release pulses of photons that are detected by an adjacent photomultiplier tube. Liquid solutions of stilbene or *p*-terphenyl in xylene or toluene are also good scintillators for beta particles, and if these solutions are loaded with boron or cadmium compounds, they will even detect neutrons. Commercially available solutions are often called "scintillation cocktails."

For the measurement of gamma rays, large crystals of sodium iodide are

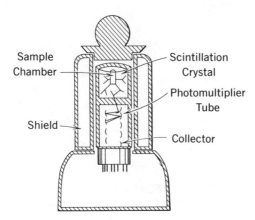

Figure 23–6 Schematic diagram of a scintillation counter and auxiliary apparatus.

preferred. The crystal is activated with about 1% of thallium iodide. Unfortunately, the crystals are highly hygroscopic and must be protected from the atmospheric moisture. A typical scintillation counter is illustrated in Figure 23–6.

Scintillation counters have very high counting efficiencies and very short dead times (0.25 μsec for NaI crystals and less than 0.01 μsec for stilbene crystals). Another very important feature is that the height of the pulse is proportional to the energy of the original particle.

Semiconductor Detectors.   Diode devices fabricated from lithium-doped germanium have improved resolution in gamma ray counting by ten-fold. A thin layer of lithium is vapor-deposited on a crystal of highly purified *p*-type germanium. Then the temperature is raised and a portion of the lithium atoms diffuse into the central portion of the crystal. Such a crystal operates much like an ionization tube. A gamma ray passing through the crystal creates "electron-hole" pairs that can be collected as a current pulse, as shown in Figure 23–7. A high sensitivity is achieved because each electron-hole pair requires only 2.9 eV of energy. In order to reduce thermal noise and obtain the optimum signal-to-noise ratio, the crystal

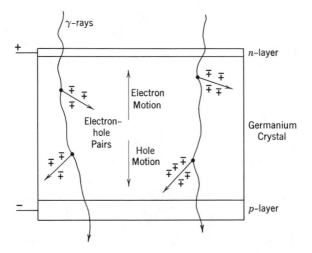

Figure 23–7 Schematic cross section of a lithium-doped germanium detector. Electron-hole pairs are formed by ionizing collisions in the central lithium-drifted region. Electrons and holes are subsequently collected by the outer layers.

temperature is maintained at $-196°C$ in a liquid nitrogen bath. These detectors are now widely used in gamma ray and X-ray spectrometry.

Scalers.   Radioactive disintegrations often occur at very high rates, faster than any mechanical counter could operate. A scaling circuit is used in such a way that only every other count is passed along. A number of these circuits in series thus produces only $1/2^n$ counts for each original pulse, where n is the number of scaling units, or the scaling factor. The number of events withheld is indicated on a series of glow lamps, with a mechanical counter measuring only the scaled-down signals. More drastic (and also more convenient) scaling is accomplished by circuits which pass along only every tenth count, the so-called "decade scalers." Electric timers may be incorporated in the scaling unit which automatically start and stop for the counting interval, or may stop the counting at a predetermined time or total number of counts.

Pulse-Height Discriminators.   If the amplitude (or height) of the pulse is proportional to the energy of the original ionizing particle, the measurement of the pulse height may be useful in identifying the nature of the radiation.

A single-channel analyzer incorporates a device that will pass signals only if they exceed a certain minimum size; those signals below the threshold are dissipated and not counted. A similar circuit can also exclude those signals that exceed a specific maximum value. Thus, a "window" is created, and only those signals within a fixed amplitude are counted. Now, if this "window" can be moved along the energy spectrum, it is possible to obtain a radiation spectrum; that is, a plot of the distribution of number of counts as a function of energy of the radiation received by the counter.

Multiple-channel analyzers are still more versatile. By means of suitable circuitry they can sort out the pulses on the basis of size and send each pulse into a particular counting device that counts only those pulses of a given range of energy. Thus the whole spectrum is obtained at once. Analyzers containing 100 to 8192 such channels are now available.

23–5   COUNTING STATISTICS, ERRORS, AND CORRECTIONS

The meaningful counting of radioactivity is not as simple as it may seem. There are a number of complications inherent in the phenomenon itself as well as specific problems within the instruments.

Background Radiation.   Even in a laboratory not contaminated by previous radiochemical work, there will be small amounts of radioactivity from naturally occurring isotopes in the surroundings and from cosmic radiation. Even with the best shielding, the cosmic effect is not entirely eliminated, nor are materials of construction completely free of radioactivity. The background counting rate must be determined and subtracted from the measured counting rate, and if the two rates are comparable, the "true" rate of the sample will be subject to large errors. Background rates of 30 cpm are not uncommon for beta counters.

Extremely "low-level" counters may be surrounded by auxiliary counters which can eliminate the background signals by a "coincidence" circuit, wherein the signal from the auxiliary counter cancels a coincident signal in the main counter.

Coincidence Corrections.   During the time when the counter is "dead," that is, recovering from a previous discharge, any new entering particle will be lost.

Especially with Geiger counters, this loss of count may be significant and can be corrected for.

If τ is the dead-time after each count and R_o is the observed counting rate in counts per sec, then the total dead-time per sec is τR_o. If R^* is the "true" counting rate which would be observed if all particles were counted, then the loss because of "coincidence" is

$$R^* - R_o = \tau R_o R^*$$

or

$$R^* = \frac{R_o}{1 - \tau R_o} \tag{23-7}$$

If the dead-time of the counter is not known, it cannot be determined from Equation 23–7, which contains two unknowns, R^* and τ. To determine τ, we may employ two sources with observed rates R_1, R_2, and R_{12} when counted separately and "stacked." The sum of R_1^* and R_2^* must equal $R_{12}^* + R_B^*$ because the background is included twice when R_1 and R_2 are measured separately:

$$R_1^* + R_2^* = R_{12}^* + R_B^* \tag{23-8}$$

Since R_B is much less than the other rates, we can ignore its coincidence correction. Substituting from Equation 23–7 into Equation 23–8, we obtain

$$\frac{R_1}{1 - R_1\tau} + \frac{R_2}{1 - R_2\tau} = \frac{R_{12}}{1 - R_{12}\tau} + R_B$$

Next, clear the fractions and discard terms in τ^2 and $R_B\tau$, which leads to a practical equation:

$$\tau \simeq \frac{R_1 + R_2 - R_{12} - R_B}{2R_1 R_2} \tag{23-9}$$

Counting Geometry. Radiation generally escapes from the sample in all directions as shown in Figure 23–8 and only those particles actually directed toward the counter can be counted. Obviously, the arrangement of counter and source including the distance between the two should be reproducible so that the same fraction of particles will always be counted. With proportional and

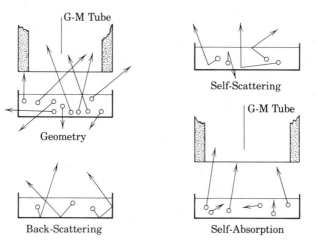

Figure 23–8 The fate of radiation from a sample, showing losses due to geometry, self-scattering, back-scattering, and self-absorption.

scintillation counters, the sample is often placed within the counter, which improves the counting efficiency enormously.

It should be clear that we rarely count all of the disintegrations and therefore, when we speak of "activity" of a sample, we should distinguish between the activity we measure (counts/sec) and the true activity (disintegrations/sec). Systems calibrated with a sample of known activity are often used. In any event, measurements must be made in a consistent manner.

Self-Scattering and Self-Absorption. Beta particles in particular are subject to deflection and/or absorption by the sample itself. Both of these effects are reduced by using a thinner sample. If the sample is a thin layer, these effects can be neglected, or if the sample is very thick, the effects become independent of thickness. But at intermediate thickness, the self-absorption effect follows a Beer's law dependence on thickness and the correction is more difficult to apply.

Back-Scattering. Some of the beta particles that would escape through the back of the sample (opposite to the counter) may be reflected by the material of the sample holder. This phenomenon is called "back-scattering," and adds to the measured counting rate. The amount of back-scattering depends on the nature of the material in the sample holder and its thickness. The error is most easily reduced by using the same type of sample holder for all comparable measurements.

Counting Statistics. Although the half-life of a given sample of radioactive material can be measured with considerable accuracy because of the large numbers of nuclides involved, the individual disintegrations occur in a random fashion. It is impossible to predict when a given nuclide will disintegrate.

As a typical example, we might obtain the following number of counts in successive 1-minute intervals:

Minute	cpm	Minute	cpm
1	93	6	104
2	95	7	100
3	110	8	102
4	101	9	95
5	96	10	104
Avg. for 1st 5 min.	99	Avg. for 2nd 5 min.	101

Avg. for 10 min. = 100 cpm

The counts per minute vary from 93 to 110, and one would have to be very lucky indeed to pick the right minute in which the average rate would be obtained. The average for the first five minutes varies from that of the second five, and the ten-minute average is still different but more likely to be correct. By "correct," we mean the average rate we would observe if we could count for an infinitely long time. But this would be quite the wrong thing to do (even if we had the time), because the rate must change with time according to the decay law (which leads to yet another type of correction!).

Assuming that individual events are random, we can apply statistical laws to predict the reliability of our measurements (see Section 24–4). In a large series of measurements of the same quantity, 68% of the measurements have a deviation from the mean no greater than one "standard deviation," given the symbol σ.

Thus the standard deviation is a measure of the scatter of a set of observations around their mean value, and it indicates the probable reliability of a single measurement. In activity measurements, if the number of counts, N, is > 100 and the half-life is much greater than the counting time, the standard deviation is equal to the square root of the total number of counts taken:

$$\sigma = \sqrt{N} \qquad (23\text{--}10)$$

The standard deviation must increase with the number of counts, but the relative standard deviation decreases with the number of counts

$$\text{rel. std. dev.} = \frac{\sigma}{N} = \frac{\sqrt{N}}{N} = \frac{1}{\sqrt{N}}$$

or

$$\% \text{ rel. std. dev.} = \frac{100}{\sqrt{N}} \qquad (23\text{--}11)$$

The standard deviation, which can be computed from a single counting run, tells us how confident we can be of our answer. There is a 68% chance (a confidence level of 68%) that our answer is within 1σ of the mean, a 50% chance that our answer is within 0.68σ of the true mean, a 95% chance that our answer is within 2σ of the true mean. The standard deviation of the counting rate is:

$$\sigma_R = \frac{\sqrt{N}}{t} = \sqrt{\frac{R}{t}} \qquad (23\text{--}12)$$

where R is the counting rate.

For example, if a sample gave 6400 counts in 2 min., the standard deviation is

$$\sigma = \sqrt{6400} = 80 \text{ counts}$$

or the relative standard deviation is

$$100/\sqrt{6400} = 1.25\%$$

and the standard deviation in the counting rate is

$$\sigma_R = \sqrt{\frac{6400/2}{2}} = 40 \text{ cpm}$$

Thus we have a 68% level of confidence that the true rate is within ± 40 of 3200 cpm. If we should prefer to be 95% confident that our answer is within $\pm 1.25\%$ of the true rate, $2\sigma_R$ must equal 0.0125:

$$2\sigma_R = \frac{2}{\sqrt{N}} = 0.0125$$

from which $\sqrt{N} = 160$ and $N = 25{,}600$ counts. Thus we must take at least 25,600 counts which will take 8 minutes.

The background count is also random in nature and subject to the same error treatment. When two rates are added or subtracted the net standard deviation is given by

$$\sigma_R = \sqrt{\sigma_A^2 + \sigma_B^2} \qquad (23\text{--}13)$$

where A represents the sample activity and B the background. When the background rate is much smaller than the sample rate, errors in the background rate will be negligible compared to the error in the sample rate. Therefore, it would not pay to spend much time counting the background. On the other hand, if the two rates are of the same order of magnitude, the time spent counting the background is just as important as the time spent on the sample itself. The optimum way to divide your time between background and sample is

$$\frac{t_B}{t_A} = \sqrt{\frac{R_B}{R_A}} \qquad (23\text{–}14)$$

Change of Rate with Time.   The counting rate, like the number of remaining radioactive nuclides, must decay exponentially with time. Thus, if the counting time is too long, the rate will change appreciably while you are counting. If the counting time is less than one-tenth of the half-life, this effect is negligible. A correction can be made if necessary.

Mixtures of Radioisotopes.   Unless the mixture is extremely potent, for example, the interior of a nuclear reactor, we can assume that each nuclide behaves independently of all others, no matter what kind. The total activity then is the sum of the individual activities. We should note, however, that counters are not equally sensitive to all kinds of radiation so that sensitivity factors must be applied to each type of activity. Some counters, as we have seen, are inherently discriminating, so that some kinds of activity may be completely eliminated from the observation.

If more than one kind of activity is being counted, it is unlikely that each will have the same half-life. The shortest lived isotope will disappear first and there will be a noticeable change in slope of the decay curve if the two half-lives vary by more than a factor of two. The decay curve for a mixture of two isotopes with half-lives of 48 min and 8 hr is shown in Figure 23–9. The composite curve can be analyzed by extrapolating the final straight portion back to zero time and

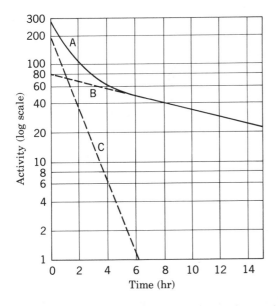

Figure 23–9   A composite decay curve of a mixture of two isotopes. A, total activity. B, extrapolated activity of isotope with half-life of 8 hours. C, hypothetical curve for activity of isotope with half-life of 48 minutes. [From Friedlander, Kennedy, and Miller, *Nuclear and Radiochemistry*, Wiley, 1964, p. 70.]

subtracting the activity of the long-lived isotope from the total observed activity. Three components can be handled in similar fashion, but experimental uncertainties make more than three very difficult. Whenever possible, it is much easier to use discrimination to handle complex mixtures. As a last resort, one must employ chemical separations.

23-6   SOME APPLICATIONS OF RADIOISOTOPES

Tracers. Except for the very lightest elements (lighter than carbon), radioisotopes are chemically identical with stable isotopes. Thus they may be used to "tag" a compound in a way that will be undetectable until the nuclide disintegrates. The history of the tagged compound and/or its successors can be followed through a chemical reaction, an industrial process, a biological process, or even a geological process. The "tag" must be an atom that is not exchangeable with the surroundings; for example, tritium would be useless as a tag on an acid if it were inserted in the carboxyl group.

Many organic and biochemical compounds are available with ^{14}C or tritium "tags" in known positions. Compounds that are not available can be synthesized by using tagged reagents as starting materials. The use of tracers is limited primarily by the imagination of the user. In biological systems, of course, we must consider possible damage to the surroundings. When necessary, and especially for carbon, oxygen, nitrogen, and hydrogen, stable heavy isotopes may be inserted as tags and detected later by mass spectrometry.

Activation Analysis. Radioactive isotopes of nearly all elements can be created artificially by bombardment with slow neutrons. In principle, any sample can be "activated" by placing it in a high flux of neutrons such as in an atomic pile (nuclear reactor). Each of the isotopes in the sample has a characteristic efficiency for capturing slow (thermal) neutrons, called its *thermal neutron capture cross-section*. Those nuclides that capture a neutron become a heavier isotope of the same element. In general, the isotope created will be unstable and will disintegrate with a characteristic half-life.

The amount of activity produced is a function of the neutron flux, ϕ, the neutron capture cross-section, σ, the time of activation, t_a, and the half-life of the radioactive product, $t_{1/2}$.

$$A = N\sigma\phi\left[1 - \exp\left(-\frac{0.693 t_a}{t_{1/2}}\right)\right] \qquad (23\text{-}15)$$

The expression within the brackets is called the *saturation factor*. Its value increases from zero initially ($t_a = 0$) to a maximum of unity ($t_a = \infty$) when the rate at which the "activated" isotopes are disintegrating equals the rate at which they are formed. For example, if a sample is activated for a period of time equal to 6 half-lives, $t_a = 6 t_{1/2}$, the saturation factor is 0.98, sufficiently close to unity so that further activation is pointless.

In general, the sample contains many kinds of isotopes, each of which captures neutrons in its own characteristic way. The measured activity produced is the sum of the individual activities for each type of isotope. Thus, the initial activity measured at time zero after activation is the summation of a number of equations like Equation 23-15. The decay curve (activity vs. time) is also the summation of the decay curves for each radioactive isotope produced (see Figure 23-9).

Activation analysis presents a challenge in selecting the optimum conditions for each type of sample. Too little activation will decrease the sensitivity, but too much may produce too high an activity for convenient measurement or it may activate so many elements that the decay curve is too complex to interpret. If the analysis calls for the determination of only one element, the conditions should be optimized for that element. For complex mixtures where it is impossible to interpret the decay curve because of interferences, a preliminary separation may be necessary either before or after activation. Decay curves, such as Figure 23–9, represent *total* activity. Complex curves can often be simplified by means of discrimination circuits (pulse height analyzers discussed above) that allow one type of radiation to be measured selectively.

Not all elements are amenable to activation analysis. The limiting sensitivity varies with the cross-section, the half-life of the isotope produced, the neutron flux, and the time available for activation as well as the counting equipment.

Comparisons of the sensitivities of some common techniques for trace analysis of selected elements are given in Table 23–2. Tables of this kind are interesting but should not be taken too seriously. The data represent the best that is possible and do not refer to a common sample. Each sample presents its own problems. The effect of interfering elements differs for each technique. Nevertheless, such a table will offer a good starting point in selecting the best method. For example, if you are required to determine traces of dysprosium, by all means choose a nuclear activation analysis.

Neutron sources are now widely available. Mixtures of radium and beryllium contained in a block of paraffin produce fluxes up to 10^7 neutrons/cm^2/sec. More active mixtures of radioisotopes are also used. The most potent source is, of course, a nuclear reactor that may produce a usable flux of 10^{14} neutrons/cm^2/sec. Because the exact flux at the sample is not usually known, quantitative analysis (which in principle could be based on Equation 23–15) is almost always done by

Table 23–2 Absolute Detection Limits in Nanograms

Element	Absorption Spectrophot.	Atomic Absorption	Flame Spectroscopy	Neutron Activation	Spark Source Mass Spectrometry
Ag	5	1	3	0.01	0.2
Al	0.5	100	4	1	0.02
Ba	100	30	3	5	0.2
Bi	600	8	100	50	0.2
Cd	3	0.3	20	5	0.3
Cu	2	0.3	1	0.1	0.08
Dy	—	300	4	0.0001	0.5
Eu	4000	100	1	0.005	0.2
Fe	200	3	3	5000	0.05
K	200	0.3	0.01	5	0.03
Mn	5	0.5	0.5	0.005	0.05
Na	8000	0.4	0.01	0.5	0.02
Ni	4	2	5	5	0.07
Pb	6	4	0.5	1000	0.3
Re	200	1000	100	0.05	0.2
Si	100	6000	700	5	0.03
Th	50	—	8000	5	0.2

[From George H. Morrison, *Trace Analysis*, Wiley-Interscience, New York, 1965, Table 2, pp. 10–13.]

comparing the activity produced in the sample with the activity produced in a standard sample of known composition treated identically and simultaneously.

Analysis by Isotope Dilution. For many complex samples, there may be no suitable quantitative method to isolate a particular component. If the interferences are difficult to remove or nullify, a method called isotope dilution analysis may provide the answer. A radioactive form of the pure substance to be determined is required. This is "diluted" in the original sample, and even if only a small portion of the component can be recovered in a pure form, then a measurement of the reduced activity will give the percent yield of the separation process.

Suppose a sample contains iron, but interferences invalidate other methods. We add W_0 g of iron as $^{59}FeCl_3$ (not necessarily a pure isotope) that has a specific activity A_0. The sample is mixed so that the ^{59}Fe is equally distributed throughout the sample. A portion of the total iron is then separated (precipitation, extraction, etc.) and isolated in a pure weighable form. If the original sample contained W_1 g of iron, then the fraction of the initial activity found in this portion is the dilution ratio, $W_0/(W_0 + W_1)$, or

$$A_0 \frac{W_0}{W_0 + W_1} = A_1$$

where A_1 is the specific activity of the isolated portion. It follows that

$$W_1 = W_0\left(\frac{A_0}{A_1} - 1\right) \tag{23-16}$$

This method has been valuable for the analysis of otherwise intractable organic and biochemical mixtures as well as geochemical and archeological samples.

23-7   RADIATION SAFETY

Although experiments normally done in elementary laboratory work present no more radiation hazard than luminous watch dials or normal cosmic radiation, we must call attention to some potential hazards. Contamination is especially insidious because there may be no apparent symptoms—the damage may show up in your offspring.

Inhalation of radioactive material is to be particularly avoided because the lungs are extremely sensitive and the contamination is difficult to remove. Ingestion may not be as bad because stomach pumps are rather effective if used in time. Absorption through the skin is also nasty, especially through cuts and blemishes and around the eyes. The hands can tolerate 20 times as much radiation as the abdominal region. One must be concerned with both the familiar radioactive or physical half-life and also the biological half-life; that is, how fast the contamination is eliminated from the body by normal biological functions.

Sensible precautions include:

1. Use a survey instrument to determine the actual level of activity present and thus the need for further precautions.
2. Use hoods and dry boxes whenever possible.
3. Use trays or absorbent paper to contain possible spills.
4. Use gloves, tongs, etc., to avoid contact with the skin.
5. Keep as much distance as practicable between source and personnel.
6. Use adequate shielding if necessary.

7. Store all radioactive materials in shielded vaults when not in use.

8. Warn other personnel or visitors whenever danger is present.

9. Don't panic. If you are in difficulty, get help.

10. For any work with more than a few microcuries of activity, consult a more complete Safety Manual before beginning to work.

QUESTIONS AND PROBLEMS

23–1. Calculate the weight of (a) 2 millicuries of ^{32}P, (b) 10 microcuries of ^{131}I.

$Ans.$ (a) 7.0×10^{-9} g

23–2. What fraction of the activity remains in the two samples of Problem 23–1 after 10 days? $Ans.$ ^{32}P, 61.5%

23–3. If the dead-time of a Geiger counter is 200 μsec, what is its counting efficiency for counting rates of 2000, 200, and 20 counts per sec? $Ans.$ 60%, 96%, 99.6%

23–4. The beta-activity of a sample was followed for a period of three hours, during which the following data were taken. Plot the decay curve and analyze the curve for: number of components, half-lives, identity of nuclides, and initial activity of each component.

Time, min.	Activity, cpm	log	Time, min.	Activity, cpm	log
10	540	2.73	80	55	1.74
20	315	2.50	90	50	1.70
30	200	2.30	100	45	1.65
40	138	2.14	120	38	1.58
50	99	2.00	140	30	1.48
60	80	1.90	160	27	1.43
70	65	1.81	180	24	1.38

23–5. Two ml of a solution containing Co^{+2} and Cu^{+2} are passed through an ion exchange column. The Co^{+2} fraction is collected in a flask and 0.1000 g of radioactive tagged $^{60}CoCl_2$ (specific activity $= 5000$ counts/min-g) is added to the flask. An excess of sulfide ion is then added to precipitate the Co^{+2} as CoS. A 0.500-g portion of the CoS is isolated and counted, giving an activity of 275 counts/min (corrected for background and coincidence). What is the concentration of Co^{+2} in the original solution?

$Ans.$ 6.6 mg/ml

23–6. A certain sample gives 450 counts/min. What is the standard deviation in the counting rate if the sample is counted for 15 min.? $Ans.$ 5.5 cpm

23–7. The observed (uncorrected) counting rate for a sample is 38.0 counts/min while under the same conditions the background rate is 15.0 counts/min. If both background and sample are counted for 20 min, what is the relative standard deviation of the corrected activity? $Ans.$ 7.1%

23–8. The dead-time of a counter is determined by measuring sample A, 10,060 counts/min; sample B, 10,950 counts/min; sample A and B together, 20,310 counts/min. Calculate the value of the dead-time. $Ans.$ 190 μsec

23–9. Some wood samples are discovered in a geological land-cut in which there is evidence of glaciation having occurred at some time in the past. The carbon-14 present in the samples decomposes at a rate of 7.00 atoms per minute per gram of wood. Carbon-14 in fresh wood decomposes at a rate of 15.30 atoms per minute per gram of wood. Carbon-14 is a weak β-emitter with a half-life of 5720 years. How many years have elapsed since glaciation occurred in the area under study? $Ans.$ $t = 6.46 \times 10^{3}$ yr

REFERENCES

G. R. Choppin, *Experimental Nuclear Chemistry*, Prentice-Hall, Englewood Cliffs, N.J., 1961.

G. Friedlander, J. W. Kennedy, and J. M. Miller, *Nuclear and Radiochemistry*, 2nd ed., Wiley, New York, 1964.

R. T. Overman and H. M. Clark, *Radioisotope Techniques*, McGraw-Hill, New York, 1959.

G. T. Morrison, ed., *Trace Analysis*, Wiley-Interscience, New York, 1965.

J. Krugers, ed., *Instrumentation in Applied Nuclear Chemistry*, Plenum Press, New York, 1973.

EVALUATION AND PROCESSING OF ANALYTICAL DATA

24

STATISTICAL TREATMENT OF DATA

24–1   INTRODUCTION

No analysis is complete until the results have been calculated and properly reported. Clearly, the report should indicate the best value obtained, but it should also reflect in some succinct fashion the probable accuracy or reliability of the result. For example, if the problem is to determine the weight of a five cent coin, the answer might be variously reported as: 5 g, 5.06 g, 5.0625 g, 5.0625 ± 0.0002, or that there is a 90% probability that the true value falls within the range 5.0625 ± 0.0010 g. Each of these answers conveys a somewhat different degree of reliability and should be consistent with the type and condition of the balance used, as well as the technique of the operator.

A single result gives no information regarding its reliability and therefore is considered unacceptable in scientific literature. However, in commercial laboratories, where time is at a premium, many routine analyses are done only once—a practice that indicates a degree of confidence in an experienced analyst. In general, an experiment must be repeated until consistent results are obtained. The confidence we have in the repeatability of experiments is the cornerstone for scientific progress.

The treatment of analytical results follows the same logic as the treatment of all experimental measurements. Over the years, scientists have formalized the nomenclature and the mathematical expression of a set of replicate measurements to convey the maximum of information about the reliability of a result. We present here a brief summary of some of the commonly used terms.

24–2   SIGNIFICANT FIGURES

The uncertainty in a quantity may be expressed by the number of significant figures used. A weight given as 5 g implies that a rough balance was used and that the weight was probably between 4 and 6 g. Likewise, 5.0625 g indicates a true weight between 5.062 and 5.063 g. As a general rule, the significant figures include all those known to be correct plus one doubtful figure. The "zero" is an exception since it may be used as a significant figure or merely to locate the decimal point.

Numbers should always be written so that the number of significant figures is unambiguous; that is, 1.23×10^5, not 123,000 for only three significant figures.

Expression	No. of Significant Figs.
5.0625	5
5062.5	5
5.06250	6
123,000	3 (or 6)
1.23×10^5	3
1.23000×10^5	6
0.001230	4

Computation Rules. In addition or subtraction of a set of numbers, all are rounded to the same number of decimal places.

Addition	Subtraction
1.6375	$1.32 \times 10^{-2} = 13.2 \times 10^{-3}$
75.2	$-6.72 \times 10^{-3} \simeq\ \ 6.7 \times 10^{-3}$
6.002	6.5×10^{-3}
82.8	

In multiplication or division, the result should contain the same number of significant figures as that term with the fewest number of significant figures.

$$A = \pi r^2 = 3.14159 \times 3.82^2 = 45.8$$
$$C = A/\epsilon b = 1.20/33176 \times 1.001 = 3.61 \times 10^{-5}$$

With the popularity of electronic calculators, there is a great temptation to ignore the rules for significant figures. At the push of a few buttons, $\pi \times 3.82^2$ becomes 45.84337664! Obviously the calculator cannot improve the accuracy of the original data. Likewise, it would be pointless to weigh a 10-g sample to the nearest 0.1 mg if the buret in a subsequent titration will be read only to three significant figures.

24–3 TYPES OF ERRORS

The True Value. If you are asked to determine the number of pages in this book or the number of dollars in your wallet, you can undoubtedly arrive at the true (or accurate) answer on the first attempt. However, the number of atoms in a mole is not known precisely because the number is derived from experimental measurements, each of which contains at least a small degree of error. Unlike counting the number of "things," the true value of a physical measurement is never known with absolute certainty. It is simply a value that has been "accepted." It may be the average of the best results obtained by experienced workers in many laboratories using different techniques, such as the International Atomic Weights that are accepted by the International Union of Pure and Applied Chemistry; or it may be the average of student results obtained on a particular unknown sample. In any event, it is the best value available for whatever purpose is intended. It is understood that a better value may be subsequently determined.

Expressions of Error. The difference between a given measurement and the true value is the *absolute error* of that measurement. Since the importance of the error may depend on the magnitude of the quantity, it is useful to relate the two quantities as a ratio called the *relative error*.

$$\text{Relative error} = \frac{\text{measured value} - \text{true value}}{\text{true value}} \qquad (24\text{-}1)$$

Sources of Errors.   It would be hopeless to list all the possible sources of error, but those errors that can be assigned definite causes are called *determinate* or *systematic* errors. In principle, these errors can be either eliminated by improved procedures, or corrected for if they are reproducible. The procedure of *calibration*, if properly carried out, presumably makes allowance for all systematic errors.

A group of unsuspected and nonreproducible errors remains that are beyond control; for example, temperature fluctuation, electronic noise, mechanical vibrations, friction, spillage, reading errors, contamination, variations in sample composition, etc. These errors occur randomly and are called *random* or *indeterminate errors*. Their random nature allows us to use the powerful techniques of statistics and the laws of probability.

24-4   NORMAL DISTRIBUTION OF RANDOM ERRORS

Consider a set of data obtained by an analyst who is evaluating a new method for determining tetraethyl lead in gasoline. The results (g of Pb/gal of gasoline) for the first 20 trials are given in the order obtained in Table 24-1. The list of results (Column x) means very little until they are evaluated.

Table 24-1   Data Obtained for the Determination of Lead in Gasoline, g Pb/gal

No.	x	$|x - \bar{x}|$	$(x - \bar{x})^2$
1	4.20	0.04	0.0016
2	4.28	0.04	0.0016
3	4.45	0.21	0.0441
4	4.17	0.07	0.0049
5	4.30	0.06	0.0036
6	4.22	0.02	0.0004
7	4.24	0	0
8	4.14	0.10	0.0100
9	4.08	0.16	0.0256
10	4.26	0.02	0.0004
11	4.23	0.01	0.0001
12	4.18	0.06	0.0036
13	4.31	0.07	0.0049
14	4.23	0.01	0.0001
15	4.38	0.14	0.0196
16	4.21	0.03	0.0009
17	4.12	0.12	0.0144
18	4.25	0.01	0.0001
19	4.33	0.09	0.0081
20	4.26	0.02	0.0004
$\bar{x}$	4.24		
$\bar{d}$		0.06	
s			0.087

Mean. For a set of n results, the arithmetic mean, designated $\bar{x}$, is:

$$\bar{x} = \frac{\Sigma x_i}{n} \tag{24-2}$$

As more and more results are taken, $\bar{x}$ approaches μ, the true value. For example, in the data set of Table 24–1, $\bar{x}$ for the first 5 determinations is 4.28, for the first 10 it is 4.21, and for the entire set of 20, $\bar{x} = 4.24$. Although the mean has not shifted greatly, we do have more confidence in the latter value.

Average Deviation from the Mean. Once the mean is established, the variability of the results can be expressed in terms of the *average deviation*, $\bar{d}$:

$$\bar{d} = \frac{\Sigma |x_i - \bar{x}|}{n} \tag{24-3}$$

Since high and low results are equally significant, the absolute value of the deviation is the important quantity. For our 20 results in Table 24–1, $\bar{d} = 0.06$ and it would be proper to express the answer as 4.24 ± 0.06 g/gal, provided the meaning of "$\pm$" is understood. For the first 5 results the answer would be 4.28 ± 0.08 g/gal.

Frequency Distribution. A more meaningful pattern appears when we group the results into intervals:

Interval	No. of Results	Interval	No. of Results
4.06 to 4.10	1	4.26 to 4.30	4
4.11 4.15	2	4.31 4.35	2
4.16 4.20	3	4.36 4.40	1
4.21 4.25	6	4.41 4.45	1

It is now fairly evident that, barring systematic errors, the true value is probably between 4.21 and 4.30 g/gal. The *range* of values of the entire set is from 4.08 to 4.45, but 55% of the values occur within the smaller range, 4.21 to 4.30.

Normal Error Curve. The results, as grouped above, are presented as a *histogram* in Figure 24–1. The distribution shows that random errors are equally probable in both directions, and that small errors occur more frequently than large errors. Another hundred, or even a thousand additional determinations would not change the shape of the distribution curve by very much. In the limit, using very small intervals and an infinite number of results, the curve becomes the *normal error curve* shown in Figure 24–2, also called a *Gaussian distribution* after the German mathematician, Karl Gauss. This curve can be described completely by just two parameters: the position of its midpoint, μ, and the half-width, σ measured between the points of inflection (which always occur at a vertical distance of $0.607 \times$ height). The mathematical equation for the curve is:

$$y = \frac{\exp[-(x - \mu)^2/2\sigma^2]}{\sigma\sqrt{2\pi}} \tag{24-4}$$

Although we can never construct this curve from actual data (an infinite number of points would be required), the shape can usually be inferred from a limited number of results (10 or more). However, there always remains a degree of uncertainty that additional results might change the position of the curve and give

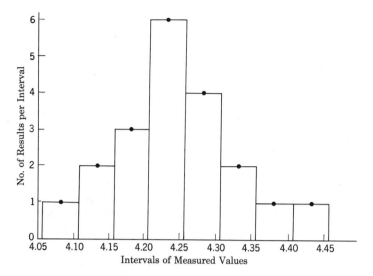

Figure 24–1   Histogram of results from Table 24–1.

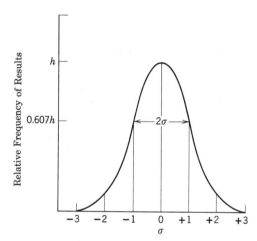

Figure 24–2   Normal error curve.

us a better answer. Fortunately the theory of statistics allows us to make a number of deductions from limited sets of data.

Standard Deviation.   The average deviation is not amenable to statistical interpretation; therefore it is more useful to determine the *standard deviation*, defined as σ in Equation 24–4. An alternate definition of σ is

$$\sigma = \sqrt{\frac{\Sigma(x_i - \bar{x})^2}{n}} \tag{24–5}$$

provided the number of determinations is large ($n > 50$). The square of the standard deviation, σ^2, known as the *variance*, is a convenient way to describe the population distribution.

In general, our frequency distribution is based on a relatively small sample of a much larger population. Thus, although the parameters are calculated from the sample distribution, the parameters required are those of the population distribution. The sample parameters are thus estimates of the population parameters, and it is found that the best estimate of the standard deviation of the population from a sample is actually given by

$$s = \sqrt{\frac{\Sigma(x_i - \bar{x})^2}{n - 1}} \tag{24-6}$$

For our 20 results, $s = 0.087$, somewhat larger than the average deviation. The first five results give $s = 0.12$.

One of the properties of the normal error curve is that 68% of the area under the curve falls within the region bounded by $\mu - \sigma$ and $\mu + \sigma$. In other words, with a large number of results, 68% of them will fall within one standard deviation from the mean, or there is a 68% probability that a single determination will fall within this range. For 2σ, the probability is 95% and for 3σ the probability is 99.7%.

Distribution of Sample Means. If samples of size n are taken from a population distribution, the means of these samples will be distributed normally (provided that $n \geq 4$). The distribution of these means has a mean value of μ, that of the population distribution, and the standard deviation (the standard error of the mean) is given by

$$s_m = \frac{\sigma}{\sqrt{n}}$$

24-5   STATISTICAL TREATMENT OF SMALL DATA SETS

Reliability of an Average Value. In general, we compute an average value from a small set of results and it may depart considerably from the true mean value. Thus, the probabilities given above are not reliable. W. S. Gossett, an English chemist who used the pen name of Student, developed an expression known as *Student's t*, which is used in many kinds of statistical interpretations.

Gossett defined t as:

$$\pm t = (\bar{x} - \mu)\frac{\sqrt{n}}{s} \tag{24-7}$$

In general, working with small sets of data, we do not have reliable values for μ or s. Therefore t-values cannot be computed, but are obtained from a table such as Table 24-2, that takes into account the possible variation of the value of $\bar{x}$ and μ and the uncertainty caused by using s instead of σ.

Equation 24-7 becomes more useful when rearranged:

$$\mu = \bar{x} \pm \frac{ts}{\sqrt{n}} \tag{24-8}$$

Equation 24-8 defines the limits around the average value within which the true mean lies with a given level of certainty. In practice, one chooses a level of certainty and then computes the *confidence limits*. For example, let us choose a 95% level of certainty for the first five determinations in Table 24-1, for which $\bar{x} = 4.28$ and $s = 0.12$. From Table 24-2, $t = 2.78$; therefore

$$\mu = 4.28 \pm \frac{2.78 \times 0.12}{\sqrt{5}} = 4.28 \pm 0.15$$

Thus with 95% confidence we can state from the first five values that the true

Table 24–2 Values of Student's t

No. of Results	Probability Level		
	90%	95%	99%
2	6.31	12.71	127.32
3	2.92	4.30	9.92
4	2.35	3.18	5.84
5	2.13	2.78	4.60
6	2.01	2.57	4.03
7	1.94	2.45	3.71
8	1.89	2.36	3.50
9	1.86	2.31	3.35
10	1.83	2.26	3.25
20	1.72	2.09	2.84

value lies within the limits 4.13 and 4.43. From the entire set of 20 determinations, the limits become (also 95% confidence)

$$4.24 \pm \frac{2.1 \times 0.087}{\sqrt{20}} = 4.24 \pm 0.04$$

If we relax the confidence level to 90%, the confidence limits become narrowed to 4.24 ± 0.03, and at 99% confidence they are widened to 4.24 ± 0.05.

Rejection of a Suspect Value. In dealing with less than 10 results, there is often a single result which deviates from the mean far more than any of the others. If there is a known cause for the error, the result should of course be rejected. Frequently the cause is uncertain and a judgment must be made whether to include the result as valid or to reject it. The Q-test is one of the most reliable of the objective tests available.

$$Q = \frac{\text{suspect value} - \text{nearest value}}{\text{largest value} - \text{smallest value}} \tag{24–9}$$

A Q-value is determined from the data and compared to a tabulated value, such as offered by Table 24–3, which refers to a 90% confidence level. If Q (experimental) $> Q$ (table), the value can be rejected with 90% confidence. Of the first five results given above (Table 24–1), 4.45 might be considered suspect.

$$Q = \frac{4.45 - 4.30}{4.45 - 4.17} = 0.54$$

In Table 24–3, $Q = 0.64$ for 5 results and 90% confidence. Thus, the result should not be rejected.

Table 24–3 Q-test Values for 90% Confidence

n	Q	n	Q
3	0.94	7	0.51
4	0.76	8	0.47
5	0.64	9	0.44
6	0.56	10	0.41

24-6   CONTROL CHARTS

Let us suppose that the method used to obtain the data in Table 24–1 is accepted for routine use. The particular sample used may be retained as a "standard sample" and run as a daily check on laboratory performance. It is informative to plot the results of this daily check analysis on a *control chart*, as in Figure 24–3 The chart consists of a central line at the best available value (in this case at the mean of the 20 earlier determinations), and two pairs of limit lines at the *inner* and *outer control limits*. The inner limits are often set at $\pm 2s$ and a value outside these limits is a warning signal that something is wrong. The outer limits are often set at $\pm 3s$ and a value outside these limits indicate that the procedure is no longer in *statistical control.* Trends are also very evident and may help to locate the cause as for example the rising trend from days 6 through 14 might have resulted from slow evaporation of the standard sample or slow deterioration of a reagent.

Control charts are frequently used as aids to quality control in manufacturing or processing. Any measure of quality may be plotted sequentially. Sometimes it is desirable to eliminate the effect of short-term fluctuations by averaging two or more consecutive data points. For example, the dashed line in Figure 24–3 represents the average of the last three points for any given time. Otherwise an unnecessarily drastic correction might be applied.

There are numerous variations of the use of control charts for the statistical monitoring and control of quality—more complete discussions are available in the references below.

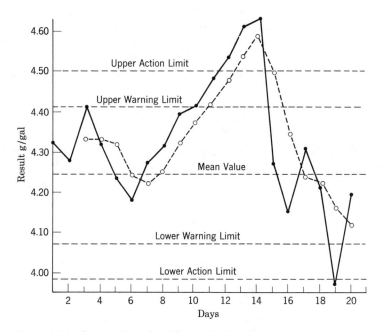

Figure 24–3   Control chart for daily check of analysis of standard gasoline samples.

QUESTIONS AND PROBLEMS

24–1. Express the results of the following calculations with the appropriate number of significant figures.

(a) $17.593 + 3.4756 - 0.0459 + 1.75$

(b) $\dfrac{\pi \times 75.3^3}{7.77 \times 10^{-5}}$

(c) $\dfrac{-1.75 \times 10^{-5} + \sqrt{(1.75 \times 10^{-5})^2 + 4 \times 1.75 \times 10^{-9}}}{2.000}$

(d) $5.8 \times 10^{-6} \times \dfrac{0.1000 - 2 \times 10^{-4}}{0.1044 + 2 \times 10^{-4}}$

Ans. (a) 22.77, (b) 1.726×10^{10}, (c) 6.80×10^{-5}, (d) 5.5×10^{-6}

24–2. Calculate the formula weight of $Pd(NO_3)_2$ to the proper number of significant figures.

Ans. 230.4

24–3. To check his technique, an analyst was given a sample of pure calcium carbonate to analyze for Ca. If he obtained 39.78%, what were his absolute and relative errors?

Ans. -0.26% and -0.65% (6.5 ppt)

24–4. A bronze sample weighing about 4 g is to be analyzed spectrophotometrically for manganese. If the absorbance can be read to the nearest 0.001 absorbance unit, how accurately should the sample be weighed? Assume that the volume of the solution will be adjusted to give maximum accuracy in the absorbance reading. *Ans.* ± 0.01 g

24–5. A manufacturer of light bulbs claims that they will last an average of 1200 hours with a standard deviation of 100 hours. (a) How many bulbs of a lot of 400 would be expected to burn for more than 1400 hours? (b) If 30 of the bulbs burn out in less than 1000 hours, does this invalidate the claim? *Ans.* (a) 10 (b) Yes

24–6. The determination of chromium in an ore sample gave the following results: 28.53%, 28.47%, 28.72%, 28.39% and 28.64%. Calculate (a) the mean, (b) the average deviation, (c) the standard deviation, and (d) the 95% confidence limits.

Ans. (a) 28.55%, (b) 0.10, (c) 0.13, (d) $28.55 \pm 0.16\%$

24–7. Replicate determinations of the % Ni in a coin gave: 24.92%, 25.07%, 25.51%, 24.65%, and 24.95%. Should any of these results be discarded?

REFERENCES

E. L. Grant, *Statistical Quality Control*, 2nd ed., McGraw-Hill, New York, 1952.

E. B. Wilson, Jr., *An Introduction to Scientific Research*, McGraw-Hill, New York, 1952, p. 263.

G. Wernimont, *Ind. Eng. Chem., Anal. Ed.*, **18**, 587 (1946).

J. A. Mitchell, *Ind. Eng. Chem., Anal. Ed.*, **19**, 961 (1947).

25

DATA PROCESSING

25-1   INTRODUCTION

Thirty years ago the deflection of a galvanometer was the only means of
obtaining a readout from an infrared spectrometer; the galvanometer deflection at
a series of discrete wavelengths had to be plotted manually to provide a complete
record of an infrared spectrum. Twenty years ago the introduction of the pen
recorder enabled a permanent record of a spectrum to be obtained automatically
as a continuous function of wavelength. Ten years ago digital computers began to
be used in conjunction with analytical instruments, and since then the use of
computers with these instruments has increased rapidly. In this chapter we shall
discuss simple readout systems, examine how the digital computer is used with
different instruments, and then consider simple operational amplifier circuits.

25-2   INDICATING AND RECORDING DEVICES

Meters.   A simple meter is one of the commonest output systems for analytical
instruments. It is adequate where only a few readings are required from each
sample, provided that a high degree of precision or a large number of samples is
not involved. Common examples are laboratory pH meters, oxygen meters,
colorimeters, and atomic absorption spectrometers.

Digital Indicators.   As an alternative to the meter, a digital indicator can be used
to give a reading equivalent to that provided by the meter, as shown in Figure
25–1. The digital display system is preferable because it is clearer and eliminates
errors that may be made in reading the meter, it occupies less space, and it may be
more accurate. At present it is somewhat more expensive but this situation could
well change in favor of digital indicators in the near future.

Pen Recorders.   Where more detailed information is required, such as that
provided by an infrared or NMR spectrometer, a pen recorder is necessary. This
will usually be of the potentiometric type, that is, an automatic recording
potentiometer. In this device, illustrated in Figure 25–2, the input voltage signal is

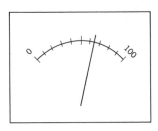

Figure 25–1   Simple meter and the
equivalent digital indicator.

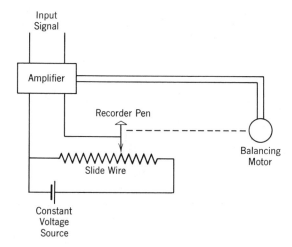

Input
Signal

Amplifier

Recorder Pen

Balancing
Motor

Slide Wire

Constant
Voltage
Source

Figure 25–2   Potentiometric recorder.

compared with a reference voltage generated by a moving contact on the potentiometer slide wire. The difference is converted to an a.c. signal, and the amplified output is fed to a phase-sensitive motor. This moves the sliding contact to balance the input and reference voltages. Since a null-balance system is used (see Chapter 10), the response of the recorder is independent of the linearity of the amplifier.

Typical specifications for such a recorder are from 0–1 mV up to 0–1 V for full-scale deflection, an accuracy of 1 part in 400 and resolution of 1 part in 1000, with chart widths between 10 and 30 cm, and a response speed of from 2 sec to 0.25 sec for full-scale deflection. The recorder may be a separate unit or it may be an integral part of the instrument with which it is used. In the latter case there will be mechanical linkages common to both the recorder and the instrument. For example, in a spectrophotometer the optical wedge and recorder pen positions may be directly linked. The recorder paper is usually driven at a pre-selected speed by a stepping-motor, and different paper speeds are obtained by the use of a frequency divider network that drives the stepping-motor. Alternatively, the paper may be fixed and the pen mounted on a moving slidewire carriage (see Figure 10–11). Another possibility is for both axes (X and Y) to represent independent input voltages.

Normally a single-pen recorder is used to represent only one variable, but it can be used for more than one item of information if these are displayed sequentially (see Figure 26–9). A recorder with two or more separate pens is necessary for the simultaneous representation of more than one variable. It is unusual for a continuously recording potentiometric recorder to have more than two pens and a separate slidewire drive unit is required for each pen, but some recorders used for process control applications may have up to twelve or more pens. The record then consists of a series of lines made up of discrete points. Separate lines are identified by using numbers, as shown in Figures 25–3, or different shapes or colors. These recorders are relatively slow since the printing time is typically 30 seconds per point.

The output from a spectrophotometer is normally recorded in terms of percent transmittance (%T). However, the concentration of an absorbing species is generally proportional to the absorbance, $-\log T$. In the past, some recorders have incorporated a logarithmically wound slide wire (or its equivalent) to provide a recording that is linear in concentration. It is now more usual to generate $\log T$ by electronic means before the result is fed to the recorder.

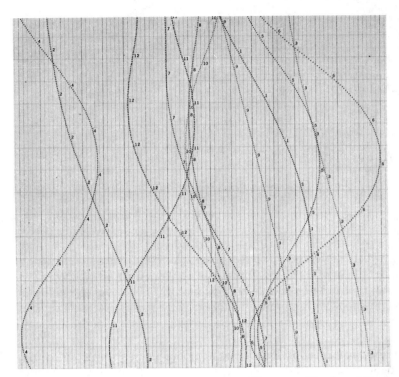

Figure 25–3 Multi-channel pen recorder output. [Courtesy of Leeds and Northrup.]

Galvanometer Recorders or Oscillographs.   The inertia of the pen carriage limits the speed of response of a potentiometric recorder. Consequently, other recording techniques are necessary for pen speeds in excess of approximately 50 cm/sec. In some galvanometer recorders the pen assembly is attached to the galvanometer suspension. However, for highest recording speeds the pen is eliminated altogether by using a reflecting galvanometer with an ultraviolet radiation source and UV-sensitive recording paper. This type of recorder is commonly used for mass spectrometry, and several separate galvanometers in the one recorder provide overlapping sensitivity ranges. An example of the record produced by such an instrument is given in Figure 25–4. Recording speeds of up to 1000 m/sec can be achieved with light-beam oscillographs but the clarity of the resulting spectrum generally leaves much to be desired. This has resulted in the development of a number of ink-writing oscillographs in which a relatively high writing speed is achieved by forcing a stream of ink out of the pen under pressure. A much clearer and more permanent record results but the maximum recording speed is limited to about 50 m/sec.

The recording of transient signals can also be achieved by the use of a storage oscilloscope and an oscilloscope camera. The storage oscilloscope will retain on

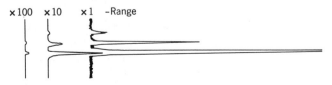

Figure 25–4 Part of a mass-spectrum obtained with a 3-channel oscillographic recorder. [Courtesy of DuPont Instruments.]

the screen for several hours the trace of a signal moving at up to 5000 m/sec. This trace can then be photographed at leisure.

25–3 INTEGRATORS

The determination of the area under individual peaks is of particular importance in n.m.r. spectroscopy and in gas chromatography. The areas of the individual peaks in an n.m.r. spectrum provide a measure of the relative number of protons (or other nuclei) of a particular type in the sample molecule. In gas chromatography the area of an individual peak in the chromatogram is proportional to the amount of that particular component in the sample. The area is a more reliable measure than peak height because it is independent of the shape of the peak and of minor changes in the column operating conditions. However, it should be noted that the area figure is not a direct measure of the component concentration. There is a proportionality factor that is dependent on both the sample component and the detector. This will be discussed further in conjunction with digital integrators.

Ball-and-Disc Integrator. A diagram of a ball-and-disc integrator is given in Figure 25–5. It consists of a flat disc driven at constant speed by a synchronous motor, two balls whose speed of rotation is determined by their position on the rotating disc, and a counter. The balls are mechanically linked to the recorder pen carriage so that they are in the center of the disc when the recorder pen is at zero and near the periphery of the disc when the recorder pen is at the top end of the recorder scale. Their speed of rotation is proportional to their distance from the center of the disc. Since the top ball drives a counter, the count rate is proportional to the distance of the recorder pen from the zero point. The total count generated as the recorder pen traverses a peak is therefore proportional to the area under the peak. This type of integrator has been used extensively for the measurement of gas chromatographic peak areas, but it is unsuitable for the measurement of very narrow peaks because of the inertia inherent in the system.

Electronic Analog Integrator. In a typical n.m.r. spectrometer, integration is done electronically and the integral is recorded on the same chart as the original spectrum. The circuit of a simple electronic integrator using an *operational amplifier* is shown in Figure 25–6. Operational amplifiers, so called because they are used in analog computers to carry out various mathematical operations, are characterized by high gain, high stability, and low current drain. The basic operational amplifier is a three-terminal device, with two inputs and an output, and is represented by the triangular symbol shown in the figure. The amplifier provides a positive output to one input and an inverted (or negative) output to the other, all referred to a common ground. It is therefore said to have a differential input, but in Figure 25–6 the positive input has been grounded. Normally, feedback between output and input is provided by a resistor, but if this is replaced by a capacitor the

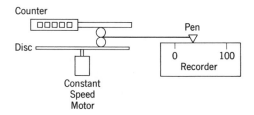

Figure 25–5 Ball-and-disc integrator.

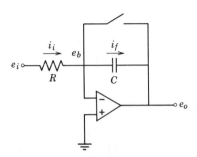

Figure 25–6 Simple electronic integrator using an operational amplifier.

device acts as an integrator. If the input current to the amplifier is negligible we can write

$$i_i = \frac{e_i - e_b}{R} = i_f \qquad (25\text{–}1)$$

The voltage drop across the capacitor is given by

$$(e_b - e_o) = \frac{q}{C} = \frac{1}{C}\int dq = \frac{1}{C}\int i_f\, dt \qquad (25\text{–}2)$$

since current is the rate of change of charge, dq/dt. Designating the amplifier gain by A, we can replace e_b by $-e_o/A$, and using Equations 25–1 and 25–2 we obtain

$$-e_o/A - e_o = \frac{1}{RC}\int (e_i + e_o/A)\, dt \qquad (25\text{–}3)$$

Because A is large, typically 10^4 to 10^5, this reduces to

$$e_o \simeq -\frac{1}{RC}\int_0^t e_i\, dt + E_0(t = 0) \qquad (25\text{–}4)$$

where the product RC is the integrator time constant. E_0 is the integrator output at time $t = 0$, and would be due to integration of earlier input signals. The integrator can be reset to zero by momentarily closing the switch in Figure 25–6, thereby discharging the capacitor.

Digital Integrators. A simple form of digital integrator consists of a voltage-to-frequency converter and an electronic counter that counts the number of impulses. In comparison with the ball-and-disc integrator, the voltage-to-frequency converter has three important advantages. First, there are no mechanical components and a very high count rate can be achieved. Second, the signal to be integrated is fed directly into this device quite independently of the recorder. The resultant area figures will therefore be free from errors due to malfunctioning of the recorder or inability of the recorder to accurately follow the peak. Third, a high degree of accuracy ($\sim 0.01\%$) can be obtained in the conversion of voltage to frequency, and this is reflected in the overall accuracy obtained with this type of integrator. In addition, the fact that this is entirely an electronic device offers the possibility of carrying out other operations on the input data. For gas chromatographic applications the two most important additional operations are the separation of the areas due to overlapping peaks and compensation for baseline drift.

A simplified schematic diagram of a digital integrator for gas chromatography is shown in Figure 25–7. The functions of the various components are outlined below. Initial amplification of the input signal may be required to match the signal to the operating range of the voltage-to-frequency converter. The input amplifier also

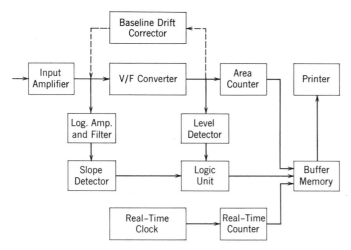

Figure 25–7 Simplified schematic diagram of a digital integrator for gas chromatography.

isolates the chromatograph from the integrator. During the emergence of a peak, the pulses generated by the voltage-to-frequency converter are fed to the digital area counter, and the total number of pulses accumulated during the peak is a measure of the peak area. At the end of the peak the total count is transferred to the buffer memory and the counter is re-set to zero. The use of a buffer memory allows the counter to clear rapidly (approximately 10 μsec) so that there is little interference with the storage of data from the next peak, whereas the operation of the electromechanical printer is relatively slow (50 msec/digit). Retention times are printed out by means of a clock that generates a pulse every second and an associated counter to provide a total count corresponding to the retention time. This is also fed to the printer via the buffer memory.

Integration and retention time measurement are controlled by a *logic system*. The signal from the input amplifier goes via a logarithmic amplifier and filter to a differentiator. The onset of a peak, and therefore the start of integration, is indicated by a rise in the output from the differentiator above the value set by the *slope detector*, as shown in Figure 25–8b. The use of the logarithmic amplifier

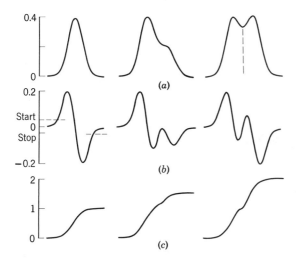

Figure 25–8 First derivative (b) and integral (c) curves for various peak profiles.

ensures that, within reason, the output from the differentiator is independent of the absolute magnitude of the input peak. At the end of the peak the derivative returns to zero, and this provides the signal to stop integration and print out the peak area. A problem arises with some partially resolved peaks, as shown by the second curve in Figure 25-8b, because the slope detector may incorrectly indicate that the peak is finished. This is overcome by the addition of a *level detector*, Figure 25-7, which provides an indication of the input signal level to the logic unit and thereby allows integration to continue. The logic circuits also find the position of the peak maximum for printout of the retention time. Where two peaks are only partially separated, the integrator normally uses the vertical from the minimum to the baseline (broken line in Figure 25-8a) as the dividing line for the allocation of peak areas. Provided that the two peaks are of similar height, this has little effect on the accuracy. However, substantial errors will arise in the measurement of a trace component on the tail of a major peak. Some integrators have been designed to handle this problem by calculating the area between the minor peak and the tangent to the major peak over that section of the curve occupied by the minor peak, as shown in Figure 25-9.

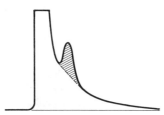

Figure 25-9   Tangent baseline correction.

In addition to the measurement of peak areas and retention times, digital integrators generally include circuitry to offset *baseline drift*. This drift can be distinguished from the onset of a peak by means of the slope detector. In the absence of a peak the rate of drift can be monitored and a signal fed back in opposition to the input signal to produce a steady baseline. In the presence of a peak the automatic drift correction can either be suspended or it can be allowed to continue at the rate just prior to the start of the peak. Correction should be suspended if the direction of drift varies randomly with time, but continued if it is consistently in one direction.

Computing Integrators.   Until recently the major defect of digital integrators has been the fact that the area figures they produce are not corrected for differences in the detector response to different components in the sample. However, several computing integrators are now available that will accept detector calibration factors and then print out a complete analysis in terms of component percentages. The insertion of detector calibration factors and the calculation of the final analysis results can also be done separately with a programmable calculator, but this requires the manual transfer of the area figures to the calculator.

A computing integrator can be interfaced to a number of chromatographs but can handle the output from only one at a time. If it is necessary to process the data from a number of chromatographs simultaneously, a separate integrator is required for each chromatograph. Alternatively, a digital computer can be used to handle a number of chromatographs virtually simultaneously on a time-shared basis.

25–4 DIGITAL COMPUTERS

In comparison with special devices such as the computing integrator, the general purpose digital computer offers a greater degree of flexibility and the ability to handle very large quantities of information. Furthermore, because of its speed of operation and storage capacity, a computer is generally capable of handling more than one instrument at a time. However, to do this it must have facilities that are not found in many computers designed purely for batch processing. In the following discussion we shall first briefly look at the computer itself and then consider its use with specific analytical instruments.

The computer consists of a group of devices that store and carry out mathematical operations on *binary numbers*. Binary numbers are represented by a series of 0's and 1's, and each 0 or 1 in the number represents the absence or presence of a particular power of two. This is in comparison with decimal numbers that represent a series of powers of ten. Thus, the binary number 1101 represents $1 \times 2^3 + 1 \times 2^2 + 0 \times 2^1 + 1 \times 2^0$, which in decimal form is 13 or $1 \times 10^1 + 3 \times 10^0$. Binary numbers or *words* in the computer may represent either numbers or computer operating instructions, and are stored in the various computer memories.

Control of the system, and mathematical operations, are carried out within the *central processing unit* (*C.P.U.*) that interacts directly with the core memory, Figure 25–10a. The rate at which operations are carried out by the C.P.U. is determined largely by the time taken to transfer data into and out of the core memory, about 1 μsec, and is controlled by a clock that generates a series of

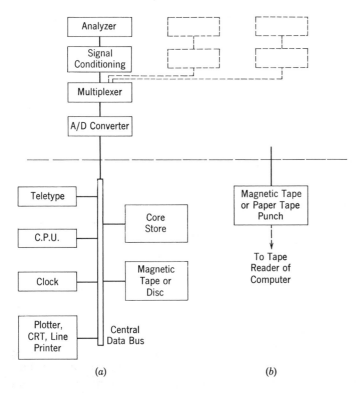

(a)   (b)

Figure 25–10 On-line (a) and off-line (b) computer systems.

pulses within the C.P.U. The actual operations are specified by a set of instructions, the *computer program*, which is stored in the core memory. The core memory might contain from 4096 (4K) to 32,768 (32K) words. Word size depends on the particular computer, but would typically be either 12 or 16 binary digits (*bits*). For floating point arithmetic, two words would be used to represent each number. The first word and part of the second will represent the mantissa (less than 1) and the remainder represents the exponent. Because of the cost of core memory, additional storage is provided by using magnetic tape or revolving magnetic discs, which might contain 100K or more words. However, mathematical operations on the data stored in these can only be carried out by first transferring the data (in appropriate sized blocks) into the core memory.

Addition processes in the computer are relatively simple and the addition of two numbers might take only a few microseconds. Subtraction is slightly slower because several operations are required. Multiplication and division may take from 10 μsec to 1 msec, depending on whether special equipment is included for speeding up these processes.

The computer operator normally controls the computer by means of a *Teletype*. This is an electric typewriter that converts alphabetical and numerical (alphanumeric) characters into groups of electrical impulses. However, the Teletype and similar devices provide a relatively slow means for the output of data from the computer, being capable of printing only 10 to 30 characters per second. The *line printer*, a more expensive output system, will print an entire line of up to 132 characters in one operation at printing rates of 300 or more lines per minute. Intermediate between these two are printers that have been developed mainly for use with the smaller computers commonly used with analytical instruments. These printers achieve fast printing rates by the use of thermal or electrostatic printing techniques. Large screen oscilloscopes can be used to display both alphanumeric and graphical data, but a separate *hard-copy* device is then required to produce a permanent record. Digitally controlled graph plotters can also be used with the computer.

The analytical instrument is connected to the computer by means of a separate interface card. This will include an *analog-to-digital converter* to convert the analog signal from the instrument into digital form. If the computer is being used to control the instrument, a *digital-to-analog converter* may also be required to convert the digital signal from the computer into analog form. The rate of generation of data from the instrument may be controlled by pulses from a digital clock, which would then also be included as part of the interface.

The data bus consists of a series of wires that carry the signals to and from the computer (16 wires from a 16-bit computer), together with a number of control lines. In addition, there will be a number of device selection lines whereby the computer can select the piece of peripheral equipment to which it is to transmit data. The need for control wires arises because of the speed of operation of the computer. Thus, the computer might be able to carry out 1 million operations per second, whereas the Teletype takes 0.1 sec to print one character. The computer can output the binary representation of the character to be printed by the Teletype into a buffer memory. It can then continue to perform other operations until the Teletype sends a signal back to the computer advising that the character has been printed. The next character is then placed in the buffer memory, and so on. When a number of devices are connected to the computer a *priority interrupt system* is required. In this way the external devices can temporarily stop other computer operations so that data can be transferred to the computer. Furthermore, the existence of a priority system allows the faster external devices to have priority in data transfer over the slower devices.

The development of programs to carry out data transfer from a number of instruments, plus the appropriate mathematical operations on these data, involve a considerable amount of work. However, a number of suitable programs for specific computers are available from some of the computer manufacturers.

25–5   COMPUTERIZED GAS CHROMATOGRAPHY

Figure 25–10a shows the basic components of an *on-line* computer system in which a number of instruments, in this case chromatographs, are linked directly to the computer. This can be contrasted with the *off-line* system shown in Figure 25–10b in which the computer is not physically connected to the analyzers. An on-line system in which the processing of data is carried out at a speed comparable to that of the input of data is said to be a *real-time* system.

Data Sampling.   The first step in processing data from the chromatograph is to provide the computer with a representation of the chromatogram. This is done by sampling the output of the chromatograph at fixed time intervals, as shown in Figure 25–11. These analog voltages which represent the peak are converted into digital form by an analog-to-digital (A/D) converter before being fed into the computer. Some form of signal conditioning may be interposed between the analyzer and the A/D converter, for example, to provide amplification of the signal or to filter out extraneous noise. Typical gas chromatographic peaks are in the order of 1 sec wide, for which a sampling rate of 10 samples per second is generally adequate (depending on the signal-to-noise ratio). At this sampling rate a 20-minute chromatogram will produce 12,000 data points. Either these must all be stored in memory for subsequent processing, or the processing must be carried out by the computer between taking data samples. The latter greatly decreases the amount of data to be held in memory at any one time.

Multiplexing.   Since computer operations are carried out many orders of magnitude faster than the above sampling rate, it is possible to interface a number of chromatographs to the one computer. This is done by feeding the signal from each chromatograph to the A/D converter via a *multiplexer*. The latter is a high speed switch which sequentially samples the output from each chromato-graph. The computer identifies the signals coming from each chromatograph and then processes them separately.

Data Processing.   To determine the concentration of each component in the sample injected into the gas chromatograph, the computer must calculate the area

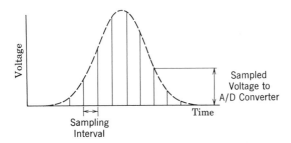

Figure 25–11   Computer sampling of a chromatographic peak.

under each peak. This may involve digital smoothing of the input data, correction for baseline drift, and curve-fitting for the calculation of the areas of partly resolved peaks. For a simple curve, as shown in Figure 25–11, the area can be determined simply by adding together the values of the sampled voltages across the curve. The computer must also determine the retention time corresponding to each peak maximum, and it can then identify each sample component from retention time data previously fed into it. The top of the peak is recognizable to within 1.5 sampling intervals by the point at which the sampled voltage starts to decrease. If a more accurate measurement of the peak center is required it can be obtained by fitting a parabola to the top of the curve. Once the components have been identified the computer can apply the appropriate detector response factor to each peak and calculate the percentage of each component in the sample. This may be done either on the basis of normalization of the results (total percentages must add to 100%) or by comparison with an internal standard (added to the sample) or an external standard (injected separately).

The next step is to prepare a report which might resemble that shown in Figure 25–12. To identify the sample components it is necessary to specify both the retention time (or relative retention time) for each component and the acceptable range or tolerance about the specified value. Over a period the retention times will drift because of aging of the column, and it is possible to program the computer so that it continually updates the stored retention times to allow for this drift.

Alternative Computer Systems. The computer used for the above applications might be capable of handling up to 40 chromatographs, and the capital and operating costs could then be spread over these 40 instruments. To justify the use of the same computer for a small number of chromatographs would be much more difficult and an alternative approach might be preferable. The smoothing of data, correction for baseline drift, calculation of peak areas, and determination of the retention time of each peak involves a considerable amount of computing. However, all these operations can be carried out by a digital integrator. One alternative therefore is to use a small computer just to carry out the final data processing and to equip each chromatograph with its own digital integrator. This approach reduces the cost of the central processing unit and the size of the computer memory required, but each chromatograph added to the system must have its own integrator.

There are other alternatives. For example, the output from each chromatograph could be fed via the A/D converter to a magnetic tape or paper tape punch, as

```
INSTRUMENT: GC4
METHOD: N/TC/1000
DATE: 3/4/73
TITLE: IGM/REACTOR 3
RUN NUMBER: 5
```

NO	COMPOUND	WT%	TIME	RET	%TOL	AREA	FACTOR
1	BENZENE	12.54	50	51	5	731	1002
2	TOLUENE	51.26	118	118	5	2942	1164
3	?	1.90	173			87	1000
4	ETHYLBENZENE	32.56	239	230	3	1849	1327
5	M-XYLENE	0.99	269	265	3	56	1327
6	U-XYLENE	0.76	315	320	5	43	1327

```
               100.01
```

Figure 25–12 Typical computer printout for a gas chromatographic analysis.

shown in Figure 25–10b. The tape would then be taken to a remote batch-processing computer for data reduction. This might mean that an existing computer system could be used, thereby saving the cost of the on-line computer, but there would then be a considerable delay in obtaining the analysis results.

Programmable calculators, actually a form of mini-computer, are finding increasing use for the processing of data from instruments such as chromatographs and clinical analyzers. Extended memory capacity and printers and other graphical devices are now available for use with a number of these calculators. However, they are usually slow and do not have the capability of handling the output from more than one instrument at a time. The combination of a programmable calculator and digital integrator produces, in effect, a computing integrator, but this system requires manual transfer of data from the integrator to the calculator.

Advantages of the Computer.   In addition to a saving in manpower requirements, the computer offers a number of other advantages. More effective use can be made of existing chromatographs, and tests have shown that the results produced by the computer are more accurate and precise. Typically, a standard deviation of better than 0.5% can be obtained by the use of a computer or digital integrator, compared with 1 to 5% by other methods. Where a large number of samples is involved, the cost per analysis may be much lower, and a further benefit is the increased amount of information obtainable through the use of the computer. One example of the latter occurs in the analysis of partially resolved peaks where the computer can carry out a curve-fitting procedure to obtain individual component percentages. Another is the use of a computer in conjunction with a gas chromatograph for simulated distillation. This is an analytical technique that uses a temperature-programmed gas chromatograph to determine the boiling range distribution of a sample. Simulated distillation is commonly used in the petroleum industry because many petroleum products are marketed with boiling range specifications. The computer might also be used to actually control the operation of the chromatograph, for example, sample injection, column switching, and so on.

25–6 COMPUTERIZED MASS SPECTROMETRY

The use of computers with mass spectrometers is increasing rapidly because of the large amount of data that can be generated by high resolution instruments or when a mass spectrometer is used in conjunction with a gas chromatograph. A typical gas chromatogram might contain about twenty peaks and it is usual to run three mass spectra on each peak, one on the rising side of the peak, one near the center, and one on the tail of the peak. It could take only ten minutes to generate the mass spectra, but a week or more to handle the resulting data manually.

Data Processing.   We shall consider first the use of computers for data processing. As with the chromatograph, the first step is to sample the output from the mass spectrometer to obtain the intensity of each peak and its mass. Because of the rate at which information may be generated by a mass spectrometer, data sampling rates of up to 40,000 samples/sec may be necessary. This is several orders of magnitude greater than that required for the input of gas chromatographic data. The mass scale is calibrated initially by the use of a standard compound, such as perfluorokerosene. The computer generates a polynomial relationship between the known mass numbers in the standard and the mass spectrometer magnetic field, measured with a magnetic field sensor, or the scan time, depending on the type of mass spectrometer. Subsequently, the mass corresponding to the center of each peak in the sample spectrum may be obtained by a magnetic-field to mass conversion or by a scan-time to mass conversion.

The peak center can be determined in a number of ways. An analog differentiator can be used that produces a zero signal at the top of the peak, and this in turn provides a signal to the computer. Alternatively, the peak center can be determined digitally in a similar way, or by calculating the center of the peak width at half-height, or by calculating the centroid of the peak. With a high resolution mass spectrometer the peaks are very narrow and occupy only a small fraction of the total baseline. A threshold intensity value is therefore set, below which any signals are ignored, and this considerably reduces the amount of data processing required.

The next step is to reduce the resulting mass spectrum to a form acceptable to the mass spectroscopist. It is usually the practice to normalize the mass spectrum by putting the intensity of the most intense or base peak equal to 100 and scaling all other peaks accordingly. The mass scale is also linearized and the resulting mass spectrum can then be printed out either in tabular form or as a line-plot. The latter can be printed out on a line-printer or plotter, or displayed on a cathode ray tube as shown in Figure 25–13.

The line-plot is the normal form of presentation for low resolution spectra but is unsatisfactory for high resolution spectra because several peaks may be separated by only a few milli-mass units. An *element map* is therefore commonly used for high resolution mass spectra. The element map lists in each column those ions having the same number of hetero atoms. In the example in Figure 25–14 the ions in Column 1 contain neither nitrogen nor oxygen, and the relative intensity of each ion on a logarithmic scale is indicated by the number of dashes. Ions in Column 2 contain one oxygen atom, and those in Column 3 contain two oxygen atoms. The molecular formula is always found at the bottom right-hand corner. It should be

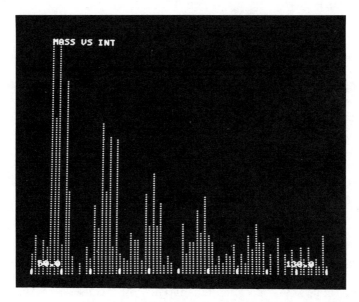

Figure 25–13 CRT display of a low resolution mass spectrum. [Courtesy of Digital Equipment Corporation.]

	NO 00	NO 01	NO 02
93	------ C 7H 9		
95	------- C 7H11		
97		------- C 6H 9	
104	---- C 8H 8		
105	------- C 8H 9		
106	---- C 8H10		
107	------- C 8H11	----- C 7H 7	
108	---- C 8H12		
109	-- C 8H13	-------- C 7H 9	
110		---- C 7H10	
111		----- C 7H11	
117	---- C 9H 9		
119	----- C 9H 11		
120	----- C 9H12		
121	----- C 9H13	---- C 8H 9	
122	----- C 9H14	----- C 8H10	
123		----- C 8H11	
124		--------C 8H12	
125		----- C 8H13	
128	----- C10H 8		
129	----- C10H 9		
131	--- C10H11		
133	----- C10H13	---- C 9H 9	
135	------ C10H15	---- C 9H11	
136		------- C 9H12	
143	--- C11H11		
144	---- C11H12		
146	---- C11H14		
147	---- C11H15	----- C10H11	
148	-------- C11H16	---- C10H12	
149	-- C11H17	------ C10H13	
150		------ C10H14	
159	---- C12H15	-- C11H11	
162		----- C11H14	
163		----- C11H15	
171	--- C13H15		
173	---- C13H17	--- C12H13	
175		--- C12H15	
229		---- C16H21	
271			-- C18H23
286			---------C19H26

Figure 25-14 Element map of androstene diketone. [From J. R. Chapman, *Chemistry in Britain*, 5, 564 (1969).]

pointed out, however, that even element maps become unwieldly for some compounds. For example, the element map for a compound containing four nitrogen atoms and six oxygen atoms would require thirty-five columns!

Compound Identification. The generation of an element map by the computer involves the determination of the mass of each peak to a high degree of accuracy. The elemental composition of each peak can then be obtained by having the computer generate all possible combinations of the elements (normally limited to C, H, O, and N) that could correspond within a few milli-mass units to that particular mass, and then selecting the best fit. This generally completes the initial data processing, although other output forms are sometimes used for special purposes. One example is the generation of specific-ion chromatograms, such as that shown in Figure 25-15. In this figure the ion currents corresponding to masses 458 and 460 have been plotted as a function of time, showing that one peak in the chromatogram actually contains two unresolved sample components.

The next step is the identification of the compound, and this is usually done off-line because of the computer time required. Computer methods for the identification of mass spectra fall into three main areas: library searching, methods employing empirical breakdown rules, and a learning-machine approach.

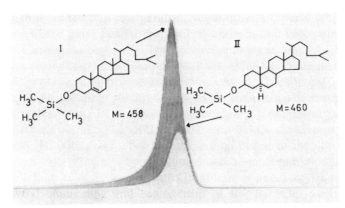

Figure 25-15 Specific-ion gas chromatogram. [Courtesy of LKB-Produkter AB, Stockholm, Sweden.]

Probably the most obvious, and perhaps the most straightforward, method for the identification of an unknown mass spectrum is to compare it with a library of reference spectra. There are, however, two main problems with this approach. The first is the fact that the spectrum of an unknown compound is frequently not amongst the many thousand spectra that are available in, for example, the API (American Petroleum Institute) or ASTM (American Society for Testing Materials) reference collections. Furthermore, considerable care is necessary in using collections of spectra derived from different sources. This is because the relative peak intensities obtained for the same compound on two different mass spectrometers may vary by a factor of over 100 to 1 from one end of the spectrum to the other! Research workers who do use library search methods generally build up their own libraries covering a narrow range of compounds in which they are specifically interested. The second major problem is the large amount of data processing required in searching a file containing some thousands of spectra. One way to overcome this is to use only the most intense peaks in each mass spectrum so that the search time is considerably reduced. A variation of this is to select the two most intense peaks in each successive group of 14 mass numbers, the latter being chosen because it corresponds to one CH_2 increment. This also helps to minimize the mass discrimination effects that make the comparison of spectra obtained with different types of mass spectrometers difficult.

A more general approach, which is not limited to compounds whose spectra are readily available, is to employ the methods adopted by the practising mass spectroscopist. He uses empirical rules that have been formulated to explain the observed breakdown patterns of many compounds. Individual mass spectroscopists may use certain intuitive steps in identifying a mass spectrum which have not yet been included in present computer-based identification routines, but the computer still has one major advantage over its human counterpart. That is its thoroughness. It will consider all possibilities exhaustively and will do this without becoming fatigued. There is mounting evidence to show that programs can now be written that provide more accurate, and of course much faster, interpretation of spectra than is achieved by most experienced mass spectroscopists.

One of the most interesting of these fragmentation pathway approaches is the *Heuristic DENDRAL* program developed by Lederberg and his co-workers at Stanford University. The first step in this procedure is to determine, from the low resolution mass spectrum, the functional group present in the molecule. For some classes of compounds, such as ethers and aliphatic amines, n.m.r. data are also required to determine the number of methyl groups. The next step is to generate a complete but nonredundant list of all stable structural isomers by the use of the DENDRAL program. The n.m.r. data are used to truncate this list. The mass numbers and peak intensities that would occur in the breakdown pattern for each of the remaining isomers are then predicted using empirical breakdown rules. The actual spectrum and the predicted spectra are compared and those isomers that most nearly fit the experimental data are ranked in priority order. A high degree of success has been obtained with this program and work is currently in progress to extend it to the more difficult polyfunctional and cyclic compounds.

Another approach is to use the computer as a *learning machine*, or trainable pattern classifier. In this method the computer is trained to recognize common features in a series of spectra. It does so, not by looking at isolated peaks, but by calculating a discriminant function that is used to weight each peak in the unknown spectrum. Training is carried out by repeatedly presenting to the computer a set of reference spectra until there is no significant improvement in its ability to produce the correct response. Different discriminant functions are generated for different features, for example, for the presence or absence of oxygen, for the number of carbon atoms present in the molecule, and so on.

Control of the Mass Spectrometer. In scanning a mass spectrum a substantial amount of time is wasted by looking at sections of the spectrum that do not provide useful information. This includes the sections between peaks and the sections on the sides of the peaks. For a quadrupole mass spectrometer, the normal continuous scanning can be replaced by a stepwise scanning system controlled by the computer. At each step the mass spectrometer voltage is varied to shift the spectrum position to that corresponding with the center of the next mass number. The output signal from the mass spectrometer is then integrated for a preset time period, thereby increasing the output signal and hence the sensitivity. However, the scan time can still be short since the number of steps in the scan is equal only to the number of atomic mass units in the spectrum range.

25–7 COMPUTERIZED N.M.R. SPECTROSCOPY

Time-Averaging. Probably the most familiar application of computers in n.m.r. spectroscopy is time-averaging, whereby successive spectra run on the same sample are added together to increase the signal-to-noise ratio. The signal intensity increases in direct proportion to the number of scans, n, whereas random noise tends to average out and increases only as $\sqrt{n}$. The signal-to-noise ratio is therefore enhanced by the ratio $n/\sqrt{n}$, that is, by $\sqrt{n}$. With a stable n.m.r. and a multi-channel integrator, commonly known as a CAT (*Computer of Average Transients*), samples can be run for extended periods (frequently overnight) to substantially increase the signal-to-noise ratio. It may be thought that the same result would be easier to obtain in n.m.r. spectroscopy by increasing the r.f. transmitter power and reducing the scan speed. However, this is impractical because the increased power raises the nuclei to the higher state faster than relaxation processes can return them to the lower state. Without an excess of nuclei in the lower energy state there is no net absorption of energy and the absorption spectrum disappears.

Fast Fourier Transform. More recently, the technique of pulsed Fourier-transform n.m.r. spectroscopy has been developed, mainly to enable ^{13}C spectra to be obtained on a routine basis. This nucleus has an abundance of only 1% and an inherently lower sensitivity than protons because of the closeness of the upper and lower energy levels. Its overall sensitivity in natural abundance is only 10^{-4} of that of protons. In the conventional n.m.r. experiment the frequency (or field) is varied as a function of time and at any one instant the response to only one narrow range of frequencies is being observed. This is very inefficient and an obvious improvement would be to have a large number of fixed-frequency transmitters and receivers that simultaneously cover the required spectral range. However, to cover the ^{13}C chemical shift range of 5000 Hz at 1 Hz intervals would require 5000 of these—which is impractical. Nevertheless, the same effect can be achieved in a somewhat different way.

A single frequency that is represented in time as a cosine wave of indefinite extent, Figure 25–16a, can also be represented in the frequency domain as a single line (Dirac δ function) in an otherwise zero-valued spectrum, Figure 25–16b. As shown in Figure 25–17, a series of equally spaced r.f. pulses is equivalent to a series of discrete frequencies and the spacing between these frequencies is equal to the pulse repetition rate. The intensities of the various frequency components depend on the duration of the individual r.f. pulses, with the distribution becoming flatter as the pulse length shortens. The use of a series of r.f. pulses thus has the same effect as a large number of separate transmitters.

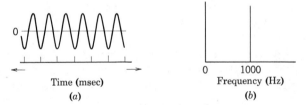

Figure 25–16   Pictorial representation of the Fourier transform (b) of a 1000 Hz cosine wave (a).

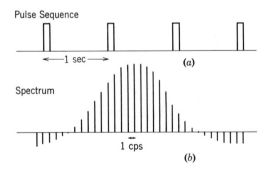

Figure 25–17   Pulse sequence (a) used in Fourier transform spectroscopy to simultaneously excite the entire spectrum, and the corresponding frequency domain representation (b). [From R. R. Ernst, *Adv. Mag. Resonance*, **2**, 109 (1966).]

Fortunately an equivalent number of separate receivers is not required because the free induction decay signal, which occurs after each pulse, is the Fourier transform of the steady state spectrum as shown in Figure 25–18. The transformation procedure is carried out digitally and may be viewed as a matrix operation involving the array of points that represents the spectrum. To obtain the steady state spectrum without undue delay the Fast Fourier Transform is used. This, in effect, considerably reduces the size of the arrays that must be handled by the

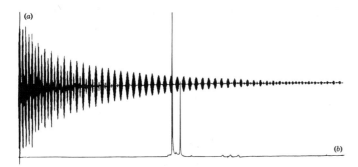

Figure 25–18 Free induction decay signal (a) and the Fourier transform spectrum (b) of (CH₂=CH)₄Si. [From G. C. Levy and G. L. Nelson, *Carbon-13 Nuclear Magnetic Resonance for Organic Chemists*, Wiley-Interscience, New York, 1972, p. 14.]

computer and the number of operations required. In practice, the sum of the n.m.r. responses to a series of r.f. pulses is stored in the computer (time-averaged) and the Fourier transform of this enhanced signal is then calculated.

Although pulse techniques have been used mainly for ^{13}C spectra, they are of most value with systems for which the spin-lattice relaxation time is of the same order as the spin-spin relaxation time, for example, protons. An example of the results obtainable for protons is given in Figure 25–19 which shows a 10-fold improvement in the signal-to-noise ratio when compared with the equivalent single scan spectrum. Conventional time-averaging would take 100 times as long to achieve the same result.

Simulation of n.m.r. Spectra. Another application of the digital computer in n.m.r. spectroscopy is to generate theoretical spectra from given chemical shift and spin-spin coupling constants. Initially an approximate set of parameters can be used and these can be successfully refined until the spectrum produced by the computer matches that obtained experimentally. The technique can be used both for the determination of shift and coupling constants and for the identification of unknown compounds.

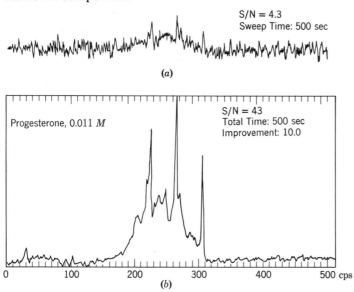

Figure 25–19 Single scan (a) and Fourier transform n.m.r. spectrum (b) of a 0.01 molar solution of progesterone in hexafluorobenzene. [From R. R. Ernst, *Adv. Mag. Resonance* **2**, 117 (1966).]

25-8   OTHER ANALYTICAL APPLICATIONS OF DIGITAL COMPUTERS

Digital computers are now being used extensively for the processing of analytical data from many different instruments, for the interpretation of results, and for instrument control. One of the more important applications is the determination of crystal structures from X-ray diffraction data. Computer-controlled diffractometers are becoming increasingly common. In the medical area, small digital computers are an integral part of some clinical analyzers, and computers are also being used to assist doctors in diagnostic procedures.

Digital averaging and smoothing techniques can be applied to the output obtained from most instruments to reduce high frequency noise, and deconvolution procedures can be used to increase the effective instrument resolution. A simple example of a deconvolution operation is provided by compensation for the broadening of gas chromatographic peaks due to a finite detector time constant, τ. In this case the true peak shape $E(t)$, a function of time, t, is broadened and skewed to give the output function $D(t)$, and the input and output functions are related by

$$D(t) + \tau \frac{dD(t)}{dt} = E(t) \qquad (25\text{--}5)$$

The original function can thus be obtained by adding the output signal and a portion of its first derivative. This is shown in Figure 25–20 in which the upper curve is the actual output signal produced by the gas chromatograph and the lower curve is the corrected response. This particular example was in fact obtained by using an analog differentiator, and analog techniques have also been used in the past for sharpening the peaks obtained from low resolution mass spectrometers. However, the digital computer is more suitable than analog methods for complex deconvolution operations because of its inherent flexibility in the handling of data. Digital computers have been used to enhance the resolution of gas chromatographs, mass spectrometers, and n.m.r. spectrometers.

As discussed earlier in this chapter, the efficiency of a conventional n.m.r. spectrometer is relatively low because at any one time the instrument response is confined to that due to one narrow band of exciting frequencies, and this situation can be improved by the use of pulse techniques and a digital computer. A similar problem occurs with an infrared spectrometer in which the light-gathering efficiency is severely restricted by narrow entrance and exit slits. Once again the efficiency can be improved by Fourier transform techniques, or by the Hadamard transform that allows the use of entrance and exit masks instead of single slits. In either case, a digital computer is required to obtain the final spectrum.

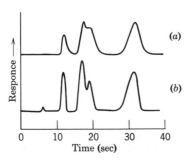

Figure 25–20   Actual (a) and deconvoluted (b) gas chromatograms. [Reprinted with permission from J. W. Ashley and C. N. Reilley, *Anal. Chem.*, **37**, 626 (1965). Copyright by the American Chemical Society.]

25-9   OPERATIONAL AMPLIFIERS

Operational amplifier circuits are used in three main areas: in processing data from instruments, in instrument control systems (particularly for electrochemistry), and in analog computers. We have already briefly looked at the operational amplifier in considering signal integration. The operational amplifier has a high gain, typically 10^4 to 10^5, high stability, low input current drain, and low output impedance. Voltage inputs can be applied to both input terminals, in which case the output is proportional to the difference between the two input signals. Alternatively, the signal can be applied to one input. In Figure 25–21a the signal is applied to the inverting input via a resistor R_1 and the noninverting input is grounded. (Power supply connections to the operational amplifier are not shown.)

Since the gain of the amplifier is high, and its output is limited to about ± 10 V, the differential input signal must be very small. As a result, the voltage at the inverting input to the amplifier in Figure 25–21a must be very close to ground and is often referred to as *virtual ground*. Because of this, and the low input current drain, we can write

$$i_i = i_f \quad \text{or} \quad e_i/R_1 = -e_o/R_f \tag{25-6}$$

so that

$$e_o = -(R_f/R_1)e_i \tag{25-7}$$

This circuit is an *inverter* because the sign of the input signal is reversed. Its gain is determined by the ratio R_f/R_1 (provided that this value is considerably less than the gain of the operational amplifier itself). If the feedback resistor is replaced by an electrochemical cell, the cell current is controlled because of Equation 25–6. The cell current can then be varied by changing the value of the input voltage to the circuit. If in Figure 25–21a the input resistor is deleted we have a *current-to-voltage converter*, since

$$i_i = -e_o/R_f \quad \text{or} \quad e_o = -R_f i_i \tag{25-8}$$

In Figure 25–21b there are a number of input signals, each connected to a separate input resistor. For this circuit we have

$$\frac{e_1}{R_1} + \frac{e_2}{R_2} + \frac{e_3}{R_3} = -\frac{e_o}{R_f} \quad \text{or} \quad e_o = -R_f(e_1/R_1 + e_2/R_2 + e_3/R_3) \tag{25-9}$$

This circuit is referred to as a *summer*. If all the resistors are equal the operation is simply the addition of all the input voltages, together with the overall sign change.

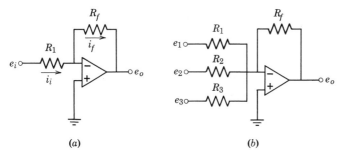

(a)   (b)

Figure 25–21   Inverter (a) and summer (b) operational amplifier circuits.

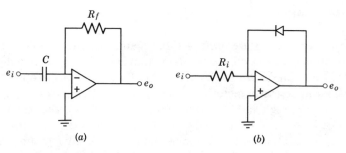

Figure 25–22 Differentiator (a) and logarithmic (b) circuits.

By employing other circuit components we obtain different circuit characteristics. In Figure 25–6 the integrator circuit was shown. Interchanging the resistor and condenser, as in Figure 25–22a, produces a *differentiator*. For this circuit,

$$C\, de_i/dt = -\,e_o/R_f \quad \text{or} \quad e_o = -\,r_f C\, de/dt. \tag{25–10}$$

In Figure 25–22b a diode is used in the feedback circuit. Provided that the voltage across the diode is greater than about 100 mV, the voltage-current relationship for the diode is

$$e = B \log (i/i_0) \tag{25–11}$$

where B is 0.059 V and i_0 is a constant which depends on the particular diode and is of the order of 10^{-11} A. For this circuit,

$$e_o = B \log (i_f/i_0)$$

$$= B \log \left(-\frac{e_i}{R_1 i_0}\right) \tag{25–12}$$

Note that, for the circuit to operate, the input voltage must be negative. By reversing the resistor and diode, and interchanging the direction of the diode, an exponential or antilogarithmic converter is obtained.

Multiplication is one of the more difficult functions to implement with operational amplifiers, and is usually produced by the *quarter-square multiplier*. In this, the relationship

$$xy = \tfrac{1}{4}[(x+y)^2 - (x-y)^2] \tag{25–13}$$

is used, where x and y are the variables whose product is required. Squaring is achieved by special diode function generators in which the square law relationship between input and output is approximated by a series of linear segments. The multiplier is therefore bought as a separate element from the normal operational amplifier. By using a multiplier in the feedback circuit of an operational amplifier, a *divider circuit* is produced. This is shown in Figure 25–23, in which

$$\frac{e_1}{R_1} = -\frac{e_0 e_2}{R_f} \quad \text{or} \quad e_0 = -\frac{e_1}{e_2}\frac{R_f}{R_1}. \tag{25–14}$$

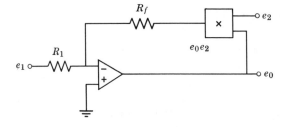

Figure 25–23 Divider circuit using a multiplier in the operational amplifier feedback circuit.

25–10   ANALOG AND HYBRID COMPUTERS

Although digital computers excel at the handling of vast quantities of numerical data, they are relatively inefficient at solving differential equations. This problem frequently arises in kinetics studies.

An analog computer consists essentially of a number of operational amplifiers that can add, subtract, amplify, and integrate. These are interconnected in such a way as to represent, or simulate, the problem to be solved. A simple example is provided by the solution of the second order differential equation:

$$\frac{d^2X}{dt^2} + k_1 \frac{dX}{dt} + k_2 X = f(t) \tag{25–15}$$

where $f(t)$ is a forcing function. The summer, integrator, and inverter connections required to solve this problem are shown in Figure 25–24. The output of each integrator is the integral of the function input to it, but with the sign reversed. An analysis of the integrator circuit has already been given. Likewise, the summer (top left-hand corner) that sums the functions input to it, and the inverter (in the center loop) reverse the sign of their respective inputs; at the same time these components can also be used as amplifiers. Constants, provided they are less than unity, can be handled by potentiometers, and other analog components can be used for multiplication and for generating special functions. In the example given, the terms generated by the integrators and inverter are brought together at the summer where it can be seen that (with the change of sign in the summer)

$$\frac{d^2X}{dt^2} = f(t) - k_1 \frac{dX}{dt} - k_2 X \tag{25–16}$$

and this satisfies the initial equation. After the various components have been

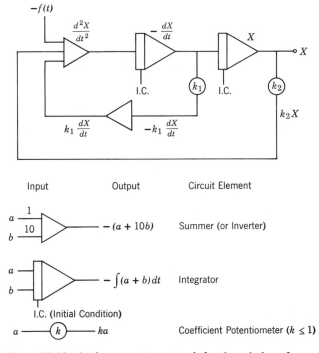

Figure 25–24   Analog computer network for the solution of a second-order differential equation.

connected, generally by means of a patchboard, the potentiometer coefficients are set and voltages are applied to each of the integrators to represent the initial conditions of the problem. The system is then set into operation by the simultaneous closing of a series of switches. The output, X, is recorded on an oscilloscope or pen recorder.

An example of the network required for the solution of the rate equations which represent the consecutive first-order reactions

$$A \xrightarrow{k_1} B \xrightarrow{k_2} C$$

is shown in Figure 25–25. The rate expressions are

$$-\frac{d[A]}{dt} = k_1[A] \tag{25-17}$$

$$\frac{d[B]}{dt} = k_1[A] - k_2[B] \tag{25-18}$$

$$\frac{d[C]}{dt} = k_2[B] \tag{25-19}$$

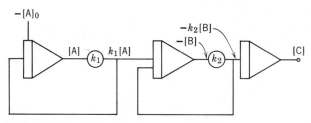

Figure 25–25 Analog computer network for the simulation of two consecutive first-order reactions.

and $[A]_0$ is the initial concentration of A. Typical results are given in Figure 25–26.

The examples given above are relatively simple; more complicated exercises, for example, the simulation of part of an industrial chemical plant, could require several hundred analog components.

The most time-consuming part of analog computing is setting up the network interconnections, scaling the problem so as not to overload the amplifiers, and then accurately setting the coefficient potentiometers. The actual solution of the

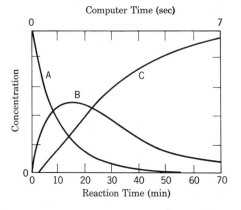

Figure 25–26 Typical results obtained for the kinetic system represented by Figure 25–25.

differential equations is very rapid, typically a fraction of a second. However, the inefficiency in the initial setting-up of the problem can be overcome with the help of a digital computer. The analog-digital combination is known as a *hybrid computer*. It combines the ability of the digital computer to handle large quantities of data with the analog computer's ability to solve differential equations very rapidly. The digital computer also provides logic facilities. This combination thereby provides a far more powerful device for simulation purposes than either computer used independently.

QUESTIONS AND PROBLEMS

Note: Assume that all peaks are Gaussian in shape.

25–1. Calculate the maximum pen speed required to accurately reproduce:
  (a) A gas chromatogram, in which the minimum peak half-width is 2 sec and the maximum peak height is 20 cm,
  (b) A mass spectrum, produced by a quadrupole mass spectrometer, in which adjacent peaks are just separated (10% valley), the scanning speed is 100 amu per second, and the maximum peak height is 10 cm.
  What type of recorder would be suitable for each of these applications?

  Ans. (a) 14.3, (b) 2968 cm/sec

25–2. The accuracy of integration achieved by a digital integrator can be adversely affected by inaccuracies in locating the onset and end of a peak. If integration is started when the peak rises to 1% of its maximum value and stopped when it falls below this value, what error results if:
  (a) The total area between these two points is measured
  (b) Only that area above the 1% height limit is measured?   *Ans.* (a) 0.24, (b) 2.7%

25–3. If a Gaussian peak is sampled at intervals of 1 standard deviation (σ) over a range of $\pm 2\sigma$ from the peak center, what is the error in the area as determined by
  (a) Summation of the sampled values
  (b) Use of the trapezoidal rule
  (c) Use of Simpson's rule?   *Ans.* (a) 0.91, (b) 6.3, (c) 5.3% negative

25–4. If the limits in Question 25–3 are extended to $\pm 3\sigma$, what results are obtained?

  Ans. (a) 0.03, (b) 0.47% negative, (c) 0.15% positive

25–5. A quadrupole mass spectrometer scans from mass 12 to mass 600 in 1 second, adjacent peaks being just separated (10% valley). If individual peaks above the 0.1% level are to be represented by 10 sampled points, what analog-to-digital conversion rate is required?   *Ans.* 3.9 KHz

25–6. In time-averaging of n.m.r. spectra, the sampled voltages that represent the spectrum are stored in individual locations in the computer memory, each location corresponding to a predetermined value of the sweep field. For a 60 MHz instrument that has a resolution (minimum peak half-width) of 0.3 Hz, how many memory locations would be needed for the full range of proton n.m.r. spectra if we allocate eight storage locations for these peaks and the measurement accuracy is 0.1% of the peak maximum? (Note that this "core" store requirement does not include that needed for the computer operating instructions.)   *Ans.* 5090 for a range of 600 Hz

25–7. Computers operate with a binary number system in which each factor of 2 is represented by one binary digit (or bit). To adequately represent an analog signal which has a signal-to-noise ratio of 30 to 1 requires a computer "word" length of at least 5 bits. If in a time-averaging experiment the input signal has a maximum signal-to-noise ratio of 30 to 1, and 1000 spectra are averaged, what computer word length is required inside the computer?   *Ans.* 15 bits

25–8. Devise an analog computer network for solution of the equation

$$\tau^2 \frac{d^2X}{dt^2} + 2\zeta\tau \frac{dX}{dt} + X = f(t)$$

that represents a damped second-order system, for example, the oscillation of a weight attached to a spring. For this exercise, multiplication can be represented by the symbol given below.

REFERENCES

G. W. Ewing and H. A. Ashworth, *The Laboratory Recorder*, Plenum, New York, 1975.

J. S. Mattson, H. B. Mark, and H. C. MacDonald, eds., *Computer Fundamentals for Chemists*, Dekker, New York, 1973.

S. P. Perone and D. O. Jones, *Digital Computers in Scientific Instrumentation: Applications to Chemistry*, McGraw-Hill, New York, 1973.

B. H. Vassos and G. W. Ewing, *Analog and Digital Electronics for Scientists*, Wiley-Interscience, New York, 1972.

A. Weissberger and B. W. Rossiter, eds., *Physical Methods of Chemistry, Part 1B*, Wiley-Interscience, New York, 1971.

R. C. Weyrick, *Fundamentals of Analog Computers*, Prentice-Hall, Englewood Cliffs, New Jersey, 1969.

*P. Hepple, *The Applications of Computer Techniques in Chemical Research*, Institute of Petroleum, London, 1972.

*C. F. Klopfenstein and C. L. Wilkins, eds., *Computers in Chemical and Biochemical Research*, Academic Press, New York, 1972.

*G. W. A. Milne (ed.), *Mass Spectrometry: Techniques and Applications*, Wiley-Interscience, New York, 1971.

*W. T. Wipke et al., eds., *Computer Representation and Manipulation of Chemical Information*, Wiley, New York, 1974.

H. A. Strobel, *Chemical Instrumentation*, 2nd ed., Addison-Wesley, Reading, Mass., 1973.

*Selected applications of computers.

AUTOMATIC AND PROCESS ANALYZERS

26

AUTOMATIC ANALYZERS

26–1 INTRODUCTION

In this chapter we discuss automatic analyzers of the type used mainly for clinical analyses. Process analyzers, which are used for the monitoring and control of chemical processes, are examined in the next chapter. Typical applications of automatic analyzers include the routine determination of glucose, albumin, cholesterol, creatinine, bilirubin, and SGOT (serum glutamate–oxaloacetate transaminase). The concentration of these and other compounds in the blood serum or plasma provides an indication of normal or abnormal health. For example, both glucose and cholesterol concentrations are high in uncontrolled diabetes, creatinine levels are high in renal disease, and bilirubin is high in jaundice.

Early forms of automatic analyzers were designed for the repeated analysis of samples taken from a single process stream, and these instruments duplicated as exactly as possible the volumetric titration procedures used by practicing analysts. The normal sequence of operations would be: add a measured quantity of sample to a titration vessel, add other reagents if necessary, titrate to the end-point, discard solution, wash out the titration vessel, and then repeat these procedures with a new sample. However, this repeated cycle of operations with a single reaction vessel is very time-consuming.

The need to analyze large numbers of samples in clinical laboratories has resulted in renewed interest in the development of automatic analyzers, and we discuss several different approaches to the solution of this problem. However, these instruments generally have two things in common. One is the use of specific color-forming reagents so that the compound of interest can be determined colorimetrically rather than by titration procedures. The second common feature is that the analysis steps are carried out sequentially as the sample moves through the instrument, whereas in earlier instruments the sample was held in a stationary vessel during the analysis. In three of the five analyzers discussed, the sample is placed in a disposable sample container to which the various reagents are added; a semi-continuous approach is adopted with the fourth instrument, and the fifth instrument uses a multiple sample container that is rinsed out between samples.

26–2 BASIC AUTOMATIC ANALYZER

The basic sequence of steps used in most of these instruments is shown in Figure 26–1. In this diagram a number of samples are held in a rotating sample tray. The first step is the transfer of a fixed aliquot of one of the samples to a sample cup. This is done with an automatic pipet and eliminates the need for any manual measurement of the size of the initial sample. The cup with the sample aliquot then moves to the next position where predetermined amounts of reagents are added. The next steps are stirring and, if necessary, filtration. The sample cup now moves into an incubator to develop the color characteristic of the sample component of interest. After a sufficient time has been allowed for color development, the sample is transferred to a colorimeter and the intensity of the color is measured at a fixed wavelength. This provides a measure of the concentration of the sample component. Because each step in the analysis is a sequential one, there may be a number of discrete samples progressing through the instrument at any given time; it is not necessary to wait for one sample to pass right through the instrument before another sample is taken from the sample tray.

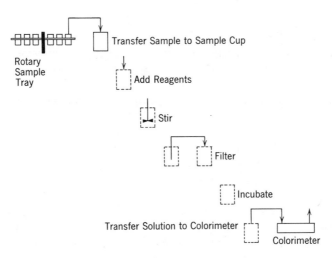

Figure 26–1 Basic sequence of operations used in automatic analyzers.

The instrument is normally set up to analyze for the same component in a number of samples. However, there is a large degree of flexibility in the arrangement just described, and different components can be determined by the use of different reagents. In some cases it may be necessary to add extra components, for example, to provide for the addition of further reagents after the filtration step. This is facilitated by constructing the instrument on a modular basis and, at any one time, using only those modules required for that specific analysis.

The methods used for the various steps, such as reagent addition, stirring, and filtration, depend on the manufacturer of the instrument. The instruments that are now briefly described achieve the same objective but by rather different methods. Different manufacturers may also use different reactions for determination of the same compound, but the present discussion covers only the differences in the instrumentation itself.

26–3   BECKMAN "DSA 560"

In the Beckman Discrete Sample Analyzer, Figure 26–2, the samples are initially placed in a rotary sample tray that holds up to 40 samples. An analysis starts with the automatic transfer of about 20 μl of a sample into a 5-hole disposable sample cup, called a "Q-cup." As this moves through the instrument, reagents and diluents are added by means of automatic pumps. (These are so designed that the sample and reagents never come into actual contact with them, thus preventing contamination and corrosion problems.) To stir the sample a jet of air is directed at an angle into the sample cup and this causes the solution in the cup to rotate rapidly. Filtration is accomplished by means of the vacuum filtration system shown in Figure 26–3. The filtration module drops a small filter hat onto the precipitated sample in the first cavity of the Q-cup. On applying vacuum, the liquid is drawn up into the filter hat and solid material remains behind on the outside of the filter. The filtered solution is then transferred to another section of the same Q-cup. The incubator block, across which the Q-cups travel, is heated by a circulating ethylene glycol mixture, and its temperature is adjustable between 35° and 65°C. The sample and (optional) reference solution are then transferred to a dual-channel colorimeter equipped with interchangeable interference filters covering the range 340 to 700 nm. The output from the colorimeter is presented in

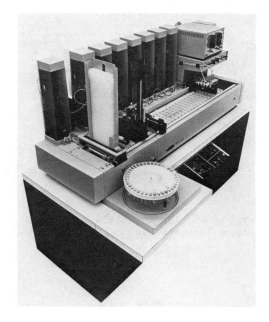

Figure 26-2   Beckman Discrete Sample Analyzer, DSA 560. [Courtesy of Beckman Instruments Inc.]

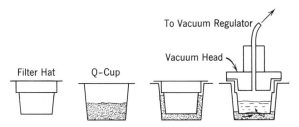

Figure 26–3   Filtration system used in the Beckman DSA 560. [Courtesy of Beckman Instruments Inc.]

concentration units and is displayed on a pen recorder or by means of an optional digital output system. A sample rate of between 120 and 160 samples per hour can be achieved with this instrument.

26–4   LJUNGBERG "AUTOLAB"

In the Autolab, shown diagrammatically in Figure 26–4, samples are placed in cylindrical cups that are linked together to form a flexible chain. In the sampling unit this sample chain runs parallel to the continuous analysis chain made up of a similar set of containers. A predetermined sample aliquot is transferred from the sample cup to one of the analysis cups and appropriate reagents are added to the analysis cup by means of up to five automatic dispensing syringes. The solution is then stirred with a small electric stirrer that is lowered into the cup. The stirrer is rinsed between samples. There is no provision for filtration, and with some analyses it is necessary to carry out a preliminary precipitation step before the main analysis sequence. The precipitate is removed by means of a manually operated centrifuge. The incubator, if used, is a shallow water-bath with a preset temperature of 37°C. The output from the colorimeter is displayed on a digital display unit in terms of either concentration (mg/100 ml) or extinction (mole/kg) units and this, together with the sample number, is also printed out. The instrument will handle up to 240 samples per hour.

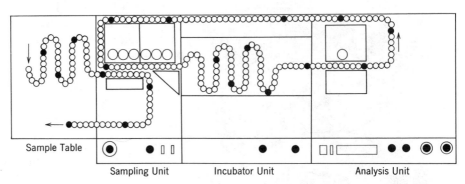

Figure 26–4   Layout of four of the basic modules of the Ljungberg AutoLab (top view). [Courtesy of A. B. Lars Ljungberg and Company, Stockholm, Sweden.]

26–5   DUPONT "ACA"

The two previous instruments, and other similar instruments, have been designed for the sequential analysis of the same component in a large number of samples. To analyze for a different sample component may involve a complete change in the reagents and analysis sequence.

The DuPont Automatic Clinical Analyzer overcomes this problem by the use of a disposable plastic test pack. This contains the reagents appropriate for the particular analysis, serves as the reaction vessel during passage through the analyzer, and is then formed into a cuvet for the final colorimetric measurement. One of these packs is shown in Figure 26–5a. The reagents are in tablet form in the upper section of the pack.

To carry out an analysis the sample is placed in a sample cup and inserted into the analyzer. This is followed by the appropriate reagent pack, or series of packs

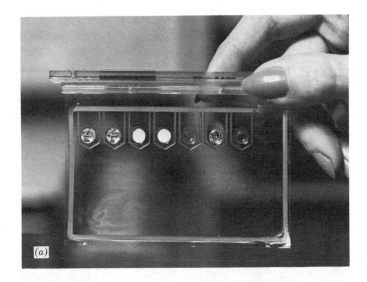

Figure 26–5 Typical reagent pack used in the DuPont Automatic Clinical Analyzer—at the start of the analysis (a), and after forming the cuvet (b). [Courtesy of DuPont Instruments.]

if more than one analysis is to be carried out on the sample. As the sample and reagent pack traverse the instrument, a predetermined aliquot of sample is transferred from the sample cup to the reagent pack, diluents are added, and the pack is thermostatted at 37°C. The reagents are released by clamping the top of the pack and then applying pressure to the sample/diluent solution to force open the temporary seals of the reagent compartments. Mixing is accomplished by vibration of the pack with an electromechanical vibrator. After the reaction is complete, the pack is placed in a combined die and photometer and one section of the pack is formed into a cuvet for the spectrophotometer by hydraulic pressure. The result is shown in Figure 26–5b, in which the pressure-formed cuvet can be seen on the left-hand side and a smaller cavity to take excess fluid on the right-hand side.

The photometer operating conditions (e.g., filter selection) and other variable parameters in the system are controlled by a built-in computer that obtains its information from a binary code on the top of each reagent pack. The top of the pack may also contain a short chromatographic column to effect protein removal, ion exchange, or gel filtration for specific analyses.

The results of the analyses carried out on each sample, and sample identification data, are printed out photographically on a report slip. Currently, some 22 separate tests can be carried out on this analyzer and it will handle up to 97 tests per hour with the first result available 7 minutes after insertion of the sample and test pack.

26–6   TECHNICON "AUTOANALYZER"

In the Technicon AutoAnalyzer—probably the most popular automatic analyzer developed to date—an alternative approach is adopted. Instead of taking a fixed aliquot of each sample for the analysis, a constant flow of sample is used. The first application for this instrument was the determination of urea nitrogen in blood, and a diagram of the instrument being used for this purpose is shown in Figure 26–6.

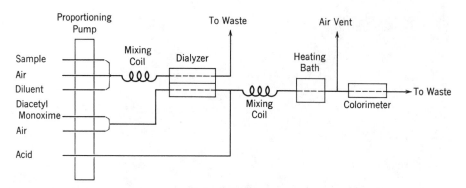

Figure 26–6 Flow diagram for the determination of blood urea nitrogen with the Technicon AutoAnalyzer.

There have been a number of changes made in the chemical system used for this particular analysis (as well as for other analyses), mainly to improve quantitative results. In the first system developed, the urea was incubated at 55°C with a buffered urease solution, and the resultant ammonia was picked up by a water stream in the dialyzer. To this stream was added a solution of Nessler's reagent that produced a yellow color in proportion to the concentration of urea in the

$$
\begin{array}{c}
\text{CH}_3\!-\!\text{C}\!=\!\text{O} \qquad \text{H}_2\text{N} \\
\qquad\qquad | \qquad\qquad + \qquad\qquad \diagdown\!\!\text{C}\!=\!\text{O} \rightarrow \\
\text{CH}_3\!-\!\text{C}\!=\!\text{O} \qquad \text{H}_2\text{N}
\end{array}
\qquad
\begin{array}{c}
\text{CH}_3\!-\!\text{C}\!=\!\text{N} \\
\qquad\qquad | \qquad\qquad \diagdown\!\!\text{C}\!=\!\text{O} + 2\text{H}_2\text{O} \\
\text{CH}_3\!-\!\text{C}\!=\!\text{N}
\end{array}
$$

$$\text{H}_2\text{N}\cdot\text{NH}\cdot\text{CS}\cdot\text{NH}_2 \downarrow$$

$$
\begin{array}{c}
\text{CH}_3\!-\!\text{C}\!=\!\text{N} \\
\qquad\qquad | \qquad\qquad \diagdown\!\!\text{C}\!=\!\text{N}\!-\!\text{NH}\cdot\text{CS}\cdot\text{NH}_2 + \text{H}_2\text{O} \\
\text{CH}_3\!-\!\text{C}\!=\!\text{N}
\end{array}
$$

sample. In the latest version, diacetyl monoxime hydrolyzes in acid solution to form diacetyl. The diacetyl reacts with the urea to produce a yellow diazine derivative and the color is intensified by the addition of thiosemicarbazide that forms a red triazine derivative.

The reaction rate is speeded up by heating the solution to 90°C. The light transmitted is then measured at 520 nm in a 15-mm flowcell. In Figure 26–6 the diacetyl monoxime solution contains thiosemicarbazide and the acid solution is 20% sulfuric acid containing ferric chloride and phosphoric acid. Chloride ions sensitize the reaction and phosphate ions increase the reproducibility of the reaction.

Proportioning Pump. The heart of the AutoAnalyzer is the proportioning pump. This is a peristaltic pump which, in the latest version, has up to twenty-eight separate flexible pumping tubes. The sample and reagents are forced through the tubes by means of transverse rollers that continuously squeeze the fluids ahead of them, as shown in Figure 26–7. Different flow rates can be obtained by using tubes of different internal diameter. Since the ratios of the volumetric rates of reagents and sample remain constant, there is no need for measuring out fixed aliquots, and automatic pipets are therefore not required in this instrument.

Sample Mixing. Sample and reagents are mixed by passing them through a series of horizontal glass coils. The claim is that as the mixture flows through each loop, the heavier fluid flows down through the lighter, thereby ensuring a homogeneous solution.

Dialyzer. In place of filtration, a dialyzer is generally used. This consists of two matching grooves cut in a pair of transparent plastic plates. The plates are clamped together with the grooves separated by a cellophane membrane. The sample flows through the upper half, and diffusible constituents pass through the cellophane semi-permeable membrane into the flowing stream of reagent in the lower half. The amount that diffuses through the membrane is proportional to the concentration but depends on the sample component. For example, approximately 10% of glucose might pass through the membrane, whereas the figure would be nearer 50% for urea. Since the dialysis rate is temperature dependent, the membrane assembly is housed in a temperature controlled water bath.

Color Development and Measurement. Color development occurs in a heating bath normally set at 95°C (or 37°C for enzyme reactions and bacterial activity tests). The color-forming reaction may be relatively slow and the stream containing the sample and reagents therefore flows through a glass helix, approximately 33 m long, which is immersed in the bath. The time taken to traverse the helix is normally about 5 minutes. From here the stream flows to the colorimeter, and the concentration of the sample component is continuously displayed on a pen recorder. As an alternative to the colorimeter a fluorimeter is used for some clinical analyses and a flame photometer for the determination of sodium and potassium.

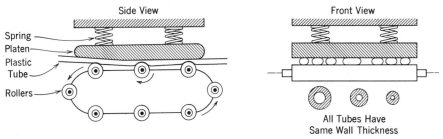

Figure 26–7 AutoAnalyzer proportioning pump. [Courtesy of Technicon Instruments Corp.]

Air Segmentation of Samples and Reagents.   As with the other instruments described above, the AutoAnalyzer is used for the sequential analysis of a number of samples. These are placed in a rotary sample tray and, as the tray rotates, a metal probe dips into each sample cup and aspirates the sample for a fixed time interval. Between 0.2 and 0.6 ml of sample is normally used, depending on the speed of rotation of the sample tray (20, 40, or 60 samples per hour). The probe then lifts out of the cup, aspirates air for one second, and goes into a wash receptacle. After that it lifts out of the wash solution, aspirates air again, and then moves into the next sample cup. Each sample is thus separated by segments of air and by a wash solution. Air is also introduced into the fluid lines to segment the reagent streams. This segmentation of the fluid streams provides a wiping action that removes traces of residual material from the tube walls, thereby further preventing cross-contamination between samples. However, the air bubbles must be removed before the fluid stream enters the colorimeter.

Multi-Analyzer Systems.   The fact that the proportioning pump can take up to twenty-eight tubes makes it possible to carry out up to three analyses simultaneously. Calibration results for BUN (blood urea nitrogen) and glucose, run simultaneously on the same samples but using two colorimeters and a two-pen recorder, are shown in Figure 26–8. The standards for BUN range from 10 to 100 mg % in multiples of 10, and those for glucose range from 50 to 500 mg % in multiples of 50. Each segment of the trace shows the rise in signal output as the sample enters the colorimeter, a plateau corresponding to the true concentration of the sample, and a decrease in the signal corresponding to wash-out of the sample. To maximize the sample throughput, the length of the horizontal plateau is normally reduced and the curves will then be sharper than those shown here.

By adding additional channels it is possible to perform other analyses simultaneously on the same sample and this has resulted in the development of the Technicon SMA12/60. This instrument analyzes individual serum samples for 12 selected biochemical parameters at the rate of 60 samples per hour. Any 12 of the following biochemical tests can be selected: albumin, alkaline phosphatase,

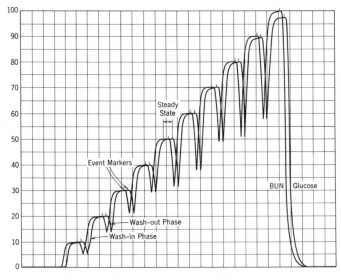

Figure 26–8   AutoAnalyzer calibration curves for blood urea nitrogen and glucose. [Courtesy of Technicon Instruments Corp.]

bilirubin, calcium, carbon dioxide, cholesterol, chloride, creatinine, glucose, inorganic phosphate, LDH (lactic dehydrogenase), potassium, SGOT, protein, urea nitrogen, uric acid. Potassium and sodium are measured by means of a continuous flame photometer in place of the colorimeter cell.

The results for each sample are presented on precalibrated paper as a "Serum Chemistry Graph." An example is shown in Figure 26–9. Normal ranges for each parameter are shaded, so that no intermediate data processing is required and the physician can see any abnormalities at a glance.

Similar instruments have been developed for the hematology laboratory. The SMA4 analyzes undiluted whole blood samples for RBC (red blood count), WBC (white blood count), hemoglobin, and hematocrit. The SMA7 is similar but carries

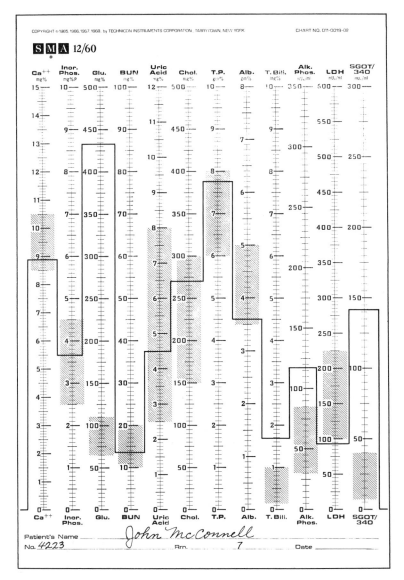

Figure 26–9 Technicon serum chemistry graph of a patient with severe diabetes mellitus. The following biochemical abnormalities can be seen: elevated SGOT, alkaline phosphatase, glucose, and bilirubin. [Courtesy of Technicon Instruments Corp.]

out additional calculations to provide MCV (mean corpuscular volume), MCH (mean corpuscular hemoglobin), and MCHC (MCH concentration).

The basic AutoAnalyzer has also been used for routine inorganic analyses, particularly for trace materials in boiler feedwater, soft drinks, and plating solutions, and for fertilizer analyses.

26–7   ELECTRO-NUCLEONICS "GeMSAEC" SYSTEM*

The analyzers described above are all of the sequential type, in which one sample follows another through the instrument. It is also possible to analyze a number of samples simultaneously for a single component, as in the GeMSAEC analyzer.

In this instrument a flat rotor with a number of spectrophotometer cuvets at its periphery is mounted on the drive shaft of a centrifuge. On top of the rotor is a transfer disc containing the samples and reagents. A cross-section through this assembly is shown in Figure 26–10. As the disc is rotated, the samples and reagents are moved by centrifugal force over a barrier and into the mixing chamber, and from there via a capillary into the appropriate cuvets. Mixing is achieved by rapidly changing the speed of rotation of the rotor. After reaction is complete, the sample transmittance is measured by means of a light beam that scans all the cuvets in turn as they pass the stationary light source. A typical oscillographic display of transmittance data produced by a series of serum protein standards is shown in Figure 26–11.

The instrument can also be used as a kinetic enzyme analyzer by the incorporation of a small computer. Enzymes, which are proteins, catalyze a large number of reactions in living organisms. Enzyme levels in the blood are normally

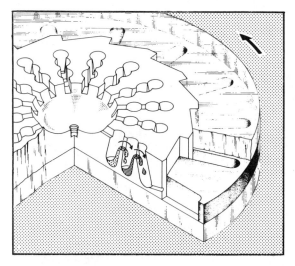

Figure 26–10 GeMSAEC rotor assembly, showing sample transfer during rotation. [Reprinted with permission from C. D. Scott and C. A. Burtis, *Anal. Chem.*, **45**, 327A (1973). Copyright by the American Chemical Society.]

* An acronym derived from the National Institute of General Medical Sciences and the U.S. Atomic Energy Commission.

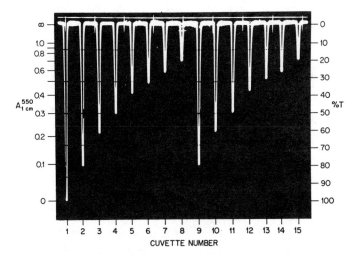

Figure 26–11 Calibration data for serum protein standards as displayed on the GeMSAEC oscilloscope. [From D. W. Hatcher and N. G. Anderson, *Am. J. Clin. Path.*, **52**, 645 (1969). Copyright 1969, The Williams and Wilkins Co., Baltimore.]

limited to a relatively narrow range and changes in these levels provide an indication of a number of disorders. A given enzyme will catalyze a reaction with a specific substrate for which we can write

$$E + S \rightleftharpoons X \rightleftharpoons E + P$$

where E is the enzyme, S is the substrate, X is an enzyme-substrate complex, and P is the reaction product. When the concentration of the substrate is large compared with the enzyme concentration, the forward reaction proceeds at close to the maximum rate and this rate is determined by the activity of the enzyme. This situation will continue until the concentration of substrate has been substantially reduced or until the product has an effect on the overall reaction rate. The rate of formation of product in the early part of the reaction will therefore be constant and provides a measure of enzyme activity. The computer will accumulate the data obtained from successive measurements of the solution absorbance, convert these to concentrations, select the linear portion of the reaction curve, and from that determine the enzyme activity.

QUESTIONS AND PROBLEMS

26–1. Hydrazine, which is added to boiler feedwater to remove traces of oxygen, can be determined by measuring the intensity of the red-orange azine formed by its reaction with *p*-dimethylamino benzaldehyde and $1N$ HCl. Show the AutoAnalyzer diagram for this analysis.

26–2. Creatinine can be determined by its reaction with alkaline picrate that results in a red-colored complex. Critical conditions in this reaction include the time and temperature of reaction, the degree of agitation during reaction time, and the wavelength. A saturated solution of picric acid and $0.6M$ sodium hydroxide are used as the main reagents. The effect of noncreatinine substances can be minimized by dialysis of the sample from a solution containing 1.8% sodium chloride into distilled water. Show the AutoAnalyzer diagram for this reaction.

26–3. The concentration of urobilinogen in urine increases with hemolytic anemia and decreases with jaundice. Urobilinogen can be determined by its reaction with Ehrlich's Aldehyde Reagent (0.7 g of paradimethylaminobenzaldehyde in 150 ml conc. HCl and 100 ml distilled water) to produce a red color. A sodium acetate solution, added 15 seconds after the urine and aldehyde reagent are added together, stops the reaction and intensifies the color complex. Show how this reaction would be carried out with one of the discrete type analyzers.

REFERENCES

E. C. Barton et al., eds., "Advances in Automated Analyses—Technicon International Congress, 1969" (3 volumes), Mediad Inc., White Plains, N.Y., 1970. A series of papers on applications of the AutoAnalyzer.

R. Robinson, *Clinical Chemistry and Automation*, Griffin, London, 1971. A discussion of accuracy in clinical analyses.

W. L. White, M. M. Erickson, and S. C. Stevens, *Practical Automation for the Clinical Laboratory*, 2nd ed., C. V. Mosby Co., St. Louis, 1972. Contains details of a number of clinical analyzers.

P. L. Wolf, D. Williams, T. Tsudaka, and L. Acosta, *Methods and Techniques in Clinical Chemistry*, Wiley-Interscience, New York, 1972. Provides details of the chemistry of common clinical laboratory procedures.

J. T. van Gemert, "Automated Wet Chemical Analyzers and their Applications," *Talanta*, **20**, 1045 (1973). Reviews instruments, modifications and applications in clinical chemistry, pharmacy, agricultural and food chemistry, environmental chemistry, and miscellaneous applications.

D. B. Roodyn, *Automated Enzyme Assays*, North-Holland, Amsterdam, 1970.

27

PROCESS ANALYZERS

27-1 INTRODUCTION

The product of a manufacturing process must normally conform to a detailed product specification. In the chemical industry this specification limits the amount of impurities the product may contain, and close control of the process is required to ensure that the final product is within specification. In many processes this is achieved mainly through the control of pressure, temperature, and flow rate at various points in the process. However, particularly where a range of products is being produced, it may also be necessary to control the chemical composition of the process stream at various points in the plant.

Laboratory analyses are generally adequate to provide a check on raw materials, but the delay inherent in taking a sample, transporting it to the laboratory, analyzing it, and then taking action if the result is out of specification, is frequently too great for adequate control to be achieved at intermediate stages in the process. It is then necessary to install an analyzer on the plant, close to the sampling point. It may also be necessary to make the analyzer part of a closed loop system in which the output from the analyzer is used to automatically control one of the process variables. This situation is becoming more frequent because of the increasing complexity and tightness of design of modern chemical plants and the continuing need for improved product quality and reduced operating costs.

Process analyzers may also be installed on a plant because of safety considerations, or to obtain improved raw material usage, increased plant throughput, reduced labor or utility usage, or more efficient by-product or waste recovery.

27-2 SPECIFIC AND NONSPECIFIC ANALYZERS

Process analyzers can be divided broadly into two classes: specific and nonspecific. In nonspecific analyzers some bulk property of the process stream is measured, for example, viscosity or thermal conductivity. This is generally useful as an analytical method only for binary mixtures but, if applicable, it is often the simplest, cheapest, and most reliable method. An exception to the binary mixture limitation occurs where the contribution from one component in the mixture differs markedly from that of all other components. The position is particularly favorable when its contribution is considerably greater than that of the other components. This occurs in the measurement of hydrogen by thermal conductivity and of oxygen by direct paramagnetic methods, as shown in Table 27-1.

Specific analyzers are capable of measuring one or more individual components in a sample mixture, and a number of these instruments are discussed in this chapter. The gas chromatograph and some colorimetric analyzers can be used to measure more than one component in the same sample. The infrared analyzer and

Table 27–1   Bulk Properties of Common Gases

	Thermal Conductivity[a] (100°C, air = 1.00)	Magnetic Susceptibility[b] (oxygen = 100.0)
Hydrogen	7.10	− 0.1
Oxygen	1.01	+ 100.0
Nitrogen	1.00	− 0.4
Argon	0.70	− 0.6
Carbon monoxide	0.96	− 0.4
Carbon dioxide	0.70	− 0.6
Methane	1.45	− 0.4
Ethane	0.97	− 0.8
Ethylene	0.98	− 0.4

[a] E. R. Weaver, in *Physical Methods of Chemical Analysis*, (W. G. Berl, ed.), Vol. II, Academic Press, New York, 1951.
[b] Beckman Instruments Inc., Bulletin 0-4016.

mass spectrometer can also be used for multi-component applications but process versions of these instruments have been largely replaced by the gas chromatograph.

To obtain a complete analysis of a sample, it may be necessary to use a number of different instruments. For example, in air pollution studies nondispersive infrared analyzers might be used for carbon monoxide, carbon dioxide, and nitric oxide, a UV filter photometer for nitrogen dioxide, and a gas chromatograph for individual hydrocarbons. Additional analyzers might also be required for the determination of ozone (by chemiluminescence), sulfur dioxide (coulometry), and mercury vapor (UV absorption).

27–3   TYPICAL PROCESS ANALYZER

In principle, any laboratory instrument can be adapted for process use, but in practice special instruments are required to meet plant requirements. The need for continuous analyses has also resulted in the development of a number of analyzers that are not normally found in the research laboratory.

Because process analyzers must function unattended for extended periods of time they must be more reliable and more stable than the corresponding laboratory instruments. Furthermore, chemical plants provide a relatively hostile environment for analytical instruments. On the one hand the analyzer must be protected from corrosive materials within the plant. On the other, the plant must be protected from any malfunction by the analyzer. Generally the major problem is that created by flammable process materials and most process analyzers must operate in a flameproof area.

A block diagram of a typical process analyzer installation is shown in Figure 27–1. It consists of:

1. A sample conditioning system usually located on the plant.
2. The analyzer, which is usually located in an enclosed analyzer house, and has associated with it any necessary gas supplies, reagents, and calibration samples.

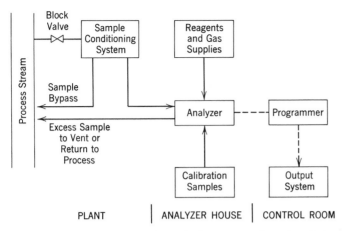

Figure 27–1 Block diagram of a typical process analyzer installation.

3. The output system, usually located in the plant control room. A programming (or sequencing) unit, which controls the operation of the analyzer, is also required for repetitive instruments, such as the chromatograph.

27–4 SAMPLE HANDLING SYSTEM

Successful results from a process analyzer depend to a large extent on the sampling system. This must present to the analyzer a representative sample in an acceptable form, with a minimum of delay and at a suitable flow rate. The location of the sample point must be chosen carefully to ensure that the sample is representative of the process stream. If the process stream is hot, the sample line will probably require heating to avoid loss of some of the sample components by condensation in the line. Heating is provided by steam jacketing the line or by steam tracing, that is, by running a steam line adjacent to the sample line.

The function of the sample conditioning system is to treat the sample so that it is acceptable to the analyzer. For the simple case of a gas under positive pressure, the sampling system may be no more than a filter and pressure reducer. In general, however, additional components may be necessary for heating or cooling the sample, drying it, removing entrained gas or liquid droplets, and for scrubbing out corrosive or other unwanted substances. At the same time it is important to ensure that the components of interest are not removed by the sample handling system. When sample streams need extensive clean-up, the sample conditioning system may resemble a small chemical processing plant.

The sample flow rate required depends on the particular analyzer but it would typically be about 0.1 to 1 l/min. This determines the size of the sample conditioning system. To reduce the time required for the sample to travel from the sampling point to the analyzer, a much larger sample flow is normally taken from the process stream and most of this is returned to a lower pressure point in the process via the sample bypass line shown in Figure 27–1. Delays in the sampling system are also minimized by making the sample lines as short and narrow as possible.

An example of a sampling system designed for the analysis of HCl in a process stream containing hydrocarbons, HCl vapor, and solid catalyst is given in Figure 27–2. The sample is initially cleaned by passing it through a glass-wool filter. After

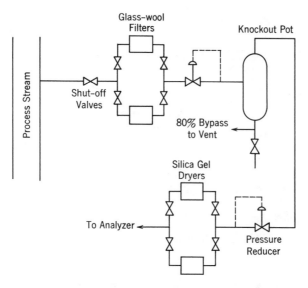

Figure 27–2 Sample handling system for the analysis of HCl in a process stream containing hydrocarbons, HCl vapor and solid catalyst. [After W. C. Schall, *Chem. Eng.*, **69** (10), 166 (1962).]

pressure reduction, approximately 80% of the sample is bypassed through the bottom of a knockout pot to vent and thence to an HCl absorber. The remainder, after a further reduction in pressure, passes to a silica gel dryer and thence to the analyzer. The installation of filters and dryers in parallel should be noted. This allows these components to be replaced without interfering with the continued operation of the system.

27–5 ANALYZER CALIBRATION

Calibration standards can be purchased from commercial suppliers or made up in the laboratory. However, most process analyzers are calibrated by comparison with laboratory analyses of samples taken from the process stream. This eliminates the problems involved in making up synthetic calibration samples but, for an instantaneous comparison, the laboratory sample must be taken from the same point and at the same time as that for the process analyzer. Failure to do so can well produce discordant results, as shown by Table 27–2. If a sufficiently large

Table 27–2 Comparison of Process Acid Analyzer Results and Laboratory Results

	Laboratory Sampling Point	
	50 feet away	*At analyzer*
Process analyzer, mean % H_2SO_4	91.7	91.8
Laboratory analyses, mean % H_2SO_4	90.9	91.7
Average difference	0.8	0.1
Standard deviation of differences	1.2	0.3

Data from J. W. Loveland, *Instr. Tech.*, **14**, 57 (May 1967).

process sample is taken, it can be used subsequently (without further laboratory analyses) as the calibration standard for the process analyzer. Gas samples containing less than 1% of selected impurities can be prepared by using permeation tubes or diffusion tubes. If a liquid is sealed in a plastic tube, it will permeate slowly but at a constant rate (after an induction period of some days) through the tube. The tube can be calibrated by measuring its weight loss over several days, and this method can be used for very volatile materials such as sulfur dioxide or propane. Some typical results are shown in Figure 27–3. For less volatile materials (e.g., iso-octane) an open vertical tube partly filled with the liquid sample can be used. The rate of loss of sample from this tube is controlled by the vapor pressure of the sample and its diffusion coefficient.

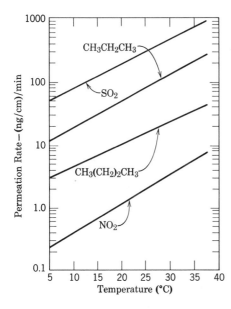

Figure 27–3 Permeation rates as a function of temperature for various gases. [From G. O. Nelson, *Controlled Test Atmospheres*, Ann Arbor Science Publishers, Ann Arbor, Mich., 1971, p. 139.]

27–6 PROCESS GAS CHROMATOGRAPH

The gas chromatograph is one of the most important process analyzers and also provides an excellent example of a complete analyzer system. In addition to the basic components already discussed in Chapter 7 the process chromatograph requires an automatic sample valve for the repetitive injection of a fixed quantity of sample and a programmer to control the overall operation of the instrument. Another feature of many process chromatographs is the use of multiple columns to enable complex samples to be analyzed.

Sample Valve. The principle of the automatic sample valve is shown diagrammatically in Figure 27–4. In the version shown here there are two discs. The upper disc contains inlet and outlet tubes and the sample loop or tube (shaded in the diagram) that determines the size of the sample, usually about 0.1 to 1 ml of gas. This disc is normally made of stainless steel and its bottom face is polished optically flat. The lower disc is of PTFE, and seals against the polished face of the upper disc. It has three channels cut into its surface that link together the inlet, outlet, and sample tubes. In position (a) the flowing sample fills the sample loop. In position (b) the lower disc has been rotated 60°, thereby interchanging the tubing

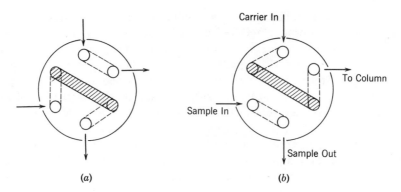

(a)   (b)

Figure 27–4 Operating principle of the gas sampling valve: filling the sample loop (a) and injecting the sample onto the column (b).

connections and introducing the sample trapped in the sample loop into the carrier gas line. In practice these valves may utilize either a rotary or sliding action, or the channels may be formed by a pneumatically operated flexible diaphragm. With liquids, very small sample volumes are involved, only a few μl. For this reason, liquid samples are usually vaporized and a vapor sample is injected into the column by means of a heated gas sample valve. Vaporization of the sample without fractionation of the sample components can be achieved by passing it through a heated capillary tube. By vaporizing the sample close to the sampling point the time lag normally present when transmitting liquid samples is also reduced.

  Columns and Detectors.   The characteristics of chromatograph columns used for process instruments must remain stable for an extended period of time. Consequently some column operating conditions that are acceptable in the laboratory are unsuitable for process chromatographs. The use of multiple column systems in process chromatographs is quite common, particularly for the removal of unwanted high-boiling components. This procedure, which employs a stripper column, is illustrated in Figure 27–5. At the start of the analysis the stripper and analysis columns are in series (Figure 27–5a) and this configuration is retained until all of the wanted components have passed through the stripper and into the analysis column. The carrier gas flow through the stripper column is then reversed (Figure 27–5b) and the heavier sample components are backflushed to vent. At the

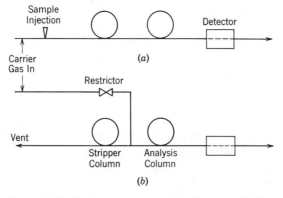

Figure 27–5 Dual column system for the removal of unwanted heavy sample components: at the start of the analysis (a), and during the back-flush cycle (b).

same time the carrier gas flow through the analysis column is maintained constant by introducing a flow restrictor into the carrier gas line in place of the stripper column. Column switching systems, involving up to three or more columns, may be required for the rapid analysis of samples containing a large number of components or for the separation of important trace impurities.

As with laboratory instruments, the two detectors most commonly used in process gas chromatographs are the thermal conductivity detector and the flame ionization detector. The latter provides the sensitivity required to analyze for trace quantities of hydrocarbons and can also be used for carbon monoxide and carbon dioxide at the ppm level by converting these compounds to methane over a nickel methanation catalyst before detection. The oven and detector assembly is either housed in a flameproof case or is purged with a continuous flow of air to meet plant safety requirements.

Programmer and Readout Systems.   Column switching and repetitive injection of the sample are controlled by a programmer. This sequencing or timing device also presets the signal attenuation for each peak, controls the operation of the recorder, and automatically re-zeros the baseline prior to each analysis. Various programming devices have been developed but those most commonly used with process analyzers employ either manually set timing cams or a multichannel rotating disc and photoelectric pickup head, together with associated switches. The relationship between the chromatogram and the programmer timing functions is shown in Figure 27–6. The normal chromatogram is unsuitable for the repetitive display of data and it is usually converted into bar-graph form as shown in the figure. This is done by keeping the recorder paper stationary during the emergence of each peak, and stepping it on a fixed distance between each peak. Alternatively, a multipoint recorder can be used, with a different color or shaped point for each component in the sample.

Applications of Process Chromatographs.   Process gas chromatography is used widely for the analysis of petroleum and petrochemical products because of its ability to provide detailed information about these materials. Other applications include the monitoring of annealing furnace and blast furnace atmospheres, air separation processes, ammonia synthesis, helium in natural gas, and sulfur recovery. Process gas chromatographs can also be used in the research laboratory to monitor research reactors.

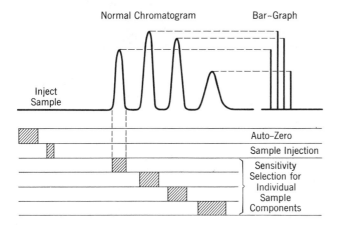

Figure 27–6   Relationship between the chromatogram and the programmer timing functions.

27-7   OXYGEN ANALYZERS

Methods for the determination of oxygen include automatic titration, electrical conductivity, electrochemical cells of various types, gas chromatography, heat of reaction, mass spectrometry, paramagnetic and thermomagnetic oxygen analyzers, and photometric methods. The particular method employed depends to a large extent on whether the measurement is for oxygen in the gas phase or for dissolved oxygen and on the detection limit required.

Electrical Conductivity.   One of the most sensitive methods for dissolved oxygen is based on the reaction

$$4Tl + O_2 + 2H_2O \rightarrow 4Tl^+ + 4OH^-$$

and the subsequent determination of the ion concentrations by measurement of the electrical conductivity of the solution. In earlier instruments the reaction of oxygen with nitric oxide was used for the same purpose, but the need to replace the gas cylinder makes the nitric oxide method less convenient. On the other hand, the toxicity of thallium creates a potential problem. In practice the sample is split into two streams, one of which goes via a reaction tube containing thallium to a sensing conductivity cell, and the other via a dummy column to a reference cell. This instrument is used mainly for the determination of oxygen in boiler feedwater. A boiler operating at 400 to 600 psi might have a feedwater specification of less than 10 ppb (parts per billion) of dissolved oxygen, while one operating at 1000 psi would have a specification figure of only 1 to 2 ppb. The full-scale ranges of this instrument cover from 0 to 10 ppb up to 0 to 1000 ppb, with 95% response being obtained in three minutes. (1 ppb dissolved oxygen = 1 μg/l.)

Electrochemical Cells.   Several electrochemical cell systems are used in commercial oxygen analyzers. An example is shown in Figure 27–7, which is basically a polarographic oxygen electrode. The silver anode and gold cathode are separated from the sample by a thin PTFE membrane that is permeable to oxygen. With a potential of 0.8 V applied between the anode and cathode, the following reactions take place in the presence of oxygen

$$\text{Cathode:}\quad O_2 + 2H_2O + 4e^- \rightarrow 4OH^-$$
$$\text{Anode:}\quad 4Ag + 4Cl^- \rightarrow 4AgCl + 4e^-$$

The resulting current flow is proportional to the partial pressure of oxygen in the sample. Because the rate of diffusion of oxygen through the membrane varies by about 3% per °C, a thermistor is built in to automatically adjust the output signal to compensate for temperature changes. The speed of response of the sensor is also determined by the oxygen diffusion rate; typically 90% of the final response occurs within 10 seconds.

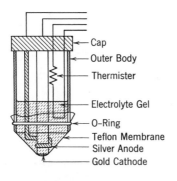

Figure 27–7   Polarographic oxygen electrode. [Courtesy of Beckman Instruments Inc.]

Cap
Outer Body
Thermister
Electrolyte Gel
O–Ring
Teflon Membrane
Silver Anode
Gold Cathode

This device can be used for both dissolved oxygen and gaseous oxygen, but the sample must be free from other gases that are reduced at approximately 0.8 V. These include sulfur dioxide, chlorine, bromine, iodine, and oxides of nitrogen. Since the sensor consumes a small amount of oxygen, and diffusion of oxygen through liquids is relatively slow, a liquid sample must flow continually past the sensor for valid results to be obtained. With a static liquid sample the oxygen supply would be gradually depleted. The device will cover the ranges 0 to 2.5 ppm up to 0 to 100 ppm of dissolved oxygen or 0 to 5% up to 0 to 100% (v/v) of gaseous oxygen.

Heat of Reaction. A relatively simple method for the determination of gaseous oxygen (or, conversely, hydrogen) is by measurement of the heat of reaction of oxygen and hydrogen over a palladium catalyst. The oxygen-containing sample, and additional electrolytically generated hydrogen if required, is first passed through a drying chamber and an active-charcoal purifier. The temperature of the gas is then measured before and after reaction, and the amount of heat generated is directly proportional to the concentration of oxygen in the sample. This type of instrument is suitable for the measurement of oxygen in gases over the range from 0 to 200 ppm up to 0 to 1%. However, some sample components (e.g., carbon monoxide) can adversely affect the palladium catalyst. Unsaturated hydrocarbons are hydrogenated and the heat developed by this reaction also interferes with the determination of oxygen or hydrogen.

Paramagnetic Analyzer. The paramagnetic properties of oxygen were discovered by Faraday in 1851. He found that a hollow glass sphere filled with oxygen and supported at the end of a horizontal rod by silk fibers was attracted by a magnet. The only other common paramagnetic gases are nitric oxide and nitrogen dioxide (see Table 27–1). In one form of paramagnetic oxygen analyzer, a precision hollow glass dumbbell is suspended in a symmetrical nonuniform magnetic field as shown in Figure 27–8. The dumbbell is made conductive with a rhodium coating and potentials are applied to it and to two adjacent electrodes. The magnetic force acting on the dumbbell is proportional to the difference between its volume magnetic susceptibility and that of the surrounding gas. In the absence of oxygen, the dumbbell tends to take up a position away from the most intense part of the field, since it is slightly diamagnetic. The presence of oxygen in the surrounding gas increases the magnetic force on the dumbbell and tends to rotate it further away from the magnet pole pieces. This deflection is sensed by an optical system, and a restoring torque is applied by changing the potential on the dumbbell. (Alternatively, an electromagnetic restoring torque can be generated by passing a current through a small coil wound on the dumbbell.) The change in potential or current required to return the dumbbell to its original position is proportional to the oxygen

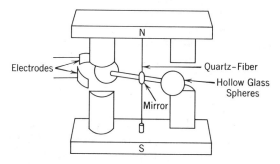

Figure 27–8 Paramagnetic oxygen analyzer.

concentration. Typical ranges for this instrument vary from 0 to 1000 ppm up to 100% and it can be calibrated by using air. The instrument is relatively robust but care is necessary to avoid a sudden large change in the sample flow rate that could destroy the dumbbell assembly.

Thermomagnetic Analyzer. Another instrument that relies on the paramagnetic properties of oxygen is the thermomagnetic analyzer. In one version, Figure 27–9, there are two heated filaments forming part of a Wheatstone bridge. One of these filaments is placed in a magnetic field. In the absence of oxygen, the sample diffuses equally into each of the cell cavities and the bridge is balanced. Any oxygen in the sample is initially attracted toward the magnetic field, but as the oxygen becomes warmer due to heating by the filament its paramagnetism decreases and it is displaced by cooler oxygen from the flowing sample. This induced convection or magnetic wind in one side of the bridge cools the measuring filament and unbalances the bridge. Thermostatting of the instrument is required, but automatic compensation for pressure variations is provided by the use of an additional pressure-compensating cell. These instruments are more robust but do not have the sensitivity of the dumbbell arrangement. They are normally available with ranges from 0–5% to 0–25% oxygen, and are calibrated directly with air.

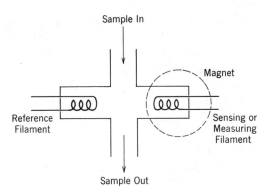

Figure 27–9 Thermomagnetic oxygen analyzer.

27–8 INFRARED ANALYZERS

Infrared analyzers used for process applications are quite different from their laboratory counterparts. Mechanically their design is much simpler and they do not employ prisms or gratings, that is they are nondispersive instruments. Optical filters are used to isolate the required wavelength region, and pure gases sensitize the instrument to one particular component in the sample. There are two basic designs of nondispersive infrared analyzer: positive-filtering and negative-filtering.

Positive-Filtering Analyzer. In the positive-filtering infrared gas analyzer the detector consists of a sealed chamber filled with the sample component of interest. The optical arrangement is as shown in Figure 27–10, which incorporates the dual chamber detector arrangement due to Luft. The two halves of the detector are separated by a flexible metal diaphragm. This diaphragm and an adjacent metal plate or grid form an electrical capacitor. Because the detector is filled with a specific sample component it absorbs radiation characteristic of that compound. The absorption of this radiation by one side of the detector raises the gas

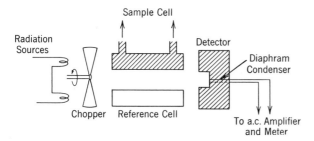

Figure 27–10   Positive-filtering Luft-type infrared gas analyzer.

temperature and pressure and distends the diaphragm, thereby changing the capacitance.

When both the reference and sample cells are filled with a nonabsorbing gas, an equal amount of radiation is absorbed by each half of the detector and the pressure on each side of the diaphragm is equal. This is the balance or zero position. When the component of interest is introduced into the sample cell, the amount of detectable radiation reaching the sensing side of the detector is reduced. As a result of the higher pressure in the reference side of the detector, the diaphragm distends toward the sensing side of the detector and this changes the capacitance. A chopper, located between the radiation sources and the sample and reference cells, chops both radiation beams simultaneously at approximately 10 Hz. The resulting changes in capacitance modulate a radio-frequency signal, and the demodulated signal is amplified and fed to an indicating meter and/or a recorder. Any long-term difference in pressure between the two halves of the detector is avoided by means of a small interconnecting channel. This allows gradual pressure variations due to temperature differentials to be equalized without affecting the diaphragm movement arising from the 10 Hz pressure pulses.

Negative-Filtering Analyzer.   In a negative-filtering analyzer, Figure 27–11, a sensitizing cell that completely absorbs radiation corresponding to the component of interest is placed in front of one half of the detector. In the absence of this particular component in the sample there is therefore a difference in the radiation intensity measured by the two halves of the detector (which in this example is a differential thermopile). When this component is present in the sample the intensity of the characteristic radiation passing through the sample cell is reduced. This affects only that side of the detector preceded by the chamber containing the nonabsorbing gas, and the difference signal between the two halves of the detector is reduced.

Because infrared absorption bands are not unique for each compound, there may

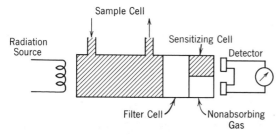

Figure 27–11   Negative-filtering infrared gas analyzer.

be unwanted absorption by other components in the sample that interferes with the measurement. It is frequently possible to compensate for this by placing the unwanted compounds in a filter cell, for example, as shown in Figure 27–11. In this way the signals characteristic of the interfering compounds are removed and are not measured by the detector. This procedure can also be used with the positive-filtering analyzer but it is claimed that the negative-filtering system provides better compensation. Best results, however, are obtained with more complex instruments that employ a combination of the two methods.

Applications of Infrared Gas Analyzers. One of the major applications for the infrared gas analyzer (IRGA) is the determination of carbon monoxide and carbon dioxide. Continuous analyses for these components are required for the control of ammonia synthesis plants, boilers (flue gas analysis for combustion efficiency), blast furnaces, and metal treatment plants. Other important applications of the IRGA are in air pollution studies and vehicle exhaust emission control. The range required varies from 0 to 100 ppm up to 0 to 100% depending on the particular application. An IRGA can also be used to determine ammonia, anesthetics, sulfur dioxide, and water vapor. Infrared analyzers can be used for liquid samples (e.g., for water in liquid hydrocarbons or other organic solvents), but this application involves cell path lengths of less than 1 mm, compared with 20 to 100 cm for gas samples. For most organic analyses the infrared analyzer has been replaced by the gas chromatograph.

The absorption bands of carbon monoxide and carbon dioxide are reasonably well separated and there is no problem in sensitizing the instrument to just one of these gases. This is shown in Figure 27–12 which depicts the situation where the detector of a positive-filtering IRGA has been sensitized to carbon dioxide by filling with this gas. In the ammonia synthesis plant application, the major sample components are hydrogen and nitrogen which do not absorb in this region of the spectrum. Methane, which is present in small concentrations in plants that are based on the cracking of hydrocarbon feedstocks, has absorption bands that are also well separated from those of carbon monoxide and carbon dioxide.

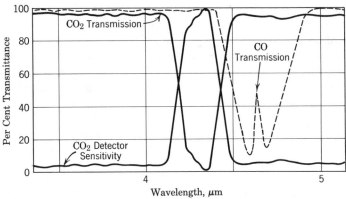

Figure 27–12 Selective response of an IRGA detector sensitized to carbon dioxide. [Courtesy of Beckman Instruments Inc.]

27–9   MOISTURE ANALYZERS

There are approximately twenty different methods for the automatic determination of moisture in solids, liquids, and gases. These include automatic titration procedures for liquids and infrared absorption for gases. Electrical capacitance measurements can be used to continuously monitor the water content of sheet material, and microwave absorption for granular solids. We shall concern ourselves mainly with the measurement of moisture in gases.

Electrolytic Hygrometer. The commonest process moisture analyzer for gases is the electrolytic hygrometer. In this instrument there is an electrolytic cell which consists of two closely spaced platinum or rhodium wires, approximately 0.3 mm apart, wound on a cylindrical former and coated with a film of phosphorous pentoxide. This film absorbs water from the gas sample, and the absorbed water is continuously electrolyzed by a dc voltage applied between the two wires. At a constant sample flow rate, and with all the water removed from the sample stream, the electrolysis current is a direct measure of the water content of the sample gas. The instrument can therefore be calibrated directly by the application of Faraday's law. In spite of the fact that the response of the electrolytic hygrometer depends on the electrolysis of absorbed moisture, approximately 90% of the final response is reached in 1 minute with a sample flow rate of 100 ml/min, provided that the change in moisture content is not too large.

Rhodium is preferred as the electrode material because it minimizes the errors that can occur due to recombination of the electrolytically generated oxygen and hydrogen in streams containing high concentrations of hydrogen. It also increases the operating life of the cell at high moisture concentrations. The instrument is suitable for measuring moisture over the range 0 to 10 to 0 to 1000 ppm in inert gases, hydrocarbon gases, and even in chlorine. However, it cannot be used with basic gases such as ammonia or with certain organic materials such as alcohols and acetone. The phosphorous pentoxide film also catalyzes polymerization of some unsaturated organics, particularly conjugated dienes such as butadiene. To overcome this problem, a gas chromatograph can be used in conjunction with the electrolytic hygrometer to carry out a preliminary separation of the unwanted sample components.

Piezo-electric Sorption Hygrometer. A more recent instrument that can also be used for the measurement of moisture in gases is the piezo-electric sorption hygrometer. The sensing element in this instrument is a radio-frequency (9 MHz) quartz crystal coated with a hygroscopic material and located in a temperature-controlled oven. When the coating absorbs moisture from the surrounding sample gas the crystal gains in weight and the frequency of oscillation decreases. The sensitivity of this device is similar to that of the electrolytic hygrometer but it is not as specific. However, interference from other compounds can be minimized by a suitable choice of crystal coating or, conversely, the coating can be selected to make the instrument sensitive to materials other than water.

The principle employed in this instrument has also been used as the basis of an ultrasensitive microbalance for the continuous measurement of particulate matter in the atmosphere.

Calibration and Use of Moisture Analyzers. In measuring low concentrations of moisture in gas samples, considerable care is necessary to exclude atmospheric moisture. Rubber and many plastics are permeable to moisture and therefore cannot be used. Small leaks in the system cannot be tolerated because atmospheric moisture will back-diffuse into the system, even against a high sample pressure. This occurs because the moisture diffusion rate is determined by the difference between the partial pressures of water vapor in the sample and in the surrounding atmosphere and not by the difference in the total pressures. Similar considerations also apply to oxygen analyzers. Moisture is adsorbed readily by many materials, and desorption can give rise to errors that persist for extended periods of time. This means that ceramic or fiber filter elements, which adsorb large amounts of water, cannot be used in the sampling system. Stainless steel (which is preferred) or other materials that do not adsorb moisture to any great extent (e.g., nickel or PTFE) must be used.

Calibration of moisture analyzers other than the electrolytic hygrometer presents a problem, but samples containing known low concentrations of water vapor can be produced by the catalytic conversion of hydrogen and oxygen to water. A moisture-free gas containing the appropriate quantity of hydrogen plus excess oxygen, for example, 100 ppm hydrogen and 1 to 2% oxygen in nitrogen, will remain stable in a gas cylinder for an extended period and can be used as required.

Corrosion Rate Measurements. One of the main reasons for wanting to measure the moisture content of both gases and liquids is to reduce corrosion rates. Alternatively, the rate of corrosion of a metal specimen placed in the sample stream can be used as an indirect measurement of the moisture content of the process stream. If a metal wire is used as the specimen, it can be made part of a Wheatstone bridge circuit. As the wire corrodes, its cross-section decreases and its resistance increases. The rate at which the bridge output changes is therefore a measure of the corrosion rate. By choosing a wire material such as mild steel, which corrodes at a much greater rate than the normal chemical plant materials, this method can be made quite sensitive to small changes in the moisture content of the process stream.

27–10 COLORIMETRIC ANALYZERS

The selective determination of a number of materials can be effected by the use of specific color-forming reactions. This approach is used extensively in automatic clinical analyzers for many compounds of biological importance. It can also be used for continuously monitoring for trace quantities of inorganic materials such as iron, copper, chlorine, fluoride, phosphate, and sulfite.

The problem of adding preselected quantities of reagents to the sample can be overcome in a number of ways. The use of automatic pipets or of a peristaltic pump has been discussed in Chapter 26. Another possibility for a continuous flow system is to control the rates of sample and reagent addition by means of restrictors of predetermined flow characteristics.

The construction of a simple continuous analyzer based on this principle is shown in Figure 27–13. This particular example is of a water hardness monitor in which a mixed reagent containing EDTA, a buffer, and an indicator is added to the sample. The sample flow rate in the reaction cell is controlled by means of the sample metering capillary and by maintaining a constant liquid level in the cell. Excess sample flows over the top of the cell to waste. The reagent rate is controlled by the reagent metering capillary and by the reagent level. The latter may be kept relatively constant by using a squat container for the reagent. Alternatively, more precise control can be obtained by means of a float valve that maintains the liquid level constant at a point intermediate between the reagent container and the reaction cell. The controlled sample flow discharges into a colorimeter cell directly below the reaction cell. When the hardness level increases above a predetermined value the solution color changes from blue to red and this triggers an alarm. The alarm point can be set at between 1 and 100 ppm by choice of the appropriate strength reagents.

A few reactions result in the emission of radiation, that is, chemiluminescence,

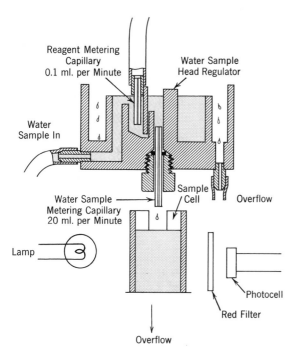

Figure 27–13   Reaction cell and colorimeter in a simple water hardness monitor. [Courtesy Hach Chemical Co.]

and this has been used recently for the determination of several air pollutants. One example is the determination of NO, which is reacted with ozone.

$$NO + O_3 \rightarrow NO_2^* + O_2$$
$$NO_2^* \rightarrow NO_2 + h\nu.$$

The intensity of the light emitted is directly proportional to the concentration of NO. Provision is usually made for the determination of NO_2 by first converting it to NO by means of a high temperature coil and then measuring NO_x, the sum of NO and NO_2. Instruments available cover the ranges 0.01 to 10,000 ppm full scale. In another analyzer, ozone in the atmosphere is monitored by the chemiluminescence resulting from its reaction with ethylene. Ranges available vary from 0.01 to 10 ppm full scale.

QUESTIONS AND PROBLEMS

27–1. Show diagrammatically the components you would use in a sample conditioning system designed to handle a 300 psi 200°C process gas saturated with water and containing suspended catalyst particles. The analysis is for methane, using an IRGA. (See W. C. Schall, *Chem. Eng.*, **69**, 166 [May 14, 1962].)

27–2. A 6 in. long permeation tube is to be used to calibrate a sulfur dioxide analyzer that requires a sample flow rate of 1.00 l/min. If the standard required is 2.00 ± 0.05 μg of sulfur dioxide per liter of air, at what temperature and within what limits would the permeation tube need to be thermostatted? From Figure 27–3 the permeation rates for this particular tube are 50 and 300 ng/min per cm length of tube at 5 and 25°C, respectively.   *Ans.* 15.7 ± 0.3°C

27–3. A gas mixture containing hydrogen, oxygen, nitrogen, carbon monoxide, methane, and carbon dioxide can be analyzed by gas chromatography by using two columns in series. The first column contains silica gel that separates carbon dioxide from the remainder of the sample. The carbon dioxide is measured by a thermal conductivity cell and is then removed by an absorption tube before the other sample components are separated on a molecular sieve column in the order given above.

    (a) Prepare a diagram of the chromatographic system.

    (b) Show the relationship between the programmer timing functions and the chromatogram for this particular analysis (cf. Figure 27–6).

27–4. What is the difference between a positive-filtering and a negative-filtering IRGA?

27–5. What current is generated by an electrolytic hygrometer for a moisture concentration of 10 ppm by weight in air, and a sample flow rate of 100 ml/min at 20°C and 760 torr?

Ans. 216 μA

REFERENCES

K. J. Clevett, *Handbook of Process Stream Analysis,* Ellis Horwood, Chichester, U.K., 1974.

I. G. McWilliam, "Process Control by Gas Chromatography" in J. C. Giddings and R. A. Keller (eds.), *Advances in Chromatography,* Vol. 7, Dekker, New York, 1968, p. 163.

W. C. Schall, "Selecting Sensible Sampling Systems", *Chem. Eng.,* **69**, 157 (May 14, 1962).

A. Verdin, *Gas Analysis Instrumentation,* Macmillan, London, 1973.

28

PROCESS CONTROL

28–1 INTRODUCTION

The reason for installing analyzers on a plant is to provide closer control of the process. If the process is relatively stable, the analysis results can be used to tell a plant operator what action to take to manually correct for any process disturbances. However, because of the number of variables, or the rapidity with which the process responds to disturbances, it is often necessary to use some form of automatic control.

The common variables that are controlled are flow rate, pressure, level, and temperature. We shall first consider these before discussing several examples of the control of product composition.

28–2 FEEDBACK CONTROL

Let us consider as an example of a simple process—the control of water temperature. One method of doing this, as in a bath or a shower, is the direct mixing of hot and cold water. This is shown in Figure 28–1a. The objective is to maintain the temperature of the water emerging from the mixing system constant in spite of variations in the hot water temperature. (We shall assume initially that the hot water flow rate remains constant.) It is obvious that we must measure the temperature of the emerging water and include a valve in the cold water line to control the cold water flow rate, as shown in Figure 28–1b. If the emerging water temperature is too high we can add more cold water; if the temperature is too low we can reduce the cold water flow rate. This can be done manually, but manual control is tedious and may not be fast enough to cope with sudden changes in the hot water temperature.

To provide automatic control, we require a *controller* that accepts the signal from the temperature measuring device, compares it with the required value (the *set-point*), and provides an appropriate signal to adjust the cold water control valve. Since the *primary measuring element*, in this case the temperature sensor, may be physically separated from the controller, we also need a *transmitter* to transmit the signal from the measuring element to the controller. If the final control element is a valve, as it is in our example, a *valve positioner* will be required to provide sufficient energy to operate the valve.

The primary element and the valve positioner are frequently not shown as separate elements on process flow sheets, and our temperature control system can be depicted as in Figure 28–1c. (To avoid congestion on a flow sheet, transmitters are also frequently not explicitly included.) This is a feedback control system because a signal from the primary measuring element at the output of the process is fed back to a point at the input to the process, in this case the cold water control valve. The symbols T, P, L, and F are used on process flow sheets to represent

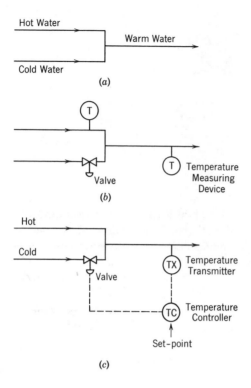

Figure 28–1 Temperature control system.

temperature, pressure, level, and flow, respectively. Likewise, I, R, and C represent indicator (usually just a meter), recorder and controller. Thus PI represents a pressure indicator and FRC a flow recorder-controller. The latter both controls the flow and provides a continuous record of the controlled flow rate. X is used to represent a transmitter, and A is an analyzer. The individual components that make up the complete control loop are shown in block diagram form in Figure 28–2. In our case the *controlled variable* is the warm water temperature and the *manipulated variable* is the cold water flow rate.

Referring again to Figure 28–1b, we shall assume that the hot water flow rate and temperature are F_H and T_H, respectively, and the cold water flow rate and

Figure 28–2 The control loop.

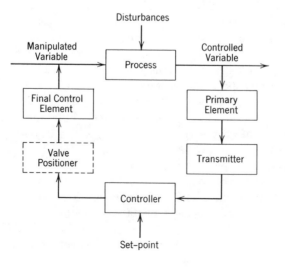

temperature are F_C and T_C. Since the amount of heat into the system is equal to the amount of heat out of the system, and assuming that the specific heat of water is 1.0, we can write

$$F_H T_H + F_C T_C = (F_H + F_C) T_M \tag{28-1}$$

where T_M is the temperature of the emerging water. If a change ΔT_H occurs in the hot water temperature, there will be a corresponding change in the warm water temperature from T_M to $T_M + \Delta T_M$. We can therefore write

$$F_H (T_H + \Delta T_H) + F_C T_C = (F_H + F_C)(T_M + \Delta T_M). \tag{28-2}$$

Subtracting Equation 28–1 from Equation 28–2, we have

$$F_H \Delta T_H = (F_H + F_C) \Delta T_M \quad \text{or} \quad \Delta T_M = \frac{F_H \Delta T_H}{F_H + F_C} \tag{28-3}$$

In the system in Figure 28–1c we have introduced a controller, and we shall assume that it increases the cold water flow rate in direct proportion to the increase in product temperature, that is,

$$\Delta F_C = G \Delta T_M \tag{28-4}$$

where G is the gain of the controller. For this system we can write

$$F_H (T_H + \Delta T_H) + (F_C + \Delta F_C) T_C = (F_H + F_C + \Delta F_C)(T_M + \Delta T_M) \tag{28-5}$$

Subtracting Equation 28–1 from Equation 28–5, and ignoring the small term $\Delta F_C \Delta T_M$, we get

$$F_H \Delta T_H + \Delta F_C T_C \simeq \Delta F_C T_M + F_H \Delta T_M + F_C \Delta T_M$$

Replacing ΔF_C by $G \Delta T_M$ (Equation 28–4), and rearranging the terms, gives

$$\Delta T_M \simeq \frac{F_H \Delta T_H}{(F_H + F_C) + G(T_M - T_C)} \tag{28-6}$$

This shows that introducing the controller into the system has reduced the error in the final temperature (cf. Equation 28–6 with 28–3) and this error becomes smaller as the gain of the controller is increased. Unfortunately, the gain of the controller cannot be increased indefinitely, because the dynamic response of the process can lead to instability. However, the resulting error, or *offset*, can be eliminated by a controller that includes integral action. Stability and offset are discussed in Section 28–5.

28-3 DYNAMIC RESPONSE OF THE SYSTEM

First-Order Systems. The ideal response of the measured variable (the output of a measuring instrument or transmitter) to a step change in a variable at the input to a process is an instantaneous step change. In practice the response is likely to be similar to that shown in Figure 28–3, in which t is time and τ is the time constant of the system.

This curve is characteristic of the temperature of a bare thermocouple immersed in hot water, or of the change in concentration of a stirred tank due to a step change in inlet concentration. Consider the situation shown in Figure 28–4 in which a solution of dissolved salt flows at a constant rate, F, into a rapidly stirred tank of constant holdup volume, V. Let the initial concentration of salt in the liquid entering the tank be C_i and that in the tank at any time, t, be C. If a step change in the inlet

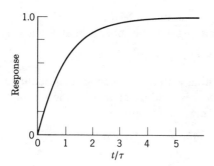

Figure 28–3 Response of a first-order system to a step input change.

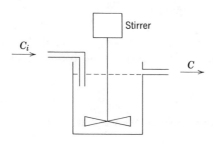

Figure 28–4 Mixing process in a stirred tank.

concentration to a new value C_f occurs at time $t = 0$, the difference between the flow rate of salt in and the flow rate of salt out must equal the rate of accumulation of salt in the tank, and we can write (ignoring density changes)

$$FC_f - FC = V\frac{dC}{dt}. \tag{28-7}$$

This equation can be solved by using the integrating factor $\exp(tF/V)$ and the solution is

$$C = C_i + (C_f - C_i)[1 - \exp(-tF/V)] \tag{28-8}$$

This is shown by Figure 28–3 if the response limits vary between $C_i(0)$ and $C_f(1.0)$. It is of the same form as the equation obtained for the charging of an electrical capacitor with the time constant τ being given by V/F. Such a system is referred to as a *first-order lag system*.

   Higher-Order Systems.   Measuring systems which combine a series of elements, for example, a thermocouple placed in a thermocouple well, or more complex processes, generally require a higher order differential equation to represent their response to an input function. The system can generally be represented by a second-order differential equation of the form

$$\tau^2\frac{d^2Y}{dt^2} + 2\zeta\tau\frac{dY}{dt} + Y = X(t) \tag{28-9}$$

where $X(t)$ is the driving force and ζ is a damping factor. The solution of this equation depends on the value of ζ and, for a step input, is shown in Figure 28–5. Most process components, with the exception of certain chemical reactors, are overdamped ($\zeta > 1$). The underdamped situation ($\zeta < 1$) is most commonly found in automatically controlled systems, as a result of the effect of the controller on the system response.

   Dead-Time.   The response of the system shown in Figure 28–1b to a step increase in the hot water temperature will be of the form shown in Figure 28–6. The sigmoid-shaped section of the curve arises because of the need to bring the outlet

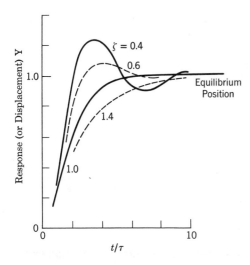

Figure 28–5 Response of second-order systems with different damping factors (ζ) to a step input change.

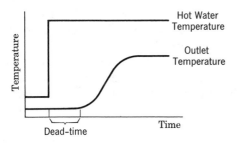

Figure 28–6 Response of the system depicted in Figure 28–1(b) to a step change in hot water temperature.

water pipes up to the new outlet temperature. However, there is also a dead-time in the system (sometimes known as *distance-velocity lag* or *transportation lag*). In this case it is due to the finite time required for the water to flow from the hot water temperature measuring device to the warm water temperature measuring device. This dead-time effect also occurs in the sampling line used to transport a sample from the process stream to a process analyzer. A similar time-delay effect occurs with an instrument such as the gas chromatograph in which the analysis results are only obtained some time after the injection of the sample.

The slow response of a system to an input disturbance, as shown in Figures 28–3 and 28–5, does create a control problem. However, the major disadvantage of feedback control is its inability to cope rapidly with changes in a system in which dead-time is significant, and this can easily lead to instability.

28–4 CASCADE AND FEEDFORWARD CONTROL

Cascade Control. In Figure 28–7 we have a heat exchanger in which a flowing liquid is heated by means of a steam-heated coil. We shall assume that the rate of flow of the liquid product remains constant but that disturbances can occur in the steam supply. These disturbances may be due to varying demands by other processes for steam which is supplied by a central boiler. The objective is to maintain the temperature of the liquid emerging from the heat exchanger constant in spite of the steam supply fluctuations.

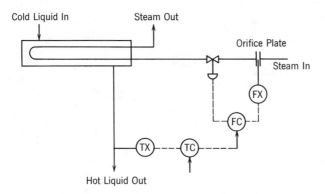

Figure 28–7   Cascade control of a heat exchanger.

A simple solution to this problem is to measure the liquid outlet temperature and use this to control the steam flow rate. However, there are two problems. One is the dead-time which is the time taken for the liquid to be transported from the heating element to the temperature sensor. The other is the fact that disturbances may occur very quickly in the steam supply, but the final equilibrium temperature will be reached relatively slowly because of the combined effect of the heat capacities of the liquid and of the metal components of the plant. It is therefore evident that, in the feedback configuration, corrective action can only be taken well after the initial disturbance has occurred. During this time the steam flow rate change may have corrected itself, so that the corrective action taken by the controller would then be inappropriate.

One method of overcoming this problem is to employ the *cascade control system* shown in Figure 28–7. In this arrangement there is a *primary* (or master) temperature control loop and a *secondary* (or slave) flow control loop. The secondary loop is relatively fast-acting and controls the steam flow rate. (The orifice plate is a constriction in the pipeline and the pressure difference developed across this constriction is proportional to the square of the flow rate.) The slower-acting primary loop is used to adjust the set-point of the steam flow controller to compensate for any residual error in the liquid temperature.

With many analyzers a cascade control system is commonly used, and the analyzer forms part of the master control loop. The cascade system not only makes allowances for the slow response speed of the analyzer but also ensures that failure or malfunction of the analyzer does not completely disrupt the control system, since a limit can be put on the allowable set-point change.

Feedforward Control.   A similar time delay problem arises in Figure 28–7 if the process liquid flow rate or input temperature changes, that is, the resulting temperature change at the outlet will only be sensed some time after the input flow rate or temperature changes. We can overcome this problem by measuring the input flow rate or temperature and using this signal to reset the steam flow rate, as shown in Figure 28–8. This is called a *feedforward control system* because it makes corrections on the basis of information obtained at the input to the process. However, to use this form of control we must have precise knowledge about the effect of both the input disturbance and the manipulated variable (in theory steam flow rate but in practice valve position) on the controlled variable (product temperature).

Feedback control is used more commonly than feedforward because the

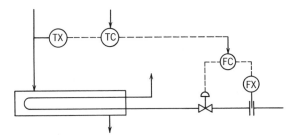

Figure 28–8 Feedforward control of a heat exchanger.

feedback system (a) requires little knowledge about the internal workings of the process other than what is the proper variable to control, and (b) continuously updates the signal from the controller as the process changes, due both to short-term disturbances and to long-term changes such as fouling of flow passages. In some cases a combination of feedforward and feedback loops is advantageous since feedforward provides rapid response and feedback can be used to compensate for any long-term variations at the end of the process.

28–5   CONTROLLER CHARACTERISTICS

On–Off Control.   The simplest control system is on–off control, in which the value of the controlled variable is allowed to fluctuate between upper and lower limits. Examples are commonly found in level control systems in which the exact level is not critical. On–off control can also be used for the control of temperature when precise control is not required or when the heat capacity of the system is relatively large.

Controllers used for general process applications are more complex and can respond in three ways. The basic control mode is proportional control, to which can be added integral and/or derivative action.

Proportional Control.   In proportional control, Figure 28–9, the difference between the signal generated by the primary element and that provided by the controller set-point is amplified and fed to the final control element. The output signal is thus proportional to the error in the value of the measured variable and the proportionality factor is determined by the gain of the amplifier. The constant p_0 in Figure 28–9 determines the initial steady state position ($e = 0$) of the final control element, for example, a valve.

It is important to realize that proportional control implies a fixed relationship between the measured variable and the controller response, and consequently between the measured variable and the position or effect of the final control element. This has already been shown by Equation 28–4 which applies to the proportional controller used in our first example. To make a change in the cold water flow rate there must be a net error in the warm water temperature and this is

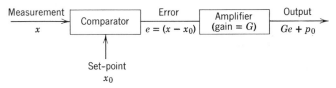

Figure 28–9   Proportional controller.

calculated, for a change in hot water temperature only, by Equation 28–6. Any change in the process that results in a change in the controlled variable is referred to as a *load change*. In this example, a change in either the water inlet temperature or flow rate is a load change. Proportional action provides partial compensation for load changes, but not perfect compensation because an error signal must persist for a new valve setting to be maintained. This is shown in Figure 28–10 in (*P*) which represents the response of a specific system with proportional control to a unit-step load change.

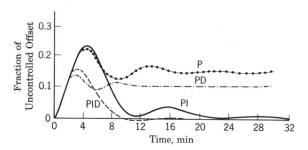

Figure 28–10   Controlled response of a system, for various modes of control. [From D. R. Coughanowr and L. B. Koppel, *Process Systems Analysis and Control*, McGraw-Hill, New York, 1965, p. 310. Used with permission of McGraw-Hill Book Co.]

Closer control should be achieved by increasing the controller gain, as shown by Equation 28–6, but too high a gain may well lead to instability. In deriving Equation 28–6 we have considered only the static response of the system. However the dynamic response of the system to a step input change will be similar to that shown in Figure 28–6. If we consider a disturbance that occurs sinusoidally (e.g., the hot water temperature changes in Figure 28–1b), the corrective action should ideally exactly oppose the effect of the disturbance.

In practice, because of the plant characteristics, there will be a phase lag between the disturbance and the sensing of the effect due to the disturbance, as shown in Figure 28–11. The phase lag is the separation between the disturbance and response curves measured in degrees (approximately 50° in Figure 28–11). This leads to a corresponding error in the timing of the feedback signal from the controller. The higher the frequency of the disturbance, the greater will be the phase lag and the smaller the output signal. This is shown by the *Bode diagram* in Figure 28–12. Provided that the phase lag is small, adequate control can be achieved by feedback control. However, as the gain of the system is increased, we may well find that the ratio of output response to input response is greater than 1 when the phase lag is

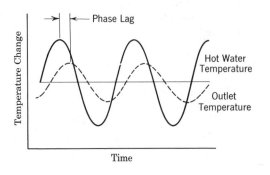

Figure 28–11   Response of the system in Figure 28–1(b) to a sinusoidally varying disturbance.

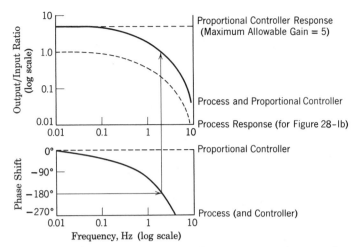

Figure 28-12 Bode diagram for the system in Figure 28-1(c), showing (arrows) the gain limitation for stability.

180°. In that case the controller response will add to the input signal, rather than oppose it, and the disturbance rapidly becomes larger and larger, that is, control of the process is lost. Feedforward control cannot lead to instability in this way, but if feedforward control is used incorrectly it can result in very poor control. The determination of the Bode diagram for a section of a process plant could take a considerable amount of time. However, if the diagrams for separate units are available, the overall response is obtained by multiplication of the output–input ratio for each component and addition of the phase shift values.

Other Controller Modes. The residual error or offset that is inherent with proportional action can be removed by adding *integral* action to the controller. With integral action (also termed *reset* action) the time integral of the error is determined and this is applied to correct the deviation. Hence, as long as there is a difference between the measured variable and the set-point, the controller will continue to correct the situation. The controller response is now represented by the sum of the proportional action and integral action, that is, by

$$\text{output} = G \text{ (gain)} \left(e \text{ (error)} + \frac{1}{IAT} \int e \, dt \right) + p_0 \qquad (28\text{--}10)$$

where *IAT* is the *integral action time*. The reciprocal $1/IAT$ is the *reset rate* (repeats per minute) or the number of times per minute that the proportional response is repeated by the reset action. The combined effect of proportional and integral action is shown in Figure 28–10 (*PI*).

Derivative (or *rate*) action is used in order to partially compensate for rapid changes in the process and therefore to improve the dynamic response of the control system. With derivative action an output signal proportional to the first derivative of the error signal is obtained. When combined with proportional action, we have

$$\text{output} = G \left(e + DAT \frac{de}{dt} \right) + p_0 \qquad (28\text{--}11)$$

where *DAT* is the *derivative action time*. The more rapid compensation for load changes obtained with derivative action is shown in Figure 28–10 (*PD*).

In addition to the above, we can combine proportional with both integral and derivative action. This provides rapid compensation for load changes and elimination of offset. However, it is more difficult to tune the three-term controller, that is, to choose the appropriate values of the three coefficients, G, IAT, and DAT, than it is to tune a two-term controller. A further problem that arises in tuning the controller is that there may be interaction between the three control modes, that is, the adjustment of one of the three parameters may affect the values of the others.

The controller modes that are used depend on the specific application. For example, flow changes may be very fast with effective time constants of less than 1 second. However, rate action is undesirable because it would tend to amplify the high-frequency noise that is present, due to stream turbulence and vibration from pumps. Proportional control and reset action is generally used, with a low gain and fast reset—that is, with a low value of both G and IAT in Equation 28–10.

Pressure control systems are generally characterized by a large capacity and small time constant, the latter being only slightly longer than for flow control. Proportional control is often the only mode used in pressure control, but reset action can be used where offset cannot be tolerated. Rate action is generally not used because it is not required in a system having a large capacity and a small time constant, and therefore some degree of self-regulation.

Level control has characteristics similar to pressure control, but some systems may also have considerable dead-time. In many cases precise control of level is not important and an on–off control system may be adequate.

Because of the nature of heat transfer, temperature control systems are usually characterized by large process heat capacity, slow response, and often a large dead-time. For precise temperature control a combination of proportional, reset, and derivative action is generally used. When dead-time is significant, cascade control on the faster variable is necessary, as in the example given in Figure 28–7.

There are a number of methods used for the adjustment of controller settings, and the appropriate settings depend on the process characteristics. In one method, the response of the control loop to a small step input is observed at low proportional gain with proportional control only. The step input is repeated at higher gains until the system produces continuous cycling. The proportional gain at that point is termed the *ultimate gain*, G_u, and the period of cycling (minutes per cycle) is the *ultimate period*, P_u. (The phase shift at this point is 180° and the gain of the system as a whole is one, as shown by the Bode diagram in Figure 28–12. The ultimate gain is therefore the reciprocal of the process gain at this point.) The proportional gain of the controller is then set at $0.5G_u$. In a system that can be represented by a single time constant, τ, and a dead-time, L, the appropriate integral action time depends on the value of L/τ. It varies from $2P_u$ for $L/\tau = 0.2$ to $0.6P_u$ for $L/\tau = 1.0$ and $0.31P_u$ for $L/\tau \geqslant 10$.

Pneumatic Versus Electrical Control Systems. In the past, most process control systems have been pneumatically operated. This means that the controller operation and the transmission of signals from the primary measuring element to the controller, and thence to the final control element, is all carried out by means of changes in air pressure. One of the main reasons for using a pneumatically operated system is its suitability in hazardous areas. In complete contrast is an electrical control system that could accidentally produce a spark in an area containing flammable vapors and thereby destroy an entire chemical plant. However, solid-state electrical control systems have now been developed in which it is impossible to generate sufficient energy for a spark to be produced. These electrical systems are termed *intrinsically safe*. Electrical control systems have a number of advantages over pneumatic systems. They are more directly compatible with many primary elements and with the increasing use of digital computers in control systems. They offer greater accuracy and suitability even where elementary calculations are involved, for example, for the control of electric power, or of flow which is proportional to the square root of the pressure differential across an orifice plate, or of pH which is a logarithmic function of concentration. In addition, electrical signals can be transmitted over longer distances than pneumatic signals. However, pneumatically operated control valves are commonly used in control systems, and these are cheaper than the equivalent electrically operated valves, so that a pneumatic control system has an advantage in this respect.

Pneumatic control systems normally operate with an air pressure range of 3 to 15 psig (pounds per square inch gauge, that is, psi above atmospheric pressure). The way in which a pneumatic controller supplies the necessary power to move a control valve is shown in Figure 28–13. Filtered air, at a pressure of about 20 psig, is admitted to the controller through a restriction that is about one-third the size of the nozzle opening; this air supply also goes directly to a relay valve. Close to the nozzle is a movable baffle or flapper whose position is controlled by the primary measuring element. When the flapper is against the nozzle it blocks the flow of air, and the pressure above the flexible diaphragm in the relay valve increases. The resulting downward movement of the diaphragm, and of a flapper attached to it, opens the relay valve and allows the supply air to flow directly to the control valve. Conversely, the movement of the controller flapper away from the nozzle releases the air pressure above the relay diaphragm and the upward force of the spring closes the relay. The air pressure at the valve will then drop because of a continual bleed of air at the control valve. A movement of the controller flapper of only a few thousandths of an inch is sufficient to change the output pressure from 3 to 15 psig.

The system just described is much too sensitive to be used without modification for proportional control, but is used directly for on–off control. Proportional control is achieved by mounting the nozzle, or the pivot point for the flapper, on a bellows so that its position varies with the output pressure. As a result the output pressure changes linearly with the motion of the flapper. Integral and derivative action can be added by suitably interconnecting the nozzle and bellows via adjustable restrictors.

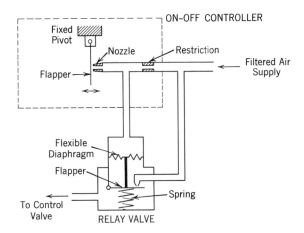

Figure 28–13 Pneumatic controller operation.

28–6   APPLICATIONS OF CONTROL SYSTEMS

Flow and Pressure Control in Gas Chromatography.   Under isothermal conditions, a constant flow rate through the gas chromatograph column can be achieved by maintaining the pressure constant at the inlet to the column. Figure 28–14a shows the basic construction of a pressure regulator. The controlled pressure is determined by the pressure above the flexible diaphragm, and this can be varied by means of the pressure-adjusting knob. In control terms, this varies the set-point of the pressure controller. If the outlet pressure tends to rise above the pre-set value, it will force the diaphragm upward and thereby close the orifice. This prevents any further gas passing from the high-pressure side of the orifice to the controlled pressure side and, since gas is continually flowing from the outlet to the column, the outlet pressure will fall. If it falls below the controlled value, the diaphragm will move downward. This opens the orifice and allows more gas to pass through from the high pressure supply. In practice, a balanced position is achieved with the orifice open just enough to keep the controlled pressure constant. It should be noted that the operation of the pressure controller is dependent on a continual flow of gas through it. As the gas flow rate decreases, it becomes more difficult to maintain a precisely controlled outlet pressure. Sometimes the tapered cone tends to stick in the orifice and a larger pressure differential than normal is required to open the orifice again. This produces an oscillating outlet pressure and the effect may be visible on the chromatogram as an oscillating baseline.

When temperature programming is used in gas chromatography there are advantages in being able to maintain the mass flow rate of gas through the column constant, in spite of the change in column temperature. If the pressure controller just described is used, the increase in gas viscosity with temperature will result in a continually diminishing flow rate through the column as its temperature increases. This problem can be overcome by modifying the pressure controller as shown in Figure 28–14b. The effect of this arrangement is to maintain a constant pressure

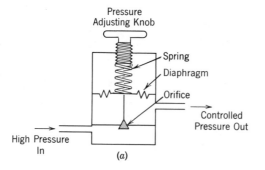

Figure 28–14   Pressure controller (a), and mass flow controller (b).

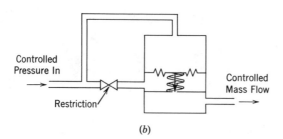

drop across the restriction so that the flow rate through the restriction remains constant. The outlet pressure is free to vary to compensate for changes in the carrier gas viscosity as the temperature of the column changes, and the mass flow rate can still be changed by changing the restriction, normally a needle valve.

Gas Chromatograph Oven and Temperature Control. The design of an oven for a gas chromatograph provides an interesting example of some of the problems inherent in temperature control systems. Ideally we want the oven to have a large heat capacity because this makes it easier to control. However, if the oven is to be temperature programmed it must be possible to change the temperature relatively rapidly. For this reason, and for general convenience, most chromatograph ovens are now of the air circulation type. An example of the construction of such an oven is shown in Figure 28–15. The limiting factor in using an air bath of this type is the poor heat transfer from the heater to the circulating air and from there to the chromatograph columns. The position is made as favorable as possible by using a bare wire heating element and very rapid air circulation to create turbulence. This improves the rate of heat transfer and minimizes temperature variations within the oven.

We are trying to control the column temperature, and it would seem logical to employ a temperature sensor that is embedded in the column or strapped to the wall of the column. This, however, is inconvenient if the column is likely to be changed, which is usually the case, and the air temperature is therefore sensed by a separate thermocouple, a thermistor, or a platinum resistance thermometer. The resistance of the latter varies practically linearly with temperature, which simplifies the design of the temperature controller, and the thermal mass of the resistance thermometer casing to some extent simulates the column tubing. If the thermal mass of the temperature sensor is too small, or if it is placed too close to the heating element, the heater will be turned off prematurely and the column will reach temperature equilibrium too slowly. Conversely, if the thermal mass of the temperature sensor is too large, the heater will not be turned off (or on) soon enough and the column temperature will overshoot the preset value. Since this will occur each time it approaches the preset value, irrespective of whether the column is heating or cooling, it may continue to cycle around the preset value indefinitely.

In a system with a large thermal capacity, or in which the heater power closely matches the load requirements, on–off control can be used. However, the former is not the case with an air bath and proportional control is necessary because the amount of heat required to maintain the oven temperature constant near room temperature is far less than that required at, say, 200°C. This is partly due to the fact that these ovens have a minimum of insulation so that they can be cooled rapidly

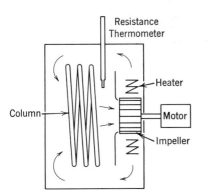

Figure 28–15 Construction of a typical gas chromatograph oven.

after completion of a temperature programmed run. Cooling is normally achieved by passing air at room temperature through the oven, without recirculation. A high degree of temperature stability is obtained by using a proportional band of only a few degrees. (Proportional band is a measure of the gain of the system and in this case is the temperature change required to turn the heater from fully on to fully off.) Any further narrowing of the band, which would be more equivalent to on–off control, would probably lead to instability because of the thermal lags in the system. A narrow proportional band corresponds to a high gain, and the amount of offset at different temperature settings will be small and can be neglected. Integral action is therefore not required, and derivative action is inappropriate because of the thermal lag in the column.

Proportional control of the power input to the oven is achieved by means of a Triac, which is an electronic switch. The a.c. supply is fed to the oven heater via this switch, and pulses generated by the oven temperature controller turn on the Triac, and hence the heater, for a predetermined part of each half-cycle. This is shown by the shaded sections in Figure 28–16. The amount of power produced by the heater is controlled in proportion to the demands of the process by changing the position in the cycle at which the Triac is switched on. It is switched off automatically as the heater voltage passes through zero. The above situation is somewhat unusual in that proportional control is achieved by means of an on–off switching system. However, this method of control is possible because the heater is supplied by an a.c. current and its frequency and the switching rate are both much faster than the thermal time constant of the heater.

pH Control of Effluent. The pH meter is the commonest analytical process instrument because of its simplicity and relatively high degree of reliability. However, it is a somewhat peculiar device because its response is not linear with concentration but varies as the logarithm of the concentration. At 25°C the glass electrode gives a voltage output of 59 mV per pH unit, that is, 59 mV for each 10-fold change in free hydrogen ion concentration. The voltage generated varies somewhat with temperature, but compensation is provided for this by inserting into the flowing sample stream a resistance element to measure temperature, in addition to the reference and sensing electrodes. Since the internal resistance of the glass electrode is high (about 30 MΩ) there is insufficient power available from the electrode to operate the controller, and a high-impedance current amplifier is required between the electrode assembly and the controller. The amplifier is normally mounted close to the electrodes to minimize electrical pickup.

We shall take as our example an effluent disposal system that must handle the waste products from a variety of chemical processes. To conform with local regulations the output from the disposal system must be limited in pH to between 6 and 8, but the incoming fluid streams may vary between 1 and 13. If the disposal system has a very large capacity, some of the extreme fluctuations may be averaged out and a single neutralization step might be possible. However, the effluent flow

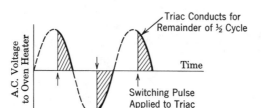

Figure 28–16 Proportional power input by means of electronic switching.

rate may well be so high that the capacity required is quite impractical. We are therefore left with a concentration range of 10^5 to 1 to cover, by the addition of either alkali (pH 1 to 6) or of acid (pH 13 to 8). Since the maximum range of a control valve for the accurate addition of reagent is about 30 to 1, a number of separate reagent addition valves must be used to cover the required range. Four valves, of different sizes or handling reagents of different concentrations, will provide a combined range (allowing for some overlap) of approximately $(25)^4$ or 4×10^5 so that a total of eight valves is required, four for the addition of acid and four for alkali.

A neutralization system designed by simulating the process with an analog computer is shown in Figure 28–17. The first tank is simply a mixing tank that averages out some of the sharper fluctuations in the effluent pH. The first neutralization stage incorporates four reagent addition valves to reduce the pH range to between 4 and 11, the particular valve to use at any one time being determined by the pH of the solution entering the second tank. To provide a linear relationship between the amount of reagent required and the valve position, the pH reading is converted to ion concentration by means of an anti-log converter. A strong acid–strong base reaction is assumed. The in-line mixer and neutralizer provides a small mixing time constant, and the additional line capacity allows the turbulence due to mixing to be transformed to laminar flow before the pH is measured. Because this is a fast process and the dead-time is small (about 1.5 sec), a three-term controller was used. The in-line mixer is followed by a mixing tank which evens out any residual pH fluctuations arising from the first neutralization step. The second in-line neutralization stage provides a finer degree of control to complete the neutralization, and this is again followed by a mixing tank. The inlet and outlet pH values are monitored by separate pH meters, linked to a two-point recorder, to check the overall performance of the system. Feedback rather than feedforward control is used in this example because of possible variations in the concentrations of the neutralization reagents.

Gas Chromatograph Control of a Distillation Column. For binary systems, temperature and composition are uniquely related, provided that pressure is held constant, and most distillation columns are under temperature and pressure control. However, for multicomponent systems temperature is not uniquely related to composition and it is therefore important to use an analyzer to control the

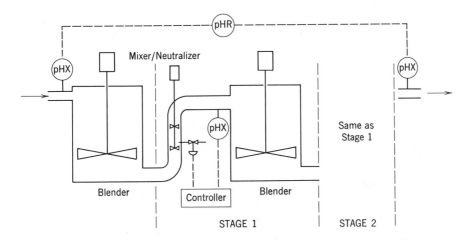

Figure 28–17 Effluent neutralization system. [After W. B. Field, *ISA Journal*, **6**, 44 (1959).]

composition of the product leaving the distillation column. This in fact is the major application for closed-loop control by gas chromatography.

The introduction of a chromatograph into a process control system generates a large time delay. This may arise partly because the sample needs to be pretreated before it is fed to the analyzer, but even without that delay the instrument itself will produce a measurement only at discrete time intervals. With an analyzer whose sampling period, or analysis time, is relatively long compared with the process time constant, good control cannot be expected. However, when the sampling period is similar to the process time constant and disturbances occur relatively slowly, the system response may be considerably better than with continuous control. This is because there is then sufficient time for the effect of the previous correction to manifest itself before a new sample is taken for analysis.

In a distillation column it is necessary to control either the overhead product (distillate) flow rate or the bottom product flow rate, Figure 28–18. If the overhead product is impure, it contains too much of the material that should be part of the bottom product. The correction is therefore to decrease the overhead product rate. This may be done either directly with a control valve in the product line, or indirectly by changing the reflux rate or the steam flow rate. Control of reflux flow rate is frequently used, but the overall response of the system is then slower than if product or steam flow rate had been controlled. One basic arrangement that has been used for the control of distillate composition is shown in Figure 28–18.

For the control of the overhead product the sample to be analyzed should be taken from a point between the condenser and the reflux accumulator; a more rapid response to process fluctuations would be expected if the sample was taken directly from the top of the column, but it has been found that large concentration gradients occur on the top tray of the column and in the overhead vapor. Sampling the product taken from the reflux accumulator introduces too large a delay because of the additional liquid capacity that is introduced into the system.

In the example given in Figure 28–18, short-term steam fluctuations are controlled by the reboiler steam flow control loop and longer term variations by the chromatograph resetting the flow controller set-point. Not shown in the figure are the flow control systems on the overhead and bottom products (normally by means of level control on the accumulator and reboiler respectively) and direct flow control on the reflux and feed supply. Note that direct flow control of both the

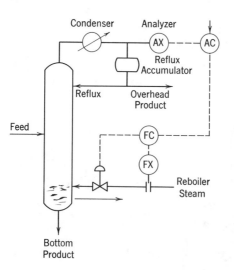

Figure 28–18   Closed loop control of a distillation column by gas chromatography.

overhead product and bottom product cannot be used, because the sum of these flows would not exactly equal the input feed rate. Pressure control at the top of the column is often achieved by controlling the cooling water flow rate to the condenser.

The feed rate to a distillation column is often determined by upstream processing equipment such as reactors or other distillation columns. This problem can be overcome by using a feedforward control system in which the steam flow rate is varied in proportion to the input feed rate and an analyzer is used to adjust the product flow rate.

QUESTIONS AND PROBLEMS

28–1. Using a person taking a shower as an example, draw a diagram of a feedback temperature control system. What is the primary measuring element? What is the valve positioner?

28–2. What determines the dead-time in the system in Question 28–1?

28–3. In the temperature control system in Question 28–1 a proportional controller, with gain G, is installed. This controls the hot water flow rate as a result of measurement of the temperature of the water emerging from the shower, the cold water flow rate F_C and temperature T_C remaining constant. The hot water flow rate and temperature are initially set at F_H and T_H respectively, and the resulting shower temperature is T_M. If the inlet hot water temperature changes by ΔT_H, show that the resulting temperature offset ΔT_M is given by

$$\Delta T_M \simeq \frac{F_H \Delta T_H}{(F_C + F_H) + G(T_H - T_M)}$$

28–4. If two thermocouples are immersed in hot water, the thermocouple of lower mass responds more quickly. Why is that?

28–5. An analyzer sampling system contains a sampling line, 20 m long and 4 mm internal diameter, and a filter and dryer each with a total volume of 200 ml. If the sample flow rate is 1.00 l/min, and a step change occurs in the concentration of the sample at the inlet to the line, how long does an analyzer with a time constant of 0.5 min take to register (a) 60% and (b) 95% of the concentration change? Assume that no mixing occurs in the line and that mixing in the filter and dryer is instantaneous. Note that for most analyzer systems the time constants (and dead-times) of individual components can be considered to be additive, with the entire system then being handled as a single first-order lag system with dead-time.   *Ans.* 1.07, 2.95 min

28–6. Draw the temperature versus time graph which results from a step change in the column temperature set-point for a chromatograph oven in which:
 (a) The temperature sensor is located too close to the heater.
 (b) The thermal mass of the temperature sensor is much greater than that of the column.

REFERENCES

N. H. Ceaglske, *Automatic Process Control for Chemical Engineers*, Wiley, New York, 1956.

P. Harriott, *Process Control*, McGraw-Hill, New York, 1964.

E. F. Johnson, *Automatic Process Control*, McGraw-Hill, New York, 1967.

A. Pollard, *Process Control*, Heinemann Educational Books, London, 1971.

E. I. Lowe and A. E. Hidden, *Computer Control in Process Industries*, Peter Peregrinus, London, 1971.

D. E. Smith, C. E. Borchers and R. J. Loyd, "Automatic Control" in A. Weissberger and B. W. Rossiter (eds.), *Techniques of Chemistry*, Vol. 1, Part 1B, Chapter 8, Wiley-Interscience, New York, 1971.

J. T. Miller, *The Revised Course in Industrial Instrument Technology*, United Trade Press, London, 1964. Details of process instruments.

INDEX

Absorbance, 135
Absorption cell, 153, 171, 521, 523, 541
Absorption coefficients, for gamma rays, 470
  for infrared, 171
  terminology, 135
  for X rays, 273
Absorption edge, 273
  table of, 270
Absorptivity, 135
Acetic acid, titrations in, 432
Acids and bases, acidity constants of, 416
  amino, 448
  definition of, 413
  dissociation of, 416
  dissociation constants of, 420, 439, 449
  hydrolysis of, 417
  indicators for, 429, 434
  monoprotic equilibria, 419
  polyprotic equilibria, 438
  relative strength of, 416
  titrations, aqueous, 427
    nonaqueous, 431
  see also Carboxylic acids
Action limit, 492
Activation analysis, 480
  detection limits, 481
Activity, radiochemical, 468
Activity coefficients, 64, 371
ADP-ATP system, 396
Adsorbents, 56, 107
Adsorption, 54
  isotherm, 55
Alcohols, infrared bands, 206
  mass spectra of, 340
Aldehydes, infrared bands, 209
  mass spectra of, 336
  ultraviolet bands, 232
Alkanes, infrared bands, 201

  mass spectra of, 334
  vapor pressure of, 10
Alkenes, infrared bands, 202
  mass spectra of, 335
Alkynes, infrared bands, 203
Alpha particles, 466, 469
Alumina, 57
Amides, infrared bands, 213
Amines, basicity of, 447
  in complex ions, 456
  infrared bands, 213
  mass spectra of, 342
Amino acids, dissociation of, 448
  separation of, 68
Amphiprotic solvents, 414
Amphoteric substance, 414
Amplifier, 158. See also Operational amplifier
Analyzers, see Automatic analyzers; and Process
      analyzers
Anhydrides, infrared bands, 213
Anisotropy, magnetic, 287
Anodic stripping analysis, 411
Appearance potential, 330
Area measurement, 103, 284, 497, 500
Aromatics, infrared bands, 202
  mass spectra of, 335
  ultraviolet bands, 233
Atomic absorption spectroscopy, 243
  detection limits, 263, 481
  quantitative analysis, 259
  sample handling, 259
  sensitivity of, 264
Atomic fluorescence spectroscopy, 265
  detection limits, 263
Atomization, flame, 245
  nonflame, 251
Atomizer, carbon rod, 251
Automatic analyzers, 519

basic unit, 520
Beckman "DSA 560," 521
DuPont "ACA," 522
for enzymes, 528
Electro-Nucleonics "GeMSAEC," 528
Ljundberg "Autolab," 522
multianalyzer systems, 526
Technicon "Autoanalyzer," 524
Autoprotolysis, 414
Auxochrome, 228
Azeotropic mixture, 24

Back e.m.f., 401
Back scattering, 476
Back titration, 463
Base peak, 328
Bases, definition of, 413
  dissociation constants of, 420
  see also Acids and bases
Bathochromic effect, 228
Beer's law, 171, 273, 470
  derivation of, 134
  deviations from, 138
Bending vibration, 169
Benzoic acid, distribution behavior of, 29
Beta filter, 275
Beta particles, 466, 469
Beynon's tables, 332
Binary numbers, 501
Binomial distribution, 36
Biochemical formal potential, 376
Bio-Gel, 70
Bode diagram, 554
Bohr principle, 167
Boiling point, 8
  of common solvents, 31
Bragg's equation, 276
Brockmann scale, 57
Brønsted theory, 413
Bubble-cap column, 20
Buffers, 420, 423
  metal ion, 464
Buffer capacity, 426
Burners, 248
Burning velocity, 248

Camphor, phase diagram of, 13
Capacity factor, 45
Carbon-13, in n.m.r., 305, 509
Carboxylic acids, infrared bands, 211
  ultraviolet bands, 231
  see also Acids and bases
Carrier gases, 86, 96, 98. *See also* Gases
Cascade control, 551
Cells, absorption, 153, 171, 521, 523, 541
  electrochemical, 362, 538
  n.m.r., 284
Chelate effect, 457
Chelates, 456
Chemical shift, 285
  tables of, 290

Chopper, 161, 244
Chromatogram, 45
Chromatography, 41
  adsorbents for, 56, 107
  apparatus, 60, 78, 81, 82, 85, 535
  column efficiency, 48
  columns for, 88, 536
  computerized, 503
  control of distillation column, 561
  deconvolution of peaks, 505, 512
  definition of, 41
  detectors for, 59, 60, 67, 73, 82, 92, 537
  flow control, 558
  gas-liquid, 85, 503
  gas-solid, 106
  gel, 70
  high performance liquid, 60
  ideal, 46
  ion exchange, 62
  ligand exchange, 69
  liquid-liquid partition, 59
  liquid phases for, 90
  liquid-solid adsorption, 54
  with mass spectrometer, 327
  nonideal, 46, 73
  packing columns, 58, 67, 73, 89
  paper, 76
  peak distortion, 56
  plate theory, 47
  preparative scale, 109
  pressure control, 558
  process, 535
  programmed temperature, 105
  qualitative analysis, 97
  quantitative analysis, 102
  rate theory, 48
  resolution, 50
  retention parameters, 45, 76, 98
  reverse phase, 59, 80
  sample valves, 87, 535
  solid supports, 60, 89
  solvents for, 57, 80
  temperature control, 559
  theory of, 44, 64, 71
  thin-layer, 81
  types of, 42
  see also specific entries
Chromophores, definition of, 226
  ultraviolet bands, 229
Chromosorb, 89
Clapeyron equation, 8
Clinical analyzers, *see* Automatic analyzers
Coincidence correction, 475
Colthup chart, 174
Compton effect, 470
Computers, analog, 515
  of average transients (CAT), 509
  central processing unit, 501
  control of instruments, 503
  digital, 501
  Fourier transform, 509

hybrid, 517
  learning machine, 509
  programs for, 502, 508
  time-averaging, 509
Conditional formation constant, 458
Conductivity, electrical, 362, 538
Confidence limits, 490
Conjugate acid (base), 414
Control charts, 492
Control loop, 548
Converter, A/D, 502
Coolidge tube, 267
Coordination compounds, 454
Copper-ammonia system, 454
Copper-trien system, 456
Cornu prism, 150
Correlation chart, for infrared, 173
  for n.m.r., 293
  for ultraviolet, 229
Corrosion rate measurement, 544
Coulometric titrations, 403
Counters, for radioactivity, 471
Counting geometry, 476
Counting rate, 476
Counting statistics, 477
Coupled reactions, 394
Coupling, in chromatography, 88
  in n.m.r., 292
Coupling constant, 292
  table of, 298
Craig process, 34
Crystals, for X-ray diffraction, 271
Curie, 468
Cyclic voltammetry, 411

Dead-time, 472, 475
  in process control, 550
Debye-Hückel limiting law, 372
Debye-Scherrer camera, 276
Decay constant, 467
Deconvolution of peaks, 505, 512
Derivative action, 555
Detection limits, definition of, 264
  comparative tables of, 263, 481
Detectors, differential, 93
  electron capture, 97
  evaluation of, 92
  flame ionization, 93
  gas chromatography, 92
  Geiger counter, 472
  hydrogen flame, 93
  infrared, 158
  integral, 93
  ionization chamber, 471
  level, 500
  liquid chromatography, 60
  mass spectrometer, 325
  photographic, 272, 277, 325, 358, 471
  proportional counter, 471
  radioactivity, 471
  scintillation counter, 473

semiconductor, 474
  slope, 499
  spectrophotometric, 154
  thermal conductivity, 95
  ultraviolet, 155
  visible, 155
  X rays, 272
Deviation, in counting statistics, 478
  from the mean, 488
  standard, 478
Dialyzer, 525
Diamagnetic shielding, 285
Dielectric constants, of common solvents, 31, 415
  effect on acidity, 415
Dienes, ultraviolet bands, 230
Differentiating solvents, 418, 432
Diffraction, gratings, 125, 152
  of X rays, 275
Diffusion current, 408
Discriminators, 475
Disintegration process, 466
Dispersion of light, 123, 125, 148
Dissociation constants, of amino acids, 449
  of monoprotic acids, 420
  of monoprotic bases, 420
  of polyprotic acids, 439
Distillation methods, 19
  control by gas chromatography, 561
Distribution coefficient, 28, 36, 44, 55, 64, 77, 100
Distribution of species, 439, 459
Distribution ratio, 29, 64
Diynes, ultraviolet bands, 232
Double bond equivalent, 176

EDTA complexes, 457
  applications of, 463
  distribution of species, 459
  formation constants, 457
  titrations with, 460
Efficiency of a column, 48
Electrochemical cell, 362, 538
Electrodes, calcium, 382
  dropping mercury, 406
  glass, 379
  for hygrometers, 543
  indicator, 377
  mercury-EDTA, 563
  platinum, 378
  for potentiometry, 363
  quinhydrone, 377
  saturated calomel, 378
  sign conventions, 365
  silicone rubber, 381
  silver chloride, 379
  solid state, 382
  for spark source, 354
Electrode potential, 365
Electrolysis, 400, 543
  controlled current, 401
  controlled potential, 403

Electromagnetic radiation, 115
  absorption of, 117, 133, 243, 273
  diffraction of, 125
  dispersion of, 123, 150
  emission of, 117, 143, 243, 256, 267
  nature of, 115
  refraction of, 122
  spectrum of, 118
  units of, 119
  see also Detectors; Radiation sources; and individual techniques
Electromotive force, 365
Electron microprobe analysis, 278
Electrophoresis, 82
Element map, 506
Eluotropic series, 58
Emission lines, 254, 269
Energy levels, of an atom, 117, 270
  of a molecule, 117, 166, 227
Equilibrium, in acid-base systems, 416, 419, 430, 442
  adsorption, 55
  complex ion, 455, 458
  effects on Beer's law, 140
  extraction, 28
  in flames, 245
  ion exchange, 64
  liquid-gas, 8, 17
  solid-gas, 11
  solid-liquid, 13
Equilibrium constants, determination of, 386
Errors, absolute, 486
  normal distribution of, 487
  photometric, 142
  random, 487
  relative, 486
  systematic, 487
Esters, infrared bands, 210
  mass spectra of, 339
Ethanol-water system, azeotropic behavior of, 24
Ethers, infrared bands, 209
  mass spectra of, 341
Ethylenediamine, 447
Ethylenediaminetetraacetic acid, see EDTA complexes
Eutectic point, 15
Exclusion limit, 70
Exposure, photographic detector, 353
Extraction, 28
  continuous countercurrent, 38
  by Craig method, 34
  of mixtures, 33
  simple, 29
  successive, 30

Faraday's law, 402
Feedback control, 547
Feedforward control, 552
Fenske equation, 21
Filtration, in autoanalyzers, 521
Fingerprint region, 173

Flame emission spectroscopy, 243
  detection limits, 263, 481
  quantitative analysis, 256, 259
  see also Atomic absorption entries
Flames, characteristics of, 246
  recommended types, 247
  spectra of, 247, 256
  temperature of, 248
Flow control, 558
Fluorescence, 121
  atomic, 265
  X-ray, 271
Force constant, 166
  table of, 168
Formal potentials, 375
Formation constants, 455
  conditional, 458
  definition of, 455
  of EDTA complexes, 457
Fourier transform, 509
Fractional distillation, 19, 561
Fractionation of polymers, 74
Fragments, table of, 344
Freezing point, of common solvents, 31
  depression constant, 14
  of mixtures, 13
  see also Melting point
Frequency, definition of, 115
  overtone, 167
Freundlich isotherm, 55
Furnace, atomic absorption, 251

Gamma rays, 466, 470
Gases, carrier, 86, 96, 98
  for flame spectroscopy, 246
  infrared analyzers for, 540
  infrared cells for, 171, 541
  magnetic susceptibility of, 532
  sample valves for, 87, 535
  thermal conductivity of, 96, 532
Gaussian distribution, 488
Geiger counter, 472
Glass electrode, 379
Glycine, 448
Goniometer, 277
Gradient elution, 57
Grating, 125, 152

Half-cell, 362
Half-life, 467
Half-wave potentials, 409
Heat exchange, 552
Heat of fusion, 6
Heat of reaction, 539
Heat of sublimation, 11
Heat of vaporization, 8
Height equivalent to a theoretical plate, see HETP
Heisenberg uncertainty principle, 283
Henderson-Hasselbalch equations, 421
Henry's law, 18
Heptane-hexane, vapor pressure-composition, 17

Heterocyclics, ultraviolet bands, 233
HETP, 23, 48, 72, 88
Hexane-heptane system, vapor pressure-composition, 17
Hollow cathode lamp, 253
Hooke's law, 167
Hydrogen electrode, 366, 378
Hydrolysis, 417
Hydronium ion, 419
Hygrometer, electrolytic, 543
  piezo-electric, 543
Hygrostat, 66
Hypsochromic effect, 228

Ideal chromatography, 46
Ideal solution, 17
Immunoelectrophoresis, 83
Indicator electrode, 377
Indicators, acid-base, 429
  compleximetric, 462
  meters, 494
  nonaqueous titrations, 434
  redox, 393
Infrared analyzers, 540
Infrared spectroscopy, 165
  absorption band tables, 201
  band description, 171
  cells for, 153, 171, 541
  correlation chart, 173
  effect of slit width, 173
  instruments, 162, 540
  prism materials, 151
  sample preparation, 170
  sources, 148
  theory of, 166
  vibrational modes, 169
  wavelength calibration, 152
Infrared spectrum, of acetophenone, 182, 196
  of allyl alcohol, 188
  of benzene, 179
  of *n*-butyl acetate, 183
  of *n*-butyl-methyl ether, 181
  of cyclohexane, 179
  of diethyl carbonate, 198
  of ethyl benzoate, 183
  of ethyl formate, 200
  of *n*-hexanal, 182
  of *n*-hexane, 178
  of *n*-hexanoic acid, 184
  of *n*-hexanol, 180
  of hexene-1, 179
  of hexylamine, 181
  of mesitylene, 189
  of 1-methoxybut-1-en-3-yne, 192
  of 2-methylcyclohexanone, 194
  of 2-methylpentane, 178
  of nujol, 170
  of oct-1-yne, 185
  of phenol, 180
  of *trans*-stilbene, 190
Injectors, sample, 87, 535

Integral action, 555
Integration of area, 103, 284, 497, 500
Integrators, analog, 497
  ball-and-disc, 497
  computing, 500
  digital, 494, 498
Interference, background radiation, 475
  effect on ion selective electrodes, 381
  in flame spectroscopy, 256, 262
  masking of, 464
  removal of, 68
  in radiation, 125
Internal conversion, 121
Intersystem crossing, 121
Ion exchange resins, 62
Ionic strength, 372
Ionization, of acids, 415
  chemical, 318
  electron impact, 316
  field, 320
  in flames, 245
  from radioactivity, 469
  repression of, 259
Ionization chamber, 471
Ion-pair, 469
Ion selective electrodes, 380
Isoelectric point, 450
Isotope dilution, 482
Isotopes, in mass spectrometry, 331
  table of radioactive, 468
Isotope labeling, 347

Junction potential, 374

Ketones, infrared bands, 207
  mass spectra of, 338
  ultraviolet bands, 232

Langmuir isotherm, 56
Learning machine, 509
Leveling solvents, 417, 432
Ligand, 454
Ligand exchange, 69
Line printer, 502
Line width, in atomic absorption, 254
  for n.m.r. absorption, 283
Liquid phases, 90
Littrow prism, 151
Logic system, 499
Luminescence, 120, 544
  intensity of, 143
Lyate ion, 418
Lyonium ion, 418

Magnetic properties of nuclei, table of, 280
Magnetic susceptibility, of common gases, 532
Magnetogyric ratio, 281
Masking agents, 464
Massmann furnace, 251
Mass spectrometry, 316
  analyzer, 321, 354

base peak, 328
  computerized, 505
  detection limits, 481
  fragments, table, 344
  element map, 506
  with gas chromatograph, 327
  inlet systems, 325, 354
  instruments, 316, 321, 353
  ion collection, 325, 354
  ion production, 316, 353
  ion separators, 321, 354
  isotope effect, 331
  isotope labeling, 347
  metastable ion peaks, 330
  molecular ion, 328
  parent peak, 328
  qualitative analysis, 343, 356
  quantitative analysis, 358
  resolution, 328
  sample handling, 325, 354
Mass spectrum, of bronze, 357
  of *n*-butylamine, 343
  of *n*-butylbenzene, 336
  of *n*-butyraldehyde, 337
  of cetyl alcohol, 341
  of dimethylamine, 343
  of *n*-dodecane, 334
  of ethyl-*sec*-butyl ether, 342
  of 1-heptene, 335
  of *n*-hexanal, 338
  of methane, 318
  of methyl butyrate, 340
  of 2-methylheptane, 335
  of *n*-octanol, 340
  of octan-4-one, 339
  of propionaldehyde, 336
McLafferty rearrangement, 337
Mean value, 488
Melting point, 5
  identification by, 16
  of mixtures, 13
  see also Freezing point
Mercury electrode, 463
Metal chelates, 456
Metal ion buffers, 464
Metastable ions, 330
Moisture analyzers, 542
Molar response factor, 103
Molecular ion, 328
Molecular orbital theory, 226
Molecular separator, 327
Molecular sieves, 107
Monochromators, 148, 254, 271
Mulls, 170
Multiplexer, 503

Nebulizers, 248, 250
Nephelometry, 127
Nernst equation, 367
Neutron capture cross section, 480
Neutrons, 470

Ninhydrin, 78
Nitrogen rule, 333
N.M.R. spectroscopy, 279
  carbon-13, 305, 509
  chemical shift table, 290
  computerized, 509
  correlation chart, 293
  coupling constants, 298
  fast Fourier transform, 509
  high resolution, 292
  instruments, 283
  integration, 284, 497
  line width, 283
  magnetic properties, 279
  nuclear relaxation, 282
  nuclear resonance, 281
  shielding effects, 285
  simulation of spectra, 511
  splitting patterns, 292
  time averaging, 509
N.M.R. spectrum, of acetaldehyde diethylacetal,
    303
  of diethyl maleate, 301
  of ethyl bromide, 294
  of 2-methylbut-3-en-1-ol, 304
  of methyl ethyl ketone, 300
  of 1-*p*-nitrobenzoyloxy-3-methylallene, 305
  of *p*-nitrophenol, 302
  of 1-nitropropane, 296
  of 1,1,2-trichloroethane, 301
  of valine, 303
Nonaqueous solvents, 431
Nonideal solutions, 24
Normal error curve, 488
Normalization, 104
  in mass spectra, 328
Nuclear magnetic moment, 279
Nuclear magnetic resonance, *see* N.M.R. *entries*
Nucleic acids, ultraviolet bands, 235

Operational amplifiers, 497, 513
  current-to-voltage, 513
  differentiator, 514
  exponential converter, 514
  integrator, 497
  inverter, 513
  multiplier, 514
  summer, 513
Overtone frequency, 167
Overvoltage, 401
Oxygen analyzers, 538

Packing a column, 58, 63, 73, 90
Paraffins, vapor pressure of, 10. *See also* Alkanes
Paramagnetic analyzer, 539
Parent peak, 328
Partition coefficient, *see* Distribution coefficient
pH, of amino acids, 448
  definition of, 419
  in monoprotic systems, 419
  in nonaqueous media, 434

operational definition, 384
  in polyprotic systems, 442
pH controller, 560
pH meter, 385
pH scale, 385
Phase diagram, 11
Phenols, infrared bands, 206
Phosphorescence, 122
Phosphoric acid, dissociation of, 438
Photoelectric effect, 470
Photometric error, 142
Photomultiplier tube, 157
Photon, properties of, 116
Phototube, 155
Planck's equation, 117
Plate theory, 47, 64
Pneumatic controllers, 557
Polarity of phases, 92
Polarized light, 129
Polarography, 406
  oxygen electrode, 538
Polymers, fractionation of, 74
Potential, electrode, 365
  half-wave, 409
Potentiometer, 383
  recording, 494
Potentiometric titration, 389
  with EDTA, 463
  iron(II) - cerium(IV), 389
Powder camera, 276
Premix burner, 249
Pressure drop correction factor, 98
Priority interrupt system, 502
Prism, 123, 150
  materials for, 151
Process analyzers, 531
  basic unit, 532
  calibration of, 534
  chemiluminescence, 544
  colorimetric, 544
  gas chromatographic, 535
  infrared, 540
  moisture, 542
  multiple column systems, 536
  oxygen analyzer, 538
  sample handling, 533
Process control, 547
  Bode diagram, 554
  cascade control, 551
  composition control with gas chromatograph, 561
  controller modes, 553
  dead time, 550
  electrical control systems, 556
  feedback, 547, 552
  feedforward, 552
  flow control, 547, 552, 556, 558
  frequency response, 554
  instability, 555
  level control, 556
  pH control, 560

pneumatic control systems, 556
  pressure control, 556, 558
  process response, 549
  temperature control, 547, 552, 559
  terminology, 547
  time constant, 549
Process controllers, 547
  derivative (or rate) action, 555
  electrical, 556
  error, 553
  gain, 549, 553
  integral (or reset) action, 555
  intrinsically safe, 556
  pneumatic, 556
  proportional, 553
  set-point, 547, 553
  setting controllers, 556
Programmed temperature, 105
Proportional control, 553
Proportional counter, 471

Q-test, 491
Quadratic formula, 422
Quadrupole spectrometer, 323
Qualitative analysis, *see specific techniques*
Quantitative analysis, *see specific techniques*
Quenching effect, 473

Radiation safety, 482
Radiation sources, for atomic absorption, 253
  gamma rays, 470
  hollow cathode lamp, 253
  infrared, 148
  radio frequency, 283
  ultraviolet, 148
  visible, 148
  X-rays, 267
Radiochemistry, 466
  activation analysis, 480
  counting statistics, 477
  detection and measurement, 471
  radioactive nuclides, 468
  rates of disintegration, 467
Raman scattering, 127
Raoult's law, 14, 17
  deviations from, 24
Rate theory, 48
Rayleigh scattering, 127
Real-time system, 503
Recorders, galvanometric, 496
  oscillographic, 496
  potentiometric, 495
Reduced mass, 167
Reference electrode, 377
Reflux ratio, 22
Refractive index, 115, 122
  of glasses, 124
  table of, 125
Rejection of a value, 491
Relative retention, 46, 100
Relative volatility, 18

Relaxation, nuclear, 282
   vibrational, 121
Reliability of a value, 490
Residual current, 408
Resolution, chromatographic, 50
   enhancement of, 512
   in mass spectrometry, 328
Resonance, nuclear, 281
Respiratory chain, 397
Retardation factor, 76
Retention index, 101
Retention parameters, 97
Retention time, 45
Retention volume, 46, 98
Ring rule, 333
Rocking vibration, 169
Rotation of light rays, 128

Safety, radiation, 482
Sample handling, *see individual techniques*
Saturated calomel electrode, 378
Saturation factor, 480
Scalers, 475
Scattering, in radioactivity, 476
   Raman, 127
   Rayleigh, 127
Scintillation counter, 473
Scissoring vibration, 169
Selectivity coefficient, for ion exchange, 64
Selectivity ratio, 380
Separation factor, 37
Sephadex, 70
Shielding parameter, 285
Sign conventions in electrochemistry, 365
Significant figures, 485
Silica gel, 57, 81
Slope detector, 499
Snell's law, 123
Solubility product, 387
Solution, ideal, 17
   nonideal, 24
Solvation, effect on dissociation, 416
   effect on ultraviolet spectra, 235
   of metal ions, 454
Solvents, absorption bands of, 155
   for acid-base titrations, 415
   amphiprotic, 414
   for atomic absorption, 257, 260
   for chromatography, 57, 80
   dielectric constants of, 31
   differentiating, 417
   for extraction, 31
   for flame analysis, 257
   inert, 433
   for infrared spectroscopy, 171
   leveling, 418
   for n.m.r., 284
   nonaqueous, 432
   for plane chromatography, 80
   role in acidity, 414
   for spectrophotometry, 153

   transparency limits of, 154
Specific absorptivity, 135
Specific activity, 468
Specific retention volume, 99
Specific rotation, 130
Spectrofluorimeter, 163
Spectrometer, mass, 353, 505
   n.m.r., 283, 509
   spark source, 353
   X-ray emission, 271
Spectrophotometers, 147
   atomic absorption, 244
   for autoanalysis, 521, 525
   commercial instruments, 161
   flame photometer, 244
   process analyzers, 544
   single versus double-beam, 160
Spectroscopy, *see individual methods*
Splitting patterns, 294
Spin-lattice relaxation, 282
Spin number, 279
Spin-spin relaxation, 282
Spin-spin splitting, 292
Splitter, sample, 87
Square-wave polarography, 411
Stability constant, 455
Standard addition, 262
Standard deviation, 478, 489
Standard hydrogen electrode, 365
Standard potential, 367
   table of, 369
Standards, for activation analysis, 481
   for atomic absorption, 257, 260
   coulometric, 405
   for gas chromatography, 104
   internal, 261
   for process control, 534
   reference electrode, 365
Standard state, 368
Statistical control, 492
Statistics, 477
Stoke's lines, 128
Stretching vibration, 169
Stripping analysis, 411
Student's *t*-value, 490
Styragel, 71
Successive approximations, 422, 443
Supercooling, 6
Supporting electrolyte, 408
Supports, chromatographic, 60, 89

Tailing of peaks, 56, 500
Teletype, 502
Temperature, control of, 548
   effect on diffusion current, 409, 538
   effect on flame emission, 256
   effects in gas chromatography, 105
   effect on vapor pressure, 9
Temperature controller, 559
Temperature programming, 105
Tetramethyl silane, 285

Theoretical plates, 21, 47
  determination from chromatogram, 47
  number in distillation column, 21, 23
Thermal conductivity of common gases, 96, 532
Thermomagnetic analyzer, 540
Thermometer correction, 7
Time-averaging, 509
Time-of-flight spectrometer, 321
Titrations, acid-base, 427
  amino acids, 451
  back, 463
  compleximetric, 455
  EDTA equilibria, 460
  iron(II) - cerium(IV), 389
  nonaqueous, 431
  photometric, 463
  polyprotic acids, 445
  potentiometric, 463
  secondary coulometric, 403
Total consumption burner, 248
Tracers, radioactive, 480
Transmittance, 135
Triethylenetetramine, 456
Triple point, 12
True value, 486
t-values, 490
Twisting vibration, 169

Ultraviolet spectroscopy, 226
  absorption bands, 229
  correlation chart, 229
  instruments, 162
  kinetic studies with, 239, 529
  molecular orbitals, 226
  qualitative analysis, 237
  quantitative analysis, 133, 136, 159
  sample handling, 153
  solvation effects, 235
  solvents for, 154
  theory of, 226
Ultraviolet spectrum, of benzene, 236
  of diethyl barbituric acid, 240
  of fluorobenzene, 236
  of methyl salicylate, 239
  of nitrophenol, 240
  of phenol, 237
  of trimethyl picrate, 240

Uncertainty principle, 283

Vanadium, stepwise oxidation, 395
Van Deemter equation, 49, 54, 88
Vapor pressure, definition of, 8
  dependence on temperature, 9
  of paraffins, 10
Variance, 489
Vibrations, types of, 169
Visible spectrophotometry, *see* Ultraviolet *entries*

Wagging vibration, 169
Warning limit, 492
Water, phase diagram of, 12
  de-ionization of, 67
  ion product of, 388, 414
  moisture analyzers, 542
  regain, 72
  removal with gels, 73
Water-ethanol system, azeotropic behavior of, 24
Wavelength, calibration methods, 152
  definition of, 115
Wavenumber, definition of, 116
Weight factor, 103
Woodward's rules, 230

X-ray crystallography, 276
X-ray spectroscopy, 267
  absorption, 273
  absorption edge, 270
  detectors, 272
  diffraction, 275
  emission, 267
  energy levels, 270
  fluorescence, 271
  instruments, 271, 277
  line spectra, 268
X-ray spectrum, continuous, 267
  of copper, 275
  of lead, 273
  line, 268, 270
  of molybdenum, 269
  of tungsten, 269

Zinc, complex ions of, 460
Zone electrophoresis, 82
Zwitterion, 448